NUMERICAL SOLUTION OF ORDINARY DIFFERENTIAL EQUATIONS

NUMERICAL SOLUTION OF ORDINARY DIFFERENTIAL EQUATIONS

Lawrence F. Shampine
Southern Methodist University

CHAPMAN & HALL
New York • London

First published in 1994 by
Chapman & Hall
One Penn Plaza
New York, NY 10119

Published in Great Britain by
Chapman & Hall
2-6 Boundary Row
London SE1 8HN

Printed in the United States of America

Library of Congress Cataloging in Publication Data

Shampine, Lawrence F.
Numerical solution of ordinary differential equations / Lawrence F. Shampine.
p. cm.
Includes bibliographical references and index.
ISBN 0-412-05151-6
1. Differential equations—Numerical solutions. I. Title.
QA372.S417 1993
515'.352—dc20 93-30156
CIP

British Library Cataloguing in Publication Data available

Please send your order for this or any**Chapman & Hall book to Chapman & Hall, 29 West 35th Street, New York, NY 10001, Attn: Customer Service Department.** You may also call our Order Department at 1-212-244-3336 or fax your purchase order to 1-800-248-4724.

For a complete listing of Chapman & Hall's titles, send your requests to **Chapman & Hall, Dept. BC, One Penn Plaza, New York, NY 10119.**

Contents

Preface

This book is an introduction to the numerical solution of the initial value problem for a system of ordinary differential equations (ODEs). The first three chapters are general in nature. Although they should be read quickly, their importance should not be underestimated. Chapter 1 provides a sense of what kinds of problems can be solved numerically and describes how typical problems can be formulated in a way that permits their solution with standard codes. It is easy to get the impression that numerical methods are more powerful than they really are, this being especially true when reading books devoted to the theory of ODEs. Chapter 2 explains some of the limitations of numerical methods. The numerical solution of an initial value problem requires more information than the mathematical statement of the problem. Chapter 3 is devoted to the computational problem and associated software issues.

Chapters 4 through 8 derive the basic numerical methods, prove their convergence, study their stability, and consider how to implement them effectively. A great many methods have been proposed, but only a few are seen in popular codes. By narrowing the focus of the book to the most important methods, it is possible to consider them rather carefully and still make minimal demands on the mathematical and computational background of the reader.

Some of the exercises appear in context and others at the end of the chapter. Generally it is worth reading the exercises for the applications and techniques they introduce. There is a collection of substantial problems at the end of the book. Experience solving non-trivial problems with state-of-the-art codes is indispensable for the person who wishes to come to a good understanding of the numerical solution of the initial value problem for ODEs (or any other area of numerical analysis, for that matter!).

References for the literature and codes cited in the text appear at the end of the book. They are supplemented by a list of books that contain comprehensive bibliographies. References to surveys of software for the initial value problem are found in Section 5 of Chapter 3 along with directions for obtaining some quality software that is free or inexpensive.

In an appendix are found some results from the theory of ODEs, vector calculus, matrix theory, and the theory of polynomial interpolation. Most of the results in this appendix would be encountered in first courses on the subject.

This book presents a perspective of the numerical solution of ordinary differential equations that results from some two decades of research and development by the author in both universities and scientific laboratories. Thanks are due Bob, Buddy, Marilyn, Kris, Przemek, Wen, and Mike for all they have taught the author about the subject. Special thanks are due Ian Gladwell for the help he has so generously provided in both his roles as friend and colleague.

1

The Mathematical Problem

The aim of this chapter is to develop a sense of what kinds of initial value problems can be solved numerically and how to prepare problems for their numerical solution. First we review the "facts of life" about the existence and uniqueness of solutions. It is convenient in the theory of ordinary differential equations (ODEs) to work with problems written in a standard form, and because the codes all expect problems to be presented in this way, we must go into this. Some basic mathematical tools are found in the appendix. The concept of "order" is fundamental to a study of the numerical solution of the initial value problem. Because it may not be familiar to the reader, the elements are developed here. Finally a series of substantial examples are taken up to show how one might be able to deal with problems that do not fit neatly into the standard theory of ODEs and their numerical solution. Some of the examples will be used throughout the book for illustrative purposes.

§1 Existence, Uniqueness, and Standard Form

We begin by considering the initial value problem itself to see what kinds of problems are meaningful and what kinds we might hope to solve numerically. Even very simple problems that can be understood with arguments from calculus show what can happen. If $F(x)$ is continuous on an interval $[a, b]$, the equation

$$\frac{dy}{dx} = F(x) \tag{1.1}$$

has a solution given by the fundamental theorem of calculus,

$$y(x) = B + \int_a^x F(t)\, dt.$$

This is a solution of the differential equation (1.1) for any value of the constant B, so to select a particular solution some additional information must be supplied. There are a number of ways this might be done. Although several will appear in this chapter, the most common, and the subject of this book, is to specify the initial value

$$y(a) = A. \tag{1.2}$$

The two requirements (1.1) and (1.2) make up an initial value problem for an ordinary differential equation. There is a solution and only one, namely

$$y(x) = A + \int_a^x F(t)\, dt.$$

In general, an initial value problem for an ordinary differential equation has the form

$$\frac{dy}{dx} = F(x, y), \qquad a \le x \le b, \qquad y(a) = A.$$

It is assumed that $F(x, y)$ is continuous in both variables. A solution is a function $y(x)$ that is continuous and has a continuous first derivative on $[a, b]$ (in symbols, $y \in C^1[a, b]$), satisfies $y(a) = A$, and satisfies

$$\frac{dy(x)}{dx} = F(x, y(x))$$

for each x in $[a, b]$. The continuity requirement on F excludes some interesting problems, including many of those of classical applied mathematics. Later some examples will be taken up to illustrate how one might be able to deal with such problems.

It is often useful in a study of the initial value problem to look at the special problems (1.1, 1.2), known as *quadrature problems*, to gain insight. Further insight can be obtained from another class of problems that can be transformed to quadrature problems:

$$y' = F(y). \tag{1.3}$$

If $F(A) \neq 0$, say $F(A) > 0$, the differential equation states that $y(x)$ increases as long as $F(y)$ remains positive. But then the inverse function $x(y)$ exists and

$$\frac{d\,x(y)}{dy} = 1 \bigg/ \frac{d\,y(x)}{dx}.$$

Accordingly, if $F(A) > 0$, the quadrature problem

$$\frac{dx}{dy} = \frac{1}{F(y)}, \qquad x(A) = a,$$

can be solved to obtain

$$x(y) = a + \int_A^y \frac{dt}{F(t)}.$$

The solution $x(y)$ has the same form when $F(A) < 0$.

As a specific example of this technique, consider

$$\frac{dy}{dx} = y, \qquad y(0) = A > 0,$$

for which

$$x(y) = \int_A^y \frac{dt}{t} = \int_1^y \frac{dt}{t} - \int_1^A \frac{dt}{t} = \ln y - \ln A = \ln(y/A).$$

Inverting the relationship leads to $y(x) = A \exp(x)$. The procedure for $A < 0$ is the same and leads to the same result. This familiar problem and solution shows that it may happen that a solution is valid on *any* interval $[a, b]$. Generally the interval on which a solution exists depends on both the differential equation and the initial condition. The problem

$$\frac{dy}{dx} = y^2, \qquad y(0) = 1$$

furnishes an example. In this case

$$x(y) = \int_1^y \frac{dt}{t^2} = 1 - \frac{1}{y},$$

hence $y(x) = 1/(1 - x)$. The solution "blows up" at $x = 1$ and past this point there is no solution of the initial value problem.

Notice that if $F(A) = 0$, then $y(x) \equiv A$ is a (constant) solution of the differential equation (1.3). The technique for solving such problems by interchanging independent and dependent variables used in the preceding examples does not apply when $F(A) = 0$, so we cannot conclude that this is the only solution of the problem. In point of fact, it *is* possible that there be other solutions. For example, if $F(y)$ is $3y^{2/3}$ and $A = 0$, it is easy to verify that $y(x) = x^3$ is another solution to the initial value problem. Indeed, there are infinitely many solutions:

EXERCISE 1. Consider the initial value problem

$$y' = 3y^{2/3}, \qquad a \le x \le b, \qquad y(a) = 0.$$

Verify that for any value of the constant γ such that $a \le \gamma \le b$, the function

$$\begin{aligned} y(x) &= 0 && a \le x \le \gamma, \\ &= (x - \gamma)^3 && \gamma \le x \le b, \end{aligned}$$

is a solution of the initial value problem. (Don't forget to verify that both $y(x)$ and its derivative are continuous on $[a, b]$.)

As another example of the difficulties that might arise in connection with the existence and uniqueness of solutions of an initial value problem, consider

$$\frac{dy}{dx} = \frac{y}{x} = F(x, y).$$

For intervals $[a, b]$ with $0 < a < b$, the function F is continuous in both variables. However, if $a = 0$, $F(x, y)$ is not continuous, or even defined, for all arguments (x, y) of interest and the problem is said to be *singular*. Singular initial value problems may have solutions. For any constant c, the function $y(x) = cx$ satisfies the differential equation of the example for all $x \neq 0$. It also satisfies the differential equation at $x = 0$ as a limit when $x \to 0$. Thus if the initial condition is $y(0) = 0$, the function $y(x) = cx$ satisfies the initial condition and satisfies the differential equation for all x, at least as a limit. In a natural sense this initial value problem has solutions, but there are infinitely many of them because c is arbitrary. What if $y(0) = A \neq 0$? There cannot then be a solution with a finite derivative at x because as $x \to 0$,

$$\frac{dy(x)}{dx} = \frac{y(x)}{x} \to \frac{A}{0} = \pm\infty.$$

Another kind of singular problem is one posed on an infinite interval. Such problems scarcely need exemplification because they arise frequently as idealizations of physical processes lasting for a "long" time or involving a distance that is "large." The fact that they are singular is perhaps more obvious when a change of independent variable is made so that the problem is posed on a finite interval. For example, if the equation

$$\frac{dy}{dx} = F(x, y)$$

is to be solved on an interval $[1, \infty)$, introduction of the new independent variable $t = x^{-1}$ converts the equation into

$$\frac{dy}{dx} = \frac{dy}{dt}\frac{dt}{dx} = -x^{-2}\frac{dy}{dt} = -t^2\frac{dy}{dt} = F(t^{-1}, y),$$

and finally into

$$\frac{dy}{dt} = -t^{-2} F(t^{-1}, y),$$

to be solved on the interval $[1, 0)$. In this form it is perhaps clearer that the problem may be singular as $t \to 0$, corresponding to $x \to \infty$.

Singular systems present both theoretical and practical difficulties. In the course of this chapter examples will be given that show they do arise naturally. It will also be shown how one might be able to prepare them for solution with standard codes for the initial value problem. An important part of the classical theory of ordinary differential equations is concerned with singularities. The theory applies classical techniques of applied mathematics such as series, perturbation, and asymptotic expansions to determine the behavior of solutions. These techniques are also useful computationally. They are not competitors for the kinds of methods studied in this book; they are complementary. Often the beginner wonders why texts on the theory of ODEs spend so much time on, say, series solution of differential equations in the neighborhood of singular points instead of just relying upon a library computer code. There is a short explanation—the codes do not even attempt to solve singular problems! This is why we must spend some time discussing how to prepare singular problems for solution by standard codes. To prepare a problem we must understand how its solutions behave near a singularity and for this we need the analytical theory of ODEs and the classical techniques of applied mathematics.

We have now seen simple examples illustrating the fundamental facts that an initial value problem may not have a solution at all, or may have exactly one solution, or may have infinitely many solutions. Furthermore, a solution may, or may not, exist throughout the interval of interest.

Problems arise more commonly as a system of equations or are put into this form. The *standard form* for the initial value problem for a system of n first-order ODEs is

$$\begin{aligned}
y_1' &= F_1(x, y_1, y_2, \ldots, y_n), & y_1(a) &= A_1,\\
y_2' &= F_2(x, y_1, y_2, \ldots, y_n), & y_2(a) &= A_2,\\
&\vdots \\
y_n' &= F_n(x, y_1, y_2, \ldots, y_n), & y_n(a) &= A_n,
\end{aligned}$$

or, in vector notation,

$$\mathbf{y}' = \mathbf{F}(x, \mathbf{y}), \qquad \mathbf{y}(a) = \mathbf{A}.$$

All vectors in this book are column vectors that are indicated by boldface type. A superscript T indicates transpose, so that $\mathbf{v}^T$ is a row vector. The formal resemblance to the case of a single equation is not misleading, a great convenience for the study of differential equations. The standard form is the one accepted by the vast majority of codes. The major exception is that there are codes intended for systems of the special form

$$\mathbf{y}'' = \mathbf{F}(x, \mathbf{y}), \qquad \mathbf{y}(a) = \mathbf{A}, \qquad \mathbf{y}'(a) = \mathbf{S}.$$

Although problems arise in diverse forms, both theory and practice make use of a standard form. It is essential to understand how to go about putting problems into this form. After all, if you cannot put a problem into the form required by a code, you cannot use the code to solve the problem. There is no one way to put a problem into standard form and a few examples will be given to illustrate how it might be done. Frequently differential equations involve derivatives higher than the first, e.g.,

$$y'' = -y, \qquad y(0) = 0, \qquad y'(0) = 1.$$

To write such problems as systems of first-order equations, the basic device is to introduce new variables. In this case, let $y_1(x)$ be $y(x)$ and $y_2(x)$ be $y'(x)$. Then

$$y_1' = y'(x) = y_2,$$
$$y_2' = y'' = -y = -y_1.$$

Along with the initial conditions this becomes a system in standard form,

$$y_1' = y_2, \qquad y_1(0) = 0,$$
$$y_2' = -y_1, \qquad y_2(0) = 1.$$

This problem is an example of the standard initial value problem for a single equation of order n,

$$y^{(n)} = F(x, y, y', y^{(2)}, \ldots, y^{(n-1)}),$$
$$y(a) = A_1, y'(a) = A_2, \ldots, y^{(n-1)}(a) = A_n.$$

The usual way to convert this problem to a system of first-order differential equations is to introduce the unknowns

$$y_1 = y, y_2 = y', \ldots, y_n = y^{(n-1)},$$

and then

$$\begin{aligned} y_1' &= y_2, & y_1(a) &= A_1, \\ y_2' &= y_3, & y_2(a) &= A_2, \\ &\vdots \\ y_n' &= F(x, y_1, y_2, \ldots, y_n), & y_n(a) &= A_n. \end{aligned}$$

Converting to standard form may not be merely a technical device. In principle, once we know the function $y(x)$, we know its derivative $y'(x)$, but the matter is not so simple in practice. Many numerical methods provide approximations to $y(x)$ only at certain values of x and do not attempt to approximate $y'(x)$ at all. Other numerical methods approximate $y(x)$ at all x and do produce an approximation to $y'(x)$, but the approximation to $y'(x)$ may not be as accurate as the approximation to $y(x)$. If the derivative is of independent interest, introducing $y'(x)$ as a new variable leads to an independent and accurate approximation to the derivative, whatever numerical scheme is used. Introducing new variables can facilitate the computation of useful auxiliary quantities. An example that will be taken up later is Blasius' similarity solution of the velocity profile for the flow past a plate of an incompressible viscous fluid. There is a differential equation to be solved for a quantity $f(\eta)$ and its derivative $f'(\eta)$. The momentum thickness is proportional to

$$\theta(\eta) = \int_0^\eta \frac{df}{d\tau}\left(1 - \frac{df}{d\tau}\right) d\tau.$$

There are many practical difficulties in computing θ if it is done after the computation of f and f'. However, it can be obtained easily by adjoining the differential equation

$$\theta' = f'(1 - f')$$

along with the initial condition $\theta(0) = 0$ to the equations for f and f' and computing θ at the same time that f and f' are computed. This device is particularly important when it is the auxiliary quantity that is the interesting one because it allows the error control to be applied directly to this quantity. For example, if we want a certain accuracy in θ, it is far from clear how much accuracy will be required of f and f' to achieve it. By solving simultaneously for θ, it is possible to specify directly the accuracy desired for θ and the code will automatically compute f and f' as accurately as required.

The basic device is to introduce new unknowns for each derivative up to one less than the highest order appearing in the equation. The equations

$$u'' + u'v' = \sin x,$$
$$v' + v + u = \cos x,$$

illustrate the manipulations that might be necessary. Unknowns $y_1 = u$, $y_2 = u'$, and $y_3 = v$ are introduced and then

$$y_1' = y_2,$$
$$y_2' = u'' = \sin x - u'\, v' = \sin x - y_2\, y_3',$$
$$y_3' = \cos x - v - u = \cos x - y_3 - y_1.$$

This is not in standard form because of the presence of y_3' on the right-hand side of the second equation. Eliminating it by means of the third equation leads to

$$y_1' = y_2,$$
$$y_2' = \sin x - y_2(\cos x - y_1 - y_3),$$
$$y_3' = \cos x - y_3 - y_1,$$

which is in standard form.

The following exercises give some practice in converting to standard form.

EXERCISE 2. Write the following equations as systems of first-order equations in standard form. Which of the original variables require initial values to specify an initial value problem when in standard form? Three common notations for derivatives are used: $(d/dt)\, y = y' = \dot{y}$

(a) $y'' - (1 - y^2)y' + y = 0$
(b) $x^3(d^3y/dx^3) + 6x^2(d^2y/dx^2) + 7x(dy/dx) = 0$
(c) $x + \sin z = \dot{x}$, $\quad \dot{z} + z = x$ (Differentiation is with respect to t here; x is a dependent variable.)
(d) $\ddot{y} + \dot{y} - \dot{v} = 0$, $\quad \ddot{y} + \dot{v} + 3y + v = 0$.

EXERCISE 3. Consider the scalar problem in standard form,

$$\frac{dy}{dx} = F(x, y).$$

Sometimes it is advantageous to introduce polar coordinates: $x = r\cos\theta$, $y = r\sin\theta$. If $y = f(x)$ is a solution of the differential equation, then in polar coordinates, $r\sin\theta = f(r\cos\theta)$. This relationship defines the function $r(\theta)$ implicitly. Find a first-order differential equation for r by differentiating the relationship and using the chain rule. Put the differential equation into standard form.

EXERCISE 4. Example 3.12 of Chapter 7 of J.D. Logan, *Applied Mathematics a Contemporary Approach* (Wiley, New York, 1987), discusses the longitudinal vibrations of a metallic rod in circumstances that lead to a shock wave. The partial differential equations that model the situation are reduced to ordinary differential equations by a similarity method. The equations

$$\frac{2y_0}{\rho_0} f \frac{df}{ds} = g - \frac{4}{3} s \frac{dg}{ds},$$
$$\frac{dg}{ds} + \frac{4}{3} s \frac{df}{ds} = \frac{2}{3} f,$$

are to be integrated numerically for $f(s)$ and $g(s)$. Here y_0 and ρ_0 are physical parameters that are constant. To integrate the system numerically, it is first necessary to put the equations into standard form. Do so.

EXERCISE 5. Write the equation

$$y'' + \nu^2 y + \varepsilon f(x, y, y') = 0$$

as a pair of first-order equations. Here ν and ε are parameters. Verify that when $\varepsilon = 0$, the solution is $y(x) = a \sin(\nu x + \phi)$ and its derivative $y'(x) = \nu a \cos(\nu x + \phi)$. In this special case the amplitude a and the phase ϕ are constants. It has been found useful in both theory and practice to write the solution for general ε in the form

$$y(x) = a(x) \sin(\nu x + \phi(x)),$$
$$y'(x) = \nu\, a(x) \cos(\nu x + \phi(x)),$$

where now the amplitude $a(x)$ and phase $\phi(x)$ are functions of x. The idea of this representation is that the amplitude and phase should vary slowly as functions of x when ε is small, even though the solution and its derivative vary rapidly when ν is large. Show that the functions $a(x)$ and $\phi(x)$ satisfy the differential equations

$$a' = -\frac{\varepsilon}{\nu} f(x, a \sin(\nu x + \phi), \nu a \cos(\nu x + \phi)) \cos(\nu x + \phi),$$

$$\phi' = \frac{\varepsilon}{a\nu} f(x, a \sin(\nu x + \phi), \nu a \cos(\nu x + \phi)) \sin(\nu x + \phi).$$

EXERCISE 6. Examples 3.13 and 3.14 of F.M. White, *Fluid Mechanics*, 2nd ed. (McGraw-Hill, New York, 1986) discuss the draining of a tank of constant cross-section area A_1 through a nozzle of area A_2. The free surface of the liquid is at a height $h(t)$ above the nozzle and moves at a velocity $V_1(t)$. The quantity of interest is the nozzle discharge velocity $V_2(t)$. For incompressible inviscid flow, the continuity equation states that $A_1\, V_1(t) = A_2\, V_2(t)$. When $A_2 \ll A_1$, a classical approximate analysis leads to the Torricelli formula $V_2(t) \approx \sqrt{2gh(t)}$, where g is the acceleration due to gravity. A more refined analysis that accounts approximately for the draining of the tank leads to the equation

$$2h\,\frac{A_1}{A_2}\,\frac{dV_2}{dt} + V_2^2\left(1 - \frac{A_1^2}{A_2^2}\right) = 2gh$$

where

$$h(t) = h(0) - \int_0^t V_1(\tau)\, d\tau.$$

It is said that this equation can be solved numerically, but to do so, you must prepare it. Obtain a differential equation for $h(t)$, eliminate $V_1(t)$ by using the continuity equation, and write the resulting system of two differential equations in standard form.

EXERCISE 7. D.A. Wells, *Theory and Problems of Lagrangian Dynamics* (Schaum's Outline Series, McGraw-Hill, New York, 1967), gives many examples of problems that need some preparation for their numerical solution. One is the motion of a dumbbell in a vertical plane. Two particles of masses m_1 and m_2 rigidly fastened to a light rod of length l move freely in the vertical x–y plane under the action of gravity. If the coordinates of the first particle are (x_1, y_1) and the angle the rod makes with the horizontal is θ, Lagrange's equations lead to

$$\begin{aligned}
(m_1 + m_2)\ddot{x}_1 - m_2\, l\, \ddot{\theta}\sin(\theta) - m_2\, l\, \dot{\theta}^2\cos(\theta) &= 0,\\
(m_1 + m_2)\ddot{y}_1 + m_2\, l\, \ddot{\theta}\cos(\theta) - m_2\, l\, \dot{\theta}^2\sin(\theta) &= -(m_1 + m_2)g,\\
m_2(l^2\,\ddot{\theta} - l\,\ddot{x}_1\sin(\theta) + l\,\ddot{y}_1\cos(\theta)) &= -m_2 g\, l\cos(\theta).
\end{aligned}$$

Here g is the acceleration due to gravity. Put this problem into standard form. If you find it convenient to use the inverse of a constant matrix, do not go to the trouble of inverting the matrix analytically.

Fortunately, it is usually easy to recognize initial value problems that have solutions. A standard *existence* result is that if $\mathbf{F}(x, \mathbf{y})$ is continuous on an open region D containing the initial point $(a, \mathbf{A})$, there is *at least* one solution $\mathbf{y}(x)$ of

$$\mathbf{y}' = \mathbf{F}(x, \mathbf{y}), \qquad \mathbf{y}(a) = \mathbf{A}$$

in the region. More details are provided in the appendix, which contains mathematical results basic to a study of the numerical solution of initial value problems. The technical assumption that D is "open" means that the initial point $(a, \mathbf{A})$ cannot be on the boundary of the region where $\mathbf{F}(x, \mathbf{y})$ is defined. It is easy to see that the result may not be true if it is. As a simple example, suppose D is the strip $-\infty < x < +\infty$, $0 \leq y \leq 1$ and $F(x, y) = +1$ for $(x, y) \in D$. Then if $(a, A) = (0, 1)$, the initial value problem has no solution because $y'(a) = +1$ requires that a solution $y(x)$ strictly increase from the initial value $y(a) = 1$, and this would take it outside the region where F is defined. This does not mean that you cannot start from a boundary—you can start from the lower boundary here—it just means that the question of existence is more complex then.

Starting on a boundary of the region of definition of $\mathbf{F}$ can present practical difficulties, too. An example will make the point. When a group of scientists at the Sandia National Laboratories was solving numerically some differential equations describing the concentrations of reactants in a liquid, they sometimes got reports from the operating system of undefined values. They consulted the author, B. Hulme, of the code they were using about this and he explained the difficulty. Negative concentrations are not physically meaningful and this translated into mathematical terms as an $\mathbf{F}$ that was not defined for negative arguments. Some reactants were not present initially, so these components of the solution were 0 at time $t = 0$—the initial values were on the boundary of the region of definition of $\mathbf{F}$. Like all differential equation solvers, Hulme's code must be able to evaluate $\mathbf{F}$ for arguments "near" its current approximation to the solution $\mathbf{y}(x)$. In the case of a component that vanishes at $t = 0$, the code could present the subroutine for $\mathbf{F}(x, \mathbf{u})$ with an argument $\mathbf{u}$ having a "small" negative value for the component. This actually happened and the operating system reported an attempt to evaluate $\mathbf{F}$ where it was not defined. When trying to solve problems with initial values on the boundary of the region of definition of $\mathbf{F}$, it must be kept in mind that even if the differential equation implies that the solution will enter the region of definition, the solver must evaluate $\mathbf{F}$ at arguments "near" the initial point and these arguments might lie outside the region of definition. Perhaps the most common situation of starting on a boundary is when the problem is posed on $a \leq x \leq b$ and $\mathbf{F}(x, \mathbf{u})$ is not defined for x less than the initial point a. Normally there is no difficulty in this situation. An existence and uniqueness theorem will be stated below and few codes will want to evaluate $\mathbf{F}$ for $x < a$.

Nearly all the problems seen in practice will satisfy the hypotheses of the existence theorem stated, at least in portions of the region of interest. However, as we have seen by example, solutions need not exist for all x.

The theorem goes on to say that each solution exists on a maximal interval $\alpha < x < \beta$ containing a and as $x \to \alpha$ or $x \to \beta$, the solution tends to the boundary of the region. In the case of the example with solution $1/(1 - x)$, the function $F = y^2$ is continuous for all arguments. As $x \to 1$, the solution tends to infinity, which in this case represents the boundary of the region. We usually suppose that $\mathbf{F}$ is continuous for all y, so a solution can fail to extend from $x = a$ to $x = b$ only if it becomes unbounded in magnitude along the way. The next exercise takes up a situation in which it is easy to show that the solution cannot become unbounded, hence extends throughout the interval of interest.

EXERCISE 8. A linear spring-mass system (without friction) is described by an initial value problem,

$$m\,\ddot{x} = -k\,x, \qquad x(0) \text{ and } \dot{x}(0) \text{ given,}$$

for the displacement x as a function of the time t. Here the mass m and the spring constant k are positive constants. Show that any solution of the differential equation satisfies

$$E = \text{constant} = \frac{m}{2}\,\dot{x}(t)^2 + \frac{k}{2}\,x(t)^2,$$

which expresses the conservation of energy. Use this law to bound the magnitude of the solution and its first derivative in terms of the given initial values and then argue that the solution exists for all t.

We have seen examples for which there is more than one solution. It is clear that such problems cannot be solved numerically in a mechanical fashion. Numerical methods require even more. Just putting the vector $\mathbf{A}$ of initial values into a computer normally introduces some error because the components A_i have to be represented by finite precision numbers α_i. It is essential, then, that when $\mathbf{A}$ is "close" to $\boldsymbol{\alpha}$, the solution $\mathbf{u}(x)$ of

$$\mathbf{u}' = \mathbf{F}(x, \mathbf{u}), \qquad \mathbf{u}(a) = \boldsymbol{\alpha}$$

be "close" to the solution $\mathbf{y}(x)$. A mathematical problem is said to be *well-posed* when "small" changes to the data, here both the initial value vector $\mathbf{A}$ and the function $\mathbf{F}(x, \mathbf{y})$, result in "small" changes to the solution. Whether a problem is well posed is a basic issue in the theory of ODEs and elementary results are stated in the appendix. Because the matter is fundamental to the numerical solution of initial value problems, we go into it now.

We need a way to measure how much $\mathbf{F}(x, \mathbf{u})$ changes when $\mathbf{u}$ is changed, but to do that we need a way to measure the size of a vector $\mathbf{v}$. A *vector norm* $\|\mathbf{v}\|$ generalizes the concept of the absolute value of a scalar to vectors and similarly a matrix norm provides a measure of the size of a matrix. The basic definitions and results for both kinds of norms and how they relate to each other are found in the appendix. Some are stated here so that they will be handy. Let the vector $\mathbf{v}$ have components v_i for $i = 1, \ldots, n$. Only two vector norms are common in the numerical solution of the initial value problem for a system of ODEs. They are the infinity, or max, norm,

$$\|\mathbf{v}\|_\infty = \max_{1\le i\le n} |v_i|,$$

and the Euclidean, or two, norm,

$$\|\mathbf{v}\|_2 = \left(\sum_{i=1}^{n} |v_i|^2\right)^{1/2}.$$

A norm that is occasionally used is the one norm,

$$\|\mathbf{v}\|_1 = \sum_{i=1}^{n} |v_i|.$$

There are *matrix norms* for $n \times n$ matrices $M = (M_{ij})$ that are *compatible* with a vector norm in the sense that

$$\|M\,\mathbf{v}\| \le \|M\|\ \|\mathbf{v}\|.$$

One such norm is the matrix norm *subordinate* to the vector norm:

$$\|M\| = \max_{x\ne o} \frac{\|M\,\mathbf{x}\|}{\|\mathbf{x}\|}.$$

In the case of the max vector norm, it is easy to compute the subordinate matrix norm from

$$\|M\|_\infty = \max_{1\le i\le n} \sum_{j=1}^{n} |M_{ij}|,$$

and in the case of the one vector norm, from

$$\|M\|_1 = \max_{1\le j\le n} \sum_{i=1}^{n} |M_{ij}|.$$

The matrix norm subordinate to the Euclidean vector norm is not readily computed. A compatible norm that is easy to compute is the Frobenius norm:

$$\|M\|_F = \left(\sum_{i=1}^{n}\sum_{j=1}^{n} |M_{ij}|^2\right)^{1/2}.$$

Returning now to measuring how much $\mathbf{F}(x, \mathbf{u})$ changes when $\mathbf{u}$ is changed, a function $\mathbf{F}$ is said to satisfy a *Lipschitz condition* with constant L in a region if

$$\|\mathbf{F}(x, \mathbf{u}) - \mathbf{F}(x, \mathbf{v})\| \leq L\|\mathbf{u} - \mathbf{v}\|$$

for all $(x, \mathbf{u})$ and $(x, \mathbf{v})$ in the region. To understand better what this means, first note that it implies continuity in the dependent variable $\mathbf{u}$—as $\mathbf{v}$ tends to $\mathbf{u}$, $\mathbf{F}(x, \mathbf{v})$ tends to $\mathbf{F}(x, \mathbf{u})$. To get more insight, let us first consider the scalar case when F has a continuous partial derivative. A mean value theorem tells us then that there is a point ξ between u and v for which

$$|F(x, u) - F(x, v)| = \left|\frac{\partial F}{\partial y}(x, \xi)(u - v)\right| = \left|\frac{\partial F}{\partial y}(x, \xi)\right| |u - v|.$$

A Lipschitz constant L provides a bound on the partial derivative here. Further, if the partial derivative is bounded in magnitude for all x in the interval and all ξ by a number L, this number will serve as a Lipschitz constant for F. Earlier we saw an example with $F = 3y^{2/3}$. It has a partial derivative $2y^{-1/3}$ that is not bounded in magnitude on any region that includes $y = 0$, hence does not satisfy a Lipschitz condition on such regions.

Linear differential equations, i.e., equations of the form

$$\mathbf{y}' = J(x)\,\mathbf{y} + \mathbf{g}(x) = \mathbf{F}(x, \mathbf{y}),$$

also help us to understand Lipschitz constants. For such equations,

$$\|\mathbf{F}(x, \mathbf{u}) - \mathbf{F}(x, \mathbf{v})\| = \|J(x)(\mathbf{u} - \mathbf{v})\| \leq \|J(x)\|\ \|\mathbf{u} - \mathbf{v}\|,$$

when the norm of the matrix $J(x)$ is compatible with the vector norm. A Lipschitz constant L can then be found as

$$L = \max_{a \leq x \leq b} \|J(x)\|.$$

This generalizes what was done for a single (scalar) equation to a system in the case of linear $\mathbf{F}$. As with a single equation, nonlinear $\mathbf{F}$ are studied with the aid of a mean value theorem. Using one taken up in the appendix we have

$$\mathbf{F}(x, \mathbf{u}) - \mathbf{F}(x, \mathbf{v}) = \mathcal{J}(\mathbf{u} - \mathbf{v}).$$

The matrix $\mathcal{J}$ here is the *Jacobian* of $\mathbf{F}$. It is the matrix that has $\partial F_i/\partial y_j$ as its (i, j) entry. Other notations for the Jacobian matrix are

$$\left(\frac{\partial F_i}{\partial y_j}\right) = \frac{\partial \mathbf{F}}{\partial \mathbf{y}} = \mathbf{F}_y.$$

In this mean value theorem, the entries of $\mathcal{J}$ are evaluated at x and *different* values of the dependent variables on the line between **u** and **v**. In the linear case, $\mathcal{J}$ is just the matrix $J(x)$. Though somewhat more complicated, the situation with vector systems is essentially the same as the scalar case.

EXERCISE 9. Airy's equation is $y'' = x\,y$. Put this linear equation into standard form. What is the Jacobian matrix $J(x)$? If the interval of interest is $0 \le a \le x \le b$, find a Lipschitz constant when the infinity norm is used and when the two norm is used. Use the subordinate matrix norm in the one case and the Frobenius matrix norm in the other.

Continuity of **F** is enough to guarantee the existence of a solution. A standard *uniqueness* result states that if **F** satisfies a Lipschitz condition, or as we say, is "Lipschitzian," then there is *at most* one solution. This follows from a stronger result about how fast two solutions can spread apart that we take up now. Suppose that in addition to the conditions of the existence theorem, **F** satisfies a Lipschitz condition with constant L. If $\mathbf{u}(x)$ and $\mathbf{v}(x)$ are any two solutions of the differential equation,

$$\mathbf{u}'(x) = \mathbf{F}(x, \mathbf{u})) \quad \text{and} \quad \mathbf{v}'(x) = \mathbf{F}(x, \mathbf{v}(x)),$$

then

$$\|\mathbf{u}(x) - \mathbf{v}(x)\| \le \|\mathbf{u}(a) - \mathbf{v}(a)\| e^{L(x-a)} \qquad \text{for} \quad x \ge a.$$

This inequality will be very important in our study of the numerical solution of the initial value problem. A sketch of this standard result can be found in the appendix. An initial value problem with Lipschitzian **F** can have at most one solution because if $\mathbf{u}(x)$ is one solution with initial value $\mathbf{u}(a) = \mathbf{A}$ and $\mathbf{v}(x)$ is another solution with $\mathbf{v}(a) = \mathbf{A}$, the inequality says that for all $x \ge a$, $\|\mathbf{u}(x) - \mathbf{v}(x)\| = 0$, hence that $\mathbf{u}(x) \equiv \mathbf{v}(x)$.

The last inequality bounds how much a change in the value of the solution at a point a can alter the solution at some $x > a$. We also need to consider the effect of changing **F**. A standard result about this is sketched in the appendix for norms that come from an inner product as, e.g., the Euclidean norm does. It is convenient to deal with changing both initial value and **F** in one inequality. If $\mathbf{u}(x)$ is a solution of

$$\mathbf{u}'(x) = \mathbf{F}(x, \mathbf{u}(x))$$

and $\mathbf{v}(x)$ is a solution of

$$\mathbf{v}'(x) = \mathbf{F}(x, \mathbf{v}(x)) + \mathbf{G}(x),$$

then

$$\|\mathbf{u}(x) - \mathbf{v}(x)\| \le \|\mathbf{u}(a) - \mathbf{v}(a)\| e^{L(x-a)} + \int_a^x \|\mathbf{G}(t)\| e^{L(x-t)}\, dt.$$

By the *classical situation* we mean that the quantity $L(b - a)$ is "not large." This inequality quantifies a fact of fundamental importance to the numerical solution of the initial value problem: In the classical situation, the initial value problem is *well posed*—small changes to the initial values $\mathbf{u}(a)$ and to the function $\mathbf{F}$ lead to small changes in the solution $\mathbf{u}(x)$.

In all that follows it is assumed that $\mathbf{F}$ is continuous in all variables and satisfies a Lipschitz condition in an appropriate region. This guarantees both the existence and uniqueness of solutions of problems with initial values in the region. Another standard result is that if $\mathbf{F}$ is continuous on $[a, b] \times R^n$ and satisfies a Lipschitz condition there, the initial value problem

$$\mathbf{y}' = \mathbf{F}(x, \mathbf{y}), \qquad a \le x \le b, \qquad \mathbf{y}(a) = \mathbf{A},$$

has a unique solution that extends all the way from a to b. Two things are different about this result: The integration is started on the boundary $x = a$ of the region where $\mathbf{F}$ is defined and it is asserted that the solution cannot "blow up" between $x = a$ and $x = b$.

§2 Order

Throughout this book we must talk about quantities tending to zero and we shall be vitally concerned with *how fast* they tend to zero. The concept of *order* of convergence quantifies this for us. The concept may already be familiar from courses in numerical or classical applied mathematics. Because some facility at using the concept is essential to understanding the numerical solution of the initial value problem, we take up now the basic definitions and properties.

A function $g_1(h)$ of a positive parameter h is said to be of order p, written as $\mathcal{O}(h^p)$ and spoken as "big oh of h to the p," if there is a constant C_1 such that for all sufficiently small h, say $0 < h \le h_1$,

$$|g_1(h)| \le C_1\, h^p.$$

The purpose of this concept is to describe how fast a function tends to 0 as h tends to 0. For this reason no attention is paid to the values of the constants C_1 and h_1 in the definition—it is the fact that such constants exist that is important. Clearly, the higher the order p, the faster $g_1(h)$ tends to zero as h does. The following properties of order follow easily from the basic definition.

If $g_1(h)$ is $\mathcal{O}(h^p)$, $g_2(h)$ is $\mathcal{O}(h^q)$, and γ is any constant, then

(1) the function $g_1(h)$ is $\mathcal{O}(h^r)$ for any $r < p$,
(2) the function $\gamma g_1(h)$ is $\mathcal{O}(h^p)$,

(3) the function $g_1(h)g_2(h)$ is $\mathbb{O}(h^{p+q})$, and
(4) the function $g_1(h) + g_2(h)$ is $\mathbb{O}(h^s)$ where $s = \min(p, q)$.

These rules say that a quantity of order p is also of order $r < p$, multiplying a quantity of order p by a constant does not change its order, multiplying quantities of orders p and q results in a quantity of order $p + q$, and adding quantities of orders p and q results in a quantity of order $s = \min(p, q)$. To see how the proofs go, let us look at property (1). By definition, there are constants h_1 and C_1 such that for all $0 < h \le h_1$,

$$|g_1(h)| \le C_1\, h^p = C_1\, h^{p-r}\, h^r \le C_1\, h_1^{p-r}\, h^r = C'\, h^r,$$

which says that $g_1(h)$ is $\mathbb{O}(h^r)$ for any $r < p$. Proving property (4) is a little more complicated. The definition of order involves an upper bound on the h that are permitted, and the bounds that appear in the definitions of the order of g_1 and g_2 might be different. So that both definitions are applicable, we take $h_3 = \min(h_1, h_2)$. Then for all $0 < h \le h_3$,

$$|g_1(h) + g_2(h)| \le |g_1(h)| + |g_2(h)| \le C_1\, h^p + C_2\, h^q.$$

Suppose that $s = \min(p, q) = q$. Then

$$|g_1(h) + g_2(h)| \le (C_1\, h^{p-q} + C_2)h^q \le (C_1\, h_3^{p-q} + C_2)h^q = C'\, h^q,$$

which says that $g_1 + g_2$ is of order q. The case of $s = p$ is handled similarly.

EXERCISE 10. Prove the other basic properties of order.

Often we shall be interested in a parameter h that might be negative. The assumption then is that there is a constant h_1 such that for all $|h| \le h_1$, we have $|g_1(h)| \le C_1\, |h|^p$. In this book the order p is typically a positive integer, but this is not necessary and there are occasions when fractional powers appear. To illustrate how negative values of the parameter arise and to obtain a collection of useful results, let us interpret Taylor series expansion using the concept of order. If $y(x)$ is any function that is continuous along with its first p derivatives in the finite interval $I = [x_0 - \varepsilon, x_0 + \varepsilon]$, Taylor's theorem with remainder tells us that for any $x \in I$,

$$y(x) = y(x_0) + y'(x_0)\delta + \cdots + \frac{1}{(p-1)!}\, y^{(p-1)}(x_0)\, \delta^{p-1} + R(\delta),$$

where

$$\delta = x - x_0, \qquad R(\delta) = \frac{1}{p!}\, y^{(p)}(\xi_x)\delta^p, \quad \text{and} \quad \xi_x \in I.$$

With the assumptions made, $|y^{(p)}(\xi)|$ is bounded on I so that $R(\delta)$ is $\mathcal{O}(\delta^p)$. The expansion is written more informally as

$$y(x) = y(x_0) + y'(x_0)\delta + \cdots + \frac{1}{(p-1)!} y^{(p-1)}(x_0)\delta^{p-1} + \mathcal{O}(\delta^p).$$

An expansion like this is used to provide an approximation to $y(x)$ for x near x_0. As $x \to x_0$, the error of this approximation tends to 0 at a rate that is $\mathcal{O}((x - x_0)^p)$. Examples used throughout the book are some approximations to exponentials:

$$e^x = 1 + x + \mathcal{O}(x^2) = 1 + x + \frac{1}{2}x^2 + \mathcal{O}(x^3) = \cdots$$

We say that $1 + x$ approximates $\exp(x)$ to order two, $1 + x + \tfrac{1}{2}x^2$ approximates $\exp(x)$ to order three, and so forth. Another important expansion is the binomial series

$$(1 + x)^\gamma = 1 + \gamma x + \frac{\gamma(\gamma - 1)}{2!}x^2 + \cdots + \frac{\gamma(\gamma - 1)(\gamma - 2)\cdots(\gamma - k + 1)}{k!}x^k + \cdots$$

which converges for $x^2 < 1$. It tells us, for example, that $1 + \tfrac{1}{2}x$ approximates $\sqrt{1 + x}$ to order two as x tends to zero.

Often we want to form composite expressions. A result general enough for our needs is that if $g(x)$ is $\mathcal{O}(x^p)$ and x is $\mathcal{O}(\delta^q)$, then $g(\delta)$ is $\mathcal{O}(\delta^{pq})$. The hypotheses stated more formally are that there are x_0 and C such that for all $|x| \le x_0$, the inequality $|g(x)| \le C|x|^p$ holds and that there are δ_0 and c such that for all $|\delta| \le \delta_0$, the inequality $|x| \le c|\delta|^q$ holds. The fact that x is $\mathcal{O}(\delta^q)$ implies that for all sufficiently small δ, we have $|x| \le x_0$. Then $|g(x)| \le C|x|^p \le C\, c^p\, |\delta|^{pq}$, which states that g is $\mathcal{O}(\delta^{pq})$.

Proving a relationship needed in the discussion of an example in the next section will show how the rules of order are used. The result is

$$(a\delta + b\delta^2 + \mathcal{O}(\delta^3))^2 = a^2\delta^2 + 2ab\delta^3 + \mathcal{O}(\delta^4).$$

Here a and b are constants, and the parameter δ tends to zero. Writing the left-hand side more formally as $(a\delta + b\delta^2 + g(\delta))^2$ and then multiplying it out in a straightforward way leads to

$$a^2\delta^2 + 2ab\delta^3 + b^2\delta^4 + 2a\delta g(\delta) + 2b\delta^2 g(\delta) + g^2(\delta) = a^2\delta^2 + 2ab\delta^3 + r(\delta).$$

Obviously the term $b^2\delta^4$ is $\mathcal{O}(\delta^4)$ and because $g(\delta)$ is $\mathcal{O}(\delta^3)$, the rules tell us that the term $2a\delta g(\delta)$ is $\mathcal{O}(\delta^4)$, $2b\delta^2 g(\delta)$ is $\mathcal{O}(\delta^5)$, and $g^2(\delta)$ is $\mathcal{O}(\delta^6)$. Now using repeatedly the rule about the order of the sum of two terms, we find that $r(\delta)$ is $\mathcal{O}(\delta^4)$, which is the result we wanted.

Another example that will be useful later is the relationship

$$\begin{aligned} T &= (1 + a\delta + b\delta^2 + \mathcal{O}(\delta^3))^\gamma \\ &= 1 + \gamma a\delta + \left(\gamma b + \frac{\gamma(\gamma - 1)}{2!} a^2\right)\delta^2 + \mathcal{O}(\delta^3) \end{aligned}$$

From the binomial series we have

$$(1 + x)^\gamma = 1 + \gamma x + \frac{\gamma(\gamma - 1)}{2!} x^2 + \mathcal{O}(x^3).$$

In this expansion we take $x = a\delta + b\delta^2 + g(\delta)$, where $g(\delta)$ is the $\mathcal{O}(\delta^3)$ term in T. According to the rules of order, x is $\mathcal{O}(\delta)$. If we let $r(x)$ be the $\mathcal{O}(x^3)$ term in the truncated binomial series, the result about composite functions tells us that r is $\mathcal{O}(\delta^3)$. Now we just substitute for x and use the rules of order for δ to get the desired result:

$$\begin{aligned} T &= (1 + a\delta + b\delta^2 + g(\delta))^\gamma \\ &= 1 + \gamma(a\delta + b\delta^2 + g(\delta)) + \frac{\gamma(\gamma - 1)}{2!}(a\delta + b\delta^2 + g(\delta))^2 + \mathcal{O}(\delta^3) \\ &= 1 + \gamma a\delta + \left(\gamma b + \frac{\gamma(\gamma - 1)}{2!} a^2\right)\delta^2 + \mathcal{O}(\delta^3). \end{aligned}$$

§3 Difficulties and Techniques for Handling Them

In this section we give a series of examples that illustrate common difficulties and show how they might be handled. Many books provide the physical context of differential equations and discuss the meaning of their numerical solutions. The *Differential Equations Laboratory Workbook* of Borrelli, Coleman, and Boyce [1992] is a good book of this kind because it has clear discussions of the physical context and nice graphical solutions obtained using a quality integrator. Our aim here is different. We are concerned not with the meaning of the equation and its solution, rather with how to integrate the initial value problem numerically. We do illustrate some of the many ways that ODEs arise, but we are mainly interested in looking at those characteristics of the problems that affect their numerical solution.

EXAMPLE 1. Problems with discontinuous $\mathbf{F}$ arise naturally in certain applications. An example will serve to illustrate further the reduction of problems to standard form and the handling of problems that are not as smooth as the codes expect. Diffusion and heat transfer in one space variable are described by an equation of the form

$$\frac{d}{dx}\left(p(x)\,\frac{dy}{dx}\right) + q(x)\,y = g(x).$$

It is assumed that $p(x) > 0$ for all x in the interval $[a, b]$ (otherwise the equation would be singular). If the coefficient $p(x)$ is differentiable, the equation can be written as

$$p(x)\,y'' + p'(x)\,y' + q(x)\,y = g(x),$$

or

$$y'' + \frac{p'(x)}{p(x)}\,y' + \frac{q(x)}{p(x)}\,y = \frac{g(x)}{p(x)}.$$

This can be put into standard form in the usual way by introducing unknowns $y_1 = y$ and $y_2 = y'$. However, converting to standard form in this way requires the derivative of $p(x)$ and this derivative may be inconvenient or impossible to obtain. Furthermore, in many physical problems the coefficients $p(x)$ and $q(x)$ are only piecewise smooth. Imagine, for example, the steady flow of heat through a stack of plates of different materials. It is usual to assume that the conductivity is constant for each material, but it will be different for different materials. Such a problem does not belong to the class of problems we investigate because the function $\mathbf{F}$ is not continuous, but it can be treated as a combination of problems that do belong to the class.

To see how we can deal with problems that involve discontinuities in the independent variable x, suppose that the coefficients are smooth on an interval $[a, b)$ and on an interval $(b, c]$. This would model, e.g., heat flowing through one material corresponding to the first interval and on through a second material corresponding to the second interval. The idea is to "tack together" solutions of two initial value problems involving coefficients that are smooth:

Problem I $\quad y_{\mathrm{I}}(a)$ given,

$$y_{\mathrm{I}}' = F_{\mathrm{I}}(x, y_{\mathrm{I}}), \qquad a \le x \le b,$$

Problem II $\quad y_{\mathrm{II}}(b) = y_{\mathrm{I}}(b),$

$$y_{\mathrm{II}}' = F_{\mathrm{II}}(x, y_{\mathrm{II}}), \qquad b \le x \le c.$$

We are supposing that the function $F_{\mathrm{I}}(x, y_{\mathrm{I}})$ is continuous and Lipschitzian on $[a, b]$ so that Problem I is in the class we are studying. Because its solution extends all the way to b, the value $y_{\mathrm{I}}(b)$ can be taken as the initial value $y_{\mathrm{II}}(b)$ for Problem II. This second problem has a function $F_{\mathrm{II}}(x, y_{\mathrm{II}})$ that we are supposing is continuous and Lipschitzian, so it is also in the class we study. As far as solving such problems with a standard code is concerned, the code just needs to "realize" that it must produce an answer $y(b)$ and start on a new problem there. More will be said about this later when we have learned more about how codes work, but the potential difficulties due to discontinuous functions should not be underestimated. It is easy to deal with the theory of discontinuities in the independent variable, but codes can have great difficulty integrating such problems. We shall see that even the theory promises difficulties when there is a lack of smoothness involving the dependent variables (the solution components).

The usual introduction of new unknowns allows us to handle layered media by splitting up the task into a succession of problems, but it can be unnatural and inconvenient. In the case of diffusion, the equation is of second order so that both $y_{\mathrm{II}}(b)$ and $y'_{\mathrm{II}}(b)$ are needed as initial conditions for the second interval. The solution $y_{\mathrm{I}}(x)$ corresponds to a temperature or a concentration. It is assumed to be continuous so that the proper value of $y_{\mathrm{II}}(b)$ is $y_{\mathrm{I}}(b)$. On the other hand, at a change of material it is usually assumed that it is the flux, $p(x)\, y'(x)$, that is continuous. This means that if p is discontinuous at b, y' must also be discontinuous. Specifically, if the flux is to be continuous, the initial value $y'_{\mathrm{II}}(b)$ must be $p(b-)y'_{\mathrm{I}}(b-)/p(b+)$. (The standard notation seen here in $p(b-)$ means the limit of $p(b - \varepsilon)$ as the positive quantity ε tends to 0 and $p(b+)$ is defined similarly.) Because $p(x)\, y'$ is to be continuous rather than y', it is more natural to introduce it as one of the unknowns. Taking then as unknowns $y_1 = y$ and $y_2 = p(x)\, y'$, we have on each interval where $p(x)$ is smooth,

$$\frac{d}{dx} y_2 = \frac{d}{dx}\left(p(x)\frac{dy}{dx}\right) = g(x) - q(x)\, y = g(x) - q(x)\, y_1$$

and

$$\frac{d}{dx} y_1 = y' = y_2/p(x).$$

These unknowns have two virtues. The more important one is that it is not necessary to differentiate $p(x)$. The other is that because the variables $y_1(x)$, $y_2(x)$ are continuous at the interface, it is easier to obtain the initial values for the second problem.

Heat flow is only one context in which equations like these arise, but in a natural way it leads us to think more about the role of initial conditions. If the equation described the flow of heat through a stack of plates, a common way to specify the solution of the differential equation would be to specify the temperatures at the two ends. Another possibility is that the stack is insulated at one end so that the flux is zero there (there is no flow of heat). This amounts to specifying the derivative of the solution at that end. The solution can be specified in more complicated ways. If heat transfer from the stack to the medium is by convection at one end, the flow of heat is usually assumed to be proportional to the difference between the temperature at the end and the temperature of the medium, a condition on the solution and its derivative. If the transfer is by radiation, it is usually assumed to be proportional to the fourth power of the temperature—a nonlinear condition on the solution. Specifying the solution of a differential equation in terms of the values of the solution and its derivatives at several points is quite common. Problems that specify the solution in such ways are called *boundary value problems*. One way to solve them is by means of solving initial value problems. Techniques for this will be discussed later, but only to make points about solving initial value problems. The numerical solution of boundary value problems is sufficiently different that a full treatment requires a book all its own. A good one is that by Ascher, Mattheij, and Russell [1988].

EXAMPLE 2. The separation of variables technique for the solution of partial differential equations gives rise to ordinary differential equations; see, for example, Haberman [1987]. More specifically, it leads to the *Sturm–Liouville problem*

$$
\begin{aligned}
(p(x)\, y')' + (\lambda\, \rho(x) - q(x))\, y &= 0, \qquad a \le x \le b,\\
y(a) \cos A - p(a)\, y'(a) \sin A &= 0,\\
y(b) \cos B - p(b)\, y'(b) \sin B &= 0.
\end{aligned}
$$

Here A and B are given numbers with $-\pi/2 \le A < \pi/2$ and $-\pi/2 \le B < \pi/2$. It is supposed that $[a, b]$ is a finite interval, that $\rho(x)$ and $q(x)$ are continuous on $[a, b]$, and that $p(x)$ is both continuous and has a continuous derivative on $[a, b]$. It is further supposed that $p(x)$ and $\rho(x)$ are positive on $[a, b]$. This defines a regular Sturm–Liouville problem; singular problems involving infinite intervals and/or p that vanish at an end point and/or q that become infinite at an end point are also important. A Sturm–Liouville problem always has the trivial solution $y(x) \equiv 0$. It is known that there are values of the parameter λ for which there is a nontrivial solution $y(x)$. Such a value λ_m is called an eigenvalue and the corresponding solution $y_m(x)$ is called the eigenfunction corresponding to λ_m. It is known that in the circumstances postulated, there is a countable sequence of real eigenvalues with $\lambda_0 < \lambda_1 < \cdots$ and eigenfunctions corresponding to different eigenvalues are orthogonal in the sense that

$$\int_a^b y_n(x)\, y_m(x)\, \rho(x)\, dx = 0 \quad \text{for} \quad n \neq m.$$

This underlies the possibility of expanding a large class of functions in terms of eigenfunctions:

$$f(x) \sim \sum_{n=0}^{\infty} c_n\, y_n(x),$$

where

$$c_n = \frac{\int_a^b f(x)\, y_n(x)\, \rho(x)\, dx}{\int_a^b y_n(x)\, y_n(x)\, \rho(x)\, dx}.$$

This expansion is fundamental to the solution of the partial differential equation. The Sturm–Liouville problem is to determine the eigenvalues, the corresponding eigenfunctions, and the normalization constants

$$\int_a^b y_n(x)\, y_n(x)\, \rho(x)\, dx.$$

Subsequently it is necessary to compute the expansion coefficients c_n and to evaluate the $y_n(x)$ at points x of interest.

This problem exemplifies the specification of a solution of a differential equation by something other than its initial values. Because of the convenience and power of codes for the initial value problem, a popular way to solve the Sturm–Liouville problem is by initial value methods. The approach is called a shooting method. The idea is to choose a nonzero value for either $y(a)$ or $y'(a)$ and then define the other quantity so as to satisfy the specified condition at a,

$$y(a) \cos A - p(a)\, y'(a) \sin A = 0.$$

If a value of λ is chosen, the differential equation can be integrated with these initial values from a to b to compute $y(b; \lambda)$. Here the dependence of this value on λ has been made explicit in the notation. This is called "*shooting* from a to b." The task is then to find a value λ such that the other boundary condition,

$$G(\lambda) = y(b; \lambda) \cos B - p(b)\, y'(b; \lambda) \sin B = 0,$$

is satisfied. This is a nonlinear algebraic equation that is to be solved by standard methods for a root λ_n of $G(\lambda) = 0$. Each evaluation of G involves the integration of an initial value problem, so the method used to solve the algebraic equation must be efficient in the sense of requiring few evaluations of G. This is a demanding task for nonlinear equation solvers because the evaluation of G is of limited accuracy, namely the accuracy obtained from the integration of the differential equation, and the more accuracy asked of the evaluation of G, the more expensive the evaluation is. What is meant by "solving an initial value problem" is clear in principle, but in practice there are several possible meanings. Here we see an example where solving the problem means approximating the solution at the single point $x = b$.

After finding an eigenvalue λ_n by shooting, the corresponding eigenfunction is $y_n(x) = y(x; \lambda_n)$. The values of the eigenfunction computed automatically in the course of computing λ_n may not correspond to the x needed for a particular purpose. Values can be obtained anywhere they are needed by integrating the differential equation again as an initial value problem. In contrast to the situation when computing an eigenvalue, "solving" the initial value problem might now mean that values of the solution are required throughout the interval of integration. Indeed, this is the case when computing the normalization constant. A convenient way to compute this quantity is to introduce a new unknown $n(x)$ by

$$n'(x) = y^2(x)\,\rho(x), \qquad n(a) = 0,$$

which is to be integrated along with $y(x)$ to obtain

$$n(b) = \int_a^b y^2(x)\,\rho(x)\,dx.$$

When $\lambda = \lambda_n$, this yields the normalization constant. A similar device provides a way to compute the expansion of a function $f(x)$. Just add another equation for the unknown $cn(x)$,

$$cn'(x) = f(x)\,y(x)\,\rho(x), \qquad cn(a) = 0,$$

and then $c_n = cn(b)/n(b)$. These are good examples of the value of introducing new variables for quantities of fundamental interest.

There are advantages to taking the flux $p(x)\,y'(x)$ as an unknown rather than $y'(x)$ when solving a Sturm–Liouville problem by shooting. It is often advantageous for this special, but important, class of problems to introduce new dependent variables. Some widely used codes are based on the Prüfer transformation that introduces the phase θ and amplitude r of the solution so that

$$y(x) = r(x)\sin\theta(x) \quad \text{and} \quad p(x)\,y'(x) = r(x)\cos\theta(x).$$

EXERCISE 11. Show that the Prüfer transformation converts the Sturm–Liouville problem into

$$\theta' = \frac{1}{p(x)}\cos^2\theta + (\lambda\,\rho(x) - q(x))\sin^2\theta,$$

$$\theta(a) = A, \qquad \theta(b) = B + n\pi,$$

along with

$$r' = \left(\frac{1}{p(x)} - (\lambda\,\rho(x) - q(x))\right)\frac{r}{2}\sin 2\theta.$$

Notice that the equation for the phase can be integrated independently of that for the amplitude and that the boundary condition is stated in terms of θ alone. This means that to compute an eigenvalue, only the first-order equation for the phase need be integrated, as contrasted to the second-order equation for y.

EXAMPLE 3. The differential equation

$$\frac{d^2y}{dt^2} = \ddot{y} = -y \tag{3.1}$$

arises in the modeling of many physical systems. For example, Rouse [1946, p. 127 ff.] models the pendulation of a column of inviscid fluid in a U-tube by this equation. (More precisely, this is the nondimensional form of the equation derived.) The sketch shows the liquid sloshing up and down in the tube. In our notation $y(t)$ measures the displacement at time t of the surface of the fluid on the right-hand side of the tube from its rest position.

To investigate the solutions of (3.1), Rouse employs the standard technique of deriving a conservation law for the energy E of the motion. If

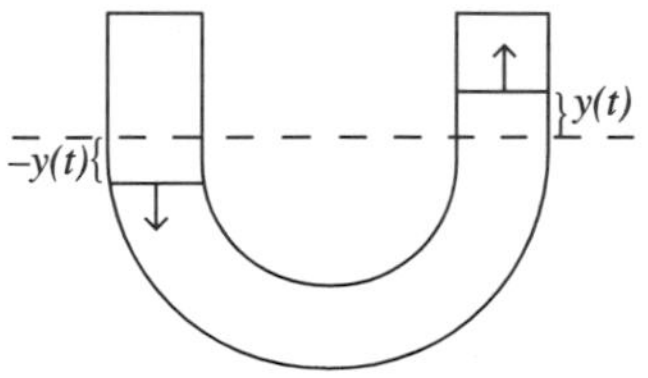

Figure 1.1 Pendulation of liquid in a U-tube.

$y(t)$ is a solution of (3.1), the energy E

$$E = \frac{1}{2}\dot{y}^2(t) + \frac{1}{2}y^2(t) \tag{3.2}$$

satisfies

$$\frac{d}{dt}E = \dot{y}(t)\,(\ddot{y}(t) + y(t)) = 0.$$

This says that E has a constant value determined by the initial conditions for $y(t)$.

EXERCISE 12. Solve (3.1) numerically. At each step compute the difference between the correct value of the energy E as obtained from the initial conditions and the approximate energy $\frac{1}{2}y_1^2 + \frac{1}{2}y_2^2$ from the approximations y_1 to the solution and y_2 to the derivative of the solution returned by your code. If the approximations are accurate, the difference must be small, but you will find that the approximate energies do not agree exactly with E. The agreement is likely to get worse as the integration proceeds.

Let us digress briefly to consider *conservation laws* more generally. These are relationships satisfied by solution components that are consequences of the form of the equation. They often express physical laws such as the conservation of energy, conservation of angular momentum, conservation of mass, charge balance, and the like. Linear conservation laws arise when $\mathbf{F}(x, \mathbf{u})$ is such that there is a constant vector $\mathbf{v}$ for which

$$\mathbf{v}^T\,\mathbf{F}(x, \mathbf{u}) \equiv 0.$$

Then if $\mathbf{y}(x)$ is a solution of the differential equation, we have

$$\mathbf{v}^T\,\mathbf{y}'(x) = \mathbf{v}^T\,\mathbf{F}(x, \mathbf{y}(x)) = 0,$$

and after integration of this equality,

$$\mathbf{v}^T\,\mathbf{y}(x) - \mathbf{v}^T\,\mathbf{y}(a) = 0.$$

This states that the linear combination of solution components $\mathbf{v}^T\,\mathbf{y}(x)$ is constant for any solution $\mathbf{y}(x)$ of the differential equation. The *Robertson problem*

$$\begin{aligned} y_1' &= -0.04y_1 + 10^4 y_2\, y_3, & y_1(0) &= 1,\\ y_2' &= 0.04y_1 - 10^4 y_2\, y_3 - 3 \times 10^7 y_2^2, & y_2(0) &= 0,\\ y_3' &= 3 \times 10^7 y_2^2, & y_3(0) &= 0, \end{aligned}$$

describing the concentrations of reactants in a certain chemical reaction was stated by Robertson [1967] and subsequently used often as a test problem. It takes just a moment to see that the solution satisfies $y_1(x) + y_2(x) + y_3(x) \equiv 1$. Interestingly, all the standard methods produce numerical approximations that satisfy all linear conservation laws, a fact that you will be asked to prove in Chapter 4.

It is important to understand what a conservation law can tell us. Suppose a solver computes $\tilde{y}_i(x) = y_i(x) + e_i(x)$ as the solution of Robertson's problem. If the numerical approximation satisfies the conservation law,

$$1 = \sum_{i=1}^{3} \tilde{y}_i(x) = \sum_{i=1}^{3} y_i(x) + \sum_{i=1}^{3} e_i(x) = 1 + \sum_{i=1}^{3} e_i(x),$$

we cannot conclude that the approximation is accurate, just that the errors are *correlated* so that

$$\sum_{i=1}^{3} e_i(x) = 0.$$

Obviously the errors can be arbitrarily large and still satisfy this relationship. Furthermore, if the approximation does not satisfy the law, say

$$\sum_{i=1}^{3} \tilde{y}_i(x) = 1 + \delta,$$

we can conclude only that

$$\sum_{i=1}^{3} e_i(x) = \delta.$$

Failure to satisfy a conservation law shows that the approximations are in error, but generally we have no idea how much. For the example at hand, we have

$$|\delta| = \left| \sum_{i=1}^{3} e_i(x) \right| \leq \sum_{i=1}^{3} |e_i(x)| \leq 3 \max_i |e_i(x)|,$$

so at least one of the components is in error by *at least* $|\delta|/3$.

Nonlinear conservation laws arise when there is a function $G(t, \mathbf{u})$ such that

$$\frac{\partial G}{\partial t}(t, \mathbf{u}) + \frac{\partial G}{\partial \mathbf{u}}(t, \mathbf{u})\, \mathbf{F}(t, \mathbf{u}) \equiv 0,$$

for then

$$\frac{d}{dt} G(t, \mathbf{u}(t)) \equiv 0$$

for any solution $\mathbf{u}(t)$ of the differential equation and

$$G(t, \mathbf{u}(t)) \equiv G(a, \mathbf{u}(a)).$$

Numerical methods generally do *not* produce solutions that satisfy *non*linear conservation laws. To fit equation (3.1) into this framework, we first write it as a first-order system with $\mathbf{u} = (u_1, u_2)^T = (y, \dot{y})^T$ and $\mathbf{F}(t, \mathbf{u}) = (u_2, -u_1)$. Then $G(t, \mathbf{u}) = \frac{1}{2}u_1^2 + \frac{1}{2}u_2^2$ satisfies

$$\frac{\partial G}{\partial t}(t, \mathbf{u}) + \frac{\partial G}{\partial \mathbf{u}}(t, \mathbf{u})\, \mathbf{F}(t, \mathbf{u}) = 0 + (u_1, u_2)\begin{pmatrix} u_2 \\ -u_1 \end{pmatrix} \equiv 0,$$

showing that the energy $E = G(t, \mathbf{u}(t))$ is constant.

Let us return now to the pendulation problem. For the value of E determined by the initial conditions for the original equation, (3.2) is a first-order differential equation for $y(t)$. Rouse puts it into standard form and investigates its solutions to determine the period of a complete pendulation. Let us consider the solution of (3.1) with initial conditions $y(0) = 0$, $\dot{y}(0) = 1$. Obviously $y(t) = \sin(t)$ furnishes the solution of the initial value problem. With these initial conditions E is $\frac{1}{2}$ and for increasing y, equation (3.2) in standard form is

$$\dot{y} = \sqrt{1 - y^2} = F(y). \tag{3.3}$$

The single initial condition for this equation is $y(0) = 0$. This is all straightforward enough, but if you attempt to solve this problem numerically, you are likely to get better acquainted with the error handling facilities of your computer system. The difficulty is that the solution we wish to compute increases to 1 at $t = \pi/2$ and this value of y is on the boundary of the region for which $F(y)$ is defined. We cannot, of course, expect the solution value of 1 to be computed exactly. If the numerical solution should be greater than 1, there will be a system failure because of an attempt to compute the square root of a negative number.

It is often the case that the person posing an initial value problem is sure on physical grounds that the solution must satisfy certain bounds. It is then easy to forget to define the equation for values of y that are not meaningful physically. A related matter is that the definition for physically meaningful values of y might extend to other y, but the behavior of solutions of the equation for nonphysical arguments might be dramatically different. Earlier we talked about starting on the edge of the region of physically meaningful values. Approaching it can also lead to disaster. In addition to the example at hand, we mention the modeling of chemical reactions taking place in a fluid. Robertson's problem is a specific example. Negative values of concentrations of reactants have no physical meaning. Often the concentration of some reactant tends to zero and it is then not unusual that a code produce an approximate concentration that is negative. Models of some reactions are defined formally for negative concentrations, but the equations are unstable, and solutions "blow up." In Chapter 8 we report numerical results of this kind for Robertson's problem. One way to cope with this kind of difficulty is to define the equation for arguments that are at least a "little" negative. For the example at hand, a simple way to obtain an $F(y)$ that is continuous for all y is to define $F(y)$ to have the value 0 when $y^2 \geq 1$.

In principle we would integrate (3.3) until our solution achieved the value 1 and then switch to the equation

$$\dot{y} = -\sqrt{1 - y^2} = G(y) \qquad \text{for} \quad t \geq \pi/2 \tag{3.4}$$

with initial condition

$$y(\pi/2) = 1. \tag{3.5}$$

However, besides the desired solution $y(t) = \sin(t)$, this initial value problem obviously has the additional solution $y(t) \equiv 1$. In point of fact, there is a family of solutions:

$$\begin{aligned} y(x) &= 1 & \pi/2 \leq t \leq \gamma \\ &= \sin(t - \gamma + \pi/2) & \gamma < t. \end{aligned}$$

To see that $y(x)$ is a solution, we must first verify that it is C^1. This is obvious except at $t = \gamma$, where it is easily verified that $y(\gamma-) = y(\gamma+)$ and $\dot{y}(\gamma-) = \dot{y}(\gamma+)$. The initial condition is obviously satisfied and it is easily verified that each of the forms for $y(x)$ given do satisfy equation (3.4). We might have expected some kind of trouble because the partial derivative of G with respect to y is not bounded as y increases to 1—the function G does not satisfy a Lipschitz condition on any region including $y = 1$. (A Lipschitz condition is sufficient for uniqueness, but it is not necessary, so an unbounded partial derivative is just a warning.)

We see from this example that problems with singular points and a failure of uniqueness are not rare. Such problems are never routine; their solutions should always be studied analytically. In this particular example, the difficulty is artificial. Although the energy equation is very useful theoretically, and sometimes numerically, here we should integrate numerically the original equation (3.1) because its solution *is* routine, whereas the energy equation is singular every time that $y^2(t) = 1$. To illustrate the handling of singular points, we consider the solution of the initial value problem (3.4, 3.5). The idea is to approximate the solution of interest by analytical means near the singular point and then to use numerical means elsewhere. Just what "analytical means" are employed will depend on the problem and the knowledge of the investigator. Often some kind of series expansion is assumed.

Frequently the physical origin of the problem tells the analyst that there is a well-behaved solution and it is just necessary to distinguish it from other possibilities. Here, for example, we expect a smooth solution $y(t)$, hence expect to be able to expand $y(t)$ near $t = \pi/2$ in a Taylor series. To illustrate a more typical approach, let us attempt to find a solution that for t near $\pi/2$ has an expansion of the form

$$y(t) = 1 + \delta(t - \pi/2)^{\mu} + \text{higher order powers of } (t - \pi/2).$$

In this expansion we have already imposed the initial condition $y(\pi/2) = 1$ and we wish to determine the constants δ and μ. To express this more carefully, we first introduce the new variable $x = t - \pi/2$ so that we are approximating y for small x. We seek a solution in the form

$$y(t) = 1 + \delta\, x^{\mu} + \mathcal{O}(x^{\mu+1}),$$

and, correspondingly, an approximation to its derivative in the form

$$\dot{y}(t) = \mu\delta\, x^{\mu-1} + \mathcal{O}(x^{\mu}).$$

Multiplying the expansion for $y(t)$ times itself leads to

$$y^2(t) = 1 + 2\delta\, x^{\mu} + \mathcal{O}(x^{\mu+1}),$$

and then a little manipulation gives

$$\begin{aligned}\sqrt{1 - y^2(t)} &= \sqrt{-2\delta\, x^{\mu} + \mathcal{O}(x^{\mu+1})},\\ &= x^{\mu/2}\sqrt{-2\delta}\,\sqrt{1 + \mathcal{O}(x)}.\end{aligned}$$

A result proved earlier using the binomial theorem allows the square root to be expanded. Then

$$\begin{aligned}\sqrt{1 - y^2(t)} &= x^{\mu/2}\sqrt{-2\delta}\,(1 + \mathcal{O}(x))\\ &= x^{\mu/2}\sqrt{-2\delta} + \mathcal{O}(x^{1+(\mu/2)}).\end{aligned}$$

Substituting this expression and the corresponding one for $\dot{y}(t)$ into the differential equation shows that if there is to be a solution $y(t)$ of the postulated form, we must have

$$\mu\delta\, x^{\mu-1} + \mathcal{O}(x^{\mu}) = x^{\mu/2}\sqrt{-2\delta} + \mathcal{O}(x^{1+(\mu/2)}).$$

If this is to hold for all t near $\pi/2$, i.e., for all sufficiently small x, it is necessary that

$$\mu - 1 = \mu/2 \quad \text{and} \quad \mu\delta = \sqrt{-2\delta},$$

hence that $\mu = 2$ and either $\delta = 0$ or $\delta = -\frac{1}{2}$. Here is where we recognize that there are two, and only two, solutions of the initial value problem.

To leading order, one is the constant 1 and the other is

$$y(t) = 1 - \frac{1}{2}(t - \pi/2)^2 + \mathcal{O}((t - \pi/2)^3).$$

(Of course, we are simply working out the Taylor series expansion of the solution $\sin(t)$ about $t = \pi/2$.)

With an expansion for $y(t)$, we can compute a value Y that approximates $y(\gamma)$ for some $\gamma > \pi/2$. We then solve (3.4) numerically for $t \geq \gamma$ with initial condition $y(\gamma) = Y$. A modern code for the numerical integration will ask you to specify the accuracy you want. To select a sensible value for the rest of the integration, we need to estimate the error in the starting value Y. If we approximate Y by the first term in the expansion, we have

$$Y - 1 \approx -\frac{1}{2}x^2 = -\frac{1}{2}(\gamma - \pi/2)^2.$$

This corresponds to approximating the error of a truncated series by the first term omitted. It will be a good approximation for all sufficiently small x. Alternatively, to say that x is small enough to neglect the $\mathcal{O}(x^3)$ terms means that we believe that the error in approximating Y by $1 - \frac{1}{2}x^2$ is rather smaller in magnitude than $|-\frac{1}{2}x^2|$. This corresponds to saying that the error of a truncated series will be rather smaller in magnitude than the last term retained. Again, this will be useful for all sufficiently small x. In this particular instance, we cannot approximate Y by 1 because the problem is still singular then, hence must adopt the second approach to assessing the error or else we must work out another term in the expansion of $y(t)$ about $\pi/2$ so that we can actually estimate the error $Y - (1 - \frac{1}{2}x^2)$ rather than (approximately) bounding it. Proceeding in this way, we move away from the singularity by analytical means and then perform the numerical integration in a region where G is smooth.

EXAMPLE 4. One of the important ways that partial differential equations are reduced to ordinary differential equations is by the assumption of a special form of solution. Traveling wave solutions furnish an example. Main [1987] discusses wave motion in a medium that is slightly dispersive and slightly nonlinear in terms of solutions $\psi(z, t)$ of the partial differential equation

$$\frac{\partial\psi}{\partial t} + c_0(1 + b\psi)\frac{\partial\psi}{\partial z} + d\left(\frac{\partial^3\psi}{\partial z^3}\right) = 0.$$

It is asked whether there are solutions that have the form of a localized hump of height ψ_{max} with ψ dropping smoothly to zero on both sides. A traveling wave solution is one for which $\psi(z, t)$ is a function of the single variable $Z = vt - z$, where the speed v is a constant that must be determined. The assumption of a traveling wave solution $\psi(Z)$ leads to

$$\frac{\partial\psi}{\partial t} = v\,\frac{d\psi}{dZ} = v\,\psi'$$

and similar expressions for the other partial derivatives. This results in an ordinary differential equation

$$(v - c_0)\psi' - bc_0\,\psi\,\psi' - d\,\psi''' = 0. \tag{3.6}$$

Following Main, this is integrated to yield

$$(v - c_0)\psi - \frac{1}{2}\,bc_0\,\psi^2 - d\,\psi'' + C = 0, \tag{3.7}$$

where C is a constant of integration. This equation is multiplied by ψ' and integrated to get

$$\frac{1}{2}(v - c_0)\psi^2 - \frac{1}{6}\,bc_0\,\psi^3 - \frac{1}{2}\,d\,\psi'^2 + C\,\psi + D = 0,$$

where D is another constant of integration. For a traveling wave solution that is an isolated hump, we need to have ψ, ψ', and ψ'' tend to 0 as Z tends to $\pm\infty$. This requires that $C = D = 0$. A little manipulation of the equation then leads to

$$(\psi')^2 = (A - B\psi)\psi^2 \tag{3.8}$$

where $A = (v - c_0)/d$ and $B = bc_0/3d$. If the solution is to have its maximum value ψ_{max} at $Z = 0$, the condition $\psi'(0) = 0$ fixes the speed v because equation (3.8) then requires that

$$A - B\,\psi_{max} = 0, \tag{3.9}$$

hence that

$$v = c_0\left(1 + \frac{1}{3}\,b\,\psi_{max}\right).$$

We refer the reader to Main's very readable book for the physical origin and meaning of the equation. This brief derivation of the traveling wave solution has been provided to illustrate one way that partial differential equations can lead to ordinary differential equations. Let us now consider the numerical solution of equation (3.8) with the initial condition $\psi(0) = \psi_{max}$. To put the equation into standard form, we must take a square root. Because we start at the top of the hump and integrate in the direction of increasing Z, the negative value of the square root is the appropriate one:

$$\psi' = -\sqrt{(A - B\psi)\,\psi^2}.$$

If we use the condition (3.9) that fixed the speed, we can write the equation as

$$\psi' = -\sqrt{B\,(\psi_{\max} - \psi)\,\psi^2}.$$

With the initial conditions specified, the integration is starting on the edge of the region where the function is defined. This is not a formality—the initial value problem obviously has a constant solution $\psi \equiv \psi_{\max}$ in addition to the one that we expect on physical grounds.

The problem at hand ought to have a smooth solution, which leads us to believe that a power series is an appropriate way to approximate the solution near the initial point. To sort out the behavior of solutions at the initial point, one term in an expansion for ψ will be found. Let us look for constants δ, μ such that

$$\psi(Z) = \psi_{\max} - \delta\, Z^{\mu} + \cdots .$$

With this assumption,

$$\psi'(Z) = -\mu\delta\, Z^{\mu-1} + \cdots$$

and

$$-\sqrt{B\,(\psi_{\max} - \psi)}\;\psi = -\sqrt{B\,(\delta\, Z^{\mu} + \cdots)}\;(\psi_{\max} - \delta\, Z^{\mu} + \cdots).$$

Substituting these expressions into the equation and equating the leading terms leads to

$$-\mu\delta\, Z^{\mu-1} + \cdots = -Z^{\mu/2}\sqrt{B\,\delta}\;\psi_{\max} + \cdots .$$

One possibility is $\delta = 0$. This corresponds to the trivial solution $\psi \equiv \psi_{\max}$. A nontrivial solution is obtained when μ is taken to be 2 and δ to be 0.25 $B\,\psi^2_{\max}$. After calculating a few more terms in the expansion of ψ near $Z = 0$, we could compute a numerical approximation to ψ at some $Z_0 > 0$. This value could then be used to start a numerical integration for $Z \geq Z_0$.

Unless one can move some distance from the origin, it is quite possible that a code will get into trouble. This is because near $Z = 0$, the solution ψ is close to its maximum value $\psi_{\max}$. Should an approximation greater than this value be formed or, less obviously, should the method need to evaluate the equation at an argument greater than this value, the computation will fail because of an attempt to take the square root of a negative number. There is, in fact, a perfectly easy way to solve this problem—just return to equation (3.6) or (3.7). The analysis allowed us to determine the speed v that permits a solution of the kind desired. With it the solution of (3.7) is routine. (Remember that $C = 0$.) The same is true of (3.6) when (3.7) is used to deduce the initial value $\psi''(0)$. This situation is not unusual. It is well to keep in mind that manipulations that facilitate analytical work may get in the way of a numerical solution. In particular, analytical treatments very often reduce systems of equations to one or a few equations of higher order. When solving the equations numerically, the equations must be reformulated as a first-order system so that a standard code can be used. The usual way of accomplishing the transformation to a first-order system may not result in variables that are as meaningful physically as the ones used to pose the original problem.

EXERCISE 13. Solve this problem numerically. There is an analytical solution,

$$\psi = C \operatorname{sech}^2 (D\, Z), \qquad \text{where } C = A/B \quad \text{and} \quad D = \frac{\sqrt{A}}{2},$$

that can be used to check your results.

EXAMPLE 5. Lighthill [1986, p. 103 ff.] presents the classical analysis of the collapse of a spherical cavity in a liquid. Because of spherical symmetry, the partial differential equation describing the problem reduces to an ordinary differential equation:

$$2\pi\, \rho\, a^3 \left(\frac{d\, a(t)}{dt} \right)^2 = \frac{4}{3}\, \pi\, (a_0^2 - a^2)\, P$$

Here $a(t)$ is the radius of the cavity at time t, $a_0 = a(0)$, ρ is the density of the fluid, and P is the (constant) pressure in the neighborhood of the cavity. As is often the case, it is useful to rewrite the problem in nondimensional form by introducing

$$A = a/a_0, \qquad q_0 = \sqrt{P/\rho}, \qquad t_0 = a_0/q_0, \quad \text{and} \quad T = t/t_0.$$

A little manipulation then results in

$$\left(\frac{dA}{dT} \right)^2 = \frac{2}{3}\, (A^{-3} - 1).$$

The initial condition becomes

$$A(0) = 1. \tag{3.10}$$

The integration is to proceed for increasing T until the bubble has radius 0—total collapse.

In standard form the differential equation is

$$\frac{dA}{dT} = -\sqrt{\frac{2}{3}(A^{-3} - 1)}. \tag{3.11}$$

This problem is singular in two ways. First notice that the initial value problem has a solution $A(T)$ that is identically 1; obviously the problem has more than one solution. This problem is rather like the preceding examples at the initial point and it can be dealt with in the same way.

EXERCISE 14. Work out the details of using an expansion to start the integration of (3.10, 3.11).

Physically we expect that the radius of the cavity decrease monotonely to zero. Clearly the equation becomes singular as A tends to zero and in particular, the derivative of A tends to $-\infty$. A sketch of the solution suggests that the curve is being described inappropriately. In such a situation a standard technique in the theory of ODEs is to interchange dependent and independent variables. Because A is monotonely decreasing, it is permissible to take it as a new independent variable and seek T in terms of A:

$$\frac{dT}{dA} = -1\Big/\sqrt{\frac{2}{3}(A^{-3} - 1)} = -A^{3/2}\Big/\sqrt{\frac{2}{3}(1 - A^3)} \tag{3.12}$$

Notice that we cannot use this new variable for the whole integration—the new equation is singular at the initial point where A is 1. We must start computing A as a function of T defined by the problem (3.10, 3.11) until at some T_0, the value $A_0 = A(T_0)$ is rather less than 1. At this point we start computing T as a function of A from equation (3.12) and the initial condition

$$T(A_0) = T_0.$$

This integration is to proceed from $A = A_0$ to $A = 0$. A quantity of special physical interest is the time at which total collapse takes place; it is the value of t when the radius a vanishes, or equivalently, it is $t_0 T$ when $A = 0$. (Recall the definition of t_0 in terms of the parameters defining the problem.)

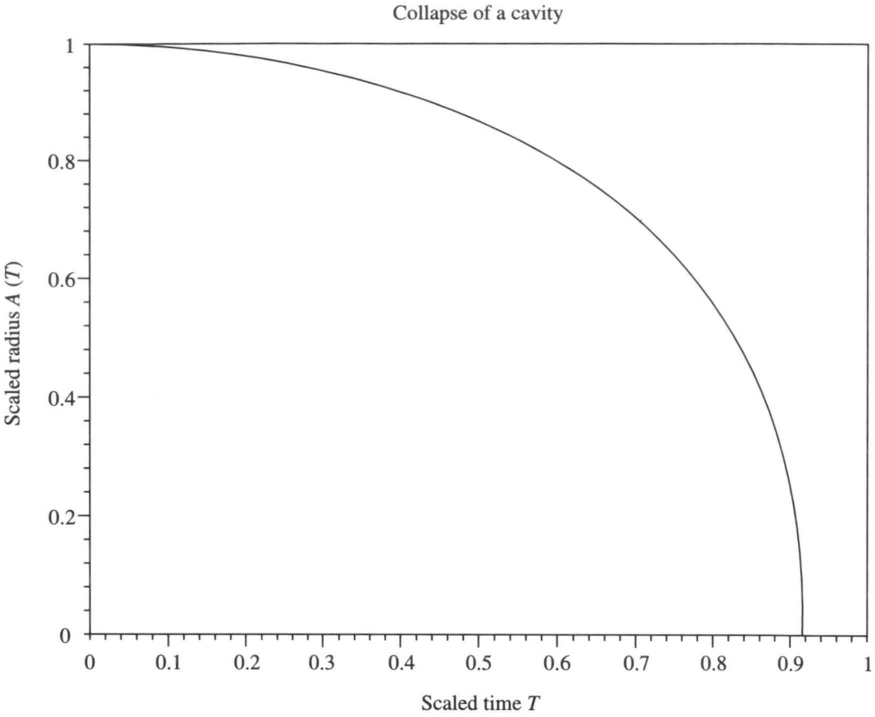

Figure 1.2

Notice that the second integration is from $A_0 > 0$ to 0. Some codes, especially old ones, assume that the integration is always in the direction of an increase of the independent variable. It is easy to introduce a change of variable to bring this about, but there is no technical reason for a code to require a specific direction and no quality code puts users to this trouble. This is a minor point. The important point is that we see here a way to locate where a solution component assumes a given value, in this instance 0. This is quite a useful application of *interchanging independent and dependent variables*. For later use we write out the technique for a general system.

Suppose we wish to solve a system of n first-order equations:

$$\begin{aligned} \frac{dy_1}{dx} &= F_1(x, y_1, \ldots, y_n), \\ &\vdots \\ \frac{dy_n}{dx} &= F_n(x, y_1, \ldots, y_n). \end{aligned}$$

If the derivative of y_j does not vanish at a point b, we can introduce y_j as an independent variable in the neighborhood of $y_j(b)$ by

$$\frac{dy_1}{dy_j} = \frac{\dfrac{dy_1}{dx}}{\dfrac{dy_j}{dx}} = \frac{F_1(x, y_1, \ldots, y_n)}{F_j(x, y_1, \ldots, y_n)},$$

$$\vdots \qquad\qquad \vdots$$

$$\frac{dx}{dy_j} = \frac{1}{\dfrac{dy_j}{dx}} = \frac{1}{F_j(x, y_1, \ldots, y_n)},$$

$$\vdots \qquad\qquad \vdots$$

$$\frac{dy_n}{dy_j} = \frac{\dfrac{dy_n}{dx}}{\dfrac{dy_j}{dx}} = \frac{F_n(x, y_1, \ldots, y_n)}{F_j(x, y_1, \ldots, y_n)}.$$

In addition to avoiding a derivative that is infinite, this change of variables can be used to control the size of all the derivatives. By selecting the component y_j for which $|F_j(b, y_1(b), \ldots, y_n(b))|$ is largest, the derivatives with respect to the new variable will be no greater than 1 in magnitude at $y_j(b)$ and of modest size near this point.

Modern codes for the initial value problem step through the interval of integration producing approximate solution values at points they find convenient or you specify. To find where a given solution component $y_j(x)$ assumes a given value ω, the integration can be monitored to determine when $y_j(x)$ passes through this value. Suppose, for example, that $y_j(b) = c < \omega$ and the next approximation $y_j(d) = e > \omega$. Introduction of y_j as the independent variable at b and integrating from $y_j = c$ to $y_j = \omega$ will then yield the desired value of x at which y_j has the specified value ω. Computation of the time of total collapse of the cavity is comparatively simple because no monitoring of the computation is necessary; we can just interchange variables as soon as it makes sense. Since Hénon [1982] proposed using the technique to compute Poincaré maps, texts on computational methods for dynamical systems have begun describing it as "Hénon's trick," but the idea is much older. The earliest reference known to the author that clearly states this technique for finding where a solution component assumes a given value is that of Wheeler [1959].

EXERCISE 15. Find the first T for which the solution $A(T)$ of (3.10, 3.11) vanishes.

EXAMPLE 6. Many important processes can be described by a pair of equations of the form

$$\frac{d}{dt}x(t) = F_1(x, y),$$

$$\frac{d}{dt}y(t) = F_2(x, y).$$

A way to think about such equations is that x and y are the coordinates of a point in the plane and the equations describe the motion of the point as a function of time. An example that is often studied is Volterra's model of the number of individuals in two populations that are assumed to conflict with one another, e.g., a predator and prey. See, for example Davis [1962, p. 101 ff.]. This model has the form

$$\frac{d}{dt}N_1(t) = aN_1 - bN_1 N_2,$$

$$\frac{d}{dt}N_2(t) = -cN_2 + dN_1 N_2,$$

where N_1 and N_2 are the number of individuals in each population and a, b, c, d are positive numbers that describe how the populations interact. A classic technique of applied mathematics is to remove the role of time by going to the *phase plane*. Here this means writing y as a function of x rather than t, so that in general

$$\frac{dy}{dx} = \frac{\frac{dy}{dt}}{\frac{dx}{dt}} = \frac{F_2(x, y)}{F_1(x, y)}.$$

Note that it is just as easy to eliminate t by writing x as a function of y. In this manner attention is focussed on the qualitative properties of the solution rather than on the solution at particular times. For example, a periodic solution is recognized by a solution curve in the phase plane that is closed.

The concept of the phase plane will be illustrated with the important *van der Pol equation*,

$$\ddot{y} - \varepsilon(1 - y^2)\dot{y} + y = 0,$$

often used as a test equation for differential equation solvers. Here ε is a positive parameter that governs the importance of the nonlinear term. There are two periodic solutions, one of which is the trivial solution $y(t) \equiv 0$.

Because all nontrivial solutions of the differential equation approach the nontrivial periodic solution as $t \to \infty$, it is called a limit cycle. When ε is "large" the approach to the limit cycle is quite rapid and the limit cycle has very sharp changes in its behavior. Figure 1.3 shows a plot of the limit cycle when $\varepsilon = 10$. The solution is plotted as a curve in three dimensions and the projections show more conventional plots. On the bottom face is seen a plot of $y(t)$ that shows the sharp changes that challenge differential equation solvers. On the back face is seen a plot of the first derivative with spikes where the solution changes rapidly. On the left face is seen a plot of $\dot{y}$ against y, a phase plane plot. The role of time disappears in such a plot and the closed curve shows that this solution is periodic.

Every modern book on the theory of ordinary differential equations devotes considerable attention to the phase plane. Problems in the phase plane interest us here for two reasons. One concerns singular problems. Viewing such equations as describing the motion of a particle, it is obvious that stationary solutions are important to understanding what kinds of

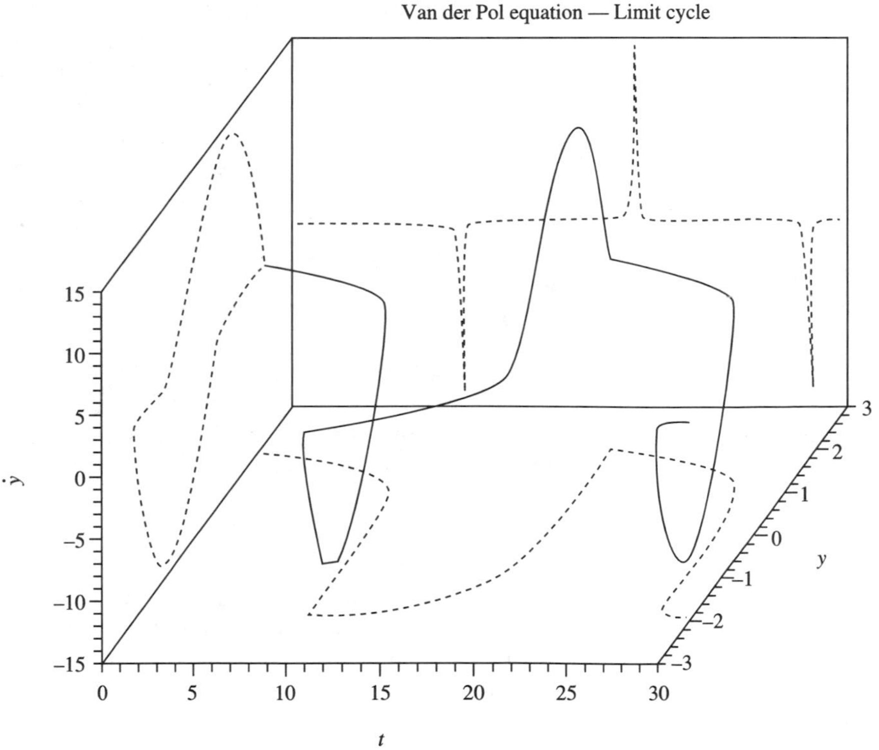

Figure 1.3

behavior are possible. A stationary solution, $x_s(t)$ and $y_s(t)$, has the coordinates as constant functions of time, hence

$$\frac{d}{dt}x_s(t) = F_1(x_s, y_s) \equiv 0,$$

$$\frac{d}{dt}y_s(t) = F_2(x_s, y_s) \equiv 0.$$

This is all easy to understand when the independent variable is the time t, but in the phase plane, it is clear that the problem is singular at a stationary point. In the theoretical analysis, attention is focussed on the singular points in the phase plane and the behavior of solution curves in the neighborhood of a singular point is sorted out by analytical means. The computation of solutions in the phase plane is important to understanding nonlinear equations, so many examples of such solutions are found plotted in all recent books on the subject. The point is that in this context, equations with singular points are to be expected and the analysis that would permit an integration to be started at a singular point is standard. Of course, it can be much easier to return temporarily to the original independent variable. Recall, for example, the equation $y' = y/x$ that was used to illustrate singular behavior. It could have originated in the pair of equations

$$\frac{dy}{dt} = y, \qquad \frac{dx}{dt} = x.$$

The general solution to these uncoupled equations is

$$y(t) = \alpha \exp(t), \qquad x(t) = \beta \exp(t),$$

for arbitrary constants α and β. If β is not zero, we then deduce that y can be written as a function of x, namely $y(x) = cx$ where $c = \alpha/\beta$. From this point of view the behavior of the singular equation is not so puzzling. The "original" system of equations is nonsingular and presents no difficulty numerically.

To explore the behavior of a system, it might well be considered unnecessary to integrate from, or to, a singular point, but there is another difficulty when integrating in the phase plane that must be recognized. Earlier it was remarked that one might write y as a function of x or vice versa. The possibility of changing the independent variable is not just a theoretical matter, when computing solutions in the phase plane it is usually necessary. The difficulty is that the motion of the particle represents a *curve* in the phase plane, but it generally does not represent a function, i.e., y is generally not a function of x for all the x of interest. This is seen in the phase plane plot of the limit cycle of van der Pol's equation. Another standard problem

$$\dot{x} = \mu x + y - x(x^2 + y^2),$$
$$\dot{y} = -x + \mu y - y(x^2 + y^2),$$

serves to emphasize the point. Here μ is a parameter. The solutions are easily understood when polar coordinates are introduced:

$$\dot{r} = r(\mu - r^2), \qquad \dot{\theta} = -1.$$

The equation for the angle θ shows that the particle is moving clockwise around the origin at a uniform rate. If the parameter μ is negative, the equation for r shows that the radial distance is decreasing, hence that the particle spirals into the origin. If the parameter μ is positive, it is observed that there is a periodic solution with the particle moving at a uniform rate clockwise on a circle about the origin of radius $\sqrt{\mu}$. Any solution curve starting at a radial distance greater than $\sqrt{\mu}$ spirals in to this circle and any curve starting closer to the origin spirals out to the circle—the circle is a limit cycle for this equation. Obviously the coordinate y is not a function of x for all x of interest. In a mechanical sense, the difficulty appears when F_1 vanishes and F_2 does not so that the derivative of y with respect to x becomes infinite—the path has a vertical tangent. One way to cope with this is to interchange independent and dependent variables as described in the last example. However, as this example makes clear, it may be necessary to interchange the variables frequently and it seems more natural to turn to a description of the curve that is coordinate-free. This is another example where a numerical difficulty has arisen because of analytical manipulations—the parametric representation in terms of t for the original system does not have this difficulty, so one remedy is to return to the original formulation. On the other hand, if it is really the path and not the speed at which it is traversed that is interesting, there are other possibilities. One is to use arc length as a parameter.

In general suppose we wish to integrate the system

$$\frac{dy_1}{dx} = F_1(x, y_1, \ldots, y_n),$$
$$\vdots \qquad\qquad \vdots$$
$$\frac{dy_n}{dx} = F_n(x, y_1, \ldots, y_n).$$

for $x \geq a$. When *arc length* s is introduced as a new independent variable, $x(s)$ becomes a dependent variable with initial value $x(0) = a$, and the system becomes

$$\frac{dy_1}{ds} = F_1(x, y_1, \ldots, y_n)/S,$$
$$\vdots \qquad \vdots$$
$$\frac{dy_n}{ds} = F_n(x, y_1, \ldots, y_n)/S,$$
$$\frac{\mathrm{dx}}{\mathrm{ds}} = 1/S,$$

where

$$S = \left(1 + \sum_{i=1}^{n} F_i^2(x, y_1, \ldots, y_n)\right)^{1/2}.$$

In this variable there is no difficulty with a derivative becoming infinite. Indeed,

$$\left(\frac{dx}{ds}\right)^2 + \sum_{i=1}^{n}\left(\frac{dy_i}{ds}\right)^2 = \left(\frac{1}{S}\right)^2 + \sum_{i=1}^{n}\left(\frac{F_i}{S}\right)^2 \equiv 1,$$

and this implies that each of the derivatives is bounded in magnitude by 1 over the whole range of integration. Some care may be required to evaluate the equation properly in floating point arithmetic. The difficulty is that if some component F_i becomes large in magnitude, its square might lead to unnecessary overflow. The usual way to handle this is to scale the expressions. To evaluate the equations, first evaluate all the F_i at the appropriate argument $x, y_1, \ldots, y_n$, and then find the largest in magnitude:

$$\zeta = \|\mathbf{F}\|_\infty = \max_i |F_i|.$$

The equations can then be evaluated in the scaled form

$$F_k/S = \frac{(F_k/\zeta)}{\left((1/\zeta)^2 + \sum_{i=1}^{n}(F_i/\zeta)^2\right)^{1/2}}.$$

This form might lead to underflows, but they are harmless and should be set to zero by the operating system of the computer. (You might have to interact with the system to bring this about.)

Notice that introducing arc length s as an independent variable has resulted in a differential equation of the form $\mathbf{u}' = \mathbf{G}(\mathbf{u})$—the independent variable s does not appear as an argument in $\mathbf{G}$. Such differential equations are said to be *autonomous*. In this way any initial value problem can be converted into an equivalent problem with a differential equation that is autonomous.

Except for possible annoyances with the exponent range of the computer used, arc length is often a very useful independent variable. It is frequently used for the computation of solutions in the phase plane. As a specific example, the TI-85 graphics calculator uses the technique for this purpose. Many problems require answers at specific values of the original independent variable. After introducing arc length as the new independent variable, answers are then required where a dependent variable assumes a given value. This task arose in Example 5 where it was handled by introducing a new independent variable. This is not an attractive possibility in the present context. For now, we just remark that some codes do have the capability of producing answers at those times when a given dependent variable assumes a given value.

EXAMPLE 7. Numerical methods for the solution of a system of algebraic equations $\mathbf{F}(\mathbf{x}) = \mathbf{0}$ typically require a good guess $\mathbf{x}^0$ for a solution if the method is to converge. Some problems are "hard" in the sense that the guess must be quite good if the method is to succeed. *Homotopy* or *continuation* methods deal with this by embedding the given problem in a family of problems. It may be that $\mathbf{F}$ depends in a natural way on a parameter λ and that for certain values of λ the equation $\mathbf{F}(\mathbf{x}, \lambda) = \mathbf{0}$ is easy to solve. If necessary, a parameter can be introduced artificially. One way this is done is to define $\mathbf{H}(\mathbf{x}, t) = \mathbf{F}(\mathbf{x}) + (t - 1)\mathbf{F}(\mathbf{x}^0)$. By construction $\mathbf{x}^0$ is a solution of $\mathbf{H}(\mathbf{x}, 0) = \mathbf{0}$ and we want the solution of $\mathbf{H}(\mathbf{x}, 1) = \mathbf{F}(\mathbf{x}) = \mathbf{0}$. In suitable circumstances an implicit function theorem says that the equation $\mathbf{H}(\mathbf{x}, t) = \mathbf{0}$ defines a function $\mathbf{x}(t)$. Davidenko [1953] exploited this to derive a differential equation for this function:

$$\mathbf{0} = \frac{d}{dt}\mathbf{H}(\mathbf{x}(t), t) = \frac{\partial \mathbf{H}(\mathbf{x}(t), t)}{\partial \mathbf{x}}\frac{d\mathbf{x}}{dt} + \frac{\partial \mathbf{H}(\mathbf{x}(t), t)}{\partial t}.$$

This is a differential equation that is not in standard form. If the Jacobian $\partial \mathbf{H}/\partial \mathbf{x}$ is not singular, the equation can be put into standard form:

$$\frac{d\mathbf{x}}{dt} = -\left(\frac{\partial \mathbf{H}(\mathbf{x}(t), t)}{\partial \mathbf{x}}\right)^{-1}\frac{\partial \mathbf{H}(\mathbf{x}(t), t)}{\partial t}.$$

The idea is to integrate this system of equations from $t = 0$ where the initial value $\mathbf{x}(0) = \mathbf{x}^0$ to $t = 1$ where the solution $\mathbf{x}(1)$ provides the solution to the given (algebraic) problem. Clearly it is necessary in this approach that the Jacobian be invertible, but not much more is needed for an implicit function theorem to apply and the scheme to be justified. Ortega and Rheinboldt [1970] provide details.

For the artificial scheme the equation is

$$\frac{\partial \mathbf{F}(\mathbf{x}(t))}{\partial \mathbf{x}} \frac{d\mathbf{x}}{dt} = -\mathbf{F}(\mathbf{x}^0), \qquad 0 \le t \le 1, \qquad \mathbf{x}(0) = \mathbf{x}^0.$$

In practice the Jacobian here should not be inverted, rather a system of linear equations solved for each value of $\mathbf{x}'$ required by the integrator. This exemplifies differential equations for which evaluation of the function defining the equation can be rather expensive when the system is large. For this reason, Watson [1979] relies upon a variable order, variable step size, Adams–Bashforth–Moulton, PECE code (all of these attributes will be explained in succeeding chapters) for the integration because it attempts to minimize the number of such evaluations.

It is opportune to note that the cost of an integration is conventionally measured by the *number of function evaluations* required. When solving problems involving complicated functions $\mathbf{F}(x, \mathbf{y})$ with the popular methods, counting the number of times $\mathbf{F}$ is evaluated is a fair measure of the cost that does not depend on the computer being used. When $\mathbf{F}$ is not complicated, the cost may vary significantly depending on which computer and even which compiler is used. Fortunately, the cost is then often small in absolute terms. Gupta [1980] provides some numerical results about the relative importance of function evaluations and the other costs called *overhead*. We shall see in Chapter 8 that when solving a special class of problems called *stiff*, the overhead can be substantial part of the total cost.

EXERCISE 16. Another way of introducing a parameter t artificially is to write $\mathbf{H}(\mathbf{x}, t) = t\,\mathbf{F}(\mathbf{x}) + (1 - t)(\mathbf{x} - \mathbf{x}^0)$. By construction $\mathbf{x}^0$ is a solution of $\mathbf{H}(\mathbf{x}, 0) = \mathbf{0}$ and we want the solution of $\mathbf{H}(\mathbf{x}, 1) = \mathbf{0}$. Derive an initial value problem for $\mathbf{x}(t)$ so that it may be integrated from $t = 0$ to $t = 1$ to obtain a solution $\mathbf{x}(1)$ of the equation $\mathbf{F}(\mathbf{x}) = \mathbf{0}$.

The focus here is on the solution of a given equation $\mathbf{F}(\mathbf{x}) = \mathbf{0}$ and the embedding is employed to facilitate the solution. There is a large body of work concerned with a closely related matter called *bifurcation theory*. Many physical problems are described by systems of differential equations that depend on a vector $\boldsymbol{\alpha}$ of parameters,

$$\frac{d\mathbf{x}}{dt} = \mathbf{f}(\mathbf{x}, \boldsymbol{\alpha}).$$

Steady-state solutions and periodic solutions are of particular interest. The former are solutions with $\mathbf{x}'(t) = \mathbf{0} = \mathbf{f}(\mathbf{x}, \boldsymbol{\alpha})$. They are constant vectors that are solutions of a set of algebraic equations depending on a vector of physical parameters $\boldsymbol{\alpha}$. How these solutions behave when the parameters are changed is often of primary interest. This is studied by fixing all the parameters but one that we shall call α. Then, as before,

$$\frac{\partial \mathbf{f}}{\partial \mathbf{x}} \frac{d\mathbf{x}}{d\alpha} = -\frac{\partial \mathbf{f}}{\partial \alpha}, \qquad \mathbf{x}(\alpha_0) \text{ given,}$$

can be solved to compute $\mathbf{x}(\alpha)$. All goes well except when the Jacobian $\mathbf{f}_x$ becomes singular at a critical point $\alpha = \alpha_c$. In this context singularities are of special interest because the qualitative behavior of the problem may change at α_c. One possibility is that the differential equation has more than one solution for $\alpha > \alpha_c$—the solution bifurcates. Numerical solution of the differential equation in such a situation is not a routine matter. Another possibility is that the derivative of $\mathbf{x}$ with respect to α becomes infinite as $\alpha \rightarrow \alpha_c$ in such a way that it compensates for the singularity of the Jacobian there with the consequence that there is a solution curve with a vertical tangent at α_c. Such a "turning point" is another example of an inappropriate choice of independent variable. Locally we need to replace α as the independent variable by one of the components of $\mathbf{x}$ or resort to arc length as an independent variable. The code DERPAR of Holodniok and Kubiček [1984] (see also Kubiček and Marek [1983]) uses arc length for dealing with turning points. Neither choice is as simple as in the examples given earlier of these techniques because the functions are evaluated by solving a system of linear equations; details can be found in the references.

As a preparatory example in Chapter 2, Seydel [1988] considers the solutions $y(\lambda)$ of

$$0 = f(y, \lambda) = y^3 - y^2(\lambda + 1.25) + y(3\lambda - 6) + \lambda^2 - 7.75\lambda + 13.75.$$

Differentiating with respect to the parameter λ leads to

$$\frac{dy}{d\lambda} = \frac{y^2 - 3y - 2\lambda + 7.75}{3y^2 - y(2\lambda + 2.5) + 3\lambda - 6}.$$

In an exercise he suggests integrating this equation from $y(2) = 1$ to the left and to the right. The partial solutions provided below make important points. The numerical integrations and the plots were all done with the MDEP package of Buchanan [1992]. (It's fun to explore the solution of ODEs with this *free*ware.)

EXERCISE 17. Integrate this equation from $\lambda = 2$ to $\lambda = 0$. One of the methods available in MDEP is a fixed step size Runge–Kutta code. Figure 1.4 was computed using this method and the default step size of 0.02. The results are not what we might expect. Another of the methods available

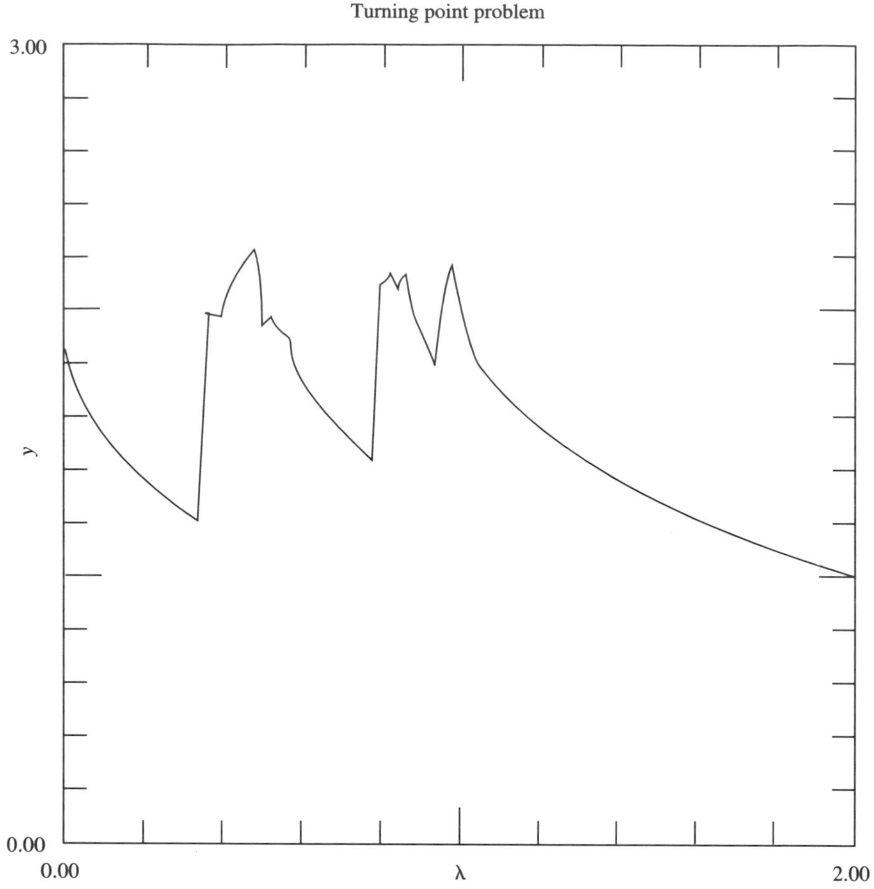

Figure 1.4

in MDEP is a variable step size Runge–Kutta code. Such a code monitors the error made at each step and adjusts the step size accordingly. When this code was used with the default maximum step size of 0.02 and default error tolerance, the code returned at $\lambda = 1$ with a solution of $y = 2$ and a message to the effect that it could no longer achieve the desired accuracy. Some people actually prefer to use fixed step codes so they will not be bothered by messages of this kind. Codes that monitor the error are trying to tell you something about the problem when they provide such messages—ignoring them is foolhardy. In this instance we are warned that the results of the fixed step computation are nonsense from $\lambda = 1$ on. To appreciate this, evaluate the differential equation at $(\lambda, y) = (1, 2)$ to understand why the variable step size code found the solution difficult to approximate. Interchange the dependent and independent variables as explained in Example 5 and solve for $\lambda(y)$. This integration is easy, even with a fixed step code. You will see an example of a turning point and understand why the fixed step size results in the original independent variable are nonsense.

EXERCISE 18. Figure 1.5 shows what happens when the integration is done from $\lambda = 2$ to $\lambda = 4$ using both the fixed step size code and the variable step size code in MDEP. Now what has happened with the fixed step size integration? It agrees nicely with the one controlling the error up to a point where it abruptly branches off from a smooth solution. The

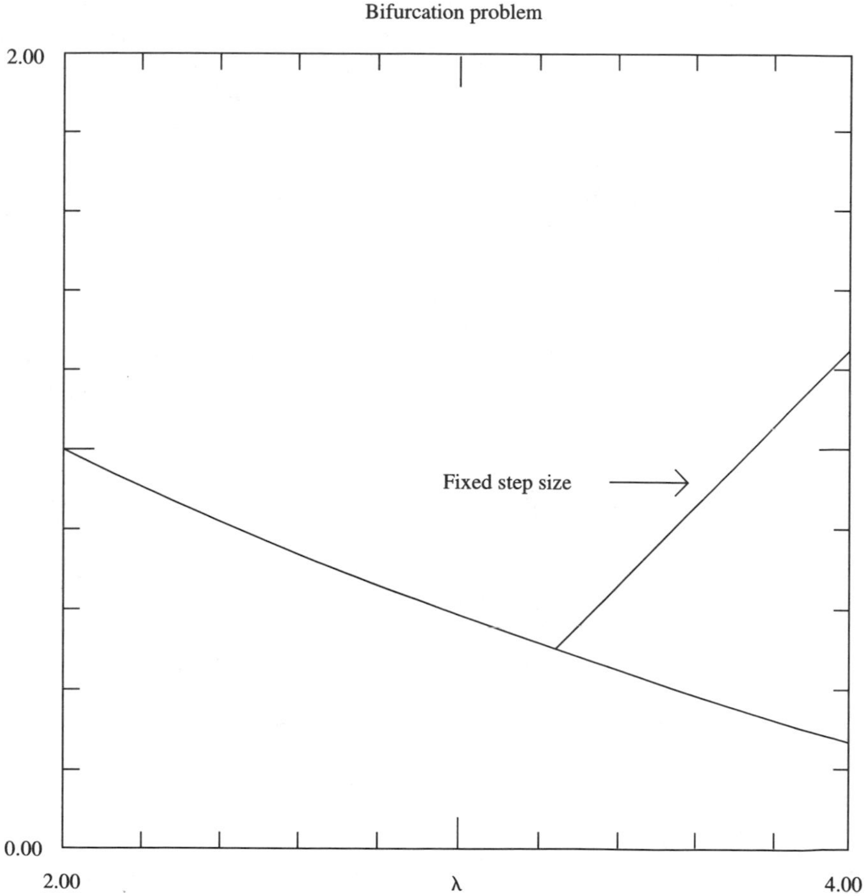

Figure 1.5

solutions separate at $(\lambda, y) = (3.25, 0.5)$. Evaluate the differential equation at this point to see that it is singular there. As in Example 3, analyze the behavior of solutions in the neighborhood of this point by looking for solutions of the form

$$y(\lambda) = 0.5 + \alpha(\lambda - 3.25) + \text{higher powers of } (\lambda - 3.25).$$

Show that there are two possible values of α. At the singular point uniqueness fails and the solution bifurcates. The kink in the numerical solution is a surprise, but this time it is the problem that has surprised us rather than the integrator.

EXAMPLE 8. Because of the availability of convenient and powerful codes for the solution of ordinary differential equations, more complicated tasks are often approximated by a system of ordinary differential equations. An example in the context of partial differential equations is *semi-discretization*, or the *method of lines*. Some of the salient ideas will be exposed here, but there is far too much to the topic to do more than present just a few ideas in the context of extremely simple examples. The recent book of Schiesser [1991] is devoted to this topic alone. Our main purpose is to show how very large systems of differential equations can arise and how characteristics of these problems can be exploited to make their solution possible. We also obtain some simple examples that will be useful later.

There are many ways to approximate a partial differential equation by a system of ordinary differential equations. Perhaps the most straightforward is to use *finite differences*. To exemplify this, suppose that we want to solve the equation

$$\frac{\partial u}{\partial t} + c\frac{\partial u}{\partial x} = g(x, t),$$

which is variously known as the *advection equation*, the *one-way wave equation*, and the *one-dimensional wave equation*. Here the constant c and the forcing function $g(x, t)$ are given and the equation is to hold for $0 \leq x \leq 1$ and $t \geq 0$. When c is positive, typical boundary conditions are

$$u(x, 0) \text{ given for } 0 \leq x \leq 1, \qquad u(0, t) \text{ given for } t \geq 0.$$

The homogeneous equation $u_t + cu_x = 0$ is easy to solve analytically. The behavior of solutions is exemplified by the particular solution $u(x, t) = 5/(1 + 100((x - t) - 0.5)^2)$ of $u_t + u_x = 0$. The data at $t = 0$ has a peak at $x = 0.5$. Figure 1.6a shows the solution plotted as a function of x for selected times t. It is seen that the peak is traveling to the right at a constant speed $c = 1$.

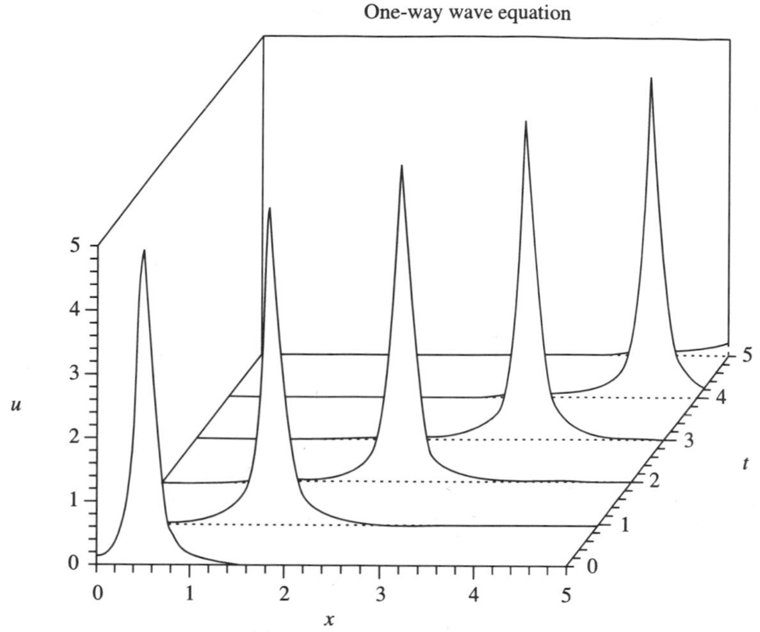

Figure 1.6a

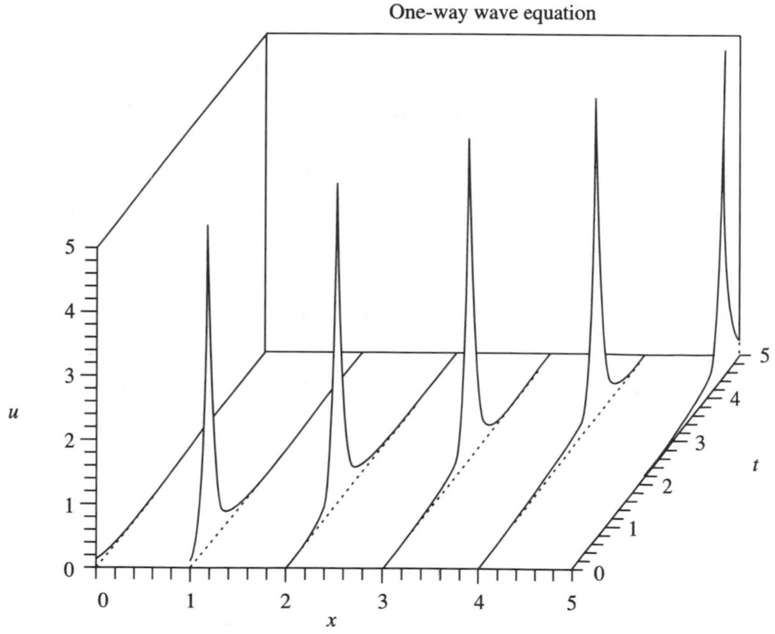

Figure 1.6b

The method of lines approximates $u(x, t)$ on a set of x values by functions of t alone. Figure 1.6b shows the example solution plotted in this way. To keep the matter simple, suppose that we choose an integer N, define a mesh spacing $\Delta = 1/N$, and approximate

$$u(i\Delta, t) \approx u_i(t) \qquad \text{for } i = 0, 1, \ldots, N.$$

The boundary condition $u(0, t)$ given provides the approximation $u_0(t)$ and the initial condition $u(x, 0)$ provides the initial value of each $u_i(t)$ as $u_i(0) = u(i\Delta, 0)$. For fixed t, the partial derivative with respect to x can be approximated by a backward difference:

$$\left.\frac{\partial u(x, t)}{\partial x}\right|_{x=i\Delta} \approx \frac{u(i\Delta, t) - u((i-1)\Delta, t)}{\Delta} \approx \frac{u_i(t) - u_{i-1}(t)}{\Delta}.$$

Using this, the partial differential equation is approximated by the system of ordinary differential equations

$$\frac{d}{dt} u_i(t) = -c\left(\frac{u_i(t) - u_{i-1}(t)}{\Delta}\right) + g(i\Delta, t) \qquad \text{for } i = 1, 2, \ldots, N.$$

To write this in terms of vectors, let $u_i(t)$ be component i of $\mathbf{u}(t)$ and let the $N \times N$ matrix

$$Q = \begin{pmatrix} 1 & 0 & \cdots & 0 \\ -1 & 1 & \cdots & 0 \\ \vdots & \ddots & \ddots & \vdots \\ 0 & \cdots & -1 & 1 \end{pmatrix}.$$

For components $i = 2, \ldots, N$ let $g(i\Delta, t)$ be component i of $\mathbf{g}(t)$ and let the first component of $\mathbf{g}(t)$ be

$$g(\Delta, t) + c\,\frac{u(0, t)}{\Delta}.$$

With this notation we have a linear system in standard form:

$$\frac{d}{dt}\mathbf{u}(t) = J\,\mathbf{u}(t) + \mathbf{g}(t).$$

Here the (constant) Jacobian J is

$$J = -\frac{c}{\Delta}\,Q.$$

Quite large systems of equations might arise in this way because Δ might have to be quite small to give a satisfactory resolution of the solution

$u(x, t)$ with respect to the spatial variable x. Although the Jacobian might be large, most of its entries are 0. This particular matrix exemplifies a class of matrices called *banded matrices*. A matrix $J = (J_{i,j})$ is said to be banded when all the nonzero elements lie in a band about the diagonal. More precisely, if $J_{i,j} = 0$ whenever $i - j > m_l$ or $j - i > m_u$, the matrix is said to have lower band width m_l, upper band width m_u, and band width $m = m_l + m_u + 1$. The example has $m_l = 1$, $m_u = 0$, and $m = 2$. Banded matrices are common in the context of semi-discretization and they provide a simple way to describe where nonzero elements can appear in the Jacobian. Note that in this definition, nothing is said about the presence of zero entries within the band, just that all entries outside the band are zero.

A more accurate approximation to the partial derivative is provided by a central difference

$$\left.\frac{\partial u(x, t)}{\partial x}\right|_{x=i\Delta} \approx \frac{u_{i+1}(t) - u_{i-1}(t)}{2\Delta}.$$

It would seem that this would allow bigger Δ for the same accuracy without any additional complication, but there is, in fact, a complication because of the boundary conditions—what do we use for $u_{N+1}(t)$? This issue is addressed more appropriately in books on the finite difference solution of partial differential equations such as Sod [1985] or Strikwerda [1989] and the interested reader may pursue the matter there. For illustrative purposes we consider a different kind of boundary condition for which there is no difficulty of this kind, namely the periodic boundary condition

$$u(0, t) = u(1, t).$$

(Of course, there is also the initial condition of $u(x, 0)$ given.) It is convenient now to define $\Delta = 1/(N + 1)$. Let us approximate $u(i\Delta, t)$ by $u_i(t)$ where

$$\frac{d}{dt} u_i(t) = -c\left(\frac{u_{i+1}(t) - u_{i-1}(t)}{2\Delta}\right) + g(i\Delta, t) \qquad \text{for } i = 1, 2, \ldots, N,$$

along with the definitions

$$u_0(t) = u_N(t), \qquad u_{N+1}(t) = u_1(t),$$

that impose periodicity. The Jacobian now has the form

$$J = -\frac{c}{2\Delta}\begin{pmatrix} 0 & 1 & 0 & \cdots & 0 & -1 \\ -1 & 0 & 1 & \cdots & 0 & 0 \\ 0 & -1 & 0 & \cdots & 0 & 0 \\ \vdots & & & & & \vdots \\ 0 & 0 & 0 & -1 & 0 & 1 \\ 1 & 0 & 0 & 0 & -1 & 0 \end{pmatrix}$$

Again the Jacobian has few entries that are not zero. If it were not for the presence of the nonzero entries $J_{1,N}$ and $J_{N,1}$ that are due to periodicity, this matrix would be a band matrix of the form called tridiagonal. A *tridiagonal matrix* is a band matrix with $m_l = 1$, $m_u = 1$, and $m = 3$. It is much easier to deal with matrices that have narrow bands than with general matrices. Fortunately, it is not necessary to treat a matrix like this one as a general matrix because there are special techniques for dealing with matrices that differ in only a few entries from a "nice" matrix like a tridiagonal one, see, e.g., Strikwerda [1989, p. 80].

Continuing with finite differences, let us now take up a problem involving two space variables x and y. Suppose we wish to solve for $0 \le x \le 1$, $0 \le y \le 1$, $t \ge 0$ the (forced) heat equation

$$\frac{\partial u}{\partial t} = \frac{\partial^2 u}{\partial x^2} + \frac{\partial^2 u}{\partial y^2} + g(x, y, t).$$

In such a case, we might not use the same mesh spacing for x and y, so let us choose integers N and M and define increments $\Delta = 1/(N+1)$ and $\delta = 1/(M+1)$. We then approximate $u(i\Delta, j\delta, t)$ by $u_{i,j}(t)$ for $0 \le i \le N+1$, $0 \le j \le M+1$. For such an equation we might have the initial condition $u(x, y, 0)$ given and boundary conditions $u(0, y, t)$, $u(1, y, t)$, $u(x, 0, t)$, and $u(x, 1, t)$ given. The boundary conditions provide us with, e.g., $u_{0,j}(t) = u(0, j\delta, t)$, and similarly they provide us with $u_{N+1,j}(t)$, $u_{i,0}(t)$, and $u_{i,M+1}(t)$. The approximations at other mesh points are to satisfy the approximating difference equations for $1 \le i \le N$, $1 \le j \le M$,

$$\frac{d\,u_{i,j}(t)}{dt} = \frac{u_{i+1,j}(t) - 2u_{i,j}(t) + u_{i-1,j}(t)}{\Delta^2} + \frac{u_{i,j+1}(t) - 2u_{i,j}(t) + u_{i,j-1}(t)}{\delta^2} + g(i\Delta, j\delta, t),$$

obtained by replacing second partial derivatives by second central differences. Indexing the approximations in this way makes their origin clear, but to solve the initial value problem we must put the system into standard form. This might be done in a number of ways. A straightforward one is to take $u_{i,j}(t)$ to be component I of $\mathbf{u}(t)$ with $I = i + (j-1)N$. Using a superscript on $\mathbf{u}(t)$ to indicate the component, indexing in this way leads to the system of equations

$$\frac{d\,\mathbf{u}^I(t)}{dt} = \frac{\mathbf{u}^{I+1}(t) - 2\,\mathbf{u}^I(t) + \mathbf{u}^{I-1}(t)}{\Delta^2} + \frac{\mathbf{u}^{I+N}(t) - 2\,\mathbf{u}^I(t) + \mathbf{u}^{I-N}(t)}{\delta^2} + g(i\Delta, j\delta, t).$$

Written in this manner it is seen that this linear system of equations has a constant Jacobian matrix that is a band matrix with $m_l = m_u = N$ and $m = 2N + 1$. Alternatively, we could define component I of $\mathbf{u}(t)$ to be $u_{i,j}(t)$ for $I = j + (i - 1)M$. Doing so results in $m_l = m_u = M$ and $m = 2M + 1$. The point is that how one chooses to index the unknowns influences the structure of the Jacobian. If, for example, we wish to minimize the width of the band, it would be better here to use the first scheme when $N \leq M$ and otherwise the second scheme.

Approximating the solution of problems involving two space variables leads to much bigger systems than those involving only one. If this had been the heat equation in one space variable

$$\frac{\partial u}{\partial t} = \frac{\partial^2}{\partial x^2} + g(x, t),$$

there would have been N unknowns and the Jacobian would have been tridiagonal. If we use the same spacing in both variables, $M = N$, for the problem with two space variables, there are N^2 unknowns. We might well be interested in solving the heat equation for three space variables and then there are N^3 unknowns. It is clear that we cannot hope to resolve the solution as well (take the mesh spacing $\Delta = 1/(N + 1)$ as small) in several space dimensions as in one because of the rapid growth of the number of unknowns. In all cases the Jacobian is a band matrix, but this is an increasingly poor description of the Jacobian as the number of space dimensions increases. This is already seen with two space dimensions because most of the entries in the band are zero.

Suppose that $u(x, t)$ is a solution of a linear partial differential equation of the form

$$\frac{\partial u}{\partial t} = \mathcal{L}u + g(x, t),$$

where $\mathcal{L}$ is a linear differential operator involving only space derivatives. Examples seen so far have

$$\mathcal{L}u = -c\frac{\partial u}{\partial x} \quad \text{and} \quad \mathcal{L}u = \frac{\partial^2 u}{\partial x^2}.$$

There are a number of approaches to semidiscretization based on approximations of the form

$$u_N(x, t) = \sum_{i=1}^{N} a_i(t)\, \phi_i(x).$$

Obviously we cannot expect that an approximate solution $u_N(x, t)$ will satisfy the equation exactly, so we ask how well it does satisfy the equation.

More specifically, we ask about the size of its residual,

$$\mathcal{R}(x) = \frac{\partial}{\partial t} u_N(x, t) - \mathcal{L}u_N(x, t) - g(x, t).$$

The idea is to choose the coefficients $a_i(t)$ in $u_N(x, t)$ so as to make the residual small in some sense. There are a variety of ways seen that are known collectively as the *method of weighted residuals*. The points we wish to make can be illustrated with *collocation*, a method that simply requires the approximation to satisfy the partial differential equation for all t at a set of selected points x_j. When the form of u_N is taken into account, this is

$$\sum_{i=1}^{N} \phi_i(x_j) \frac{d}{dt} a_i(t) = \sum_{i=1}^{N} a_i(t)[\mathcal{L}\phi_i(x)]|_{x_j} + g(x_j, t),$$

a system of ordinary differential equations. In vector notation with, e.g., $a_i(t)$ being component i of $\mathbf{a}(t)$, this is

$$M \frac{d}{dt} \mathbf{a}(t) = J\, \mathbf{a}(t) + \mathbf{g}(t).$$

Other methods of weighted residuals such as *Galerkin's method* also lead to systems of ordinary differential equations of this form.

The systems of ordinary differential equations that arise in this way are not in standard form. Provided that M^{-1} exists, systems of the general form

$$M\,\mathbf{u}' = \mathbf{f}(t, \mathbf{u}),$$

can be written in standard form as

$$\mathbf{u}' = \mathbf{F}(t, \mathbf{u}) = M^{-1}\,\mathbf{f}(t, \mathbf{u}).$$

This is fine as far as the theory goes, but it is usually not satisfactory in practice, so one of the important issues with methods of weighted residuals is how to cope with the presence of M. Sometimes the basis functions $\phi_i(x)$ are such that M is diagonal and then there is no difficulty with the inversion. Historically the basis functions were chosen to be algebraic or trigonometric polynomials and with such a choice, M is usually a full matrix. Nowadays the use of such basis functions is associated with spectral methods (Canuto et al. [1988]). In some cases there are special techniques for inverting the matrices M (or for solving the linear systems) that arise in this context. One of the important ideas of finite element methods, which are included in the Galerkin framework, is to work with basis functions that lead to banded matrices M. If the band is not too wide, the constant matrix M can be factored by Gaussian elimination in a way that takes advantage of the structure so as to be much cheaper and require much less storage than the solution of a general matrix. For extremely large systems, the fact that the percentage of nonzero entries is quite small, i.e., the matrix is *sparse*, makes iterative methods attractive for the solution of the linear systems.

One way of dealing with the presence of the matrix M is to approximate it by a diagonal matrix so as to make inversion trivial. This process of "lumping" is sometimes done even for banded matrices so as to speed up the computations and reduce the storage needed. Appendix 8 of Zienkiewicz and Taylor [1988] presents a number of procedures, especially those natural in the context of finite element methods. Ideally the approximation due to lumping would affect the numerical solution no more than the approximation due to discretization, and this may be realized in practice.

For systems of partial differential equations in one space variable, it is possible to provide very powerful and convenient software based on codes for the initial value problem for a system of ODEs. To mention just one example that is easily accessible, we cite the code of Sincovec and Madsen that discretizes by finite differences a system of partial differential equations in one space variable and time in a form that can be integrated with any suitable code for the initial value problem. The code itself is provided as "Algorithm 494 PDEONE. Solutions of Systems of Partial Differential Equations" in Sincovec and Madsen [1975b] and there is a paper, Sincovec and Madsen [1975a], that describes the use of the code by means of a number of interesting and illuminating examples. The later "Algorithm 540 PDECOL, General Collocation Software for Partial Differential Equations," Madsen and Sincovec [1979], discretizes by collocation and combines in one package the discretization and its integration. See also Keast and Muir [1991], "Algorithm 688 EPDCOL: A more efficient PDECOL code." There are many other codes for solving partial differential equations by semidiscretization; the reader can turn to Schiesser [1991] to learn more about what is available.

EXERCISE 19. The equation

$$\frac{d}{dx}\left\{(x - x^2)\frac{dy}{dx}\right\} + \lambda(1 + x)y = 0$$

is to be integrated from $x = 0$ to $x = 1$. Here λ is a parameter. The initial value $y(0) = 1$. Suppose the problem has a solution $y(x)$ that can be expanded in a Taylor series about $x = 0$.

(a) Write this equation in standard form.
(b) Find a_1 and a_2 in the Taylor series

$$y(x) = 1 + a_1 x + a_2 x^2 + \cdots .$$

(c) What value must $y'(0)$ have if $y(x)$ is to have a Taylor series at $x = 0$?

(d) Explain how to use the Taylor series to start the integration at $x = 0$. Include a discussion of the errors you make in the start.

EXERCISE 20. This is the pull-out problem for a synchronous motor with many details left out (Shampine and Gordon [1975]). The equation of interest is

$$\frac{dv}{dx} = \frac{-\lambda\, v - \sin x + \gamma}{v}$$

Here $\lambda = 0.022$ and $\gamma = 0.8$. The integration starts at a singular point x_s where there are exactly two solutions that have $v(x_s) = 0$. You are to assume that you can start the integration on the right solution. This solution behaves as sketched in Figure 1.7. The integration proceeds from x_s to the left and terminates at x_0 where $v(x_0) = 0$.

(a) Given that $\gamma > \sin x_0$, show that

$$\left.\frac{dv}{dx}\right|_{x=x0} = +\infty \qquad \text{(as sketched).}$$

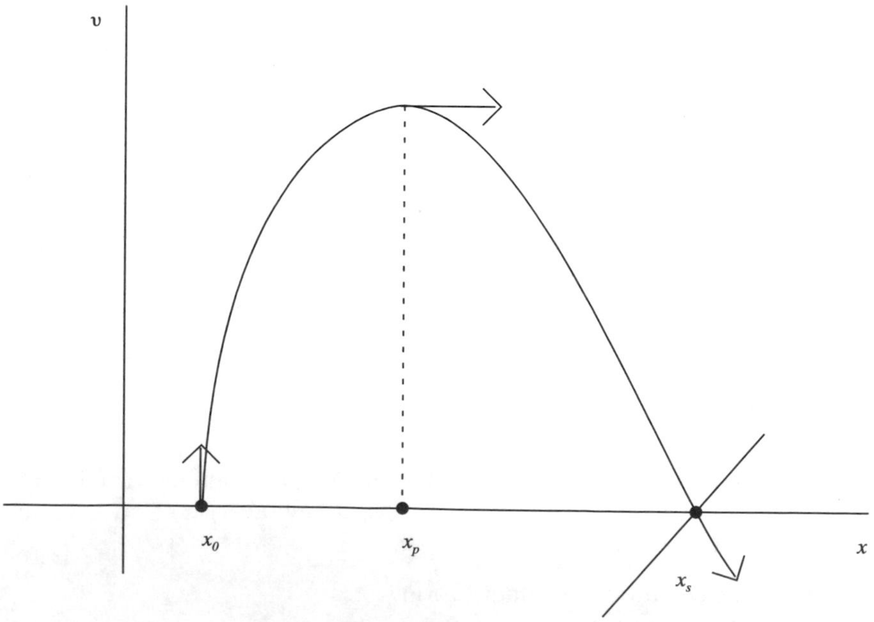

Figure 1.7

(b) There is a point x_p with $x_0 < x_p < x_s$ such that

$$\left.\frac{dv}{dx}\right|_{x=x_p} = 0 \qquad \text{(as sketched).}$$

How do you know there is such a point?

(c) Explain in detail how to compute x_0 by interchanging variables.

2

Discrete Variable Methods

Discrete variable methods for the numerical solution of

$$\mathbf{y}' = \mathbf{F}(x, \mathbf{y}), \qquad a \le x \le b, \qquad \mathbf{y}(a) = \mathbf{A} \tag{1}$$

start with the given value $\mathbf{y}_0 = \mathbf{y}(a) = \mathbf{A}$ and then produce an approximation $\mathbf{y}_1$ to $\mathbf{y}(x_1)$ at a point $x_1 > a$. This is described as advancing the integration a step of size h_0 from $x_0 = a$ to $x_1 = x_0 + h_0$. The process is repeated, successively producing approximations $\mathbf{y}_j$ to $\mathbf{y}(x_j)$ on a mesh $a = x_0 < x_1 < \ldots < x_N = b$ that spans the whole interval. Just how the integration is advanced one step is the subject of chapters that follow. What concerns us now are the implications of the approach itself as we try to answer two fundamental questions: What do the codes try to do? What kinds of problems can we hope to solve? Two ways of viewing discrete variable methods have proven useful. One focuses on the error made at each step. The other regards the numerical solution as being the exact solution of a somewhat different equation and then focuses on how different the two equations are. In both views the stability of the differential equation is fundamental. Some substantial examples are taken up to illustrate and amplify the ideas.

§1 Local Error

In the first step from a to x_1, numerical methods produce $\mathbf{y}_1$ as an approximation to the solution $\mathbf{y}(x_1)$ of the given initial value problem (1). What happens on subsequent steps is not so obvious. On the step from x_j to $x_{j+1} = x_j + h_j$, it would seem that the methods ought to try to approximate $\mathbf{y}(x_{j+1})$, but they do *not* attempt to do this, at least not directly. Some numerical methods do not "remember" results computed prior to x_j. On reaching x_j, all a code based on such a method has at its disposal is the current approximate solution $\mathbf{y}_j$ and the ability to evaluate $\mathbf{F}$. Because of this, the best it can do is to approximate the *local solution* $\mathbf{u}(x)$, the solution of the differential equation that has the value $\mathbf{y}_j$ at x_j,

$$\mathbf{u}' = \mathbf{F}(x, \mathbf{u}), \qquad \mathbf{u}(x_j) = \mathbf{y}_j.$$

Some numerical methods do remember results at a few points $x_{j-1}, \ldots, x_{j-k}$ prior to the current one, but only a few points, so that for most purposes they, too, "forget" what has transpired up to the current point. In this view of the error of discrete variable methods, codes try to control the *local error*, $\mathbf{u}(x_{j+1}) - \mathbf{y}_{j+1}$, in the step to $x_{j+1} = x_j + h_j$, rather than the *true error* or *global error*, $\mathbf{y}(x_{j+1}) - \mathbf{y}_{j+1}$. Notice that on the first step, the current approximation is the given initial value so that the local solution is the true solution and the local error is the same as the global error.

In this view of the error, there are two main issues. One is how well the numerical method approximates a solution of the differential equation over a single step and the other is the cumulative effect of the errors made at each step. The order of the method is of fundamental importance to the first issue. In this chapter the order can be defined loosely as saying that a method is of order p if the local error is of order $p + 1$ in the step size h_j:

$$\|\mathbf{u}(x_j + h_j) - \mathbf{y}_{j+1}\| \leq C\, h_j^{p+1}$$

At present all we need is that for methods of order $p \geq 1$, the smaller the step size, the more accurately the formula approximates the local solution. It is the cumulative effect of these errors that concerns us in this chapter.

In stepping from x_j to x_{j+1}, we attempt to approximate a solution $\mathbf{u}(x)$ of the differential equation. An error is made so that on the next step to x_{j+2}, we try to approximate a local solution $\mathbf{v}(x)$ that is different from $\mathbf{u}(x)$. The best we can hope to do is to approximate $\mathbf{v}(x_{j+2})$ perfectly. But then we have to ask ourselves, how close is $\mathbf{v}(x_{j+2})$ to $\mathbf{u}(x_{j+2})$, the curve we were trying to follow? This is a question about the stability of the differential equation rather than a question about the numerical method. Loosely speaking, a differential equation is said to be stable if solutions that are close at one point do not spread apart substantially thereafter, and unstable if they do.

EXAMPLE 1. Figure 2.1a shows some solutions of $y' = y^2 - x$ for $x \geq 0$. Clearly there is one solution curve that is very unstable and another that is very stable—the stability of an initial value problem depends on just which solution is being computed. In the circumstances it is easy to understand qualitatively the behavior of the solution curves. Null *isoclines* are curves $u(x)$ such that a solution $y(x)$ of the differential equation passing through the point $(x, u(x))$ does so with slope 0: $y'(x) = 0 = u(x)^2 - x$.

Obviously the null isoclines for this equation are $\pm\sqrt{x}$. A solution that passes through a point (x, y) with $y > \sqrt{x}$ has a positive slope there, which is to say that as x increases, the solution increases. If $-\sqrt{x} < y < \sqrt{x}$, the slope is negative and the solution decreases. If $y < \sqrt{x}$, the slope is positive and the solution increases. These observations tell us that as x increases, solution curves move away from the null isocline $+\sqrt{x}$ and towards the null isocline $-\sqrt{x}$. The null isoclines appear in Figure 2.1a to be solutions of the differential equation, but substitution of $u(x) = \pm\sqrt{x}$ into the equation shows that they are not. Nevertheless, Hubbard and West [1990] prove that there are solutions $y_1(x), y_2(x)$ of the differential equation that are asymptotic to the null isoclines, meaning that

$$\lim_{x\to\infty} \frac{y_1(x)}{\sqrt{x}} = 1 \quad \text{and} \quad \lim_{x\to\infty} \frac{y_2(x)}{-\sqrt{x}} = 1.$$

Figure 2.1a was computed with the MDEP package (Buchanan [1992]) using a variable step size Runge–Kutta code. Hubbard and West [1990] present in their Figure 5.4.6 some solutions computed with the three fixed step Runge–Kutta formulas available in their MacMath package. (See also Example 21.2a in Hubbard and West [1992].) On p. 243 they say ". . . first

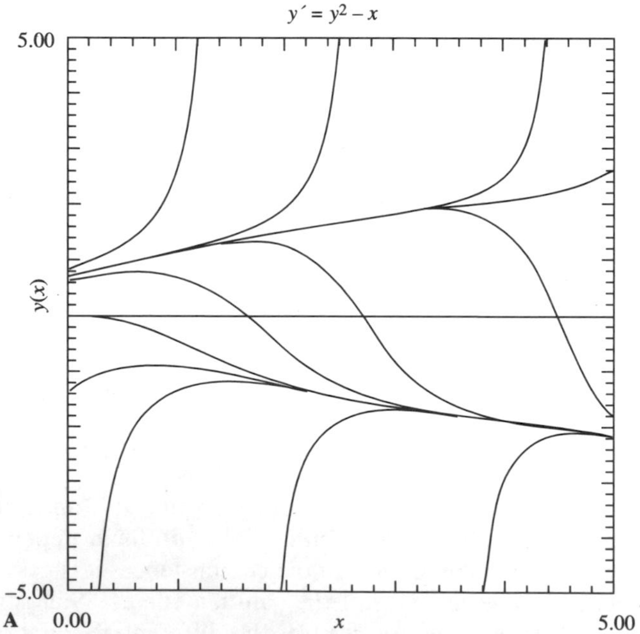

Figure 2.1a

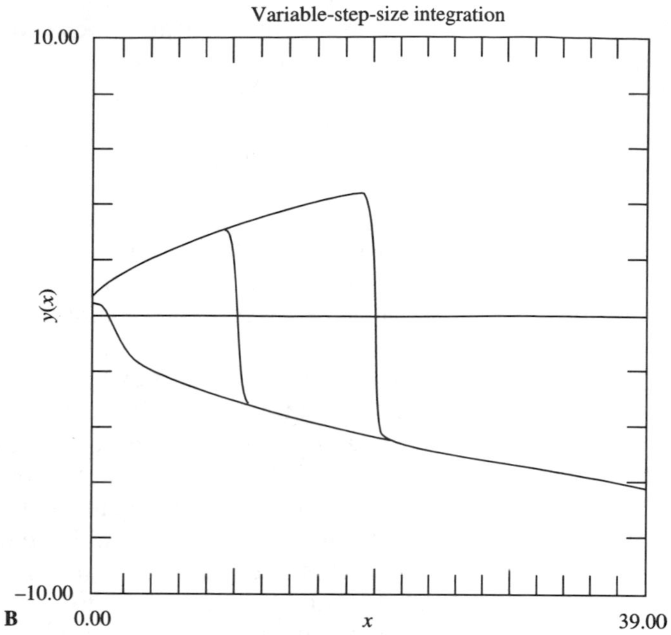

Figure 2.1b

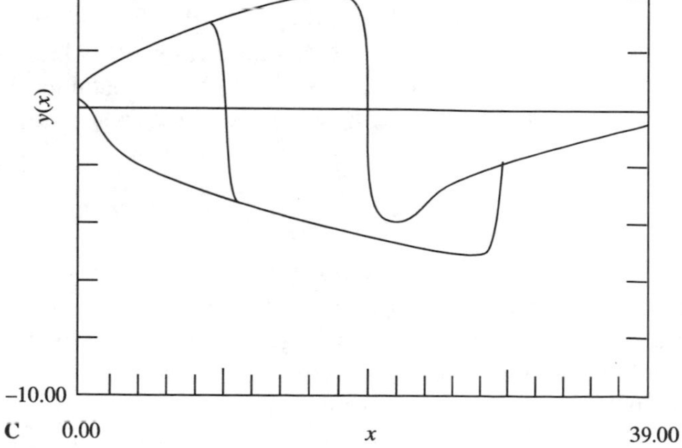

Figure 2.1c

observe that Midpoint Euler develops a really unpleasant behavior: a perfectly reasonable looking but spurious solution appears, . . .". This is the second-order Runge–Kutta formula in the package. The same thing happens with the fourth-order formula, but it occurs outside the interval of their plot. The Midpoint Euler method is not an option in MDEP, but a fixed step implementation of the fourth-order Runge–Kutta formula used in MacMath is available. Figure 2.1b shows some solutions computed using the variable step size fourth-order Runge–Kutta code in MDEP. The maximum step size was specified as 0.3. The numerical solutions have the expected behavior. Figure 2.1c shows some solutions computed with the fourth-order formula using the fixed step size 0.3. The computed solutions look reasonable, but they appear to converge to a stable solution that is qualitatively different from the stable solution of the differential equation. We shall return to this example in later chapters as we develop the tools needed to understand it. Variable step size codes monitor the local error and adjust the step size appropriately. They have two main goals. One is to get an accurate solution and the other is to accomplish the integration as inexpensively as possible. All too often people are so concerned about efficiency that they neglect the reliability of a computation. This striking example of Hubbard and West is a warning to those using codes that do not monitor the error that their numerical results might be unreliable.

In Chapter 1 it was pointed out that if $\mathbf{F}$ satisfies a Lipschitz condition with constant L, then for $x \geq x_{j+1}$,

$$\|\mathbf{u}(x) - \mathbf{v}(x)\| \leq \|\mathbf{u}(x_{j+1}) - \mathbf{v}(x_{j+1})\| \, e^{L(x - x_{j+1})}.$$

This inequality tells us that in the classical situation that $L(b - a)$ is not large, the solution of an initial value problem is moderately stable with respect to changes of the initial values. It is convenient to have a bound like this one that applies to all solutions, but, of course, it may be pessimistic for some solutions. In particular, the fact that $L(b - a)$ is large does not mean that a given solution, or even any solution, is unstable. To see an example of this, suppose that $u(x)$ and $v(x)$ are any two solutions of $y' = \mathcal{L}y$. Then

$$u(x) - v(x) = (u(x_{j+1}) - v(x_{j+1}))e^{\mathcal{L}(x - x_{j+1})}.$$

For this equation the Lipschitz constant is $L = |\mathcal{L}|$. If $L(b - a) \gg 1$ and $\mathcal{L} > 0$, the curves spread apart rapidly as x increases and the problem is unstable, but if $\mathcal{L} < 0$, the curves come together and the problem is stable.

Many people focus their attention on *the* solution $\mathbf{y}(x)$ of the given initial value problem. To understand the codes, they should focus their attention on the *set* of solutions that start out near $\mathbf{y}(x)$. At each step the numerical method tries to approximate one of these solution curves. An error is made in doing this, meaning that the code moves to an adjacent solution curve. The cumulative effect of these errors depends on the behavior of the family of solution curves. Indeed, suppose a single error is made at, say, the first step. Let $\mathbf{u}(x)$ be the solution passing through $(x_1, \mathbf{y}_1)$. If *no* further errors are made in approximating solutions of the differential equation so that each $\mathbf{y}_j = \mathbf{u}(x_j)$ for all $x_j > x_1$, the error at the end of the integration is $\mathbf{y}(b) - \mathbf{y}_N = \mathbf{y}(b) - \mathbf{u}(b)$. The question then is, if two solutions of the differential equation differ by a small amount at x_1, how much can they differ at $x = b$? When solutions spread out rapidly (the differential equation is unstable), even the one small error becomes greatly amplified. It is simply not possible to solve very unstable differential equations accurately with discrete variable methods. On the other hand, when solutions come together the effect of a small error is damped out. In the general situation the argument is repeated at each step as the numerical procedure moves from one solution curve to another nearby solution curve. Because solution curves might come together for x in one portion of $[a, b]$ and spread out in others, the behavior of the numerical errors can be complex.

The stability of an equation depends on the direction of integration. For example, in Figure 2.1 the solution $y(x) \sim +\sqrt{x}$ that is very unstable when integrated from left to right is very stable when integrated from right to left. With the MDEP program used to compute these figures and others of the kind, the user can indicate a point and ask the program to compute a solution passing through the point (a trajectory). It cannot be seen in the monochrome figures, but the program shows by means of different colors that it computes a trajectory by two integrations, one proceeding to the right and the other to the left. Because of this and the fact that stability depends on the direction of integration, initial points on the same theoretical solution curve can lead to different computed trajectories.

EXERCISE 1. A quadrature problem is a differential equation of the form

$$\frac{d}{dx} y = F(x).$$

Are such problems stable? (What is the difference between two solutions $u(x)$ and $v(x)$?) A problem of the form

$$\frac{d}{dt} u = f(u)$$

can be reduced to a quadrature problem by interchanging independent and dependent variables. Considering some of the examples presented in

Chapter 1, do the two forms always have the same stability? When might it be a good idea to reformulate such a problem as a quadrature problem?

§2 Backward Error Analysis

The other important way to view the error of discrete variable methods is called backward error analysis. Some popular methods approximate $\mathbf{y}(x)$ not just at the mesh points x_j, but at all x in the interval $[a, b]$. Methods that produce answers only at mesh points can be augmented to obtain an approximation for all x in $[a, b]$. We describe a way to do this that is not intended for numerical use, rather to understand what it means to "solve" the initial value problem. Suppose the integration has reached $(x_j, \mathbf{y}_j)$ and in the next step $\mathbf{y}_{j+1}$ is obtained as an approximation to $\mathbf{u}(x_{j+1})$. The local solution $\mathbf{u}(x)$ here is the solution of

$$\mathbf{u}' = \mathbf{F}(x, \mathbf{u}), \qquad \mathbf{u}(x_j) = \mathbf{y}_j,$$

and the local error of the step is

$$\mathbf{u}(x_{j+1}) - \mathbf{y}_{j+1} = \mathbf{le}_j.$$

The step size $h_j = x_{j+1} - x_j$. General purpose codes choose the step size, and possibly the method, so as to control the local error. Such codes might be given a tolerance τ and instructions to control the error so that

$$\|\mathbf{le}_j\| \leq h_j \, \tau$$

at each step. Alternatively, when the mesh is given (hence the step sizes), we can define τ as the smallest number for which this inequality is true for all j.

For theoretical purposes let us define $\mathbf{v}(x)$ by

$$\mathbf{v}(x) = \mathbf{u}(x) - \frac{(x - x_j)}{h_j} \mathbf{le}_j, \qquad x_j \leq x \leq x_{j+1}$$

This function interpolates the numerical solution because obviously

$$\mathbf{v}(x_j) = \mathbf{u}(x_j) = \mathbf{y}_j,$$

$$\mathbf{v}(x_{j+1}) = \mathbf{u}(x_{j+1}) - \frac{(x_{j+1} - x_j)}{h_j}(\mathbf{u}(x_{j+1}) - \mathbf{y}_{j+1}) = \mathbf{y}_{j+1}.$$

On defining $\mathbf{v}(x)$ in this manner for all subintervals $[x_j, x_{j+1}]$ we obtain a function that is continuous on the whole of $[a, b]$ and interpolates the numerical solution at the mesh points. Let us now inquire about how well this function satisfies the differential equation.

The *residual*, or *defect*, $\mathbf{r}(x)$ of a differentiable function $\mathbf{v}(x)$ measures how well it satisfies the differential equation:

$$\mathbf{r}(x) = \mathbf{v}'(x) - \mathbf{F}(x, \mathbf{v}(x)).$$

Put the other way around, the function $\mathbf{v}(x)$ is a solution of the differential equation

$$\mathbf{v}'(x) = \mathbf{F}(x, \mathbf{v}(x)) + \mathbf{r}(x),$$

a perturbed version of the equation that interests us. In the view of backward error analysis, the task given the numerical scheme is to control the size of the perturbation. Backward error analysis has proved very important to understanding the numerical solution of systems of linear algebraic equations. Just as with linear systems, satisfying an equation well does not mean that an approximate solution is close to the true solution; that depends on the stability of the problem (the term condition is used in the context of linear systems). We have some inequalities that relate the size of the perturbation $\mathbf{r}(x)$ to the difference between $\mathbf{v}(x)$ and $\mathbf{y}(x)$, but they are usually not very realistic. If the original problem is thought to be well posed on physical grounds, it is expected that if the perturbation is "small," then the solution $\mathbf{v}(x)$ will be an adequate approximation to $\mathbf{y}(x)$.

EXAMPLE 2. An example will help us understand better the role of stability in this viewpoint. It is easily verified that $y(t; \varepsilon) = (1 + \varepsilon t)\sin(t)$ is the solution of

$$y'' + y = 2\varepsilon\cos(t), \qquad y(0; \varepsilon) = 0, \qquad y'(0; \varepsilon) = 1.$$

For "small" ε, the right hand side is a perturbation of the equation that is uniformly small and on any finite interval, $y(t; \varepsilon)$ is "close" to $y(t; 0)$. On the other hand, for any given ε, if the time interval is sufficiently long, $y(t; \varepsilon)$ can differ significantly from $y(t; 0)$. Figure 2.2 shows the solutions for $\varepsilon = 0$ and $\varepsilon = 0.05$. This is the familiar phenomenon of secular growth in perturbation theory due here to resonance. In the present context, we are reminded that the definition of the classical situation involves the length of the interval as well as the Lipschitz constant—a small residual can have a large effect when the interval is large compared to the Lipschitz constant.

For x in a typical subinterval $[x_j, x_{j+1}]$, the residual of the $\mathbf{v}(x)$ just defined is

$$\mathbf{r}(x) = \mathbf{u}'(x) - \frac{1}{h_j}\mathbf{le}_j - \mathbf{F}(x, \mathbf{v}(x)) = \mathbf{F}(x, \mathbf{u}(x)) - \mathbf{F}(x, \mathbf{v}(x)) - \frac{1}{h_j}\mathbf{le}_j.$$

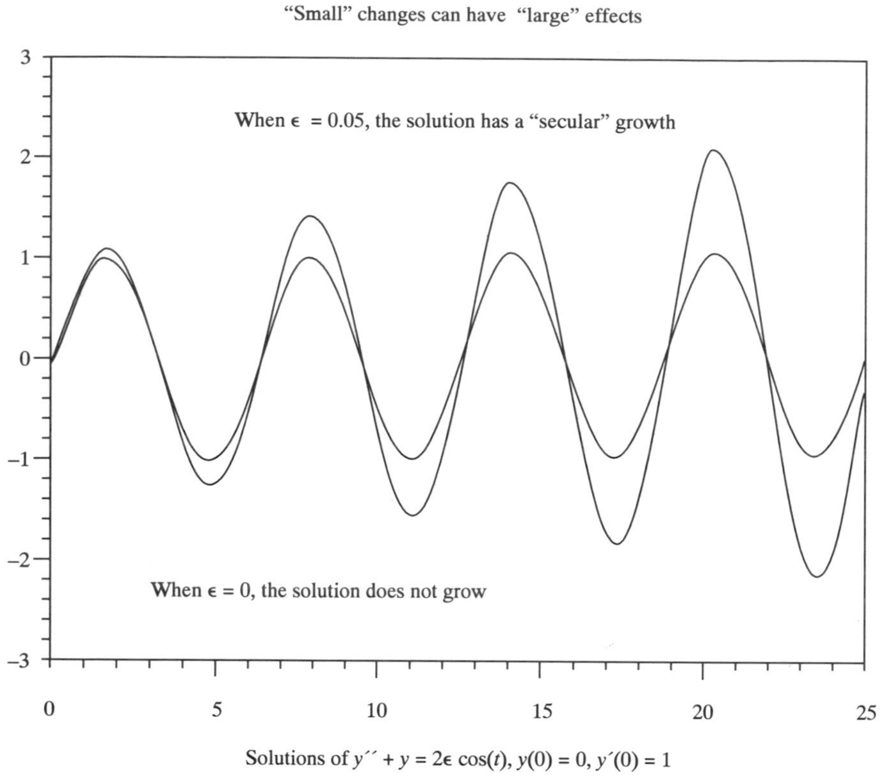

"Small" changes can have "large" effects

Solutions of $y'' + y = 2\epsilon \cos(t)$, $y(0) = 0$, $y'(0) = 1$

Figure 2.2

This implies that

$$\|\mathbf{r}(x)\| \le \frac{1}{h_j}\|\mathbf{le}_j\| + \|\mathbf{F}(x,\mathbf{u}(x)) - \mathbf{F}(x,\mathbf{u}(x) - \frac{(x-x_j)}{h_j}\mathbf{le}_j)\|$$

$$\le \frac{1}{h_j}\|\mathbf{le}_j\| + L\,\|\frac{(x-x_j)}{h_j}\mathbf{le}_j\| \le \|\mathbf{le}_j\|\left(L + \frac{1}{h_j}\right).$$

With our supposition of a bound on the local error in terms of a tolerance τ and the step size h_j, we then have

$$\|\mathbf{r}(x)\| \le \tau(1 + h_j L), \qquad x_j \le x \le x_{j+1}$$

In the classical situation that $L(b - a)$ is not large, the product Lh_j is certainly not large and is usually small compared to 1. What all this says is that if the local error divided by the step size is controlled so as to be no greater than a tolerance τ, then there is a piecewise differentiable function $\mathbf{v}(x)$ that interpolates the numerical values and satisfies the differential equation with a residual that is bounded approximately by this tolerance.

The theoretical interpolant $\mathbf{v}(x)$ is used to establish a bound on the size of the perturbation. It also serves to treat all discrete variable methods in a uniform way. However, it is not at all convenient for actually assessing the size of the residual of a computed solution. Some of the popular methods produce at each step a polynomial $\mathbf{P}(x)$ that approximates $\mathbf{y}(x)$ on all of $[x_j, x_{j+1}]$. The residual of this approximation, $\mathbf{r}(x) = \mathbf{P}'(x) - \mathbf{F}(x, \mathbf{P}(x))$, could be evaluated at selected points within the span of the step to verify that the size of the residual is acceptable.

§3 Stability

In a backward error analysis view of solving a differential equation numerically, we regard the task as one of producing a function that is the exact solution of an equation close to the given equation. In the view of local error, we always work with the given equation, but realize that an error is made at each step that moves us to a nearby solution curve. In either view, the stability of the differential equation is fundamental. In the one, the problem must be well posed with respect to changes in the equation and in the other, it must be well posed with respect to changes in initial values. Some general bounds have been given that guarantee well-posedness, but they are rarely practical. Instead, it is understanding of the problem that leads one to believe that its numerical solution is feasible. In this section some further observations are made about bounds on the effects of changes and we also look into how to approximate the effects.

When $\mathbf{F}$ satisfies a Lipschitz condition with constant L and $L(b - a)$ is not large, the problem is moderately stable and we can hope to solve the problem numerically. As we have seen by example, if $L(b - a) \gg 1$, the problem might be stable, or it might not. There is a very important class of problems called *stiff* that are stable, but have large Lipschitz constants. We must defer the discussion of these problems to Chapter 8 because it depends greatly on the numerical methods, but it is worth mentioning that we have already seen a model stiff problem. The Robertson problem

$$
\begin{aligned}
y_1' &= -0.04y_1 + 10^4 y_2y_3, & y_1(0) &= 1,\\
y_2' &= 0.04y_1 - 10^4 y_2y_3 - 3 \times 10^7 y_2^2, & y_2(0) &= 0,\\
y_3' &= 3 \times 10^7 y_2^2, & y_3(0) &= 0,
\end{aligned}
$$

which might be solved on [0, 40], is often used as an example of a stiff problem. That any Lipschitz constant must be large is easily inferred. For example,

$$\frac{\partial F_3}{\partial y_2} = 6 \times 10^7 y_2,$$

so if y_2 is of even modest size, this cannot be the classical situation for an interval of length 40. Developing a theory of stiff problems is an active area of research. Clearly something more must be said about **F** because the Lipschitz condition alone does not distinguish the unstable problem $y' = Ly$ from the stable problem $y' = -Ly$. Let us briefly take up one approach.

Suppose that the norm used to measure the size of vectors arises from an inner product, e.g., the Euclidean norm. The notation $\langle \mathbf{u}, \mathbf{v} \rangle$ will be used for the inner product of the vectors **u** and **v**. (In the case of the Euclidean inner product $\langle \mathbf{u}, \mathbf{v} \rangle = \mathbf{u}^T\mathbf{v}$.) The induced norm $\|\mathbf{u}\| = \sqrt{\langle \mathbf{u}, \mathbf{u} \rangle}$. The Cauchy–Schwarz inequality relates the inner product of two vectors and their norms:

$$|\langle \mathbf{u}, \mathbf{v} \rangle| \leq \|\mathbf{u}\| \, \|\mathbf{v}\|.$$

A Lipschitz condition on **F** and this inequality lead to

$$|\langle \mathbf{F}(x, \mathbf{u}) - \mathbf{F}(x, \mathbf{v}), \mathbf{u} - \mathbf{v} \rangle| \leq \|\mathbf{F}(x, \mathbf{u}) - \mathbf{F}(x, \mathbf{v})\| \, \|\mathbf{u} - \mathbf{v}\| \leq L \, \|\mathbf{u} - \mathbf{v}\|^2$$

or, put differently,

$$-L \, \|\mathbf{u} - \mathbf{v}\|^2 \leq \langle \mathbf{F}(x, \mathbf{u}) - \mathbf{F}(x, \mathbf{v}), \mathbf{u} - \mathbf{v} \rangle \leq L \, \|\mathbf{u} - \mathbf{v}\|^2.$$

This pair of inequalities is a consequence of the Lipschitz condition. It turns out that one of the inequalities is much more important to stability than the other. For this reason we study **F** that satisfy

$$\langle \mathbf{F}(x, \mathbf{u}) - \mathbf{F}(x, \mathbf{v}), \mathbf{u} - \mathbf{v} \rangle \leq \mathcal{L} \, \|\mathbf{u} - \mathbf{v}\|^2$$

for a constant $\mathcal{L}$. We continue to assume that **F** satisfies a Lipschitz condition, so there is no point in introducing this *one-sided Lipschitz condition* unless the constant $\mathcal{L}$ is smaller than L. By definition L is non-negative, but $\mathcal{L}$ is permitted to be negative. Problems with **F** that satisfy a one-sided Lipschitz condition with constant $\mathcal{L} < 0$ are called dissipative. The problem $y' = \mathcal{L}y$ has the Lipschitz constant $L = |\mathcal{L}|$. Now

$$\begin{aligned}\langle F(x, u) - F(x, v), u - v \rangle &= \langle \mathcal{L}u - \mathcal{L}v, u - v \rangle \\ &= \mathcal{L} \langle u - v, u - v \rangle = \mathcal{L} \, |u - v|^2.\end{aligned}$$

When $\mathcal{L} < 0$ this is a dissipative problem.

A one-sided Lipschitz condition provides more information about **F**. The proof of a stability result that takes advantage of this information is sketched in the appendix. This result says that if $\mathbf{y}(x)$ is the solution of

$$\mathbf{y}' = \mathbf{F}(x, \mathbf{y}), \qquad \mathbf{y}(a) \text{ given},$$

and $\mathbf{v}(x)$ the solution of

$$\mathbf{v}' = \mathbf{F}(x, \mathbf{v}) + \mathbf{G}(x), \qquad \mathbf{v}(a) \text{ given},$$

then

$$\|\mathbf{v}(x) - \mathbf{y}(x)\| \leq \|\mathbf{v}(a) - \mathbf{y}(a)\| e^{\mathcal{L}(x-a)} + \int_a^x \|\mathbf{G}(t)\| e^{\mathcal{L}(x-t)} \, dt$$

The constant $\mathcal{L} \leq L$, so the product $\mathcal{L}(b - a)$ is never bigger than $L(b - a)$. When it is smaller, the inequality shows that the stability is better than we would expect from the Lipschitz condition alone. When **F** is dissipative and $\mathbf{G} \equiv \mathbf{0}$, the distinction is dramatic because the inequality implies that all solutions of the differential equation come together exponentially fast.

EXERCISE 2. Stiff problems are stable problems with Lipschitz constants L such that $L(b - a) \gg 1$. Some are dissipative, but many are not. Prove that a dissipative equation cannot satisfy a linear conservation law. Because Robertson's problem satisfies a linear conservation law, this model stiff problem is not dissipative. Hint: Ask what happens in the conservation law as $x \to \infty$.

If we wish to solve a differential equation numerically by a discrete variable method, it is necessary that the problem be at least moderately stable. We consider only $\mathbf{F}(x, \mathbf{y})$ that satisfy a Lipschitz condition with constant L. In the classical situation that $L(b - a)$ is not large, the problem is moderately stable. The fact that $L(b - a)$ is large does not mean that the problem is not stable, just that the usual bound does not show the problem to be stable. If **F** satisfies a one-sided Lipschitz condition with a constant $\mathcal{L}$ such that $\mathcal{L}(b - a)$ is not large, the inequality just presented says that the problem is moderately stable. In particular, a dissipative problem is always stable, regardless of the size of L. This suggests that the analysis for the classical situation of $L(b - a)$ not large can be extended to some stable problems with large Lipschitz constants, and this is the case. Such problems are very much the subject of research at this time. There is a big gap between the theory and practice of stability. Without knowing the solution, it is not usually possible to obtain Lipschitz constants and even when this can be done, the bounds are ordinarily so crude that they are not useful. In practice, it is physical insight that leads us to think that a problem is moderately stable so that it is reasonable to attempt its solution numerically.

Bounds on the effect of an isolated error just tell us the worst that can happen. Because the behavior of solution curves is complex, so is the effect of an isolated error. For some differential equations it is possible to understand better the behavior of solutions by analytical means, hence to understand better the propagation of error. In a technical sense, linear problems are especially easy to handle and certain classes of linear problems will be important to some of our later investigations for just this reason. Linear problems are those for which $\mathbf{F}(x, \mathbf{y})$ has the special form

$$\mathbf{y}' = J(x)\mathbf{y} + \mathbf{G}(x).$$

Let $\mathbf{y}(x)$ be the solution with $\mathbf{y}(a) = \mathbf{s}$ and let $\mathbf{u}(x)$ be the solution with $\mathbf{u}(a) = \mathbf{s} + \Delta(a)$. Define the difference $\Delta(x) = \mathbf{u}(x) - \mathbf{y}(x)$. Then

$$\mathbf{u}'(x) = \mathbf{y}'(x) + \Delta'(x) = J(x)\,(\mathbf{y}(x) + \Delta(x)) + \mathbf{G}(x),$$

so that

$$\Delta'(x) = J(x)\,\Delta(x), \qquad \Delta(a) \text{ given.}$$

This gives an equation for the change in $\mathbf{y}(x)$ that is valid for all changes $\Delta(a)$ to the initial value. There is a large body of knowledge about such problems, especially when the Jacobian matrix $J(x)$ is constant, that can be useful in understanding how error propagates; examples will be given later.

For nonlinear problems we can study the effects of small changes by linearization. We begin with changes to the initial values only. With our usual assumptions about $\mathbf{F}(x, \mathbf{y})$, the problem

$$\mathbf{y}' = \mathbf{F}(x, \mathbf{y}), \qquad a \le x \le b, \qquad \mathbf{y}(a) = \mathbf{s},$$

has a unique solution $\mathbf{y}(x)$. Directing our attention to component i of the vector of initial values, let us write the solution as $\mathbf{y}(x; s_i)$. With the notation that $\mathbf{e}^{(i)}$ is column i of the identity matrix I, the vector $\mathbf{s} + \delta\mathbf{e}^{(i)}$ is the result of adding δ to component i of $\mathbf{s}$, so the solution of the problem

$$\mathbf{y}' = \mathbf{F}(x, \mathbf{y}), \qquad a \le x \le b, \qquad \mathbf{y}(a) = \mathbf{s} + \delta\mathbf{e}^{(i)}$$

is $\mathbf{y}(x; s_i + \delta)$. We can now ask whether $(\mathbf{y}(x; s_i + \delta) - \mathbf{y}(x; s_i))/\delta$ has a limit as $\delta \to 0$, i.e., whether the partial derivative $(\partial\mathbf{y}/\partial s_i)\,(x; s_i)$ exists. When it does, we can make the usual linear approximation of applied mathematics,

$$\mathbf{y}(x; s_i + \delta) \approx \mathbf{y}(x; s_i) + \frac{\partial\mathbf{y}}{\partial s_i}(x; s_i)\delta,$$

because the difference of the two sides tends to zero as δ tends to zero. If we do not worry about the existence of partial derivatives and the interchange of the order of differentiation, i.e., if we proceed formally, we find that

$$\frac{\partial}{\partial s_i}\left(\frac{d\mathbf{y}}{dx}\right) = \frac{d}{dx}\left(\frac{\partial \mathbf{y}}{\partial s_i}\right) = \frac{\partial \mathbf{F}}{\partial \mathbf{y}}(x, \mathbf{y}(x))\frac{\partial \mathbf{y}}{\partial s_i}, \qquad a \le x \le b, \qquad \frac{\partial \mathbf{y}}{\partial s_i}(a) = \mathbf{e}^{(i)}.$$

Formally $\mathbf{y}(x)$ and the partial derivative $\mathbf{v}(x) = (\partial \mathbf{y}/\partial s_i)(x)$ satisfy the initial value problem

$$\begin{aligned} \mathbf{y}' &= \mathbf{F}(x, \mathbf{y}) & \mathbf{y}(a) &= \mathbf{s}, \\ \mathbf{v}' &= \frac{\partial \mathbf{F}}{\partial \mathbf{y}}(x, \mathbf{y}(x))\,\mathbf{v} & \mathbf{v}(a) &= \mathbf{e}^{(i)}. \end{aligned}$$

When $\mathbf{F}$ is smooth enough, this formal procedure can be justified, as can the expansion

$$\mathbf{y}(x; s_i + \delta) = \mathbf{y}(x; s_i) + \frac{\partial \mathbf{y}}{\partial s_i}(x; s_i)\delta + \mathbb{O}(\delta^2).$$

It is convenient to assemble all the partial derivatives with respect to the components of the initial vector $\mathbf{s}$ into a matrix function $V(x) = ((\partial \mathbf{y}/\partial s_1)(x), \partial \mathbf{y}/\partial s_2)(x), \ldots, (\partial \mathbf{y}/\partial s_n)(x))$ and then write the equations in the more compact form

$$\begin{aligned} \mathbf{y}' &= \mathbf{F}(x, \mathbf{y}) & \mathbf{y}(a) &= \mathbf{s}, \\ V' &= \frac{\partial \mathbf{F}}{\partial \mathbf{y}}(x\ \mathbf{y}(x))V & V(a) &= I, \end{aligned}$$

involving the Jacobian of $\mathbf{F}$ and the identity matrix I. The equation for the partial derivatives is called the *equation of first variation* or the *variational equation*. The effect of changing the initial vector $\mathbf{s}$ to $\mathbf{s} + \boldsymbol{\delta}$ is to change the solution from $\mathbf{y}(x; \mathbf{s})$ to $\mathbf{y}(x; \mathbf{s} + \boldsymbol{\delta})$ where

$$\mathbf{y}(x; \mathbf{s} + \boldsymbol{\delta}) = \mathbf{y}(x; \mathbf{s}) + V(x)\boldsymbol{\delta} + \mathbb{O}(\|\boldsymbol{\delta}\|^2) \approx \mathbf{y}(x; \mathbf{s}) + V(x)\boldsymbol{\delta}.$$

To understand the effects of changing $\mathbf{F}$, suppose that it depends on a number of parameters that are collected into a vector $\mathbf{p}$. Again, formal differentiation of the problem leads to a variational equation for the partial derivatives of the solution with respect to the parameters. Now the solution $\mathbf{y}(x; \mathbf{p})$ and the matrix of partial derivatives $V(x) = ((\partial \mathbf{y}/\partial p_1)(x), (\partial \mathbf{y}/\partial p_2)(x), \ldots, (\partial \mathbf{y}/\partial p_n)(x))$ satisfy the initial value problem

$$\begin{aligned} \mathbf{y}' &= \mathbf{F}(x, \mathbf{y}, \mathbf{p}) & \mathbf{y}(a; \mathbf{p}) &= \mathbf{s}, \\ V' &= \frac{\partial \mathbf{F}}{\partial \mathbf{y}}(x, \mathbf{y}(x; \mathbf{p}), \mathbf{p})V + \frac{\partial \mathbf{F}}{\partial \mathbf{p}}(x, \mathbf{y}(x; \mathbf{p}), \mathbf{p}) & V(a) &= 0. \end{aligned}$$

The effect of changing the parameters $\mathbf{p}$ to $\mathbf{p} + \boldsymbol{\delta}$ is to change the solution from $\mathbf{y}(x; \mathbf{p})$ to $\mathbf{y}(x; \mathbf{p} + \boldsymbol{\delta})$ where

$$\mathbf{y}(x; \mathbf{p} + \boldsymbol{\delta}) = \mathbf{y}(x; \mathbf{p}) + V(x)\boldsymbol{\delta} + \mathcal{O}(\|\boldsymbol{\delta}\|^2) \approx \mathbf{y}(x; \mathbf{p}) + V(x)\,\boldsymbol{\delta}.$$

Note the forms of the two variational equations. Both are linear equations for the partial derivatives. The one for the effect of a change of initial value is a homogeneous equation with non-zero initial values and the one for the effect of a change of $\mathbf{F}$ is inhomogeneous with zero initial values. In the linear approximation being made, the effects of changing both the initial vector and a vector of parameters can be determined separately and then simply added.

§4 Examples

The examples that follow elaborate on stability and its implications for discrete variable methods. The formal statement of the initial value problem says that the integration is to go from a to b. Surprisingly, in important applications it is possible to choose the direction of integration and even to change the length of the interval of integration. A discussion of some of these situations will help us understand what is possible with discrete variable methods.

EXAMPLE 3. The excellent introductory text on ODEs by Boyce and DiPrima [1992] presents some results computed using a fixed-step Runge–Kutta code for a problem,

$$y'' = 100y, \qquad y(0) = 1, \qquad y'(0) = -10,$$

that they describe as "stiff." Following the presentation by D. Canright to a Workshop on Teaching ODEs with Computer Experiments (Harvey Mudd College, June 1992), the adaptive Runge–Kutta code in MDEP was used to obtain Figure 2.3. The solution of the initial value problem is $\exp(-10x)$ and it is clear from the plot that the numerical solution bears no resemblance to the true solution for x bigger than, say, 2.4. Because the problem is described in the literature as stiff, the Workshop participants assumed that the code had failed due to stiffness. But the problem is not stiff, and the code did not fail! This problem is unstable and the numerical results obtained are just what we should expect in the circumstances. All solutions of this equation have the form $\alpha \exp(-10x) + \beta \exp(+10x)$. The initial conditions imply that the solution sought has $\alpha = 1$ and $\beta = 0$. When a small local error is made, the code moves from the true solution $\exp(-10x)$ to a nearby solution curve with $\alpha \approx 1$ and $\beta \approx 0$. No matter

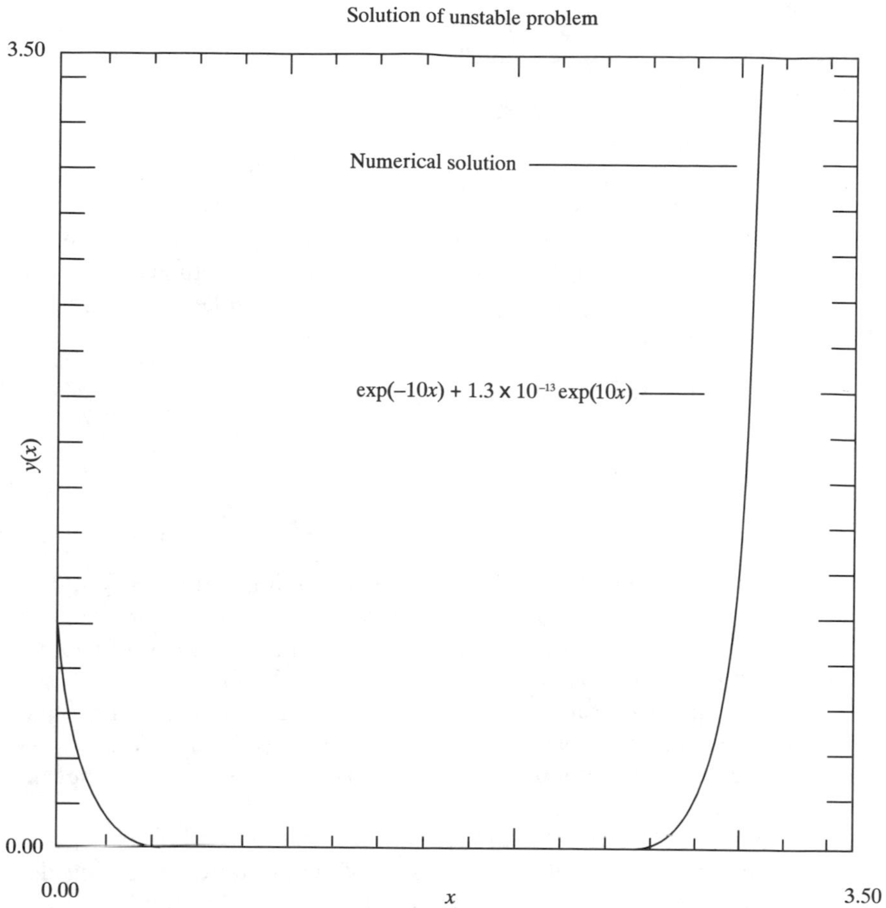

Figure 2.3

how small this error, hence no matter how small β, this new local solution diverges rapidly from the true solution. After a while, a local solution is approximately a multiple of the dominant solution $\exp(+10x)$. Using some of the nice features of MDEP, it was found that a plot of $\exp(-10x) + 1.3 \times 10^{-13}\exp(+10x)$ can scarcely be distinguished graphically from the plot of the numerical solution of the differential equation. This can be interpreted as saying that the numerical results are about the same as what one would get by making an error of about 1.3×10^{-13} in the first step and no other errors thereafter. The fact that the computed solution differs so much from the true solution is duc to the problem, not the integrator. Stiffness is a complex matter that we take up in Chapter 8. It is not easy to classify some problems, but this one is easy. It is the antithesis of a stiff problem because stiff problems are always stable, indeed, super-stable in a certain sense.

EXAMPLE 4. In Example 1 of Chapter 1 we saw how problems with discontinuities in the independent variable might be discussed by "tacking together" problems with continuous **F**. The same approach lets us see that such problems are well posed. Recall that the discussion was in terms of an example for which **F** might be discontinuous at $x = b$, but is continuous and Lipschitzian on the intervals $[a, b)$ and $(b, c]$. We broke up the original problem into two parts:

$$\begin{array}{lll} \text{Problem I} & y_I(a) \text{ given}, & \\ & y'_{\mathrm{I}} = F_{\mathrm{I}}(x, y_{\mathrm{I}}), & a \le x \le b, \\ \text{Problem II} & y_{\mathrm{II}}(b) = y_{\mathrm{I}}(b), & \\ & y'_{\mathrm{II}} = F_{\mathrm{II}}(x, y_{\mathrm{II}}), & b \le x \le c. \end{array}$$

Problem I is well-posed, so small changes to the initial value $y_{\mathrm{I}}(a)$ and to $F_{\mathrm{I}}(x, y_{\mathrm{I}})$ lead to small changes in its solution. In particular, they lead to small changes in the value $y_{\mathrm{I}}(b)$ that constitutes the initial value for Problem II. This second problem is also well posed, so small changes to its initial value and to $F_{\mathrm{II}}(x, y_{\mathrm{II}})$ lead to small changes in its solution. This informal argument tells us that the whole problem is well posed, so we might reasonably expect to solve it numerically if we take care near $x = b$ where F is discontinuous.

As was remarked in Chapter 1, the approach can be used to discuss the existence and uniqueness of problems with discontinuities involving dependent variables, but the argument just made about the stability of the problem does not go through so neatly. An example will show that the situation can be rather different. For α in $(0, 1)$ let the initial value $y(0) = \alpha$ and define the differential equation for $0 \le x \le 2$ by

$$\begin{aligned} y'(x) &= 2 - 2x \qquad y \le 1, \\ &= 1 \qquad\qquad\; y > 1. \end{aligned}$$

Both $F_1(x, y) = 2 - 2x$ and $F_2(x, y) = 1$ satisfy Lipschitz conditions (they do not even depend on y), so the existence and uniqueness of solutions in the two regions is assured. It is easily checked that as long as $y \le 1$, the solution of the initial value problem is

$$y(x; \alpha) = \alpha + 2x - x^2.$$

The equation changes, and correspondingly the solution changes, at a point $x_\alpha > 0$ where

$$1 = y(x_\alpha; \alpha) = \alpha + 2x_\alpha - x_\alpha^2.$$

This occurs at $x_\alpha = 1 - \sqrt{\alpha}$ and for $x \geq x_\alpha$, it is easily checked that the solution with $y(x_\alpha; \alpha) = 1$ is

$$y(x; \alpha) = 1 + (x - x_\alpha).$$

It was assumed that $0 < \alpha < 1$, so $x_\alpha < 1$. Because of this, the solution at, say, $x = 2$ is

$$y(2; \alpha) = 1 + (2 - x_\alpha) = 2 + \sqrt{\alpha}.$$

This expresses the value of the solution at a particular point in terms of the initial value α, so we can ask how sensitive this value is to a change in the initial value. Clearly

$$\frac{d}{d\alpha} y(2; \alpha) = \frac{1}{2\sqrt{\alpha}}.$$

This shows that for $0 < \alpha \ll 1$, the solution at $x = 2$ is very sensitive to changes in α. Figure 2.4 displays the solutions of the initial value problem for $\alpha = 0$ and $\alpha = 0.05$. It is seen that a small change of the initial value can lead to a large change in the solution. What is happening here is that the solution with initial value $\alpha = 0$ is a parabola, $y(x; 0) = 2x - x^2$, that increases to $y = 1$ at $x = 1$. This solution has a maximum at $x = 1$, so the curve grazes the boundary where there is a change in the definition of $F(x, y)$ and decreases to $y = 0$ at $x = 2$. Any solution with initial value $\alpha > 0$ crosses the boundary $y = 1$ at x_α where the equation changes and the solution increases monotonely thereafter. No matter how close α is to 0, the value $y(2; \alpha) > 1$ is very different from $y(2; 0) = 0$.

This example was contrived so as to be simple, but it illustrates a situation that is by no means rare. A real problem easily visualized is that of following the path of a particle through a pile of blocks of different materials. The differential equation describing the motion involves the material, so depends on which block the particle is in. If the particle grazes a material interface, its trajectory might depend strongly on whether it stays in one block or penetrates another. Clearly some trajectories are not stable because moving to a nearby trajectory at some time may result in quite a different path later.

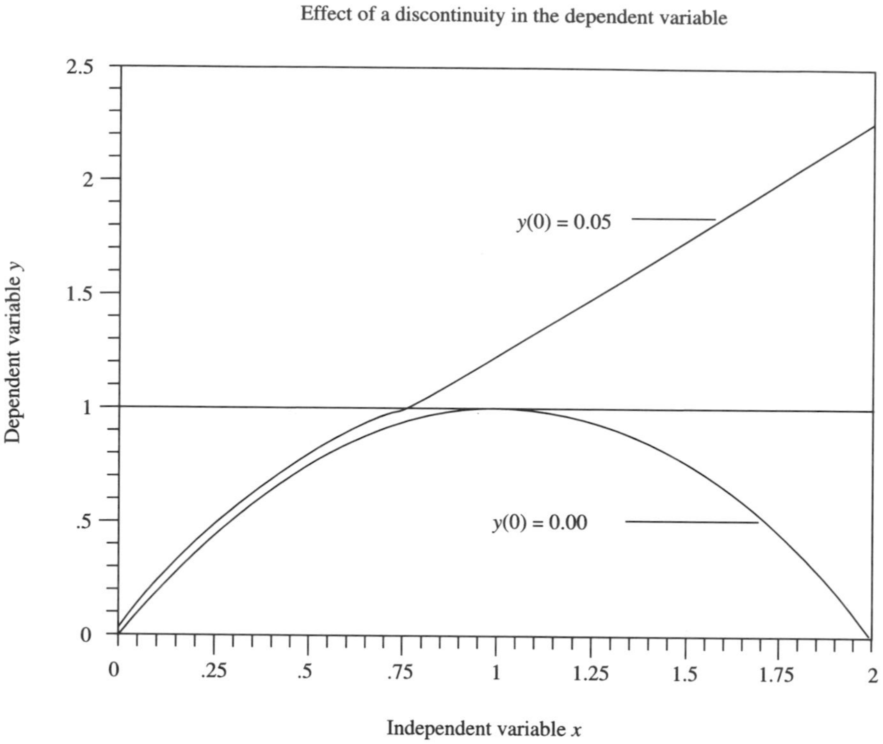

Figure 2.4

EXAMPLE 5. Important special functions are defined as solutions of Airy's equation, $y'' = x\,y$. The special function $Ai(x)$ is the solution that satisfies the initial conditions

$$Ai(0) = 3^{-2/3}/\Gamma(2/3) \approx 0.35502\ 80538\ 87817,$$

$$Ai'(0) = -3^{-1/3}/\Gamma(1/3) \approx -0.25881\ 94037\ 92807.$$

The special function $Bi(x)$ is the solution that satisfies the initial conditions $Bi(0) = \sqrt{3}\,Ai(0)$, $Bi'(0) = -\sqrt{3}\,Ai'(0)$. The two solutions are linearly independent, so all solutions of Airy's equation can be written as linear combinations of them. They are tabulated in Abramowitz and Stegun [1964], where one finds, e.g., that

$$Ai(1) = 0.13529242, \qquad Ai'(1) = -0.15914744,$$
$$Bi(1) = 1.20742359, \qquad Bi'(1) = 0.93243593.$$

For large x it is known from any asymptotic analysis that

$$Ai(x) \sim \frac{1}{2}\pi^{-1/2}x^{-1/4}\exp\left(-\frac{2}{3}x^{3/2}\right) = R(x),$$

$$Bi(x) \sim \pi^{-1/2}x^{-1/4}\exp\left(+\frac{2}{3}x^{3/2}\right) = Q(x).$$

This means that

$$\lim_{x\to\infty}\frac{Ai(x)}{R(x)} = 1 \quad \text{and} \quad \lim_{x\to\infty}\frac{Bi(x)}{Q(x)} = 1,$$

hence that for large x, $Ai(x) \approx R(x)$ and $Bi(x) \approx Q(x)$. A different way of stating this is that the relative errors of the asymptotic approximations tend to zero as x tends to infinity.

This is a very simple differential equation and its numerical solution is entirely straightforward. We computed $Ai(x)$ using codes based on each of the popular kinds of discrete variable methods taken up in this book, namely Runge–Kutta, Adams, and backward differentiation formulas. The codes were asked to produce about five correct digits at $x = 1$ and $x = 10$. The details of these computations do not matter in this chapter because we are concerned about the implications of the approach, not the methods themselves. The results were

	code A	code B	code C	$Ai(x)$
$x = 1$	.135300	.135298	.135293	.135292
$x = 10$	3742.20	3140.81	548.399	$\approx 10^{-10}$

The results at $x = 1$ are what one might expect, but those at $x = 10$ are terrible! All the codes are of the highest quality and all fail to produce acceptable results at $x = 10$; this suggests that the difficulty lies with the approach and/or the problem. Let us try to understand what is going on.

Formulation of the Airy equation as a first-order system and computation of a Lipschitz constant was set as an exercise in Chapter 1. Using, say, the infinity norm, the Lipschitz constant L on an interval $[0, b]$ is $\max(b,1)$. The computation of $Ai(x)$ on $[0, 1]$ involves an initial value problem with $L(b - a) = 1$. This is the classical situation, so we might hope for a successful numerical integration and this turns out to be the case. The computation on $[0, 10]$ has $L(b - a) = 100$. This is not the classical situation, so stability might be a problem. As was pointed out in the last section, stability is easier to study for linear equations because the difference between any two solutions of the differential equation is a solution of the homogeneous equation. Airy's equation is already homogeneous, so the stability of the problem depends on how solutions of the equation depends on how solutions of the equation itself behave. All solutions are linear combinations of $Ai(x)$ and $Bi(x)$. With approximations to these fundamental solutions valid for large x, we can sort out the behavior of solutions easily and so understand the stability of the equation.

In computing $Ai(x)$ numerically, suppose that at some point x_j we compute $y_j \approx Ai(x_j)$ and $y_j' \approx Ai'(x_j)$. From x_j on, the best we can hope to do is to compute exactly the local solution $u(x)$, the solution of Airy's equation with initial values y_j, y_j' at x_j. Now $u(x)$ is a solution of Airy's equation, so it must be a linear combination of $Ai(x)$ and $Bi(x)$: $u(x) = \alpha Ai(x) + \beta Bi(x)$. Of course we can work out what α and β are here, but all we need do is observe that if y_j and y_j' are good approximations to $Ai(x_j)$ and $Ai'(x_j)$, then α is close to 1 and β is close to 0. The solution $Ai(x)$ that we wish to approximate decays rapidly for large x. Regardless of how small β is, as x gets large, the rapidly growing term $\beta Bi(x)$ in $u(x)$ swamps the term $\alpha Ai(x)$ and the numerical result is useless. It is important to understand this. No matter how accurate the approximations at x_j are, eventually the numerical solution will be useless even if the integration is *exact* for all $x > x_j$.

Sometimes we can solve an unstable problem. Suppose it is not $Ai(x)$ that we wish to compute, but $Bi(x)$. The only difference in this case is that now α will be close to 0 and β will be close to 1 in $u(x)$. Let us say that $\beta = 1 + \varepsilon$ for a small ε. In this case the local solution $u(x)$ behaves like the solution $Bi(x)$ that we wish to compute. Specifically, as x increases,

$$\frac{u(x) - Bi(x)}{Bi(x)} = \varepsilon + \alpha \frac{Ai(x)}{Bi(x)} \to \varepsilon.$$

When computing $Bi(x)$, the introduction of a small error in a step does not prevent us from obtaining a solution accurate in a relative sense (though it does in an absolute sense). This fact is not revealed by the bounds on stability based on Lipschitz constants.

Usually we do not know so much about the solutions of the differential equation and insight is gained by approximating the equation. Here, for example, we might approximate the equation near a point (x_j, y_j, y_j') with $x_j > 0$ by an equation with constant Jacobian: $u'' = x_j u$. It is then easy to find the two linearly independent solutions

$$u_1(x) = \exp(x\, x_j^{1/2}) \quad \text{and} \quad u_2(x) = \exp(-x\, x_j^{1/2}).$$

When x_j is large, these solutions diverge very rapidly as x increases, hence the approximating problem is unstable. The argument suggests that Airy's equation will also be unstable for large x, which turns out to be the case.

The very nature of the initial value problem tells us that we cannot compute $Ai(x)$ accurately for large x by integrating the differential equation with a discrete variable method. There is a trick that allows us to do so. Besides illuminating the role of stability, it exemplifies a standard device for computing special functions such as Bessel functions by recurrence. The trick is to reverse the direction of integration. Suppose we start at some large x_0 with *arbitrary* (but non-zero) initial values y_0, y_0' and integrate to the origin. The solution of Airy's differential equation with these initial conditions is a linear combination of $Ai(x)$ and $Bi(x)$. When integrating towards the origin, it is $Ai(x)$ that strongly dominates $Bi(x)$, so whatever the initial values, a code will soon be approximating a multiple of $Ai(x)$, say $\gamma Ai(x)$. Suppose y_N approximates the solution at $x = 0$. Since $y_N \approx \gamma Ai(0)$, and we know $Ai(0)$, we can deduce the factor γ. For x_j far enough from x_0, the computed $y_j \approx \gamma Ai(x_j)$, and all we have to do is divide y_j by γ to get an accurate solution for $Ai(x_j)$.

EXERCISE 3. Integrate Airy's equation with appropriate initial conditions to compute $Ai(x)$ and $Bi(x)$ at $x = 1$ and $x = 10$. Compare your values to the tabulated values for $x = 1$ and to the asymptotic values for $x = 10$. By differentiating the Wronskian

$$W(x) = Ai(x)\, Bi'(x) - Ai'(x)\, Bi(x)$$

and using the fact that $Ai(x)$ and $Bi(x)$ are solutions of Airy's equation, show that the Wronskian is constant. This constant value can be determined from the initial conditions at $x = 0$ to be $1/\pi$. You should evaluate the Wronskian with your approximations to these four functions to see how well they satisfy this law. A discrepancy shows that at least one of the values differs from its true value (though not by how much). One way to proceed is to integrate separately two systems of two first-order equations for $Ai(x)$ and $Bi(x)$. If you do this, the typical code will produce answers at different points in the two integrations. Because the test on the Wronskian requires approximations at the same points, you will have to choose some points in advance and use the code in such a way that you get answers at these points in both integrations. A better way to proceed that allows you to test the results at every step is to compute $Ai(x)$ and $Bi(x)$ at the same time by solving a system of four first-order equations.

EXAMPLE 6. The Sturm–Liouville problem is an example of a two point *boundary value problem, BVP*, for a linear differential equation. The method of *shooting* solves the BVP by means of solving initial value problems. In principle it is possible to solve quite general BVPs for linear differential equations by shooting, but it is not always practical for reasons that illuminate our study of the errors of discrete variable methods. The reasons also help us understand some of the devices used in the practical solution of BVPs. The theory of boundary value problems for linear differential equations is based on a representation of solutions of initial value problems for such equations that is taken up in most introductions to the theory of ODEs. If $\mathbf{y}(x)$ is any solution of the system of n equations

$$\mathbf{y}' = J(x)\,\mathbf{y} + \mathbf{g}(x),$$

then it is the sum of a particular solution $\mathbf{p}(x)$ of the inhomogeneous problem,

$$\mathbf{p}'(x) = J(x)\,\mathbf{p}(x) + \mathbf{g}(x),$$

and a solution of the homogeneous problem

$$\mathbf{u}' = J(x)\,\mathbf{u}.$$

Let $\{\mathbf{u}^i(x)\}$ be a set of n solutions of the homogeneous problem. If the vectors of initial values $\{\mathbf{u}^i(a)\}$ are linearly independent, the theory says that the vectors $\{\mathbf{u}^i(x)\}$ are linearly independent for all x and that any solution $\mathbf{u}(x)$ of the homogeneous problem can be written as a linear combination of these n particular solutions. A convenient way of working with this representation is to assemble the particular solutions as the columns in the fundamental matrix $\Phi(x) = (\mathbf{u}^1(x), \mathbf{u}^2(x), \ldots, \mathbf{u}^n(x))$. A *fundamental matrix* can be defined directly as the solution of the matrix differential equation

$$\Phi' = J(x)\,\Phi, \qquad \Phi(a) \text{ (non-singular) given.}$$

There are two ways to compute a fundamental matrix with the usual codes. The obvious one is to compute the columns of the matrix, the $\mathbf{u}^i(x)$, independently. Another way is to compute all the columns at the same time by letting $\mathbf{V}$ be the vector $(\mathbf{u}^1(x), \mathbf{u}^2(x), \ldots, \mathbf{u}^n(x))^T$ of length n^2. This vector is the solution of

$$\frac{d}{dx}\mathbf{V} = D(x)\,\mathbf{V}, \qquad \mathbf{V}(a) \text{ given}$$

where D is the block diagonal matrix

$$D(x) = \operatorname{diag}\{J(x), \ldots, J(x)\}.$$

This way is usually (much) more efficient because the evaluation of $J(x)$ is a relatively expensive part of the computation and in this manner one evaluation of $J(x)$ can be used for the integration of all n columns.

Putting together these observations, any solution of a system of n linear differential equations can be computed in terms of just $n + 1$ special solutions:

$$\mathbf{y}(x) = \mathbf{p}(x) + \Phi(x)\,\mathbf{s}.$$

Here $\mathbf{p}(x)$ is any solution of the inhomogeneous equation, $\Phi(x)$ is any fundamental matrix, and the vector $\mathbf{s}$ specifies the linear combination of the $\mathbf{u}^i(x)$ yielding $\mathbf{y}(x)$. In principle, any non-singular matrix $\Phi(a)$ could be used for the initial value of the fundamental matrix. A common choice is the identity matrix. General linear two point boundary conditions have the form

$$B_a \mathbf{y}(a) + B_b \mathbf{y}(b) = \beta,$$

where B_a and B_b are given matrices and β is a given vector. The representation of solutions of the differential equation then says that the boundary value problem has a solution if, and only if,

$$(B_a\, \Phi(a) + B_b\, \Phi(b))\mathbf{s} = \beta - B_a\, \mathbf{p}(a) - B_b\, \mathbf{p}(b)$$

has a solution vector $\mathbf{s}$. As with any system of linear equations, there may be no solution, exactly one, or infinitely many solutions. In terms of the BVP, there is then either no solution, exactly one, or infinitely many. The Sturm–Liouville problem when the parameter λ is an eigenvalue is an example of infinitely many solutions because any multiple of the eigenfunction is a solution.

This all looks straightforward, but it may not be successful for reasons of stability. To see the difficulties we study some specific equations. Let us consider the homogeneous equation

$$y'' - (\alpha + \beta)y' + \alpha\beta\, y = 0$$

and a boundary value problem posed on $[0, 1]$. Provided that the parameters α and β are different, two linearly independent solutions of this equation are $\exp(\alpha x)$ and $\exp(\beta x)$. Suppose for example that $\alpha = +1$ and $\beta = +100$. It is not practical to solve this problem by shooting from $x = 0$ to $x = 1$ because the initial value problem is very unstable—the difference between two solutions can grow as fast as $\exp(100x)$. (Recall the discussion of Example 3.) Generally when we are presented an initial value problem posed on $[a, b]$, we must integrate from $x = a$ to $x = b$, and that's that. One of the things that is different about solving boundary value problems is that we get to choose the direction of integration. If we integrate from $x = 1$ to $x = 0$, the initial value problem is stable for this equation. The difference between two solutions decays so that numerical errors are even damped out.

Returning to the example problem, it is by no means true that there is always a stable direction of integration. If $\alpha = +100$ and $\beta = -100$, initial value problems are unstable in both directions of integration. The initial value problems are not well posed with respect to changes of the initial values, but this does not mean that the boundary value problem is not well posed with respect to changes of the boundary values. Although it is possible to improve the situation by multiple shooting, a technique that we take up in another example, shooting is impractical for many well posed BVPs because the initial value problems are very unstable.

Let us consider further the case $\alpha = -1$ and $\beta = -100$. The fact that the integration from 0 to 1 is stable does not mean that shooting will succeed. We need to compute a set of linearly independent solutions. The mathematical theory states that if the vectors of initial values are linearly independent, then the solution vectors will be linearly independent at other x. In practice, this need not be so. For one thing, there is nothing in the numerical methods that would cause them to maintain linear independence automatically. A fundamental difficulty is exposed when we attempt to compute the linearly independent solutions $\exp(-x) + \exp(-100x)$ and $\exp(-x) - \exp(-100x)$. The function $\exp(-100x)$ decays so fast compared to $\exp(-x)$ that computationally it is "invisible" at $x = 1$—in finite precision arithmetic both solutions are given by the floating point representation of $\exp(-1)$. This has nothing to do with the numerical solution of the initial value problem. It is the simple observation that in finite precision arithmetic, the computer representation of these two linearly independent vectors is linearly dependent. The fundamental matrix based on these solutions is nonsingular in principle, but not in practice, and the whole approach fails.

EXAMPLE 7. The solution of boundary value problems for nonlinear differential equations raises new issues, some of which we examine in the context of an example. One of the ways that partial differential equations are reduced to ordinary differential equations is by looking for similarity solutions—solutions that depend in a special way on the independent variables. An important example is Blasius' solution of the laminar flow of an incompressible fluid with small viscosity moving parallel to a semi-infinite flat plate. There are two independent variables x and y. The origin is taken at the leading edge of the plate, x measures distance along the plate, and y measures distance perpendicular to the plate. In the free stream the flow is in uniform motion parallel to the plate with constant speed U. The speed $u(x, y)$ of the flow parallel to the plate is sought in the form $u(x, y) = Uf'(\eta)$ where η is some function of x and y. It turns out that the partial differential equations and boundary conditions reduce to a boundary value problem for an ordinary differential equation when the similarity variable

$$\eta = y \sqrt{\frac{U}{2\nu x}}$$

is used. Here ν is the kinematic viscosity of the fluid. The *Blasius problem* then is to find $f(\eta)$ such that

$$\frac{d^3f}{d\eta^3} + f\frac{d^2f}{d\eta^2} = 0,$$

$$f(0) = 0, \qquad \frac{df}{d\eta}(0) = 0, \qquad \frac{df}{d\eta}(\infty) = 1.$$

The derivation of this problem is to be found in many books. The one by White [1991] discusses solution of the problem by shooting and by various techniques of classical applied mathematics. (The book also has many other examples of partial differential equations reduced to ordinary differential equations in a variety of ways along with some details of their numerical solution.)

The boundary condition at infinity arises as an idealization of a condition that holds for large η. In practice it is usual to choose a finite value B and find that solution of the differential equation that satisfies

$$f(0) = 0, \qquad \frac{df}{d\eta}(0) = 0, \qquad \frac{df}{d\eta}(B) = 1.$$

It is difficult to decide whether this results in a solution close to the one with the boundary condition at infinity. This is not an issue that is appropriate for discussion here. In practice it is usual to solve several boundary value problems with a sequence of increasing B and when the solutions appear to agree to several digits, it is presumed that they agree with the solution of the problem with $B = \infty$ to as many digits. In this way White and others have concluded that $B = 10$ is big enough. As the plot of $f'(\eta)$ displayed in Figure 2.5 shows, the function tends rapidly to a constant value so any "large" B will do.

To solve the boundary value problem on the finite interval by shooting, a value for s is chosen and the equation is integrated as an initial value problem with initial conditions

$$f(0) = 0, \qquad \frac{df}{d\eta}(0) = 0, \qquad \frac{d^2f}{d\eta^2}(0) = s$$

to obtain $f'(B; s)$. We then find an initial value s for which $f'(B; s) = 1$, i.e., we solve a nonlinear algebraic equation for the missing initial value. Here we have an equation in only one unknown that is quickly solved by, say, bisection to get a value $s \approx 0.4696$. Having determined the missing initial value, we can integrate an initial value problem to evaluate f and its derivatives wherever we like.

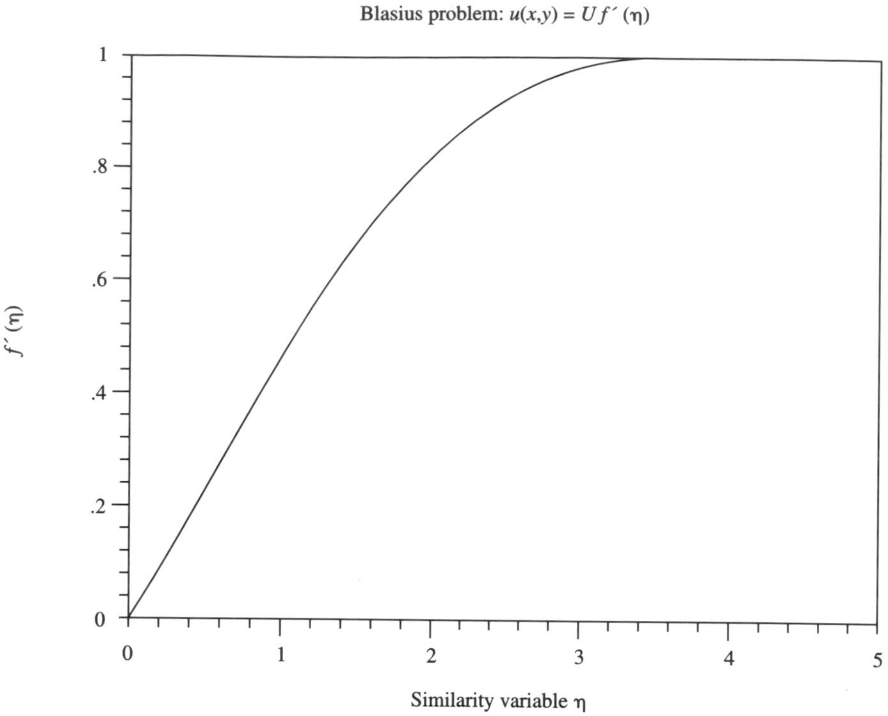

Figure 2.5

Because the solution of the Blasius problem appears as a coefficient in other equations describing fluid flow, it is important to be able to evaluate it easily. Some boundary value problems can be transformed into initial value problems because of their special form (see, e.g., Klamkin [1962, 1970], Na [1967, 1979], Rogers and Ames [1989]), and the Blasius problem is one. It is known that the first derivative of the solution of the initial value problem

$$\frac{d^3v}{dz^3} + v\,\frac{d^2v}{dz^2} = 0,$$

$$v(0) = 0, \qquad \frac{dv}{dz}(0) = 0, \qquad \frac{d^2v}{dz^2}(0) = 1,$$

increases to a limit λ^2,

$$\lim_{z\to\infty} \frac{dv}{dz} = \lambda^2 > 0.$$

It is easily verified that the solution $f(\eta)$ of the Blasius problem can be expressed in terms of $v(z)$ and λ as $\eta = \lambda z, f(\eta) = v(z)/\lambda$. By integrating $v(z)$ to a large value of z to determine λ, the solution f of the Blasius problem can be found in a single integration. However, it is more convenient to use this scheme just to compute the missing initial value for f'' and then compute f directly so as to obtain values at the desired η. This missing value is

$$\frac{d^2f}{d\eta^2}(0) = \frac{1}{\lambda^3}\frac{d^2v}{dz^2}(0) = \frac{1}{\lambda^3}.$$

EXERCISE 4. Verify that $f(\eta)$ can be expressed in terms of $v(z)$ and λ as stated and that the missing initial value $f''(0) = \lambda^{-3}$. Solve the initial value problem for $v(z)$ to determine this missing initial value. Recall that it was stated to be about 0.4696. If you are keen, solve the boundary value problem by shooting. If you have a plotting capability, use the initial value you find to compute $f'(\eta)$ and plot it to reproduce Figure 2.5.

After approximating the Blasius problem with one posed on a finite interval, it is possible to shoot in either direction. That is, initial values v_0 and v_1 could be chosen and the equation integrated from B to 0 with initial values

$$f(B) = v_0, \qquad \frac{df}{d\eta}(B) = 1, \qquad \frac{d^2f}{d\eta^2}(B) = v_2.$$

Here this is unattractive because there are two initial values to be determined such that the solution $f(\eta;\, v_0,\, v_2)$ satisfies the two simultaneous nonlinear algebraic equations

$$f(0;\, v_0,\, v_2) = 0 \quad \text{and} \quad \frac{df}{d\eta}(0;\, v_0,\, v_2) = 0,$$

and this is a harder task than solving one nonlinear algebraic equation. We would not have a choice about the matter if the integration from 0 to B were unstable and the integration from B to 0 were stable. In the case of the Blasius problem, either direction is stable enough for successful integration.

As with the initial value problem, boundary value problems arise in a great many forms and it is valuable for both the theory and practice to write them in a standard form. Moreover, there are other kinds of ODE problems that can be formulated as BVPs. For example, the Sturm–Liouville problem is not a two point boundary value problem because part of the task is to find an eigenvalue λ, but it can be written as one. Finding periodic solutions of a differential equation is a similar task that is set as an exercise. The reader can consult references such as Ascher et al. [1988] for more details about BVPs in general and conversion to standard form in particular. For our purposes two point boundary value problems of the form

$$\mathbf{y}' = \mathbf{F}(x, \mathbf{y}), \qquad \mathbf{g}(\mathbf{y}(a), \mathbf{y}(b)) = \mathbf{0}$$

will suffice. The solution of the differential equation is specified by a given algebraic relation between the values of the solution at the two boundary points a and b. A solution can be attempted by choosing a vector of initial values $\mathbf{s}$ at $x = a$, integrating the differential equation to find $\mathbf{y}(b; \mathbf{s})$, and solving the algebraic equations $\mathbf{g}(\mathbf{s}, \mathbf{y}(b; \mathbf{s})) = \mathbf{0}$ for a set of initial values, the vector $\mathbf{s}$, which will provide a solution to the boundary value problem. As we saw with the Blasius problem, we must solve a nonlinear algebraic equation for $\mathbf{s}$. This is a much more difficult task than solving a linear system, in part because the existence and uniqueness of solutions is more complicated. We do not need to go into numerical methods for the solution of nonlinear algebraic equations, but we should comment about one matter. The methods must assess how the value of $\mathbf{g}(\mathbf{s}, \mathbf{y}(b; \mathbf{s}))$ is changed when $\mathbf{s}$ is changed. For this we need to know how the value of $\mathbf{y}(b; \mathbf{s})$ is changed when $\mathbf{s}$ is changed. But that is what the variational equation gives us. Supposing that we have expressions for the partial derivatives of $\mathbf{g}(\mathbf{u}, \mathbf{v})$, differentiation of $\mathbf{g}$ with respect to $\mathbf{s}$ gives us

$$\frac{\partial \mathbf{g}}{\partial \mathbf{s}} = \frac{\partial \mathbf{g}}{\partial \mathbf{u}} + \frac{\partial \mathbf{g}}{\partial \mathbf{v}} \frac{\partial \mathbf{y}(b; \mathbf{s})}{\partial \mathbf{s}}.$$

The (matrix) derivative of the solution of the differential equation with respect to the (vector of) initial conditions $\mathbf{s}$ is the solution of

$$\frac{d}{dx}\left(\frac{\partial \mathbf{y}(x; \mathbf{s})}{\partial \mathbf{s}}\right) = \mathbf{F}_{\mathbf{y}}(x, \mathbf{y}(x; \mathbf{s}))\left(\frac{\partial \mathbf{y}(x; \mathbf{s})}{\partial \mathbf{s}}\right), \qquad \frac{\partial \mathbf{y}(a; \mathbf{s})}{\partial \mathbf{s}} = I.$$

Here I is the identity matrix and the Jacobian is evaluated at the solution $\mathbf{y}(x; \mathbf{s})$ of the initial value problem $\mathbf{y}' = \mathbf{F}(x, \mathbf{y}), \mathbf{y}(a) = \mathbf{s}$. With an analytical Jacobian, the equation of first variation can be integrated along with the shot and at the end of the integration both

$$\mathbf{y}(b; \mathbf{s}) \quad \text{and} \quad \frac{\partial \mathbf{y}(b; \mathbf{s})}{\partial \mathbf{s}}$$

will be available to the code for nonlinear algebraic equations.

We think about initial value problems being posed as an integration from a to b, so it is surprising that sometimes we actually have a choice about the direction of integration. When solving two point BVPs, the algebraic equations may be easier to deal with at one end than the other. For example, often a number of components of the solution have specified values at an end point. Nonetheless, the stability of the integration is fundamental and we must shoot in a direction that corresponds to a stable initial value problem. There is another possibility in this context that is even more surprising—we may be able to do something about the stability of the integration by shortening the interval. This is what the technique of *multiple shooting* accomplishes.

The shooting scheme we have been discussing is described more precisely as simple shooting, or single shooting. The idea of multiple shooting is to break the interval $[a, b]$ into a number of pieces $[t_0, t_1], [t_1, t_2], \ldots, [t_{N-1}, t_N]$ with $t_0 = a$ and $t_N = b$. On each of the N subintervals $[t_j, t_{j+1}]$ the differential equation is integrated with initial conditions $\mathbf{s}_j$ at t_j to determine the solution $\mathbf{y}_j(x; \mathbf{s}_j)$. These solutions of the differential equation on the various pieces will form a solution on all of $[a, b]$ when they connect properly at the breakpoints t_j:

$$\mathbf{y}_j(t_{j+1}; \mathbf{s}_j) = \mathbf{y}_{j+1}(t_{j+1}; \mathbf{s}_{j+1}) = \mathbf{s}_{j+1} \quad \text{for} \quad j = 0, 1, \ldots, N-1.$$

In terms of the pieces, the boundary conditions are

$$\mathbf{g}(\mathbf{s}_0, \mathbf{y}_{N-1}(b; \mathbf{s}_{N-1})) = \mathbf{0}.$$

This constitutes a system of algebraic equations for the N sets of unknown initial conditions $\mathbf{s}_j$. According to the bound on stability involving the Lipschitz constant, reducing the length of the interval of integration improves the stability considerably, and this often turns out to be true in practice. A key issue in practice is deciding where to introduce breakpoints. It was mentioned earlier that when solving linear BVPs, maintaining linear independence of the fundamental solutions is crucial. In that context linear independence is monitored in the course of the integration and when the numerical solutions become unsatisfactory in this regard, a breakpoint is introduced and a new set of solutions computed. Multiple shooting shifts part of the computational burden from the integrator to the scheme for solving algebraic equations. In its most extreme form multiple shooting has intervals so short that the initial value solver takes only one step per integration interval. This extreme case exemplifies what is called a boundary value approach to solving BVPs. Boundary value problems for the differential equation may be stable with respect to changes of the boundary data even when associated initial value problems are not stable with respect to such changes. By shortening the interval in multiple shooting, the boundary conditions play a more important role in determining the solution. For particular problems, shooting can be very effective; this is true, for example, of the solution of the Sturm–Liouville problem and of the Blasius problem. However, many boundary value problems are such that shooting, even multiple shooting, is unsatisfactory. The interested reader may pursue the matter in specialized texts such as Ascher et al. [1988]. Our interest in multiple shooting here is mainly in the surprising fact that it is sometimes possible to do something about the instability of an initial value problem.

EXERCISE 5. To see one example of the conversion of a task to a two point boundary value problem, suppose we want to find a periodic solution of

$$\frac{d\mathbf{y}}{dx} = \mathbf{F}(\mathbf{y})$$

of (unknown) period T, i.e., $\mathbf{y}(x + T) = \mathbf{y}(x)$ for all x. Change from the independent variable x to $t = x/T$ to make the interval [0, 1]. Then increase the size of the system by adding an equation for T so that the vector of unknowns is

$$\mathbf{w}(t) = \begin{pmatrix} \mathbf{y}(t) \\ T(t) \end{pmatrix}.$$

Remember that the period is constant—we just need to find a value for it along with a solution having this period—so a differential equation and boundary conditions for $T(t)$ are readily available. Write out the equations and boundary conditions in the form of a two point boundary value problem.

EXERCISE 6. One aspect of differential equations that receives a great deal of attention from applied mathematicians is the long-term behavior of solutions. A system of equations might, for example, have a stable periodic solution to which all nearby integral curves converge. An example of this was analyzed in Example 6 of Chapter 1 by writing the solution in polar coordinates. A classic example is van der Pol's equation, $\ddot{x} + \varepsilon(x^2 - 1)\dot{x} + x = 0$, with a positive parameter ε. The constant solution that is identically 0 is unstable and all other solutions converge to a single periodic solution, a limit cycle. To compute such a solution it is usual just to start integrating the equation and see what happens. The classical theory of the numerical solution of ODEs does not appear to offer much hope for justifying this because the theory is concerned with what happens between an initial time t_0 and a final time t_f that is finite and specified in advance. However, computation of stable periodic solutions in this way may be regarded as a shooting method for solving a boundary value problem. A periodic solution of an autonomous equation $\mathbf{y}' = F(\mathbf{y})$ with period

T satisfies $\mathbf{y}(a) = \mathbf{y}(a + T)$. Suppose for the moment that we know T and that we want to solve this boundary value problem by a shooting method. An initial value $\mathbf{s}$ is guessed for $\mathbf{y}(a)$ and the equation integrated to $a + T$ to obtain $\mathbf{y}(a + T; \mathbf{s})$. Using this value, we need to deduce an initial vector that is closer to the one defining the periodic solution. If the periodic solution is stable and the guess $\mathbf{s}$ is sufficiently good, then because the solution $\mathbf{y}(x; \mathbf{s})$ approaches the periodic solution as x increases, the value $\mathbf{y}(a + T; \mathbf{s})$ will be closer to the initial vector defining the periodic solution than $\mathbf{s}$ is. This amounts to solving the algebraic equations by the method of simple (functional) iteration. We would then shoot again from $x = a$ with this new approximation to the correct initial vector. However, the equation is autonomous, so this shot is equivalent to continuing the integration from $x = a + T$ to $x = a + 2T$ with initial vector $\mathbf{y}(a + T; \mathbf{s})$ at $x = a + T$. We see, then, that one way to implement a shooting method for the computation of a stable periodic solution of an autonomous equation amounts to just integrating for a long time. This is not the same thing as approximating the solution of the original initial value problem accurately over a long interval. It amounts to approximating repeatedly solutions of problems over an interval of length T. Notice that in this way of proceeding, it is not necessary that we know T in advance.

A standard example in perturbation theory is the approximation of the limit cycle of the van der Pol equation. For instance, Jordan and Smith [1987, p. 100] show that for small ε, the limit cycle $x(t; \varepsilon) \approx 2\cos(t + \phi)$ for a constant ϕ. Choose a small ε, arbitrary initial values $x(0)$, $\dot{x}(0)$, and compute numerically a periodic solution by integrating for a "long" time. You might, e.g., take $\varepsilon = 0.1$, $x(0) = 1$, $\dot{x}(0) = 0$, and integrate to 100π. If you have a plotting program available, a plot of the solution in the phase plane ($\dot{x}$ against x) will show the solution tending to a closed curve that represents the limit cycle. Or, if you plot $x(t)$ and $\dot{x}(t)$ for "large" t, you will see what appears to be a periodic solution of period about 2π and amplitude about 2. Alternatively, you can compare your numerical results to the asymptotic approximation as follows: If $x_n \approx \alpha \cos(t_n + \phi)$ and $\dot{x}_n \approx -\alpha \sin(t_n + \phi)$, argue that $\alpha \approx (x_n^2 + \dot{x}_n^2)^{1/2}$. Further, if $t_n = 2\pi n$ for $n = 0, 1, \ldots,$ argue that $\phi \approx \arctan(-\dot{x}_n/x_n)$. Computing α and ϕ in this way from x_n and $\dot{x}_n$ will show that the numerical amplitude α rapidly approaches a value that is approximately 2 and the numerical phase ϕ is approximately constant. The values, especially ϕ, are not exactly constant because the asymptotic approximation is only an approximation to $x(t)$.

EXERCISE 7. Seydel [1988] treats in Example 7.1 and Excrcises 1.9 and 2.8 a model used by FitzHugh to describe impulses of nerve potentials.

The differential equations are

$$\dot{y}_1 = 3\,(y_1 + y_2 - \frac{1}{3}y_1^3 + \lambda),$$

$$\dot{y}_2 = -\frac{1}{3}\,(y_1 - 0.7 + y_2).$$

Seydel's Figure 7.2 shows the spike behavior of a periodic solution when $\lambda = -1.3$. This periodic solution is computed easily by simply integrating an initial value problem for a "long" time because it is quite stable. Exercise 6 justifies the computation. Plotting the computed $y_1(t)$, as Seydel does in his figure, shows that it quickly tends to a curve that appears to be periodic. Plotting the computed solution in the phase plane, y_2 vs. y_1, shows convergence to a closed curve representing a periodic solution. Carry out this computation. If you are keen, try to determine the period of the oscillation. How difficult this is will depend on the capabilities of the software at your disposal.

EXAMPLE 7. The gravitational n-body problem is concerned with n bodies of mass $m_1, m_2, \ldots, m_n$, respectively, that move due to mutual gravitational attraction. At time t the position of the ith body is specified by the vector $\mathbf{r}_i(t)$. Assuming there are no external forces on the system, the acceleration of this body due to the gravitational attraction of the other bodies is

$$\frac{d^2}{dt^2}\mathbf{r}_i(t) = \ddot{\mathbf{r}}_i(t) = -G\sum_{\substack{j=1\\ j\neq i}}^{n}\frac{m_j}{r_{ji}^3}\mathbf{r}_{ji}.$$

Here G is the universal gravitational constant, the vector $\mathbf{r}_{ji} = \mathbf{r}_j - \mathbf{r}_i$ is the position of the jth body relative to the ith body, and r_{ji} is the relative distance $\|\mathbf{r}_{ji}\|$ (the Euclidean norm of the vector). Given the initial positions $\mathbf{r}_i(0)$ and initial velocities $\dot{\mathbf{r}}_i(0)$, this represents an initial value problem. It can be a rather large problem because there are n bodies and for each body there are three coordinates in the position vector and three coordinates in the velocity vector. Thus if the system is written as a system of first-order equations, there are $6n$ equations. This is an example of the special second-order systems mentioned in Chapter 1 that do not explicitly involve the velocity. Numerical methods that take advantage of this fact can work with only $3n$ equations.

The general n-body problem is tractable only by numerical means, but problems involving fewer bodies can be attacked analytically. In particular, it is possible to understand in a rather complete way the solution of the two-body problem. Because of this, an important tactic of applied mathematics is to approximate a situation involving several bodies by a problem involving only two bodies. Before going into this, we observe that two-body problems have a number of virtues as test problems for differential equation solvers, viz. they are important, easy to state, have a semi-analytical solution, and present a range of numerical difficulty. A family of two-body problems that are standard test problems will be useful as examples throughout this book.

As formulated in Hull et al. [1972], the *two-body problem* is

$$\ddot{x} = -x/r^3, \qquad x(0) = 1 - e, \qquad \dot{x}(0) = 0,$$

$$\ddot{y} = -y/r^3, \qquad y(0) = 0, \qquad \dot{y}(0) = \sqrt{\frac{1+e}{1-e}}$$

$$\text{where } r^2 = x^2 + y^2.$$

Here the coordinate system is chosen so that one body is fixed at the origin and $(x(t), y(t))$ is the position of the other body at times t. The orbit is an ellipse of eccentricity e. For $0 \leq e < 1$,

$$x = \cos(u) - e, \qquad \dot{x} = \frac{-\sin(u)}{1 - e\cos(u)}$$

$$y = \sqrt{1 - e^2}\sin(u), \qquad \dot{y} = \frac{\sqrt{1-e^2}\cos(u)}{1 - e\cos(u)}$$

where u is the solution of *Kepler's equation*, $u = t + e\sin(u)$, a nonlinear algebraic equation that can be solved readily using, say, Newton's method.

The standard test problems of Hull et al. [1972] are called $D1$, $D2$, $D3$, $D4$, $D5$. They correspond to $e = 0.1, 0.3, 0.5, 0.7, 0.9$, respectively. They are all posed on the interval $[0, 20]$. Figure 2.6 shows the orbit of $D5$, an ellipse of eccentricity $e = 0.9$. In this case solution of Kepler's equation gives as "true solution" the following values

$$x(20) = -1.29526625098758 \qquad \dot{x}(20) = -0.67753909247075$$
$$y(20) = 0.400393896379232 \qquad \dot{y}(20) = -0.127083815427869$$

Modern codes for the initial value problem select step sizes automatically. As described briefly in §2, at any point x in the integration, the code will select a step size small enough to yield the specified accuracy. At the same time the code will select a step size about as large as possible in order to span the interval $[a, b]$ as cheaply as possible. These matters are the subjects of chapters that follow, but some numerical results are illuminating here. On physical grounds we expect the two-body problem to be most unstable when the two bodies are closest—a small change in the orbit then

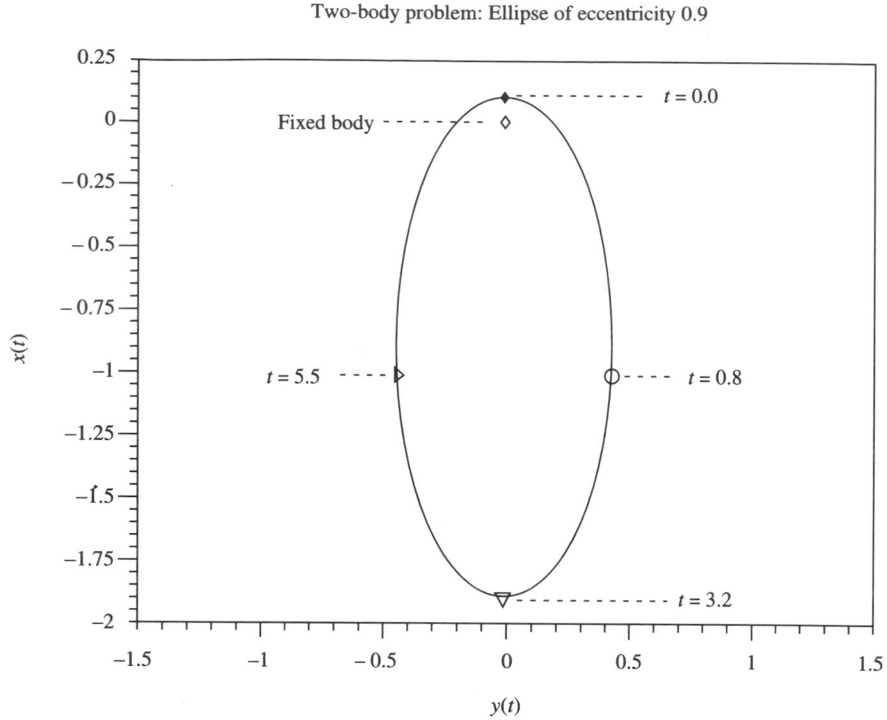

Figure 2.6

will cause relatively large changes later—and this turns out to be the case. It is also to be expected that the codes will have to use a smaller step size where the problem is unstable if it is to maintain the accuracy of the integration. Figure 2.7 shows the step sizes used by a quality code when it computed the orbit of Figure 2.6. The scale is logarithmic, so there is a considerable range of step sizes. The integration starts at the point of closest approach and it is quite clear in the graph when the bodies are again at closest approach. Even when a problem is not so unstable that it cannot be integrated, the stability of the problem can influence significantly its numerical solution by discrete variable methods.

EXERCISE 8. Show that solutions of the two-body problem satisfy two nonlinear conservation laws: The conservation of energy law is

$$\frac{1}{2}(\dot{x}^2 + \dot{y}^2) - \frac{1}{r} \qquad \text{is constant,}$$

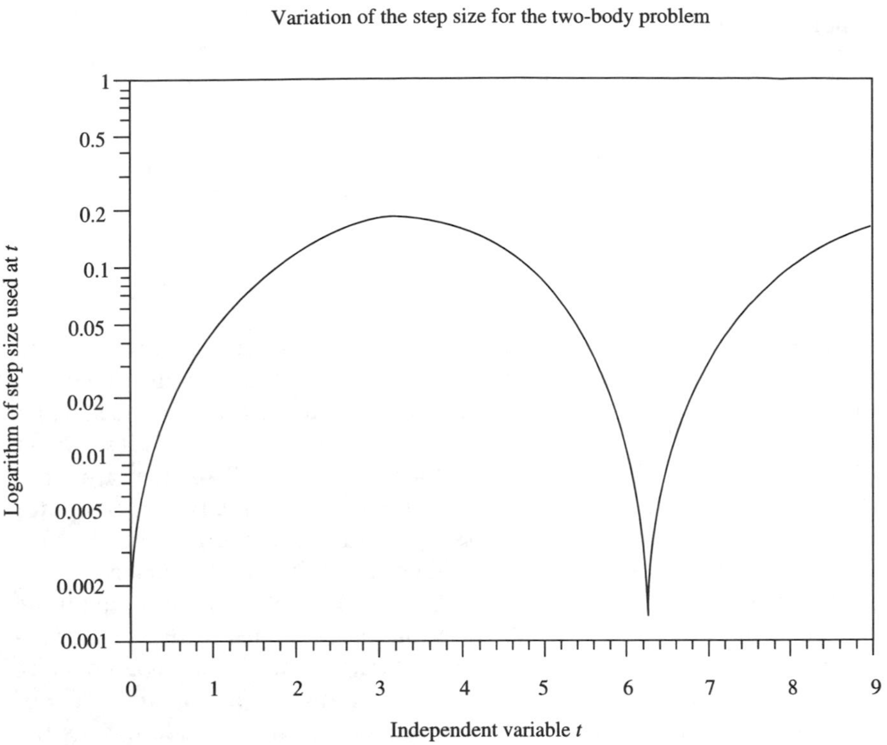

Figure 2.7

and the conservation of angular momentum law is

$$x\,\dot{y} - y\,\dot{x} \qquad \text{is constant.}$$

Integrate *D*5 with your favorite integrator and monitor how well the numerical solution satisfies these two laws. Compute the exact solution and compare it to your numerical solution. Remember that a conservation law can tell you that the numerical solution is inaccurate, but not that it is accurate.

Returning now to the use of two-body problems for understanding n-body problems, we follow the treatment in Bate et al. [1972] and suppose that m_1 is the earth and m_2 is an artificial satellite. The remaining bodies are the moon, the sun, the other planets, ... If we give our attention to the motion of the satellite about the earth, a little manipulation of the general equations leads to

$$\ddot{\mathbf{r}}_{12} = -\frac{G(m_1 + m_2)}{r_{12}^3}\mathbf{r}_{12} - G\sum_{j=3}^{n} m_j\left(\frac{1}{r_{j2}^3}\mathbf{r}_{j2} - \frac{1}{r_{j1}^3}\mathbf{r}_{j1}\right).$$

The idea is to neglect the effects of the remaining bodies and so approximate the motion of the satellite by the equation

$$\ddot{\mathbf{r}}_{12} = -\frac{G(m_1 + m_2)}{r_{12}^3}\mathbf{r}_{12}.$$

Bate et al. [1971] compare the relative accelerations for an earth satellite at 200 nautical miles due to various bodies in the solar system. The effect of the earth is 0.89. The biggest effects neglected are those of the sun and the moon, which are, respectively, 6×10^{-4} and 3.3×10^{-6}. The effect of Venus is 1.9×10^{-8} and that of other planets is still smaller. It is true that the effect of the earth dominates strongly, but with a typical computer word length, it would be quite possible to take account of other bodies. It is a scientific judgment which bodies might be safely neglected in a particular situation. Certainly it is not possible to take into account all the bodies in the universe, so whatever equations are solved, they must represent an approximation to a more exact description. This kind of modeling underlies the interpretation of solving differential equations in terms of backward error analysis. In this view, the numerical procedure finds the exact solution of a slightly different equation. If the perturbation to the equation is comparable to, or smaller than, effects neglected in the original formulation of the problem, then it is fair to say that the problem posed has been solved as accurately as is meaningful. This, however, does *not* mean that the given problem has been solved accurately. The solution is accurate only if the original problem is well posed, which is to say that the effects neglected really can be neglected. The definition of well posed here includes the time interval. As we saw earlier, after a sufficiently long time, even small perturbations might have an effect too large to ignore—we are not then in the classical situation for numerical methods. This is a fundamental difficulty when it is the long-term behavior that is interesting, as it is, for example, when studying the evolution of the solar system.

There are other aspects to the modeling that should be mentioned. Notice that one of the differential equations is singular when two bodies collide (some $r_{ij} = 0$). Treating any body as a point mass is an idealization and in appropriate circumstances the idealization, and correspondingly a differential equation, might break down. It is to be expected that numerical difficulties will arise then. For example, if the distance of the satellite from the center of mass of the earth is less than the radius of the earth itself, the equations are no longer valid! The modeling of the evolution of star clusters leads to some large-scale computations. It has been found advantageous to recognize close approach and give it special treatment. When the jth body is much closer to the ith body than any other, the relative motion can be treated as a two-body problem, hence treated in a semi-analytical way.

Another aspect of the modeling concerns the assumption that the body is a point mass. Mathematically this is possible when the density distribution of the body is spherically symmetric. The density of the earth is not spherically symmetric and Bate et al. report that this effect has a relative size of about 10^{-3} in the example quoted earlier. This effect is so large that any refined model must take it into account. In §9.7 of Bate et al. [1971], such a refined model of the distribution of mass in the earth is presented that leads to rather complicated equations of motion for the satellite. It is interesting in the present circumstances that the results of numerical integration of such models are compared to observations of real satellites and the parameters of the model adjusted with the aim of obtaining a better model of the earth's gravitational field.

3

The Computational Problem

The computational problem presented to a code is different from the mathematical problem of solving

$$\mathbf{y}' = F(x, \mathbf{y}), \qquad a \le x \le b, \qquad \mathbf{y}(a) = \mathbf{A}. \tag{1}$$

For instance, when we wish to compute a solution, we must specify what we mean by "solving the problem." Typically we must also specify how much accuracy we want. This chapter is devoted to consequences of the distinction between the mathematical problem and the computational problem and to software issues. Although there has been a considerable convergence of the designs of the better items of mathematical software, it is illuminating to see how different they can be. This chapter is only an introduction to mathematical software and it is necessarily incomplete without a deep understanding of the numerical methods. Fuller discussions that are accessible at this time are found in Shampine [1980] and Shampine and Watts [1984]. At the end of the chapter are found references to surveys of the software available and directions for obtaining some quality codes. While reading this chapter, you should study the codes available to you to see how they deal with the issues discussed here. Numerical analysis is an art as well as a science and the art can be learned only by solving problems with quality codes. Because all the general-purpose codes cited in this chapter were written in FORTRAN, issues that are specific to a programming language are discussed in terms of FORTRAN 77.

§1 Specifying the Differential Equation

When solving an initial value problem, one of the first tasks is to specify the differential equation (1). There are more possibilities than you might imagine. For instance, there are many packages that accept problems specified in physical terms and form the appropriate differential equations themselves. An exotic example is the SPEAKEASY language (Shampine [1979]) that does not even have functions in the usual sense—they are represented as arrays. Despite this, it is usual at some level in a package or language to formulate the differential equation in the form (1) and then to proceed to its solution. This is little different in a technical sense from the user specifying (1). Although our attention will be directed to problems of the form (1), it should be kept in mind that there are other *standard forms*. Some specifics will make the point. The influential early code DVDQ permits the equations in a system to have different orders (in the range one to four). There are many codes for systems of the form $\mathbf{y}'' = \mathbf{F}(x, \mathbf{y})$. As we saw in Example 8 of Chapter 1, some methods of semi-discretization of partial differential equations lead naturally to systems of the form $M(x)\mathbf{y}'(x) = \mathbf{F}(x, \mathbf{y}(x))$. The code DIVPAG of the IMSL [1989] software library accepts problems in this form with various possibilities for $M(x)$. In addition to $M(x) = I$, leading to (1), other possibilities are constant $M \neq I$ and M that depend on x. Four different structures for M are distinguished: full, band, symmetric positive definite, and band symmetric positive definite.

There is another variation on the theme of standard form. There are some technical advantages to working with $\mathbf{F}$ that do not depend explicitly on the independent variable, i.e., the equation has the form $\mathbf{y}' = \mathbf{F}(\mathbf{y})$. Such equations are said to be *autonomous*. There is no loss of generality in assuming that the equation is in autonomous form because it is easy to convert a problem of the form (1) into an equivalent problem with autonomous equation. One way presented in Example 6 of Chapter 1 is to introduce arc length as an independent variable. The usual way is to introduce a new independent variable t that is equal to x and adjoin an equation for x:

$$\frac{d\mathbf{y}}{dt} = \mathbf{F}(x, \mathbf{y}), \qquad \mathbf{y}(a) = \mathbf{A},$$

$$\frac{dx}{dt} = 1, \qquad x(a) = a.$$

The old independent variable x is now a dependent variable in the larger system. Because the new independent variable t does not appear explicitly, the larger system is in autonomous form. Some codes require the equations to be presented in autonomous form, an example being the PHASER package of Koçak [1989]. In his text Koçak describes the usual way of converting to autonomous form, but in the context of phase plane analysis, there is considerable merit to using arc length instead. As coded, the numerical methods of the package produce answers at equally spaced points in the interval of integration. If the solution should change sharply somewhere, the package must use a small step size to resolve the change accurately, but this is inefficient in those portions of the integration where the solution varies slowly. Equal spacing in arc length is better adapted to the behavior of the solution, being closely spaced in x where the solution changes rapidly and widely spaced where it changes slowly.

The most common way of specifying (1) is by means of a subroutine for **F** that when supplied x and **y**, will return $\mathbf{F}(x, \mathbf{y})$. Some solvers, e.g., DIFSUB, use a fixed name for the subroutine to evaluate **F**. Often it is desired to integrate more than one system of differential equations in a single run. This task is somewhat inconvenient when a fixed name is required of **F**. It can be accomplished by programming all the systems of equations in one subroutine with the specified, fixed name and using an integer variable along with a computed GOTO in the subroutine to select the current system. How this integer variable might be communicated to the subroutine from the calling program will be taken up shortly. There seems to be a consensus now that it is better to write different subroutines for different systems of equations, attach distinct names to them, and pass the name to the solver. A difficulty with the design is that users are prone to forget to declare the names in an EXTERNAL statement in the calling program as required by FORTRAN. This is a matter that must be made clear in the documentation.

There is some disagreement about what ought to be in the call list of a subroutine for **F**. One item that differs in popular codes is an integer NEQ specifying the number of equations. Of course, you, the user, write this subroutine and you know how many equations you have programmed, so this quantity is not obviously useful as an argument. With earlier versions of FORTRAN this quantity was needed to specify the size of the arrays for the **y** input and the $\mathbf{F}(x, \mathbf{y})$ output, but the assumed size facility of FORTRAN 77 deals with this. There are applications in which the main program must communicate some information to the subroutine, e.g., the current value of a parameter. An example arose earlier in connection with passing an integer switch for selecting which system of equations to evaluate in a single routine. Ordinarily such communication can be accomplished using COMMON, but some solvers add arguments to the call list for this purpose. As examples, the codes of DEPAC have REAL and INTEGER arrays, RPAR(*) and IPAR(*), for communicating parameters. Such arguments can also be used to pass information back to the main program. For example, some subroutines diagnose arguments for which $\mathbf{F}(x, \mathbf{y})$ is not defined. A useful measure of work is the number of times this subroutine is called by the solver. If the solver itself does not provide this number, it can be determined by incrementing a counter in the subroutine and passing the value back to the user. Neither approach to communication is entirely satisfactory: many users are uncomfortable with COMMON and arguments in the call list that are not needed for the solution of most problems are an annoyance. The authors of some codes accomplish this by *overloading* arguments in the call list for **F**. For example, DDRIV2 and LSODE allow the array holding **y** to be longer than the number of equations; the extra storage locations can be used to communicate REAL values. LSODE goes even further along these lines. In its design NEQ is an argument in the call list for **F**. This argument is treated as a *vector* with the first entry providing the number of equations and other entries being used for communication of INTEGER values. This seems an unnatural use of the argument, and Hindmarsh and his associates preferred to use arrays RPAR(*) and IPAR(*) in the later code VODE for communication rather than overload the traditional arguments.

Codes for the solution of the stiff problems taken up in Chapter 8 also require evaluation of the Jacobian matrix $J = (\partial F_i / \partial y_j)$. This is much like the evaluation of **F**, but rather more troublesome. An efficient solver will evaluate **F** much more often than J, so the evaluation of J should be done separately. When there are N equations, there are N^2 components of J to be evaluated. For even moderate N, writing a subroutine to evaluate J is tedious and prone to error. By far the most popular approach is to form the Jacobian numerically and most of the widely used solvers have this as an option. This is popular because the user does not have to get involved with the matter and numerical approximation of J is not necessarily more expensive than evaluation of analytical expressions for the partial derivatives. It is an option because it can be more expensive and, more important, because the reliable formation of the Jacobian numerically is difficult, especially with the great range of size of the components of **F** seen with stiff problems. The number of equations plays an important role for the solution of stiff problems. The cost of forming the Jacobian and the storage needed to hold it increase rapidly with the size of the system. There is also a cost of solving linear systems involving the Jacobian that increases rapidly with the size. Fortunately large Jacobian matrices are, it seems, almost always sparse. The problem then is specifying which elements of the Jacobian matrix are non-zero and what values they have. Given this information there are efficient schemes for the numerical formation of Jacobians and for the solution of the linear systems that arise. It is awkward that when solving a general sparse problem, it is not known in advance how much storage will be required. A very important special case is that of banded Jacobians. This structure occurs frequently, it is easy to specify which partial derivatives might not be zero, and the storage requirements can be specified in advance. For these reasons, most recent codes provide for banded Jacobians. Writing a subroutine to evaluate partial derivatives analytically and load them in the proper storage locations is not generally thought to be much fun, but it is sufficiently valuable to be included as an option in the solvers. If this option of using an analytical Jacobian is present in the solver, there is the matter of a name for the subroutine for J. The difficulty is that the name is an argument and some FORTRAN compilers will insist that a subroutine of this name exist even if it is not used.

There is a quite different approach to evaluating $\mathbf{F}$ called *reverse communication*. In this approach, evaluation of the equations is done in the program that calls the solver. Every time that the differential equation solver needs the value of $\mathbf{F}(x, \mathbf{y})$, it returns control to the calling program with the arguments x, $\mathbf{y}$, and a request for the value of $\mathbf{F}$. This is not entirely straightforward because the solver may need such a value at several points in the program. One way to deal with this is by means of the multiple entry facility, but this deprecated feature of FORTRAN has never been very portable and has fallen into disuse. The usual way is by means of an integer variable that the solver sets before returning to the calling program and when the solver is called again, a computed GOTO that uses the integer variable as an index transfers control to the proper point. One reason for reverse communication is to deal with problems for which evaluation of $\mathbf{F}$ is a complicated task not easily formulated as a subroutine. Also, the design is convenient when other pieces of mathematical software are being used in conjunction with a differential equation solver—the solver is more a coroutine than a subroutine. Because reverse communication leads to a more complicated software interface and involves users in matters they would ordinarily prefer to avoid, it is not common. Some kinds of problems cannot be solved easily, or perhaps at all, with the usual interface. If you do not have at your disposal a solver with reverse communication, you will have to "turn a solver inside out" (a colorful expression used to describe conversion to a reverse communication interface). Though tedious, this is not hard and it is better practice to alter the interface than to tailor the whole code to your special circumstances—as a rule you should alter quality software only when absolutely necessary and then no more than is necessary.

EXERCISE 1. A remarkable design is found in the Turbo Pascal Numerical Methods Toolbox, Borland International, Inc., Scotts Valley, CA, 1986. The procedure defined in RUNGE_S1.INC (pp. 186–195) is a fixed step size implementation of the classical four-stage, fourth-order Runge–Kutta formula. It solves a system of 10 first-order equations. The user must provide a function defining each equation in the system. There must be 10 functions in all, so if there are fewer than 10 equations in the system, it is necessary to define some functions that do nothing. Contrast this design with other codes accessible to you. Why do you think the author of the code chose this design? Is there anything about Turbo Pascal that would make such a design convenient? The programming language does make a difference to ODE codes. For example, in Turbo Pascal arrays are not communicated to procedures in the same way as in FORTRAN. It is necessary to define a variable type that specifies the size of the array, e.g., this procedure expects the user to define TNvector = **array**[0..TNRowSize] **of** Real, where TNRowSize is the number of equations in the system. The procedure takes advantage of the fact that the indices of this type range from 0 to the number of equations by storing the independent variable t in the first component $V[0]$ of a solution vector and the dependent variable $x_k(t)$ in $V[k]$.

§2 Output of the Solution

One distinction between the mathematical and the computational problem is specifying what is meant by "solving the problem." Sometimes you are interested in approximating the solution of (1) only at the end of the integration. At other times you want approximations on a set of output points in $[a, b]$. You might need them on a set of specific output points, or perhaps on any set that reveals the general behavior of the solution. For some purposes you may need to be able to approximate the solution at any point in the interval. All these situations were seen in Example 2 of Chapter 1 which discussed the solution of Sturm–Liouville problems by shooting. Which definition of solving the problem is chosen can greatly influence the cost of the computation and can even affect the accuracy of the results. All the general purpose codes proceed from a to b by a sequence of steps with the step size being chosen to yield a specified local accuracy. As a rule they take relatively short steps where the solution changes rapidly, so accepting answers where the code finds it convenient to produce them is usually a good way to ascertain the general behavior of the solution. Some methods produce answers at specific points by shortening the step size so as to land on an output point. The efficiency of such methods is degraded when the step size must be shortened often and drastically to produce output. As a byproduct of shortening the step size, more accuracy is obtained than is required. This might be regarded as a bonus, but it is not always a good thing as we shall see later. Some numerical methods produce approximate solutions not just at the end of a step, but throughout the step by means of a polynomial approximation to the solution. A good implementation then chooses the largest step size that appears to yield the desired accuracy and produces answers at specified points by evaluating the polynomial. Such an implementation is little influenced by the number and placement of output points. How to add this *interpolation capability* to methods for which it is not natural has been the object of intense research. It is important to know how a code produces answers at specific points so as to select properly a code for a given definition of solving the problem.

There are a number of practical and conceptual issues that arise in connection with output. A code with an interpolation capability must be allowed to step past an output point, but how far? Output points convey the scaling of the independent variable to the code, so the code cannot completely ignore their location. More will be said about this later. It is not always possible to step past a point, so a code with interpolation must provide for this—a complication for the user. Let us imagine integrating from, say, $x = 0$ to 1, on to 2, and then turning around and integrating back to $x = 1$. Generally the two approximate solutions at $x = 1$ will differ, and they might differ a great deal. One reason for this seen earlier is that the stability of the problem depends on the direction of integration. There are other reasons why the direction of the integration is important to the computational problem, one of which will be mentioned later in this chapter. A change of direction should be regarded as a new problem, hence the solver should be restarted. The matter is blurred in the case of codes that interpolate because when they return to the user with an answer at a specified point x_{out1}, they have an approximate solution valid throughout an interval $[x_j, x_{j+1}]$ containing x_{out1}. There is no reason why a user cannot ask for an answer at another output point x_{out2} that precedes x_{out1} but still lies within the span of the step. From the point of view of the user, the integration has reversed direction, but not from the point of view of the solver. Of course if x_{out2} precedes x_j, the integration really has been reversed, and a restart is appropriate. The Adams code ODE takes care of this matter for the user by interpolating or restarting as needed. However, experience with this code showed that users do not appreciate the implications of turning around, so when its interface was modernized to become DEABM, one of the changes made was to insist on a restart when the direction of integration is reversed.

There are two basic software designs for ODE solvers. A *step-oriented code* advances a single step in a specified direction and then returns control to the calling program. An *interval-oriented code* returns control only when it produces an answer at a specified point or interrupts the integration for a reason that it reports. Often an interval-oriented code is a driver for a lower level step-oriented code, so we begin with the more fundamental task. A step-oriented code provides a degree of control over the integration that is essential to some applications, but it is complicated for routine problems. The code must be told in which direction to go and how far by means of an input step size. For some purposes a code is simply to take the step and return, but it is better to think of this task as being accomplished by a module rather than a general purpose code. It is preferable to suppose that the code will attempt the step size input and subject the result to a test on the local error. If the step is a success, a step size appropriate for the next step is estimated and returned along with the answer. If the step is a failure, a (smaller) step size that will provide the required local accuracy is estimated and this new step size is tried. The code keeps trying until it succeeds or interrupts the computation to report some difficulty that it has recognized. Such a code will either take a step of the size input or a smaller step.

A step-oriented code is appropriate when the general behavior of a solution is to be explored by means of output at points chosen automatically by the numerical method. There are two ways to use such codes to get answers at specific points. If the code does not have interpolation, the fact that it will not take a step longer than the step size input is exploited. When the suggested step size would carry the integration past the output point, it is shortened so as to land on the output point. If the step succeeds, the desired answer is obtained. If it fails, a shorter step is taken and monitoring resumes as before. A code with interpolation could be used in the same way. In fact it must be used in this way when it is not permissible to step past the desired output point. However, when it is permissible to step past, the code takes an efficiently long step and computes an answer by interpolation. The possibility of more than one output point in the span of a step complicates the code, but this possibility is, after all, the main reason for providing the capability. The importance of the documentation of software is often greatly underestimated. The seminal code DIFSUB is a step-oriented code with interpolation. It is perfectly clear (well, it is to me!) from the prologue to the code how to interpolate and the necessity of restarting when turning around. However, the prologue to this very early code did not spell out just how to do interpolation. The author of a report on a quite substantial series of computations done with this code stepped past each output point, turned around with a restart, stepped to the output point, and then turned around with another restart to continue the integration to the next output point! This is a remarkable example of the misuse of a code; it is still more remarkable when you know that the start is a special matter in a code such as DIFSUB that is far more troublesome in both theory and practice than taking ordinary steps. To be perfectly clear about what should have been done, the user should have stepped past the output point, evaluated the interpolating polynomial at the output point, and continued on with the integration.

Interval-oriented codes are used to solve most problems. They are organized so as to return with an answer at a specific point, ready for a call specifying the next output point. There is a variation on the theme that usually allows the code to substitute for a step-oriented code. This is an optional mode, called an *intermediate output mode*, in which the code returns at each step on the way to a specified output point with the solution at the current value of the independent variable. In this mode the results are presented to the user, but the user is not permitted the kind of control of the integration available with a step-oriented code.

When the numerical method does not have an interpolation capability, the code cannot step past output points and if the output points are sufficiently close together, the step size is completely determined by the specified output points. Fortunately this situation is not common in practice. The typical situation is that the *optimal step size*, or *natural step size*, proposed by the code at each step is such that at least a few steps are taken between specified output points. In this situation, output properly handled has practically no impact on the efficiency of the code. There is another way that output affects the integration when interpolation is not possible. For a number of reasons, step size selection algorithms limit how much the step size can be increased in a single step. Unless some provision is made for output, it is possible that an output point be arbitrarily close to a mesh point, hence that an arbitrarily small step size be taken. A step size very much smaller than necessary plus a bound on the rate of increase of step size then implies a succession of steps that are smaller than necessary. The cost is put up and the results are unexpectedly accurate. One way to deal with this is to regard reducing the step size for output as a special case and on the next step return to the natural step size. A more popular approach is to prevent unduly small step sizes. Two schemes are seen, *stretching* and *look-ahead*. The latter is by far the more popular, so we present it and refer the interested reader to Shampine and Watts [1977, 1979] for a discussion of stretching. Suppose that h is the step size proposed at x_j. The code first checks whether an output point x_{out} can be reached in a step of length h. If it can, the step size is reduced so as to produce an answer at x_{out}. If it cannot be reached in a step of length h, it is asked if x_{out} can be reached in two steps of length h, i.e., whether $x_j + h < x_{\text{out}} \leq x_j + 2h$. If it can, the step size is reduced to $(x_{\text{out}} - x_j)/2$. If it cannot, no attention is paid to output. The natural step size should change slowly from step to step, so by looking ahead to an output point and adjusting the step size so as to hit it in two steps, the reduction is minimized. This simple device is very successful at ameliorating the impact of output at specified points. At an isolated output point (isolated in the sense that several natural steps are taken between output points), about the worst that can happen is two steps of about half the optimal size, which is to say, the cost is put up by at most one extra step.

There is also a software issue about specifying where output is to occur. So far we have implicitly assumed that the solver is given output points successively. Another possibility is to specify them all at once. A relatively popular way of doing this is to specify an increment Δ and ask the code to produce answers at $a, a + \Delta, a + 2\Delta, \ldots$. Some codes accept an array of values at which output is to be provided. The program cited in Exercise 1 furnishes an example of the approach. In this code the user specifies the number of steps (of constant size) and the number of answers desired. The answers returned correspond to values of the independent variable that are at roughly equal spacing throughout the interval of integration.

EXAMPLE 1. E. Spitznagel presented an exercise on two-compartment pharmacokinetics to a Workshop on Teaching ODEs with Computer Experiments (Harvey Mudd College, June 1992) that serves here to make an important point about output. Because our attention is focussed on the numerical aspects of the task, only the briefest of background is provided for this interesting exercise. Lithium carbonate is used to treat manic–depressive disorder. The range between the blood levels of the drug that are therapeutic and toxic is narrow. The drug has a half-life of about 24 hours in the blood, so there is considerable build-up over time. The first question is, with a dosage of one tablet every 12 hours, how long does it take for the medication to reach a steady state in the blood? A simple model of the transport of the drug leads to the initial value problem

$$\dot{x} = -5.6x + 48\,\text{pulsep}\left(t, \frac{1}{48}, \frac{1}{2}\right), \qquad x(0) = 0,$$

$$\dot{y} = +5.6x - 0.7y \qquad y(0) = 0.$$

The time t is measured in units of days. The function pulsep(t, w, p) defines a pulse of period p. The pulse is of unit height for $0 \le t \le w$ and zero for $w < t < p$. The term in the first differential equation involving the pulsep function describes a pill that is released uniformly over half an hour ($= 1/48$ day), and the factor of 48 accounts for 1 pill in 1/48 a unit of time.

The code DDRIV2 was used to integrate this initial value problem. Figure 3.1 shows the level of the medication in the blood, $y(t)$, over a period of 6 days. This code implements both Adams–Moulton and backward differentiation formulas. The Adams formulas were used to compute the results of the figure, but the phenomenon that interests us here does

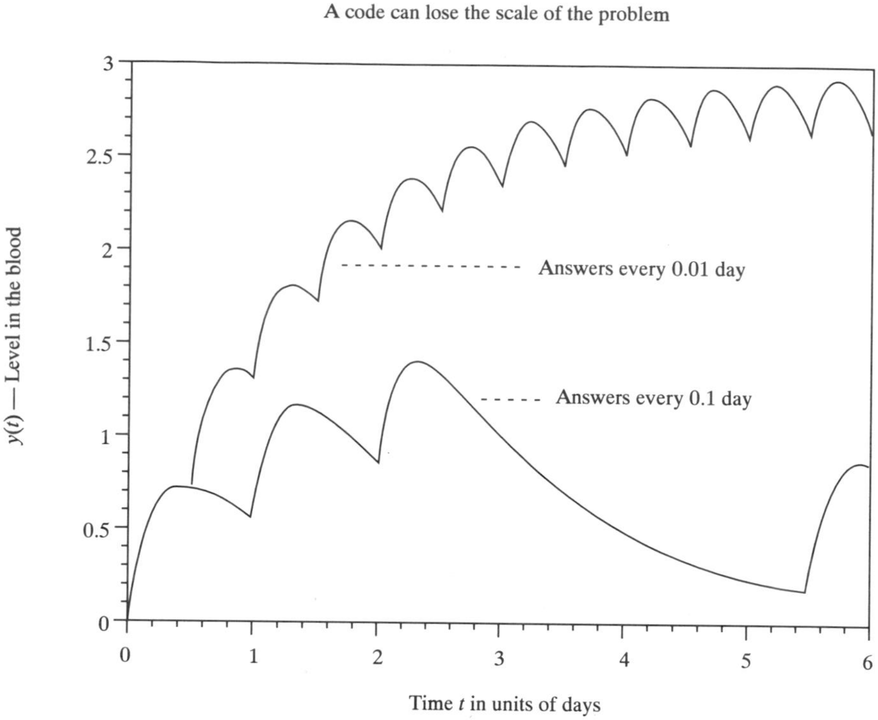

Figure 3.1

not depend on the method. This code is intended to be easy to use, so answers are provided wherever the user requests them. The two plots show what happens when answers are requested every 0.01 days and when they are requested every 0.1 days. The *only* difference between the two runs is how frequently answers are requested. Although the results are dramatically different, the code claims the same accuracy for both runs, so what is going on?

Using only a finite number of samples, a differential equation solver tries to approximate the solution throughout the interval. Clearly this is possible only if strong assumptions are made about how the solution behaves between samples, i.e., about how smooth the solution is. All methods sample at least once in a step and most sample several times. The automatic adjustment of the step size is intended to result in a step size that will yield the desired accuracy and at the same time be as big as possible. There is always a possibility that the step size selection algorithm will lose the scale of the problem and that is exactly what happens here. The code starts with a step size small enough that it "sees" the first dose by evaluating the right hand side of the differential equation during the time the pill has an effect. The equations are easy to integrate, so the code increases the step size rapidly, so rapidly that in one of the runs, the code does not "notice" the second dose because it does not evaluate **F** during the time the pill is being absorbed. After a couple of days, the integration is going so well that the code skips over a good many doses. Familiarity with step function input causes people to underestimate the strain they place upon an integrator. Because the change is discontinuous, the code has no inkling from the preceding computations that a dramatic change is about to occur and it can easily lose the scale of the problem. Though not actually necessary, codes with an interpolation capability do pay some attention to where answers are specified because generally this conveys a measure of scale. Here requesting answers every 0.1 days seems frequent, yet the absorption of a pill takes place in 0.02 ($=1/48$) days, a rather shorter time scale, and the code gets into trouble. Requesting answers every 0.01 days should be frequent enough to see the effects of the pills, and that proves to be the case.

Just how one can deal with the difficulty pointed out here depends on the code and the fact is that there is more than one way to deal with the difficulty when using DDRIV2. The code must be informed of the scale on which phenomena occur. Some codes allow the user to specify a maximum step size, a lower level subroutine called by DDRIV2 being a case in point. The trouble with this is that a maximum step size applies to the whole run, leading to a needlessly small step size when all quantities vary slowly. Codes that do not have an interpolation capability will not step past an output point, so one need only ask for answers where rapid changes are expected. Here, for example, asking for an answer when a pill is swallowed and when it has been completely digested would convey to the code the necessary scale information. Although this will usually suffice for codes that do interpolation, there is a way to be sure. Because it is not always permissible for a code to step past a certain point, all quality codes with interpolation provide the user with an option to prevent the code stepping past a specific point. The option can be used to make sure that the code "notices" sharp changes in the problem.

There is a closely related difficulty to be kept in mind. The first step is particularly difficult for the codes as they try to determine automatically the scale of the problem. The codes exploit all the information available to them, but at the initial point this information is very limited and sometimes even the best codes are fooled. The distance to the first output point is one of the things that codes use to recognize the scale. It is prudent to specify a first output point that indicates the scale on which phenomena occur near the initial point even when an answer at this point is of no interest in the application. To be concrete, if things can happen in seconds and you are interested in answers at intervals of weeks, you should tell the code to produce an answer after the first few seconds even though you have no use for this answer. Doing this will not put up the cost and it may make the difference between reliable results and nonsense.

EXERCISE 2. Spitznagel reports that with many drugs a doctor will prescribe an initial loading dose to bring the level of medication in the blood more rapidly to the desired steady state. That is, the patient takes more than one pill the first time and thereafter only one pill each time. The trick is to get to steady state more quickly without reaching levels higher than the steady state that might be toxic. Experiment with loading doses of 2, 3, and 4 tablets. Which seems to be the most satisfactory treatment?

§3 Accuracy

Practically all codes control the local error and at most attempt to estimate the true error. For this reason it is essential to understand the implications of a control of the local error for the true error of the integration. The most common way of assessing the true error is to solve a problem with one tolerance on the local error, reduce the tolerance substantially, solve the problem again, and estimate the error of the first integration by comparison. This obvious ploy can be unreliable because it can happen that the second integration is no more accurate than the first. Oddly enough, this is rather more likely to happen with a modern code that chooses the step size for reliability and efficiency than with one of the early codes that uses a fixed step size. One reason for this is that a code is obliged to keep the local error smaller than the specified tolerance, but it can, and may, choose a step size that results in a local error substantially smaller than the tolerance. Recall the situation at crude tolerances. If you ask for, say, one digit of accuracy, a code may have to work with step sizes small enough to get two digits of accuracy in order that the algorithms for local error estimation and step size adjustment be credible. Because of this, asking for two digits of accuracy will lead to results little different from those resulting from a request for one digit. Also recall that if a code does not have an interpolation capability, it will reduce the step size to obtain output at specific points. This results in more accuracy than is required. In extreme cases most of the steps are determined by the output requirements and the accuracy of the integration depends weakly, if at all, on the tolerance for a perhaps considerable range of tolerances. There are other reasons such as limiting precision that can make the second integration no more, and even less (!), accurate than the first, hence cause the ploy to fail. Despite these qualifications, this ploy of *reintegration* is useful for assessing the true error provided that very crude and very stringent tolerances are avoided. It is prudent to reduce the tolerance substantially (an order of magnitude or more) when using reintegration. Example 6 of Chapter 6 and Example 2 of Chapter 8 present results for one problem that is integrated with several codes for a wide range of tolerances. Except at the most stringent tolerances, the computations show that reducing the tolerance by an order of magnitude and integrating again provides a useful measure of the accuracy of the integration. There are a number of ways to assess the true (global) error that are more reliable because they are more adapted to the method and code. Global error assessment is a valuable capability for exploratory computations and for spot checks of a local error control. Unfortunately it is sufficiently expensive that it is not done routinely.

EXAMPLE 2. Reintegration is so plausible that it is worth presenting some numerical results from Shampine [1980] to illustrate that it can be misleading. The GEAR package was used to solve a model of the diurnal variation of a photocatalyzed reaction taken up as Example 8 in Chapter 8. Because the solution components span a wide range, a relative error control with a threshold of 10^{-20} was used. When three figures of accuracy were requested, the solution component of interest was computed to be 2×10^{-51} at the time 7.992×10^5 seconds. Quantities so small are defined by the threshold as uninteresting physically and the solution is regarded as effectively zero. "Being cautious we reduced the tolerance a *lot*—asking for six figures—and repeated the integration to get the value 10^{-61}. It seems clear that the value is zero. In fact, the value is $1.325 \times 10^{+8}$!" When the EXTRAP code was used to solve the standard test problem $y' = -0.5y^3$, $0 \leq x \leq 20$, $y(0) = 1$, it was found that the answer at $x = 20$ was the *same floating point number* for absolute error tolerances 10^{-2}, $10^{-3}, \ldots, 10^{-7}$. (Tolerances 10^{-8} and smaller give more accurate results.) When the DREBS code was used to solve the standard test problem

$$\begin{aligned} y_1' &= -y_1 + y_2, & y_1(0) &= 2,\\ y_2' &= y_1 - 2y_2 + y_3, & y_2(0) &= 0,\\ y_3' &= y_2 - y_3, & y_3(0) &= 1, \end{aligned}$$

it was found that on decreasing the tolerance from 10^{-4} to 10^{-5}, the error *increased* by a factor of 8. This is common at limiting precision, but these tolerances are far from limiting precision. Substantial exeriments (Shampine [1979c]) with a good code, DVERK, applied to routine problems showed that reducing the tolerance by an order of magnitude reduced the true error enough to get a decent estimate of the error only 85% of the time—reintegration is useful, but not truly reliable.

EXERCISE 3. Solve the two-body problem *D*5 for a range of tolerances including a crude tolerance of, say, 10^{-3} and a stringent tolerance of, say, 10^{-10}. Use reintegration to estimate the accuracy of your solutions at the end of the integration. Using the values for the true solution given in Example 7 of Chapter 2, or computing them yourself by solving Kepler's equation, determine the true error of your numerical solutions. Compare the estimates to the true error. Do the estimates represent fairly the true error? Does reintegration provide better estimates of the true error in some ranges of the tolerance than in others?

The author has written many differential equation solvers. In his experience the most annoying single issue has been unquestionably the error control. So far we have been assuming implicitly that the local error in the Ith solution component, $le(I)$, is measured in an absolute sense, that is, $|le(I)|$ is compared to a tolerance. The trouble with this is that it is not appropriate when the approximate solution $Y(I)$ is "large"' and may not even be possible. To understand why, we must take account of the fact that computers work with *finite precision arithmetic*, a fact we generally try to ignore, but sometimes cannot. Floating point arithmetic is accurate in a relative sense. Merely representing a real number $y_I(x)$ as a floating point number may result in a relative error as large as U, the unit roundoff for the computer being used. Clearly it makes no sense to ask for a numerical solution that is more accurate than the machine representation of the correct value (Shampine [1974]). Nowadays codes monitor the accuracy specified and warn the user when a relative accuracy smaller than a few units of roundoff is requested. It seems peculiar that someone might ask for so much accuracy, but it is not, in fact, rare. For one thing, most users do not know the value of U for the machine they are using and do not know how to find out. Stringent accuracy requirements are necessary for some applications and a slip is quite possible. Quality software will inform users about the system-dependent permissible values after such a slip. The most common way that an impossible accuracy is specified is by requiring an absolute error control of a solution that becomes unexpectedly large. If the approximation $Y(I)$ is to agree with $y_I(x)$ to a tolerance of ε on the absolute error,

$$|Y(I) - y_I(x)| \leq \varepsilon,$$

this corresponds to a relative error of

$$\frac{|Y(I) - y_I(x)|}{|y_I(x)|} \leq \frac{\varepsilon}{|y_I(x)|}.$$

When $|y_I(x)|$ is big enough, the absolute error test implicitly requires a relative error smaller than U, which is impossible. An absolute error control is also troublesome when a solution becomes unexpectedly small. If the solution $y_I(x)$ is smaller in magnitude than, say, $\varepsilon/2$, then any number $Y(I)$ smaller than $\varepsilon/2$ will pass the error test. What is being said, in effect, is that the solution component is not interesting when it is smaller than ε, hence no effort need be devoted to computing any correct digits. This may be exactly what you have in mind, but it should be intentional and not the result of simply not thinking about the implications of the error control.

Having mentioned the precision possible and the unit roundoff, some comments about the precision available should be made. FORTRAN specifies that both a single and a double precision be available and of course the unit roundoff depends on which is being used. The number of digits in a single precision word differs importantly among popular computers. Some provide about 14 decimal digits and this is usually adequate for the solution of ODEs. Others provide only about 7 digits and it is often found that this is inadequate. C. B. Moler used to describe jokingly the precisions available on the IBM mainframe at the University of New Mexico where he and the author worked at the time as "half precision" and "full precision" rather than "single precision" and "double precision." Anyone experienced with scientific computation on machines like this with only 7 digits in single precision will find the description to be more accurate than amusing. When working in double precision, it is easy to ruin a numerical solution without realizing it by failing to specify some constant in the routine defining the equations as double precision. The dangerous thing about this kind of slip is that the error is in the definition of the problem, not its numerical solution, so a code can report success even though the answers do not approximate well the solution of the problem intended.

When imposing an absolute control, an assumption about the scale of the solution $y_I(x)$ must be made if the tolerance is to be meaningful. If the solution becomes unexpectedly large, this kind of control is stringent and if it becomes unexpectedly small, it is lax. Many codes ask for a single tolerance that is imposed on all solution components. This is easy for the user, but implicitly assumes that all solution components are scaled the same. Often this is a bad assumption. For one thing, even if the original solution components are all scaled roughly the same, their derivatives need not be on the same scale—converting to a first-order system by introducing new unknowns for the derivatives may result in solution components with different scales. One way to deal with components scaled differently is for the code to ask for a vector of absolute error tolerances $\varepsilon(*)$ and impose the tolerance $\varepsilon(I)$ on the solution component $y_I(x)$. Another is to ask for a scalar tolerance ε and a vector YSCALE(*) and test

$$\frac{|Y(I) - y_I(\mathrm{x})|}{\mathrm{YSCALE}(I)} \leq \varepsilon.$$

This test of error relative to a scale value corresponds to an absolute error tolerance of $\varepsilon \times \mathrm{YSCALE}(I)$ on the Ith solution component. It is an attractive way to work with absolute error. The fact that vector absolute error controls are not popular simply reflects the sad fact that users often are not willing to give proper consideration to what they want in the way of accuracy. Because of this, recent software force a vector control or they emphasize a relative error control.

A relative error control, which for a tolerance ε requires that

$$\frac{|Y(I) - y_I(x)|}{|y_I(x)|} \leq \varepsilon,$$

makes it easy to test for impossible accuracies and deals with solution components that become unexpectedly large or small. An important advantage is that it is reasonable to use the same relative error tolerance on all components of the solution because the control itself takes account of different scales. For this reason, in what follows it will be assumed that a scalar relative error tolerance is specified. Unfortunately, relative error control has its defects, too. There are two issues, one technical and one of intent.

The technical defect of relative error control has two aspects. One is that it is not even defined when some $y_I(x)$ vanishes. How to provide a sensible definition of relative error then is the subject of this paragraph. As defined above, the error made in a step is controlled relative to the value of the solution at the end of the step. This is done in some codes and in others the error is taken relative to the value of the solution at the beginning of the step. Either definition can lead to disaster when a component vanishes. A more practical approach is to control the error relative to the size of the solution throughout the span of the step. In this approach a vector SIZE(I) is defined so as to represent the scale of $y_I(x)$ in the span of the step. It is perhaps surprising that simply taking SIZE(I) to be the larger, or the average, of the magnitudes of the approximate solutions at the two ends of the step will cope with most problems. Some effective numerical methods are quite expensive per step. They are competitive because they take "large" steps, meaning that the solution can change substantially in the course of a single step. The values of the solution at the ends of the step may not be representative for such methods, but as it happens, they produce approximate solutions at points within the span of the step that are not as accurate as the result at the end of the step, but accurate enough to help define the size of the solution. A case needing special attention is a solution component that underflows to zero, an event not uncommon in the context of stiff problems. The trouble is that in any reasonable definition of the size of this solution component, SIZE(I) will be (computationally) zero. Such a situation can be handled more gracefully than one might expect because the numerical method should estimate that the local error of its approximation to this solution component is zero. All one need do when writing the solver is test for a zero local error before forming SIZE(I) and testing the relative error. If the estimated local error is zero, SIZE(I) is not even formed. The first step is another special case that is more difficult to handle gracefully. Initial values are often quite special numbers and in particular, a value $y_I(a) = 0$ is not at all unusual—we have seen a number of examples already in this book. To define a reasonable value of SIZE(I), the code needs some idea of the scale of $y_I(x)$ near $x = a$. When $y_I(a) = 0$, there is no obvious scale and because the code has not yet taken a step, there is no computed approximation to $y_I(x)$ for $x \neq a$ to assist in defining a scale. Some codes resort in this situation to the first derivative $y_I'(a)$. This is a reasonable response, but unfortunately there are practical problems for which both $y_I'(a)$ and $y_I(a)$ vanish. If $y_I'(a) \neq 0$, these codes use the crude approximation $hy_I'(a)$ to the solution at the end of the first step as a measure of size. This is adequate if the initial step size is known in advance, but most general purpose codes now determine automatically the initial step size and they need a measure of size in order to estimate the h that would be used in this way of obtaining a measure of size. Because of these considerations, codes that emphasize relative error control may avoid the whole issue by not permitting pure relative error when an initial value is zero.

The other aspect of the technical defect is that even with a sensible definition of the size of the solution, numerical methods are not as effective as usual when a pure relative error control is specified and a solution component vanishes. This does not depend on the method nor on which step it happens, but the most common situation is a solution component that vanishes at the initial point a. With a tolerance ε, a relative error control on the local error requires that

$$\frac{|\text{local error}|}{|\text{size of the solution}|} \leq \varepsilon.$$

For a method of order p, the local error of a step of length h from a is $\mathcal{O}(h^{p+1})$. In the usual situation that $y(a) \neq 0$, the quantity on the left hand side here is $\mathcal{O}(h^{p+1})$, the same as an absolute error control. The situation is quite different when $y(a) = 0$. Suppose $y(x)$ has a root of multiplicity m at the initial point a, so that

$$y(x) = \frac{(x - a)^m}{m!} y^{(m)}(a) + \mathcal{O}((x - a)^{m+1})$$

where $y^{(m)}(a) \neq 0$. This shows that the "size of the solution" on $[a, a + h]$ is $\mathcal{O}(h^m)$. Now the quantity on the left hand side of the test is $\mathcal{O}(h^{p+1-m})$. It is seen that when the solution has a zero of order m and a pure relative error control is used, the effective order of the method is reduced by m. It is unfortunate that this situation is most common at the initial point because starting the integration is special in the popular variable order Adams and BDF codes. They start with a method of very low order, typically *one*, so that the reduction of order can be disastrous—an excellent reason for not permitting pure relative error when some component of $\mathbf{y}(a)$ vanishes.

The difficulty of intent alluded to earlier is that a pure relative error control specifies that all solution components be computed accurately, yet a solution component may not be interesting when it is "small." It is common that a code work very hard to compute a solution component accurate in a relative sense even though it is too small to be physically interesting. Because of this and because of the possibility that a pure relative error control fail due to a solution component vanishing, most software designs require, or at least allow, a mixed control. A control that has become quite popular in one form or another involves a scalar relative error tolerance ε and a vector THRES(I) of threshold values. The test on $Y(I)$ is

$$|Y(I) - y_I(x)| \leq \varepsilon \max(\text{SIZE}(I), \text{THRES}(I)),$$

where SIZE(I) is a measure of the magnitude of $y_I(x)$ over the span of the current step. When SIZE(I) is bigger than the threshold value THRES(I), this is a relative error test. When it is smaller, it is an absolute error test with tolerance $\varepsilon \times$ THRES(I), which is to say that it is an absolute error test with tolerance ε for a solution scaled by THRES(I). Variations on the theme have a vector AE(I) of absolute error tolerances and test either

$$|Y(I) - y_I(x)| \leq \max(\varepsilon\, \text{SIZE}(I), \text{AE}(I)),$$

or

$$|Y(I) - y_I(x)| \leq \varepsilon\, \text{SIZE}(I) + \text{AE}(I).$$

All these variations behave much the same, but the idea of a threshold below which a solution component can be neglected is perhaps the most easily understood.

There is a different way to measure size in a relative error control that has had some popularity. This is to define SIZE(I) as the biggest value of $|y_I(x)|$ seen so far in the integration. The difficulty of starting from a zero initial value remains. Other than this special case, the control is attractive to people writing codes because it is not subject to some of the difficulties of the relative error controls described so far. When a solution component is increasing, SIZE(I) is increasing and the control is a pure relative error control. When a solution component is decreasing, SIZE(I) does not change so the control amounts to an absolute error control on a solution scaled by SIZE(I). One justification for this kind of control is that when an error is made at some point, its effect generally persists. Because the effect of an error may not die out, errors of the same size should be permitted at later points. However, if a solution component decreases a great deal and then increases, the control may be unsatisfactory. This is because the error control becomes very lax as the component becomes small and the component may be computed so inaccurately that it cannot be followed accurately as it again becomes important. Other kinds of mixed error control could get into the same sort of trouble. An interesting aspect of this control is its dependence on the history of the integration. A person who does not know that DIFSUB has such a control might find the results of integrations that involve restarts or changes of direction to be very puzzling.

EXAMPLE 3. Dahlquist et al. [1982] present an interesting example known as the *knee problem*. The initial value problem is

$$\varepsilon \frac{dy}{dx} = (1 - x)y - y^2, \qquad 0 \leq x \leq 2, \qquad y(0) = 1.$$

The parameter ε satisfies $0 < \varepsilon \ll 1$. This is an example of a stiff problem arising from singular perturbation and we shall return to it later when we take up stability and stiffness. For now let us note that there are two null isoclines, $u(x) = 1 - x$ and $u(x) \equiv 0$. Arguing as in Example 1 of Chapter 2, it is seen that for $x < 1$ solutions starting near the isocline $1 - x$ approach it very rapidly and stay nearby (more precisely, within $\mathbb{O}(\varepsilon)$). The initial value problem is very stable up to $x = 1$ where the solution is nearly 0. For $x > 1$, the isocline $u(x) \equiv 0$ is very stable, so the solution of the initial value problem approaches it very rapidly and stays nearby. This isocline is obviously a solution of the differential equation and uniqueness implies that the solution of the initial value problem cannot cross it. Dahlquist et al. solved the problem using a code in the IMSL library that is a descendant of DIFSUB for several values of ε and tolerance TOL. In this code the error is taken relative to the largest value of the solution seen so far. Here the initial value is 1 and the solution is non-increasing, so what is described as a relative error control amounts to a pure absolute error control. In all cases the numerical solutions were accurate in an interval $[0, 1 - \delta]$ for some $\delta > 0$. When $\varepsilon = 10^{-4}$, the numerical solution was acceptable for TOL $= 10^{-4}$, but for TOL $= 10^{-2}$, it followed the isocline $1 - x$ for $x > 1$. Similarly, when $\varepsilon = 10^{-6}$, the numerical solution was acceptable for TOL $= 10^{-6}$, but for TOL $= 10^{-4}$, it followed the isocline $1 - x$ for $x > 1$. To make clear that the difficulty is not specific to this code, Figure 3.2 was computed using DDRIV2. Because the problem is stiff, the backward differentiation formulas were used, MINT $= 2$. The parameter $\varepsilon = 10^{-6}$ and a pure absolute error control with tolerance 10^{-4} was used. It is seen that the integration follows the isocline $1 - x$ through the x-axis down to meaningless negative values for a considerable distance before returning to the correct solution along the x-axis. The fundamental reason for this

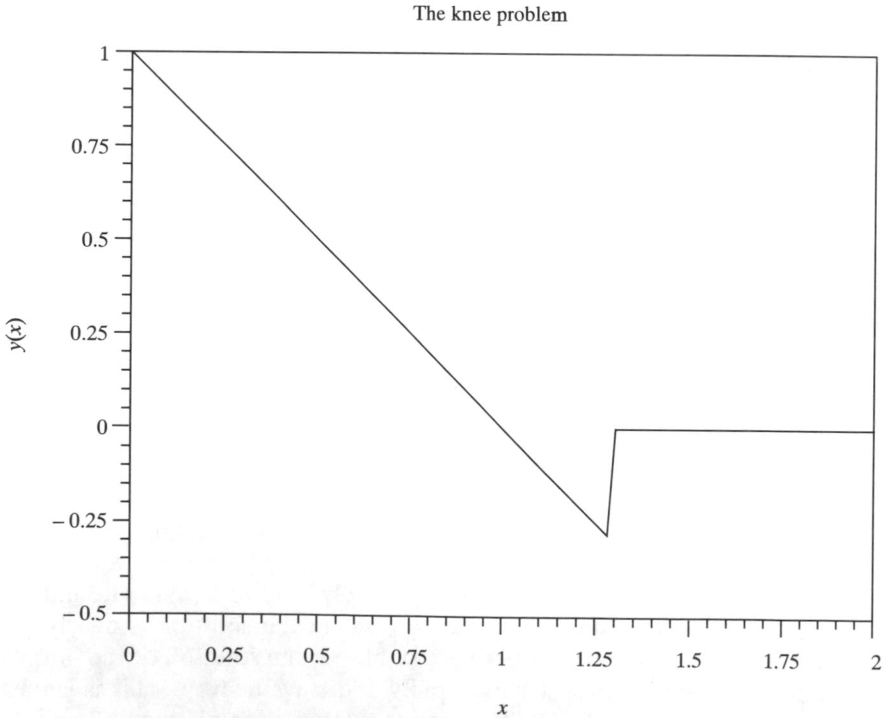

Figure 3.2

unsatisfactory numerical solution is the error control. The true solution is $\mathbb{O}(\varepsilon)$ for $x \geq 1$. When a pure absolute error control with a tolerance bigger than the solution is specified, the codes are not obliged to get any correct digits. Because of this, they do not "notice" the bend in the solution at $x = 1$. Eventually instability causes the numerical solution to leave the neighborhood of the isocline $1 - x$ and approach the stable true solution. The difficulty is aggravated greatly by the solution being so easy for the code to approximate for $x < 1$—the code accelerates so rapidly on the straightaway that when it is not told to pay close attention (by means of an appropriate error control), it fails to make the turn at $x = 1$. It is pointed out in Example 1 that output requirements convey to codes a measure of the scale of the problem. For the plot of Figure 3.2, output was requested at a constant spacing of 0.02. We might have expected this to be frequent enough to cause the code to notice the bend in the solution, but in this instance the problem is so easy to integrate that the output does not constrain the step size sufficiently. Forcing the code to take $x = 1$ as a mesh point cures the difficulty.

The issue of asking for impossible accuracy has already been raised. At stringent tolerances it may not be possible to ignore the fact that finite precision arithmetic is being used. It might be thought that asking for an accuracy that cannot be achieved on the computer being used would result in some kind of disaster, but often this is not so. It is far more common that as the tolerance is decreased to values near limiting precision, the computation becomes more expensive and the accuracy of the results actually gets worse. This is one situation in which assessment of true error of an integration by integrating with a substantially reduced tolerance will fail. It is difficult to diagnose *limiting precision*, but there is a simple test that is quite valuable. The code is to step from a mesh point x_j to $x_{j+1} = x_j + h$. If h is less than a unit roundoff in x_j, i.e., $|h| < U\,|x_j|$, then the floating point representations of x_j and x_{j+1} are the same machine numbers. Obviously, if the code predicts that a step size this small is necessary to produce the specified accuracy, limiting precision has been reached. Some popular numerical methods evaluate $\mathbf{F}(x, \mathbf{y})$ for several x in the interval $[x_j, x_j + h]$. These arguments must be different machine numbers for the analysis to have any validity. Indeed, for the analysis to be valid, the arguments must be significantly different. The simple test is to require that arguments differ in a relative sense by at least a small multiple of U, say $10U$. This simple test has been found to be of great value for the recognition of limiting precision (Shampine [1974, 1978]) and is now in wide use. As a simple example of the effects described here we cite some results from

Shampine [1979c] of integrating the two-body problem in circular motion with two codes based on Adams methods:

	DIFSUB		STEP	
Tolerance	ERROR	NFCN	ERROR	NFCN
10^{-12}	3×10^{-8}	3,487	1×10^{-9}	1,231
10^{-13}	2×10^{-9}	4,624	3×10^{-10}	1,936
10^{-14}	4×10^{-9}	13,023	1×10^{-10}	2,352
10^{-15}	3×10^{-8}	75,409	detected limiting precision	

With DIFSUB the accuracy deteriorates and the number of function evaluations, NFCN, soars as the tolerance reaches limiting precision. Because STEP uses the two tests just described, it balks at the tolerance of 10^{-15}.

EXAMPLE 4. Lastman, Wentzell, and Hindmarsh [1978] discuss the numerical solution of a problem that places great demands on the precision and the step size control. A cavitating bubble is modeled by the initial value problem

$$\frac{dy_1}{ds} = y_2, \qquad y_1(0) = 1,$$

$$\frac{dy_2}{ds} = (5\, e^{-s/s*} - 1 - 1.5\, y_2^2)/y_1 - (ay_2 + D)/y_1^2$$
$$+ (1 + D)/y_1^{3\gamma+1}, \qquad y_2(0) = 0.$$

Here $s* = 0.029/R_0$, $a = 4 \times 10^{-5}/R_0$, $D = 1.456 \times 10^{-4}/R_0$, and $\gamma = 1.4$. The quantity R_0 is taken to be 10^{-3} in most of the computations presented in the paper. In an earlier paper Wentzell found that for this value of R_0, the component $y_1(s)$ exhibits cusps. For instance, near $s = 175$ the derivative of this component changes from about -10^8 to 10^8 in a distance of only 10^{-8} units in s. The quantities of physical interest are the extrema of $y_1(s)$ and their locations. The first few values he computed and some of the values computed later by Lastman, Wentzell, and Hindmarsh [1978] are

Extrema	Wentzell	Lastman et alia
Maximum 1	$y_1 = 73.7$	$y_1 = 73.53$
	$s = 102.1$	$s = 105.29$
Minimum 1	$y_1 = 6.24 \times 10^{-5}$	$y_1 = 5.33 \times 10^{-5}$
	$s = 175.4$	$s = 174.93$
Maximum 2	$y_1 = 44.6$	$y_1 = 72.70$
	$s = 212.4$	$s = 241.45$
Minimum 2	$y_1 = 2.83 \times 10^{-3}$	$y_1 = 5.33 \times 10^{-5}$
	$s = 257.0$	$s = 307.91$

Lastman et al. obtained consistent values for these quantities in several ways. In addition to solving the problem directly, they "regularized" it by introducing new variables that change more slowly at the cusps. There were three variations of regularization and three differential equation solvers, all based on Adams–Moulton formulas. The results presented here were computed with DIFSUB using Regularization 2 with $\alpha = (3\gamma + 1)/2$.

Wentzell's computation shows a rapid damping in the motion of the bubble. He did not control the error in his integration and it is seen that an accurate integration leads to an entirely different conclusion about the motion. As Lastman et al. write, "This demonstrates the danger in integrating ODE's without some control on the single-step errors." A proper control of the error requires extraordinarily small step sizes near the minima of $y_1(s)$ but permits large step sizes near the maxima. Indeed, they report step sizes ranging over 16 orders of magnitude!

Although regularization provides a more efficient solution, a straightforward way to solve this problem is to introduce arc length as a new independent variable. It is prudent to program this in the manner explained in Example 6 of Chapter 1 in order to avoid scaling difficulties. So as to show that Runge–Kutta methods can be used, the author solved the problem with the (4, 5) pair in RKSUITE. There were a number of points of interest. One difficulty with using arc length is knowing how far to integrate to cover the interval of interest in the original independent variable. The code CT was used to advance the integration a step at a time and when the original independent variable s became greater than 310, the run was terminated. The solution y_1 has two cusps in this interval. Because of the large values attained by the derivative y_2 at the cusps, the arc length of the solution over this interval is large, about 4×10^9. The RKSUITE codes select automatically the initial step size, but the algorithm did not perform satisfactorily when CT was called with the interval $[0, 4 \times 10^9]$. Specifically, it formed an approximation to y_1 that was negative. This non-physical approximation was revealed by an "invalid exponentiation" during the run. The code allows a user to specify a bound on the initial step size and a value of 0.01 proved quite satisfactory. Some thought must be given to the error control because it is necessary to compute the important component y_1 accurately through the cusp where it gets as small as 5.3×10^{-5}. A control relative to the largest value seen so far could be used only with a very stringent tolerance because y_1 gets as big as 73 before reaching a cusp. It would be even less satisfactory for the component y_2 that gets as big as 10^8 near a cusp but decreases to 0 at a maximum. Even with an appropriate control, it was necessary to use a rather stringent tolerance to avoid non-physical approximations. With a relative error tolerance of 10^{-6} and a threshold for each component of 10^{-16}, CT integrates to $s = 310$ in

505 steps at a cost of 3952 evaluations of the equation. Despite the demands this problem places on the precision and step size control, integrating it is not expensive. Locating the cusps is easy because the code must use relatively small step sizes there. The smallest values of y_1 computed were 5.367×10^{-5} at $s = 174.93$ and 5.356×10^{-5} at $s = 307.90$.

EXAMPLE 4. Solve this problem directly with your favorite code and see what happens at the first cusp. Follow the author by integrating the problem using arc length to locate the cusps. If you are keen, look up the article of Lastman et al. and try one of their regularizations. This is as easy as using arc length and can be considerably more efficient.

It is also important not to ask for too little accuracy. This is a real temptation when the equation is defined in terms of experimental data of very limited accuracy, especially when it is appreciated that the more accuracy asked of a solver, the more expensive the integration. However, the extra cost of asking for another digit is often quite modest. Moreover, a quality code is likely to give you a couple of digits accuracy whether you ask for them or not. The issue then is the reliability of the code. All the algorithms in a modern code require some accuracy in the solution for their validity. For this reason a quality code is likely to find that it cannot reliably use a step size larger than a value yielding a couple of digits of accuracy in at least one solution component. Of course it might not succeed in recognizing that the tolerance input is too large to result in step sizes sufficiently small for a reliable solution, so it is prudent always to ask for results of at least a modest accuracy, say 1%. There are other issues. The presence of inaccurate data does not mean that an inaccurate solution is appropriate. The numerical solution of the differential equation should be sufficiently accurate that error in solving the equation does not obscure the scientific implications of error in the data. It is time once again to remark on the difference between what solvers attempt to do and what users would like for them to do. The solvers control only the local error and the implications of this for the true error that concerns users depends very much on the initial value problem itself. Most problems are moderately stable and for moderate tolerances the distinction is not particularly important in practice. However, at very relaxed tolerances, this is not so. What happens then is that the solver may do just what it is supposed to do, but produce numerical results that do not agree at all well with the true solution. Very relaxed tolerances is another situation in which assessment of true error by solving the problem again may fail. The authors of some quality mathematical software consider this issue to be so important to reliable computation that their codes require at least a modest accuracy be specified. There is a subtlety that might be noted here. The step size is chosen so that the local error of all the solution components passes a test on accuracy. Typically it is the error in only a few solution components that determines the step size. This step size is smaller than necessary for the other solution components, so they are computed more accurately than required. Because of this, it is often true that some accuracy is obtained for solution components so small that the user has defined them by means of the error control (a threshold) to be uninteresting. This is still another situation in which the accuracy in a component may be not closely related to the tolerance so that integrating again with a smaller tolerance may not provide a reliable estimate of the true error.

EXAMPLE 5. To illustrate the dangers of tolerances that are too crude, the two-body problem $D5$ was integrated with VODE using its Adams–Moulton formulas. Figure 3.3 shows the solution component $x(t)$ when computed with a relative error tolerance of 10^{-2} and with a tolerance of 10^{-12} in a mixed control with an absolute error tolerance of 10^{-14}. Asking for 2 digits of accuracy seems reasonable enough, yet it is seen that after

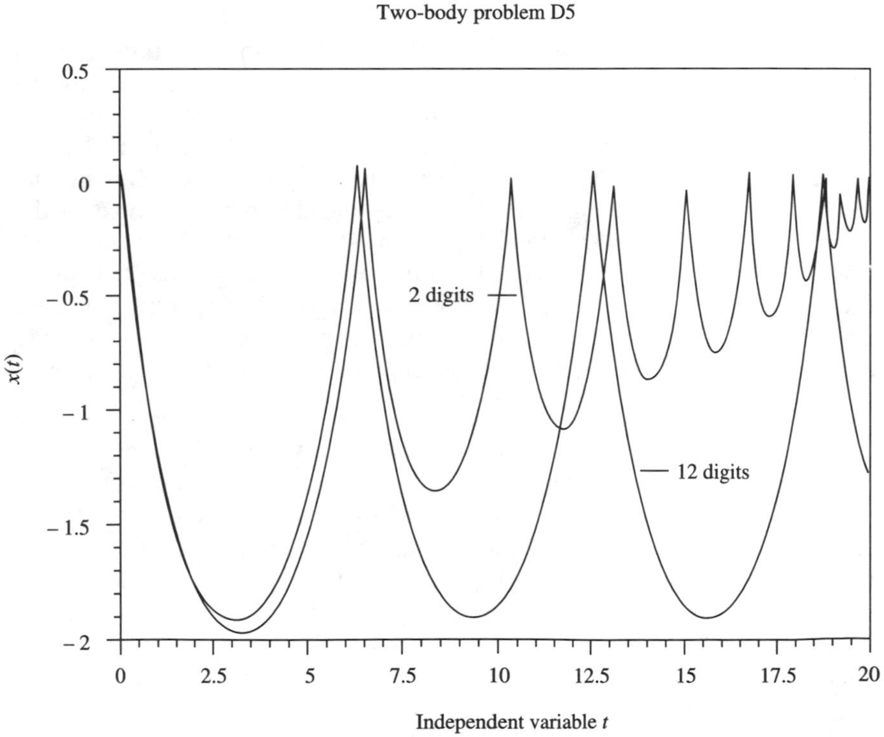

Figure 3.3

a while, the solution is not in even qualitative agreement with the solution obtained by asking for 12 digits! This is not a failure on the part of this excellent code, rather an illustration of the difference between controlling the local error and the true error for a problem that has intervals of moderate instability. Codes are "tuned" so that for routine problems the accuracy achieved is smaller than, or at least comparable to, the specified tolerance on the local error. This example makes it clear that it might be necessary to take the tolerance much smaller than the accuracy desired. This often happens with problems that exhibit some instability and with integrations that are "long" in some sense. The example makes another point. Reducing the relative error tolerance to 10^{-14} results in a solution at $t = 20$ with components having a relative error as large as 6.1×10^{-10}. The tolerance on the relative error is smaller than a unit roundoff in the double precision IEEE arithmetic used for the computations (unfortunately, VODE does not check for this), so the solution is about as accurate as we can hope to get. Single precision in this standard arithmetic is often inadequate for the solution of initial value problems and some computers have even shorter word lengths.

EXERCISE 5. Solve the two-body problem *D*5 for tolerances ranging from very crude to very stringent. Measure the cost of each integration by the number of times the differential equation is evaluated. How does the cost depend on the tolerance? You are likely to see little effect on the cost of reducing the tolerance when only a crude accuracy is desired. In a middle range of tolerances, there is ordinarily a considerable difference in the behavior of fixed and variable order codes—the cost of the latter is much less sensitive to the tolerance. What kind of code did you use? How did the cost depend on the tolerance? You are likely to find that the cost goes up and the accuracy gets worse as you approach limiting precision, but the code might fail in some manner. What did you observe? What was the best accuracy you were able to achieve?

Every writer of a code would like to make it easy to use. The code should have a prologue explaining how to use it. One aspect of making codes easy to use is keeping this prologue terse and to the point. Sometimes this is carried so far that it is not explained just what the error control *is*. It is not really possible to do sound computation without some information supplied by the user from insight or a previous attempt to integrate the problem. It appears that many users do not wish to think about what they want in an answer and some codes cater to this by making assumptions about the problem that may not be true. Some specific examples will be given to make these points concrete. It is not the author's intention to make invidious comparisons. The codes cited are of relatively high quality (if they were not, the author would not have named them), but in attempting to make life easy for users, they have also made it easy to obtain unintended results. MATLAB is a popular computational tool that the author has found very useful for a number of tasks associated with this book. It is intended to make numerical linear algebra easy. Such computations are qualitatively different from the integration of ordinary differential equations. To make the solution of ODEs resemble the solution of linear algebra problems in MATLAB, the solvers ask as little of the user as possible. Indeed, in their simplest use the solvers ODE23.M and ODE45.M do not ask the user for *any* information about the control of error—a default tolerance of 10^{-3} is used in the one case and 10^{-6} in the other. In a more sophisticated use, a tolerance TOL can be specified, but the comments in the M-files do not explain what the error control is. The same is true of the easiest-to-use Runge–Kutta code in the NAG FORTRAN Library, D02BAF. To be sure, it is "strongly recommended" that the user of this code call it with more than one value of TOL and compare the results obtained to assess their accuracy—excellent advice in the circumstances. Examination of the source code in MATLAB shows that at each step the solver estimates the error in each solution component and tests if

$$\max_I |Y(I) - y_I(x)| \leq \text{TOL} \max(\max_I |Y(I)|, 1).$$

Note that the error in a solution component is controlled relative to the size of the largest of the solution components, not the size of the component itself. Also note the presence of the number 1. The NAG library is structured in levels of codes and the level just below D02BAF, called D02BBF, not only explains what the error control is, but offers several possibilities. As it happens, the one used in the higher level code D02BAF is the same as that used in MATLAB. One problem with these easy-to-use codes is that if one component of the solution happens to be much smaller than another, the error control on this component will be very lax and the user has no inkling that this is the case. Another problem is the appearance of an absolute number unknown to the user. Unfortunately, this is by no means unusual. It prevents difficulty with a relative error control, but makes an assumption about the scale of the problem that may not be appropriate. The presence of an absolute number is not bad in itself; the bad thing is that the user does not know about it. Better practice is exemplified in the Runge–Kutta code of the IMSL library called IVPRK. Although there is a 1 used in the same way in the default option in this code, the documentation makes clear the presence of this absolute number and how to alter the value when it is not appropriate to the scale of the problem. As noted already, the second level of code in the NAG library makes the presence of the number known, but to alter it, descent to a third level of code is required.

EXAMPLE 6. To show the kind of anomalies possible when incorrect assumptions are made by the code in order to obtain a simple user interface, suppose we use ODE23.M to integrate

$$y_1' = 0, \qquad y_1(0) \text{ given},$$
$$y_2' = -\sin(x), \qquad y_2(0) = 1.$$

The solution component $y_2(x) = \cos(x)$ and the component y_1 has the constant value of $y_1(0)$. These equations are uncoupled and any reasonable integrator will integrate the first solution component without error. Because of this, the value chosen for $y_1(0)$ should have no effect on the values computed for $y_2(x)$. Using the default tolerance of 10^{-3}, it is found that 43 steps are required to integrate the problem with $y_1(0) = 1$ from $x = 0$ to $x = 10$ and the relative error in $y_2(10)$ is about 1.9×10^{-6}. In contrast, if $y_1(0) = 1000$, only 6 steps are required and the relative error in $y_2(10)$ is about 1.3×10^{-2}. What is happening in this second integration is that because the first component is bigger, the control on the error of the second component is relaxed. A different way to see the effect of the error control is to suppose that the whole problem is scaled down by a factor of 10^{-3} to become

$$y_1' = 0, \qquad y_1(0) = 10^{-3},$$
$$y_2' = -10^{-3}\sin(x), \qquad y_2(0) = 10^{-3}.$$

The solution components of this problem are 10^{-3} and $10^{-3}\cos(x)$, so if we undo the scaling by multiplication with 10^3, we get the same solution as in the first problem. This is not the case numerically. Again only 6 steps are required for solution of the scaled problem and the relative error in $\cos(10)$ (after undoing the scaling) is about 1.3×10^{-2}. Because of the embedded absolute constant of 1 in the error control, scaling the solution components down means that the error control becomes lax on both components and when the scaling is undone, we do not have an accurate result. In passing it might be mentioned that the integrators in MATLAB are examples of codes that integrate in only one direction. To integrate these equations from $x = 0$ to, say, $x = -10$ with ODE23.M, it would be necessary to change the independent variable analytically.

Let us look more deeply into these matters in the context of another code, DOPRI5. This is a code of relatively high quality and we cite it because the imprecision in the specification of its error control has a different origin—it is a research code intended to exemplify aspects of an implementation rather than serve as a production code. The prologue to the code does not explain just what the error control is. It appears to a user to have a pure relative error control with a scalar tolerance EPS. As we have seen, there are difficulties with pure relative error control that any reasonable code must deal with somehow. One difficulty is that impossible accuracy requests cannot be permitted. The code deals with this by quietly changing the input value of EPS to max(EPS, 7U). If EPS is actually increased, the user has presented the code with an impossible task. Because the code does not report that it is not attempting to solve the problem specified, on a successful integration the user has no reason to think that the desired accuracy was not achieved. This is obviously bad practice. An estimate of the true error obtained by reducing the tolerance and integrating again will certainly be incorrect if the tolerance is so small that reducing the value input does not reduce the tolerance actually attempted by the code.

In DOPRI5 the estimated local error is, in our terminology, taken relative to

$$\mathrm{SIZE}(I) = \max\left(10^{-5}, |\mathrm{YB}(I)|, |\mathrm{YE}(I)|, \frac{2U}{\mathrm{EPS}}\right).$$

Here YB(I) is the approximation to the solution at the beginning of the current step and YE(I) is the approximation at the end. It is not so easy to understand completely this definition of SIZE(I). If it were not for the last term, the definition would correspond to relative error with a threshold value of 10^{-5}. The arbitrary choice of a threshold value avoids difficulties with solution components vanishing in a way that makes no demands on the user. It is the analog of the 1 in the codes cited earlier. Often 10^{-5} is not a bad choice for non-stiff problems because the solution components are scaled so that their maximum size is about 1. That this internal choice of an absolute number for all solution components may be inappropriate for a given problem needs no further comment. The role of the last term in SIZE is more obscure. To see what it does, suppose that the solution components have magnitude no greater than 10^{-5}. As the code is presented in Hairer et al. [1987], U is 5×10^{-8}. With this value of U, for all EPS $\leq 10^{-2}$, the term $2U$/EPS is at least as big as 10^{-5} and SIZE(I) $=$ $2U$/EPS. When SIZE(I) is multiplied by EPS in the test on the local error, the divisor of EPS cancels out and the test becomes an absolute error control with fixed tolerance $2U = 10^{-7}$. The mathematical problem is independent of scale and so are almost all numerical methods (provided, of course, that the exponent range of the computer is not exceeded). The apparently pure relative error control in this code is independent of scale, too. Despite this, rescaling the problem can affect profoundly the numerical solution by this code. For example, if a problem is scaled so that all solution components are smaller than, say, 10^{-6}, the code will not attempt to compute any correct digits, regardless of the relative accuracy EPS ($\leq 10^{-2}$) that is requested.

§4 Storage Management and User Interface

A differential equation solver needs some working storage. The amount required depends on the number of equations N and the numerical method implemented. In the case of codes intended for non-stiff problems, $5N$ to $20N$ words of storage is representative. Only the largest problems present difficulties on a modern computer. The situation is different when solving stiff equations. Unless attention is paid to the structure of the Jacobian, the storage required grows like N^2 and this represents a serious expense for large systems, indeed, restricts the problems that can be solved. For this reason it is important to account for structure when solving large systems, complicating greatly the management of storage. A related matter is that a number of quantities must be retained between calls to the solver. Now that the SAVE capability is available in the FORTRAN language, this is easy to arrange.

As far as the user is concerned, the easiest way to handle storage is to leave the matter up to system routines. The PORT library (Fox et al. [1978]) is an early example. The library has a storage area that is managed dynamically to provide working storage. Generally this is very convenient. At most the user is bothered with a call to release storage. It is necessary that the system have available a substantial block of storage, but this is hardly unusual for system routines! More troublesome are problems for which there is not enough storage because the user must then get involved in determining how much storage really is needed. This is awkward for the documentation because of a natural desire not to bother users with matters that often do not concern them at all. A fundamental difficulty is that the system has to be designed for this way of managing working storage, and for stand-alone differential equation solvers this kind of support will not be available.

The simplest and most portable scheme for providing storage is to use absolute dimensions for arrays, the dimensions being chosen to accommodate the maximum number of equations allowed. The obvious difficulty is the value of this maximum. In the case of teaching packages, a limit as small as 1–8 is quite reasonable because the user types in each equation and the solution is immediately computed and plotted on a visual display unit. In scientific computation the maximum must be fairly large so as to accommodate "most" problems, but not so large that a considerable amount of storage is wasted when a small problem is solved. A value of 50 is representative of codes intended for non-stiff problems. Much smaller values are seen, e.g., the important research code DIFEX1 (Deuflhard [1983]) has 13 as the maximum number. Another difficulty is that the source code must be altered when it is necessary to solve a larger system than is provided for. This is not practical for libraries and anyway, the scheme finds no favor in libraries because it is so restrictive and does not make optimal use of storage. The author has altered source code for this reason many times. This has not always been easy because the internal documentation of the code has not always made clear which dimensions have to be changed nor the relationships between the dimensions. Advances in the FORTRAN language and modern programming technique make this both clear and easy. All one need do is specify the maximum number of equations in a PARAMETER statement and then dimension all the requisite arrays in terms of this parameter. Coded in this fashion, changing the maximum number of equations permitted is trivial.

Production-grade codes adapt the storage to the number of equations N in the system to be solved. This is not entirely straightforward. One difficulty arises in the fact that for a code to be readable, suitable names must be chosen for the quantities used. For example, a code might form a vector ERROR(*) such that ERROR(I) is an estimate of the local error in the Ith solution component. This vector would be in the call list of the subroutine and dimensioned either as having N components, with N an argument in the call list, or by an assumed size. The trouble with this approach is that in the typical code, it results in a rather long list of arguments. Users do not like such call lists, even when all they need do is allocate storage in the main program for the variables. A way out is to work with several levels of code. The code that the user sees has an array in the call list for working space, let us call it WORK(*). Instructions are given as to the length of this array in terms of the number of equations N. This code has as its principal task breaking up the work space into suitable segments for a call to a lower level code with a long call list. If, for example, the vector ERROR(*) in the lower level code is to occupy N locations starting at a pointer PTR, the lower level code is called with WORK(PTR) as the argument corresponding to ERROR. In this way the user never sees the long call list that is required for a readable code. As an example, DDRIV2 calls a lower level routine DDSTP with 34 (!) of the arguments being segments of work arrays. Robust software will test input data for legal values and internal consistency, and some checks of this kind are appropriate at the top level. One such check is on the length of WORK(*). It has been found helpful to require the user to supply the length LWORK of the array provided and check on entry that enough storage has been supplied. An annoyance with work space is dealing with the different data types. Of course it is possible to have separate work spaces for REAL, INTEGER, COMPLEX, and LOGICAL variables and some codes do provide storage areas for several types. Doing this involves the user of a code in technical issues that should not concern her, so a more popular approach is to provide only a REAL array and convert types as necessary in the interface routine. For example, a LOGICAL variable in the solver itself might be retained in WORK(*) as +1.0 for .TRUE. and −1.0 for .FALSE.. Some issues of this kind will disappear as FORTRAN 90 comes into wide use.

As we have seen in the examples, some problems involving differential equations are much more complicated to solve than others. It is hard to provide the capabilities for solving difficult problems without making the code appear formidable. This is not strictly a division between expert and novice users; an expert wants to solve easy problems easily just like anyone else. The user interface and its documentation are extremely important. When there are several codes available that purport to solve the problem at hand, it is only natural to select the one that appears to be simplest to use. Users should be skeptical about the simplest interface; it may be simple because it makes assumptions about the problem that might not be correct, or does not monitor the integration and inform the user about the problem and its solution, or . . . . Correspondingly, the person designing an interface must provide codes that are as easy as possible to use while being sufficiently powerful to solve most problems. A matter that the author feels does not get enough attention is providing *templates*. By this is meant sample programs for typical tasks. In the author's experience, when a template is available, most people simply modify it to the degree necessary to solve the problem at hand. This is by no means restricted to the inexperienced user—it is the easiest way for anyone to get going with a new code.

Two approaches to the interface problem are seen. In one approach a single code makes available all the capabilities. To keep the call list from scaring away the inexperienced, there are only a few input parameters for specifying the computational problem. These parameters can have several (perhaps many, even a great many!) possible values, which make it possible to exploit fully the capabilities available. To make it easy to solve an easy problem, the parameters have default or nominal values. With careful construction of the documentation, the person with an easy problem will have no difficulty finding out what must be done to solve the problem. It is extremely important that such a person not have to read about all the possibilities available because this is likely to lead to despair. On the other hand, a person who needs a particular capability will be sufficiently motivated to wade through the documentation to find out how to get the capability. The approach has been successful, but it is difficult to present effectively codes that have many capabilities. Unless one is very careful in the choice of input parameters, the interface can become complicated and lead to mistakes of usage. The code DIVPAG in the IMSL library mentioned earlier in connection with other standard forms of the equation makes the point. There are three possibilities for the form of the equation. There are four possibilities for the manner in which matrices are stored. There are two different kinds of numerical methods implemented. The methods involve solving nonlinear algebraic equations at each step and there are four possibilities for doing this. Clearly, it is necessary to know quite a lot about the numerical solution of ODEs in order to select properly among all these possibilities; this is a formidable user interface. The issue here is not the quality of the code itself—it is a powerful and effective code—rather one of designing software that is both capable and easy to use.

The other main approach is to have separate codes for separate kinds of problems. These are often structured in levels. For example, an interval-oriented solver often relies upon a step-oriented code that can be used independently by a person needing more control over the integration than the interval-oriented solver permits. As a case in point, the suite of Adams–Bashforth–Moulton PECE codes ODE/STEP, INTRP consists of an interval-oriented solver ODE and a step-oriented solver STEP with associated interpolation routine INTRP. Monitoring of the usage of these solvers at the Sandia National Laboratories, a large scientific laboratory, showed that practically always the interval-oriented code ODE was used. This code does have an intermediate-output mode, which helps explain why a step-oriented code was not needed more often. Investigation revealed that most people using STEP did so because ODE had only scalar absolute and relative error tolerances and a vector control is needed to deal with solution components of greatly different size. For this reason, when Shampine and Watts [1980] later designed interfaces for the DEPAC package of codes, they preferred a scalar relative error tolerance and a vector of absolute error tolerances. This design suffices for the vast majority of initial value problems. Perhaps a better example of the approach is provided by the codes of ODEPACK (Hindmarsh [1983]) that are principally intended for stiff problems. Because storage is a critical issue in this context, there are separate codes for the solution of comparatively small systems which permit Jacobians to be handled as full or banded matrices and for the solution of large systems that treat Jacobians as general sparse matrices. Only the user who needs to solve very large systems has to learn what is involved in working with sparse matrices. There are some objections to the approach of providing several codes. It can be confusing to users and it can represent a lot of code. Perhaps the most troublesome matter is one of maintenance. When it is necessary or desirable to make a change in a code, this may imply changes in a number of codes in the suite. Some care is required to be sure that the changes get made everywhere.

A variation on these themes is seen in RKSUITE. In this approach setting up the integration is separate from the integration itself. Besides being a logical distinction, this reduces greatly the complexity of the call lists. The suite has separate routines for interval-oriented and step-oriented computation. Another device employed in this suite is the use of auxiliary subroutines. For example, the step-oriented code does not pass back through the call list all the information that might be of interest to a user. A user who wants more detail about the solution process calls an auxiliary routine that extracts from a work array these details and presents them. In this way, only the person who wants more information has to look into how to obtain it.

Another issue pertinent to packages and libraries is the presence of several codes that implement different numerical methods. It is obviously valuable to make it as easy as possible to change from one method to another so that a user can determine the most appropriate one for the problem at hand. An extreme example is Gear's DIFSUB that implements both Adams–Moulton formulas and backward differentiation formulas, BDFs, in the same program. Changing from one method to the other is accomplished by setting a parameter indicating the method and taking account of the fact that specification of a stiff problem involves more than specification of a non-stiff problem. The package DEPAC includes an Adams code and a Runge–Kutta code (and a BDF code). The Adams code comes from the ODE/STEP, INTRP suite and the Runge–Kutta code comes from RKF45. Although the original codes were written with the same philosophy for use at the same laboratory, they were written at different times and involved different authors. As a consequence, they differ in their use in a number of ways, e.g., some corresponding arguments in the call lists do not have quite the same meaning in the two codes. As individual codes the designs were quite successful, but in conjunction the designs led to errors when people changed from one code to the other. This was one reason for the new design of DEPAC. Early versions of the ODE chapter of the IMSL library of mathematical software consisted of codes taken from the literature. Though modified in some respects to conform to a library standard, they were not altered so as to make them even similar in appearance to users. For example, in some cases it was not even possible to get equivalent error controls with the different solvers.

This situation is better described as a collection of codes rather than a package. It was cleaned up in a later version of the library.

EXERCISE 6. One of the popular kinds of methods for solving differential equations that we take up in Chapter 4 are explicit Runge–Kutta methods. When advancing the integration of $\mathbf{y}' = \mathbf{F}(x, \mathbf{y})$ from an approximation $\mathbf{y}_j$ of $\mathbf{y}(x_j)$ to an approximation $\mathbf{y}_{j+1}$ of $\mathbf{y}(x_j + h)$, such a method proceeds as follows:

Define $\mathbf{X}_0$, $\mathbf{Y}_0$ and evaluate $\mathbf{F}_0$ according to

$$X_0 = x_j, \qquad \mathbf{Y}_0 = \mathbf{y}_j, \qquad \mathbf{F}_0 = \mathbf{F}(X_0, \mathbf{Y}_0).$$

Then, for $k = 1, \ldots, s - 1$, define X_k, $\mathbf{Y}_k$ and evaluate $\mathbf{F}_k$ according to

$$X_k = x_j + \alpha_k h, \qquad \mathbf{Y}_k = \mathbf{y}_j + h \sum_{m=0}^{k-1} \beta_{k,m} \mathbf{F}_m, \qquad \mathbf{F}_k = \mathbf{F}(X_k, \mathbf{Y}_k).$$

The solution at the end of the step is

$$\mathbf{y}_{j+1} = \mathbf{y}_j + h \sum_{k=0}^{s-1} \gamma_k \mathbf{F}_k$$

The coefficients α_k, $\beta_{k,m}$, and γ_k define the formula. The $\mathbf{F}_k$ are called stages, so this formula is described as one of s stages.

Because the step may be rejected, the vector $\mathbf{y}_j$ must be retained. When a step is rejected, it is followed by another attempt with a smaller step size. By retaining $\mathbf{F}_0$, an evaluation of $\mathbf{F}$ can be saved on a second attempt. The usual storage management scheme forms in an auxiliary vector $\mathbf{z}$ the intermediate value $\mathbf{Y}_k$, evaluates $\mathbf{F}_k$, and stores $\mathbf{F}_k$ in a vector $\mathbf{v}_k$ for $k = 1, \ldots, s - 1$. After all the stages have been formed, the solution vector $\mathbf{y}_{j+1}$ is formed in $\mathbf{z}$. There are some other computations in the codes involving an estimate of the error of the step, but we do not go into this here. This scheme requires $s + 2$ vectors of length N. High-order Runge–Kutta formulas involve a considerable number of stages, e.g., the important (7, 8) pair of Fehlberg [1968] involves 13 stages. As we have seen in connection with solving partial differential equations by semi-discretization, the number of equations N can be large. For these reasons the codes employ a number of techniques to reduce the storage required. Two simple techniques will be taken up here; more details and a more elaborate technique are found in Shampine [1979a].

Often high-order Runge–Kutta formulas have stages that are not used in the latter parts of the computation. More precisely, there may be a stage $\mathbf{F}_m$ such that for all $k \geq M$, $\beta_{k,m} = 0$ and $\gamma_m = 0$. As soon as this stage is no longer needed, the vector used to hold it is freed up for holding later stages. A useful reduction in storage can be achieved in this way with little complication to the code. A number of authors have derived formulas that exploit this technique to minimize storage when solving partial differential equations, e.g., van der Houwen [1972]. Another vector can be saved when, as often happens, for some m, $\gamma_m = 0$. For such formulas the last stage $\mathbf{F}_{s-1}$ can be written over the stage $\mathbf{F}_m$ because it is not used in the subsequent formation of $\mathbf{y}_{j+1}$.

The 13 stage (7, 8) pair of Fehlberg [1972] has $\beta_{1,m} = 0$ for $m = 3, \ldots, 12$, $\beta_{2,m} = 0$ for $m = 5, \ldots, 12$, $\beta_{11,3} = 0$, $\beta_{11,4} = 0$, $\beta_{11,10} = 0$, $\beta_{12,10} = 0$. This is a pair of formulas and here we consider only the one with $\gamma_m = 0$ for $m = 0, \ldots, 4$ and $\gamma_{10} = 0$. Apply the techniques described to reduce the storage from the $15N$ of a straightforward implementation.

§5 Software

This text is not tied to any particular method or program; quite the contrary. It is necessary to gain experience with a variety of methods and their implementations to understand the subject. Some methods and some implementations are better than others, so references to surveys of quality software for the initial value problem are provided here. This is followed by directions for obtaining some quality software that is free or inexpensive. The author has deliberately chosen to employ a good many codes for the computations described in this book. Most can be obtained from the sources described here.

The software surveys that follow are in order of their date of publication. Naturally there is considerable overlap between the surveys and some evolution is seen as new codes were developed. The authors have different viewpoints, but all have had great experience with writing and applying software for the initial value problem.

Shampine, L. F., Watts, H. A., and Davenport, S. M., Solving Nonstiff Ordinary Differential Equations—the State of the Art, SIAM Review, 18 (1976) pp. 376–411.

Watts, H. A., Survey of Numerical Methods for Ordinary Differential Equations, pp. 127–158 in Erisman, A. M., Neves, K. W., and Dwarakanath, M. H., eds., Electric Power Problems: the Mathematical Challenge, SIAM, Philadelphia, 1980.

Gupta, G. K., Sacks-Davis, R., and Tischer, P. E., A Review of Recent Developments in Solving ODEs, ACM Computing Surveys, 17 (1985) pp. 5–47.

Section 4.1 (pp. 149–153) by R. C. Aiken and section 4.2 (pp. 153–166) by C. W. Gear, in Aiken, R. C., ed., Stiff Computation, Oxford University Press, Oxford, 1985.

Byrne, G. D., and Hindmarsh, A. C., Stiff ODE Solvers: a Review of Current and Coming Attractions, J. Comp. Phys., 70 (1987) pp. 1–62.

A particularly convenient way to acquire quality software is from netlib. Sending the message of one line, "send index", to netlib@ornl.gov results in a reply that we excerpt here:

"= = = = = How to use netlib = = = = =

This file is the reply you'll get to:
mail netlib@ornl.gov
send index
Here are examples of the various kinds of requests. . . .

The Internet address "netlib@ornl.gov" refers to a gateway machine, at Oak Ridge National Laboratory in Oak Ridge, Tennessee. This address should be understood on all the major networks.

For access from Europe, try the duplicate collection in Oslo:
Internet: netlib@nac.no
EARN/BITNET: netlib%nac.no@norunix.bitnet
X.400: s=netlib; o=nac; c=no;
EUNET/uucp: nac!netlib
For the Pacific, try netlib@draci.cs.uow.edu.au
located at the University of Wollongong, NSW, Australia. . . .

Background about netlib is in Jack J. Dongarra and Eric Grosse, Distribution of Mathematical Software Via Electronic Mail, Comm. ACM (1987) 30,403–407 and in a quarterly column published in the SIAM News and SIGNUM Newsletter. . . ." It is also possible to get information by sending the message of one line, "help", to netlib@ornl.gov. In addition to the other addresses given above for Internet, you can use netlib@research.att.com. Furthermore, there is a uucp address uunet!research!netlib to which you can direct your message.

There are a number of libraries in netlib that contain software for the initial value problem, so some searching may be necessary to locate all that is available. Netlib is probably the most convenient source for the recent code VODE of Brown, Bryne, and Hindmarsh [1989]; it is in the library 'ode'. The same is true of an important family of codes that are variations on Hindmarsh's LSODE. They form the library 'odepack.'

Some quality codes were originally published in source code form in reports or in books. For example, the source code for a widely used Runge–

Kutta code, RKF45, due to Watts and Shampine appears first in a Sandia National Laboratories report (Shampine and Watts [1976a]) and its development is described in an article and a proceedings (Shampine and Watts [1977, 1979]). In slightly modified form the source code appeared in several books, one of which is Forsythe, Malcolm, and Moler [1977]. All the codes in this last book are contained in the library 'fmm' on netlib. RFK45 is also contained in the library 'ode' on netlib. The code was one of the original codes that composed the SLATEC library (Buzbee [1984]). Its interface was modernized to become the code DERKF in a suite called DEPAC (Shampine and Watts [1980]) and this is the version of the code in the present SLATEC library. Similarly, the popular Runge–Kutta code, DVERK, of Hull, Enright, and Jackson first appeared in a University of Toronto report (Hull et al. [1976]) and subsequently in the IMSL library. After modernization of the ODE chapter of this library, the new version of the code was called DIVPRK. DVERK is available in the library 'ode' on netlib. The primary source of the suite of codes ODE/STEP, INTERP based on Adams methods is the monograph Shampine and Gordon [1975]. It was one of the original codes in the SLATEC library and was later modernized to become DEABM in DEPAC and the library. Still later it became the basis of the D02C?? codes in the NAG library. ODE is available in the library 'ode' on netlib. A seminal code, DIFSUB, due to Gear that is based on backward differentiation formulas was published in a journal (Gear [1971a, 1971b]) and also (in slightly different form) in the text Gear [1971c]. The code LSODE cited earlier is a descendant of this code. A number of journals publish programs in source code form and some make the programs available in machine-readable form. For example, the Runge–Kutta code with global error estimation, GERK (Shampine and Watts [1976a, 1976b]), is published in the ACM Transactions on Mathematical Software. The codes published in this journal are available through netlib. Some numerical analysis texts are complemented by software on a diskette. Generally this software is appropriate only for classroom or research use, but the codes that accompany the text of Kahaner, Moler, and Nash [1989] are production-grade codes. Included is a quality code, SDRIV2, for the initial value problem. The double precision version, called DDRIV2, has been used for a number of our examples. The book Schiesser [1991] is devoted to the solution of partial differential equations by semi-discretization (more specifically, the method of lines). A diskette available from the author includes both single and double precision versions of RKF45 and SDRIV2. The book and the software available from the Schiesser provide a great many example solutions of substantial problems.

Two major libraries of scientific software, IMSL and NAG, are in wide use and there has been a proliferation of smaller libraries for microcomputers. These libraries may well be providing public domain codes with minimal alterations to the user interface, error handling, storage management, and setting of machine constants to make the codes fit into the library framework. A typical computer center will have one or more libraries available and typically each library will contain a selection of codes for the initial value problem.

EXERCISE 7. One way to solve the Blasius problem requires the integration of

$$v''' + v\,v'' = 0,$$
$$v(0) = 0, \qquad v'(0) = 0, \qquad v''(0) = 1,$$

to obtain the limit

$$\lim_{z \to \infty} v'(z) = \lambda^2.$$

You are to obtain λ by integrating to $z = 10$. This is to be done with several different solvers. You are to specify a reasonable local error control and moderate tolerance(s) for your integrations. If possible, you should solve the same computational problem with all the codes. The point of this problem is for you to become acquainted with different software interfaces for the initial value problem. You are to report the differences you observe in the way important software issues are handled. In particular, include in your writeup the differences in the way storage is handled, in the form of the subroutine for evaluating the equation, in the design for the task of integrating a problem that requires an answer at only one point, and in the local error control. You are to explain the choices you made in your attempt to solve the same computational problem with all the integrators you used.

EXERCISE 8. Solutions of the predator–prey equations of Volterra,

$$\frac{d}{dt} N_1(t) = aN_1 - bN_1N_2,$$

$$\frac{d}{dt} N_2(t) = -cN_2 + dN_1N_2,$$

satisfy a nonlinear conservation law. The quantities a, b, c, d, here are constants. The law is more easily stated after the equations are simplified a little by introducing the new variables $x(t) = N_1(t)d/c$ and $y(t) = N_2(t)b/a$:

$$\frac{d}{dt} x(t) = a(x - xy),$$

$$\frac{d}{dt} y(t) = -\,c(y - xy).$$

Verify this new form and then verify that for any solution $x(t)$, $y(t)$ of these equations, the function

$$G(t, x, y) = x(t)^{-c}\, y(t)^{-a}\, e^{cx(t) + ay(t)}$$

is constant. H. T. Davis [1962, p. 101 ff.] discusses the model and presents some numerical results when $a = 2$ and $c = 1$ for initial values $x = 1$ and $y = 3$. With these initial values at $t = 0$, integrate the equations with your favorite code and test whether the numerical solution satisfies the conservation law exactly. This will help spot an inaccurate solution, but does little more than provide encouragement. Use the method of reintegration to gain more confidence in your results. If the code you use has a global error assessment, try it.

EXERCISE 9. The numerical solution of two problems with the same analytical solution may be very different and unfortunately, the effect depends on just which numerical method is used. The two-body problem

$$\begin{aligned} \ddot{x}(t) &= -\,x(t)/r \qquad x(0) = 1, \quad \dot{x}(0) = 0 \\ \ddot{y}(t) &= -\,y(t)/r \qquad y(0) = 0, \quad \dot{y}(0) = 1 \end{aligned}$$

where

$$r = (x(t)^2 + y(t)^2)^{3/2}$$

has the analytical solution $x(t) = \cos t$, $y(t) = \sin t$. Obviously this is also the solution of

$$\begin{aligned} \ddot{x}(t) &= -\,\cos t \qquad x(0) = 1, \quad \dot{x}(0) = 0 \\ \ddot{y}(t) &= -\,\sin t \qquad y(0) = 0, \quad \dot{y}(0) = 1. \end{aligned}$$

As a rule, a code based on a Runge–Kutta method will find one of these problems easier than the other. Try this out with your favorite code and monitor the number of times the code evaluates the differential equation. (If the code does not provide this information directly, put an integer variable in COMMON with your subroutine for evaluating the differential equation. Initialize this variable to 0 in the calling program and increment it by 1 in the subroutine for evaluating the equation.).

If, for example, you integrate from 0 to 3π = 3.D0*ACOS(−1.D0) with DIVPRK from the IMSL library and an absolute error tolerance of 10^{-9}, you will find that the code requires 912 evaluations of **F** for the nonlinear form and 320 for the quadrature. RKF45 does not permit pure absolute error. When used with this absolute error tolerance and the smallest relative error tolerance it permits, the code requires 1033 evaluations for the nonlinear form and 595 for the quadrature. The D02BAF code in the NAG library does not explain what kind of error control is implemented. When its tolerance is set to 10^{-9}, it requires 3089 evaluations for the nonlinear problem and 2349 for the quadrature.

These three codes all implement Runge–Kutta formulas. Evidently the form of the equation matters for such methods. The three formulas are all different and the magnitude of the effect is seen to be different. As it happens, the formulas are all of different order and at this tolerance, the higher the order, the more efficient the code for these problems.

EXERCISE 10. The author did the fixed step size computations of Exercises 17 and 18 of Chapter 1 with both MDEP and PHASER. The two packages implement the same fourth-order Runge–Kutta formula, but the user interfaces are rather different. Recall that the differential equation

$$\frac{dy}{d\lambda} = \frac{y^2 - 3y - 2\lambda + 7.75}{3y^2 - y(2\lambda + 2.5) + 3\lambda - 6}$$

is to be integrated from $\lambda = 2$ to $\lambda = 0$. As pointed out earlier, PHASER requires equations to be in autonomous form. Moreover, it does not permit integration from right to left. Try preparing this problem for integration with PHASER. No preparation of this kind is necessary when using MDEP. If PHASER is handy, carry out the integration and compare your results with those presented in Chapter 1. You will note that in order to define the equations in this package, you have to assign (fixed) symbolic names to all the numbers appearing in the equations and use fixed names for the solution components. Furthermore, if you should make mistakes in defining the equations (like the author did), you cannot edit them out; you have to start all over. Definition of an equation in MDEP is more straightforward. Numbers are entered in a natural way, the default name for the solution component can be changed if you wish, and you can make corrections to the equation at any time.

4

Basic Methods

A remarkable number of discrete variable methods have been proposed for the solution of the initial value problem

$$\mathbf{y}' = \mathbf{F}(x, \mathbf{y}), \qquad a \leq x \leq b, \qquad \mathbf{y}(a) = \mathbf{A},$$

but only a few are in wide use. One-step methods use only information gathered in the span of a single step from x_j to $x_j + h$. The very natural approach of expanding the solution in a Taylor series about x_j falls into this class. Though not in general use, the approach is valuable for understanding Runge–Kutta methods, the widely used one-step methods. Having reached a point x_j in an integration, there are (usually) previously computed solution values that can be exploited for the computation of an approximate solution at $x_j + h$. This observation is the basis of the linear multistep methods, LMMs. There are only two families of LMMs that are at all popular. Variations on the Adams formulas are one and the backward differentiation formulas, BDFs, are the other. The analysis and the use of these two kinds of formulas differ so much that it is best to consider them separately. All the methods mentioned include formulas that define the approximate solution implicitly. Using our basic assumption that **F** satisfies a Lipschitz condition, it will be shown that the usual implicit methods are well defined for all sufficiently small step sizes h. It is more complicated to evaluate an implicit formula than an explicit one, but it need not be much more expensive and implicit formulas can have advantages that make them worth the trouble.

All the popular methods can be derived in natural ways that provide insight about their strengths and weaknesses. Some feeling for this is much more important than the details that we must work out here for further investigation of the methods. One way to derive methods starts from the integrated form of the differential equation

$$\mathbf{y}(x_j + h) = \mathbf{y}(x_j) + \int_{x_j}^{x_j+h} \mathbf{F}(t, \mathbf{y}(t))\, dt$$

and generalizes what is done in elementary numerical analysis courses for the approximation of

$$\int_{x_j}^{x_j+h} f(t)\,dt,$$

the quadrature problem. Runge–Kutta methods arise from underlying quadrature formulas that use data only from $[x_j, x_{j+1}]$. Adams methods arise from underlying quadrature formulas that use data outside of $[x_j, x_{j+1}]$, specifically approximate solutions computed prior to x_j. These methods can be viewed as arising from numerical integration. Another popular family of methods, the backward differentiation formulas, BDFs, can be viewed in an analogous way as arising from numerical differentiation.

§1 One-Step Methods

There are important differences between methods that use only information gathered at the current mesh point x_j and those that "remember" information obtained earlier in the computation. This section is concerned with the former, called one-step methods, as well as some concepts fundamental to all the methods we take up. One-step methods can be introduced in a natural way by means of Taylor series and this approach may already be familiar to the reader from courses on the theory of ODEs. Though Taylor series methods can be quite practical, they are used in this book only to introduce and understand Runge–Kutta methods. Runge–Kutta methods are derived in families and it is not a simple matter to select the "best" member of a family. Indeed, how to do this has evolved considerably over the years as a better understanding of the issues of quality has been reached. These issues will be investigated in several chapters of this book. Only the issue of accuracy is taken up here and even its discussion is necessarily incomplete in this section.

§1.1 Taylor Series and Runge–Kutta Methods

If the solution $\mathbf{y}(x)$ of the initial value problem

$$\mathbf{y}' = \mathbf{F}(x, \mathbf{y}), \qquad \mathbf{y}(a) = \mathbf{A}, \tag{1.1}$$

for a system of n equations is sufficiently smooth, we can write for each component

$$y_i(x) = y_i(a) + y_i'(a)(x - a) + \frac{y_i''(a)}{2!}(x - a)^2 + \dots .$$

The first derivative in this expansion is immediately available from the differential equation because in general

$$y_i'(x) = F_i(x, y_1(x), y_2(x), \ldots, y_n(x))$$

and in particular, at the initial point

$$\mathbf{y}'(a) = \mathbf{F}(a, \mathbf{A}).$$

Differentiation of the equation for the first derivative leads by the chain rule to

$$\begin{aligned} y_i''(x) &= \frac{\partial F_i}{\partial x}(x, \mathbf{y}(x)) + \sum_{j=1}^{n} \frac{\partial F_i}{\partial y_j}(x, \mathbf{y}(x)) \frac{dy_j}{dx} \\ &= \frac{\partial F_i}{\partial x}(x, \mathbf{y}(x)) + \sum_{j=1}^{n} \frac{\partial F_i}{\partial y_j}(x, \mathbf{y}(x)) F_j(x, \mathbf{y}(x)). \end{aligned}$$

In matrix notation the value at a is

$$\mathbf{y}''(a) = \frac{\partial \mathbf{F}}{\partial x}(a, \mathbf{A}) + \frac{\partial \mathbf{F}}{\partial \mathbf{y}}(a, \mathbf{A})\, \mathbf{F}(a, \mathbf{A}),$$

where the $n \times n$ Jacobian matrix

$$\frac{\partial \mathbf{F}}{\partial \mathbf{y}} = \left(\frac{\partial F_i}{\partial y_j} \right).$$

In principle the function $\mathbf{F}$ can be differentiated repeatedly to determine the derivatives of $\mathbf{y}(x)$. The computation of higher order terms becomes complicated unless we go about it in a systematic way. The calculations are simplified by working with equations in autonomous form, $\mathbf{y}' = \mathbf{F}(\mathbf{y})$, because this eliminates the special role of the independent variable x. There is no loss of generality in supposing the equation has this form because we have seen in Chapters 1 and 3 two ways to convert general equations to autonomous form.

The notation can be simplified considerably. In this context let us denote component i of the vector $\mathbf{F}$ by F^i and denote a partial derivative with respect to component j of the solution, y^j, by a comma and a subscript. For example,

$$\text{usual notation: } \frac{\partial F_i}{\partial y_j}, \qquad \text{compact notation: } F^i_{,j}.$$

If this example is further differentiated with respect to component k of the solution, the result is denoted by

$$\text{usual notation: } \frac{\partial^2 F_i}{\partial y_j \partial y_k}, \qquad \text{compact notation: } F^i_{,j,k}.$$

In this notation, component i of the differential equation is

$$(\mathbf{y}')^i = F^i.$$

For later use we define the vector $\mathbf{D}_1^1 = \mathbf{F}$ so that

$$\mathbf{y}' = \mathbf{D}_1^1.$$

Differentiating the autonomous equation we obtain

$$(\mathbf{y}'')^i = \sum_{j=1}^{n} \frac{\partial F^i}{\partial y^j} \frac{dy^j}{dx} = \sum_{j=1}^{n} F^i_{,j} F^j.$$

The notation is simplified further by the convention that if an index is repeated in an expression, the expression is to be summed over all values of the index, here the number n of components of the differential equation. Thus we write simply

$$(\mathbf{y}'')^i = F^i_{,j} F^j = (\mathbf{D}_1^2)^i,$$

where we define another vector $\mathbf{D}_1^2$. The next derivative is

$$(\mathbf{y}^{(3)})^i = \left(\frac{d}{dx}\mathbf{y}''\right)^i = \sum_{k=1}^{n} \left(\frac{\partial F^i_{,j}}{\partial y^k}\frac{dy^k}{dx} F^j + F^i_{,j}\frac{\partial F^j}{\partial y^k}\frac{dy^k}{dx}\right)$$
$$= F^i_{,j,k} F^k F^j + F^i_{,j} F^j_{,k} F^k.$$

Correspondingly we define a pair of vectors by

$$(\mathbf{D}_1^3)^i = F^i_{,j,k} F^j F^k \quad \text{and} \quad (\mathbf{D}_2^3)^i = F^i_{,j} F^j_{,k} F^k$$

so that

$$\mathbf{y}^{(3)} = \mathbf{D}_1^3 + \mathbf{D}_2^3.$$

Expressions of the kind arising here can be manipulated by interchanging the order of products, interchanging the order of differentiation, and relabeling indices. Examples of such manipulations are

$$\begin{aligned} F^i_{,j,k} F^j F^k &= F^i_{,j,k} F^k F^j \\ &= F^i_{,k,j} F^k F^j \\ &= F^i_{,k,m} F^k F^m. \end{aligned}$$

EXERCISE 1. Show that

$$\mathbf{y}^{(4)} = \mathbf{D}_1^4 + 3\,\mathbf{D}_2^4 + \mathbf{D}_3^4 + \mathbf{D}_4^4,$$

where

$$(\mathbf{D}_1^4)^i = F^i_{,j,k,m} F^j F^k F^m, \qquad (\mathbf{D}_2^4)^i = F^i_{,j,k} F^j_{,m} F^k F^m,$$
$$(\mathbf{D}_3^4)^i = F^i_{,j} F^j_{,k,m} F^m F^k, \qquad (\mathbf{D}_4^4)^i = F^i_{,j} F^j_{,k} F^k_{,m} F^m.$$

Working out these derivatives by hand quickly ceases to be fun. Fortunately, there are now available symbol manipulation languages to carry out the drudgery for us. The $\mathbf{D}_j^i$ that appear in the Taylor series expansion are called *elementary differentials*. We do not need explicit expressions for them, just a recognition that they are sums of products of partial derivatives of components of **F**. Our interest in Taylor series is mostly theoretical, but we should mention that in their practical use the expansions are derived in a less direct way that is much more efficient, see e.g., Corliss and Chang [1982]. The key idea is to evaluate the derivatives in the course of evaluating the function rather than work out the expressions for the derivatives and evaluate them separately. Proceeding in this manner avoids repeated evaluation of common expressions.

A *Taylor series method* is obtained by truncating the Taylor series expansion of the solution. Suppose that $\mathbf{y}_j \approx \mathbf{y}(x_j)$ and we want to approximate the local solution, the solution $\mathbf{u}(x)$ of

$$\mathbf{u}' = \mathbf{F}(x, \mathbf{u}), \qquad \mathbf{u}(x_j) = \mathbf{y}_j,$$

at the next mesh point x_{j+1}. It is convenient to introduce here the step size

$$h_j = x_{j+1} - x_j,$$

or, as we often write for convenience, h. To understand the error due to truncation of the series, we write the Taylor expansion with remainder. For example, the Taylor series expansion of $\mathbf{u}(x)$ consisting of two terms with remainder is

$$\mathbf{u}(x_j + h) = \mathbf{u}(x_j) + h\,\mathbf{u}'(x_j) + \frac{h^2}{2}\,\mathbf{u}''(*).$$

In obtaining this by applying the usual expansion to the components of $\mathbf{u}(x)$ separately, the value of the independent variable in $\mathbf{u}''$ depends on the component. So as not to make the notation any more complicated, the independent variable is written as an asterisk in this and similar situations. The (forward) Euler method is based on the first two terms of the Taylor series expansion:

$$\mathbf{y}_{j+1} = \mathbf{u}(x_j) + h\,\mathbf{u}'(x_j) = \mathbf{y}_j + h\,\mathbf{F}(x_j, \mathbf{y}_j).$$

Evidently the error made in approximating $\mathbf{u}(x_{j+1})$ by $\mathbf{y}_{j+1}$ is $\mathcal{O}(h^2)$. Although it is not very accurate, the method is an option in a number of packages for teaching ODEs, e.g., MacMath, MDEP, and PHASER, no doubt because it is easy to visualize what the method does. It also plays an important role in starting an integration with the Adams methods that we take up in another section. The second order Taylor series method is based on a three-term Taylor expansion,

$$\mathbf{u}(x_j + h) = \mathbf{u}(x_j) + h\,\mathbf{u}'(x_j) + \frac{h^2}{2!}\mathbf{u}''(x_j) + \frac{h^3}{3!}\mathbf{u}^{(3)}(*),$$

and so defines

$$\mathbf{y}_{j+1} = \mathbf{y}_j + h\,\mathbf{F}(x_j, \mathbf{y}_j) + \frac{h^2}{2}\left[\frac{\partial \mathbf{F}}{\partial x}(x_j, \mathbf{y}_j) + \frac{\partial \mathbf{F}}{\partial \mathbf{y}}(x_j, \mathbf{y}_j)\,\mathbf{F}(x_j, \mathbf{y}_j)\right].$$

We see that the local error of Euler's method is $\mathcal{O}(h^2)$ and that of the second-order Taylor series method is $\mathcal{O}(h^3)$.

There are two ways to get a more accurate approximation to the local solution $\mathbf{u}(x)$. One is to use more terms in the Taylor expansion, the other is to use a smaller step size h. The expressions for the higher order derivatives are generally quite complex. For some interesting special problems the expressions are not so complicated and Taylor series of rather high order have been used. Nowadays there are programs available that make the use of high order Taylor series convenient for relatively general problems. For instance, that of Corliss and Chang [1982] might well approximate $\mathbf{y}$ by a truncated Taylor series with as many as *thirty* terms. Although Taylor series methods can be quite effective, we do not pursue them further in this book.

An *explicit one-step method* is a recipe for approximating $\mathbf{u}(x_{j+1})$ given only $\mathbf{y}_j$, h, and the ability to evaluate $\mathbf{F}$ and perhaps derivatives of $\mathbf{F}$:

$$\mathbf{y}_{j+1} = \mathbf{y}_j + h\,\boldsymbol{\Phi}(x_j, \mathbf{y}_j, \mathbf{F}, h).$$

Euler's method is an example for which the *increment function* $\boldsymbol{\Phi}(x, \mathbf{y}, \mathbf{F}, h)$ is

$$\boldsymbol{\Phi} = \mathbf{F}(x, \mathbf{y}).$$

The second order Taylor series method is an example with

$$\boldsymbol{\Phi} = \mathbf{F}(x, \mathbf{y}) + \frac{h}{2}\left[\frac{\partial \mathbf{F}}{\partial x}(x, \mathbf{y}) + \frac{\partial \mathbf{F}}{\partial \mathbf{y}}(x, \mathbf{y})\,\mathbf{F}(x, \mathbf{y})\right].$$

A one-step method is said to be of order μ for a given function $\mathbf{F}$ if there is a constant C such that

$$\|\mathbf{u}(x_j + h) - \mathbf{y}_{j+1}\| \le C\,h^{\mu+1}$$

or, stated differently,

$$\begin{aligned}\mathbf{u}(x_j + h) &= \mathbf{y}_{j+1} + \mathcal{O}(h^{\mu+1})\\ &= \mathbf{y}_j + h\,\boldsymbol{\Phi}(x_j, \mathbf{y}_j, \mathbf{F}, h) + \mathcal{O}(h^{\mu+1}).\end{aligned}$$

It is usual to speak of a method as being of order μ if it is of this order for all sufficiently smooth $\mathbf{F}$. Keep in mind that if the problem at hand is not sufficiently smooth, the accuracy of the formula will be of order lower than μ. Also, for a given $\mathbf{F}$, it is possible that a formula be of order higher than its usual order of μ—an example is given in Exercise 9.

Let us now for a moment focus on the recipe for the numerical solution $\mathbf{y}_{j+1}$. The local solution does not satisfy this recipe exactly. When it is substituted into the recipe, the discrepancy $h\tau_j$ is the *local error*:

$$\mathbf{u}(x_j + h) = \mathbf{u}(x_j) + h\,\mathbf{\Phi}(x_j, \mathbf{u}(x_j), \mathbf{F}, h) + h\tau_j.$$

(By definition $\mathbf{u}(x_j) = \mathbf{y}_j$.) The solution $\mathbf{y}(x)$ of the problem (1.1) can play the role of a local solution here. Because of the importance of this special case, local error has a special name then, namely, *local truncation error* or sometimes just *truncation error*. There is a useful way to view local truncation error. If the current solution approximation $\mathbf{y}_j$ were equal to the true solution evaluated at x_j, that is, if it were equal to $\mathbf{y}(x_j)$, the formula would produce

$$\mathbf{y}_{j+1} = \mathbf{y}(x_j) + h_j\,\mathbf{\Phi}(x_j, \mathbf{y}(x_j), \mathbf{F}, h_j)$$

as the approximation to $\mathbf{y}(x_{j+1})$. The error of this step is $\mathbf{y}(x_j + h_j) - \mathbf{y}_{j+1} = h_j\tau_j$, the local truncation error. In this view, the local truncation error is the difference between the desired value of the solution and the value produced by the recipe when the recipe is provided with data that is exact.

The first codes used a fixed formula and a fixed step size. The *classical convergence theory* models the computation by considering an integration to be a member of a sequence of integrations. In this model each integration of (1.1) from a to b uses a constant step size h and results are produced at $x_j = a + jh$ for $j = 0, 1, \ldots$. In the next chapter we shall see that for a method of order μ,

$$\max_j \|\mathbf{y}(x_j) - \mathbf{y}_j\| = \mathcal{O}(h^{\mu}),$$

i.e., the numerical values approximate the solution to order μ. The early hand computations were done in a more flexible way. The behavior of the computed solution was scrutinized. If a result did not appear to be sufficiently accurate, it was rejected and the step repeated with a smaller step size. Perhaps a different order formula was also tried. If it appeared that the results were more accurate than necessary, a larger step size would be used so as to solve the problem more efficiently. If it appeared that a formula with a different order would allow a larger step size than the current formula, the formula was changed. Modern codes do all these things. The classical convergence theory has been modified so as to model most of these procedures. Variation of order is the conspicuous exception. It is still an art, but a great deal of experience shows the plausible procedures to be effective and it is reasonable to hope that a theoretical understanding is possible.

Although we have concentrated on approximating the solution at the mesh points x_j, Taylor series methods actually provide polynomial approximations valid for all of $[x_j, x_{j+1}]$. For example, in the case of the second-order method we are approximating $\mathbf{u}(x)$ by the truncated Taylor series

$$\mathbf{p}(x) = \mathbf{y}_j + (x - x_j)\,\mathbf{F}(x_j, \mathbf{y}_j) + \frac{(x - x_j)^2}{2}\left[\frac{\partial \mathbf{F}}{\partial x}(x_j, \mathbf{y}_j) + \frac{\partial \mathbf{F}}{\partial \mathbf{y}}(x_j, \mathbf{y}_j)\,\mathbf{F}(x_j, \mathbf{y}_j)\right].$$

The Taylor polynomials arise from interpolation to the value and first few derivatives at x_j of the local solution $\mathbf{u}(x)$. Obtaining solution values between mesh points is generally called *interpolation*, even though the connection with the usual meaning of interpolation may be obscure with some of the schemes in use.

The explicit Taylor series methods use information about $\mathbf{u}(x)$ computed at the single point x_j. If $\mathbf{F}$ were to change drastically between x_j and x_{j+1}, the method might not "notice." One way to spot such changes is to expand about a tentative solution at x_{j+1} so as to obtain an approximation back at x_j and then compare this approximation to $\mathbf{u}(x_j)$. If the agreement is poor, $\mathbf{F}$ must have changed character in the span of the step. This suggests one reason for considering *implicit methods*. To be specific, let us take a step backward from $(x_{j+1}, \mathbf{y}_{j+1})$ with Euler's method. If the result of this step is to reproduce the given value $\mathbf{y}_j = \mathbf{u}(x_j)$, we would have to choose $\mathbf{y}_{j+1}$ so that

$$\mathbf{y}_j = \mathbf{y}_{j+1} - h\,\mathbf{F}(x_{j+1}, \mathbf{y}_{j+1}).$$

This is an algebraic equation for $\mathbf{y}_{j+1}$ that is generally non-linear and it is by no means obvious that $\mathbf{y}_{j+1}$ is well defined. If the function $\mathbf{F}(x, \mathbf{y})$ is linear,

$$\mathbf{F}(x, \mathbf{y}) = J(x)\,\mathbf{y} + \mathbf{g}(x),$$

the equation becomes

$$(I - h\,J(x_{j+1}))\,\mathbf{y}_{j+1} = \mathbf{y}_j + h\,\mathbf{g}(x_{j+1}).$$

Except when the matrix $I - hJ(x_{j+1})$ is singular, there is a unique solution $\mathbf{y}_{j+1}$. Clearly for "most" h the matrix is non-singular; it is certainly non-singular for all sufficiently small step sizes h. Solving a linear system to get an approximate solution $\mathbf{y}_{j+1}$ is already quite a bit of work and the matter is even worse for non-linear problems, so why bother? Right now this is not clear, with only a hint provided by the fact that an implicit method takes account of the behavior of the solution at the end of the step as well as at the beginning.

To understand why we are interested in implicit formulas, we need to raise a fundamental question—*stability*. At this time we just touch upon the matter and defer a fuller treatment to Chapter 6. For our present purpose it suffices to contrast the implicit *backward Euler method*

$$\mathbf{y}_{j+1} = \mathbf{y}_j + h\, \mathbf{F}(x_{j+1}, \mathbf{y}_{j+1})$$

and the explicit *forward Euler method*

$$\mathbf{y}_{j+1} = \mathbf{y}_j + h\, \mathbf{F}(x_j, \mathbf{y}_j).$$

Both methods come from the linear terms of a Taylor series expansion at an end of a step. From the remainder terms of the expansions, it is clear that the two methods have essentially the same accuracy for small h. In Chapter 2 we saw that the stability of a differential equation is of fundamental importance. Among other things, we were concerned with how much the value of the local solution $\mathbf{u}(x_j + h)$ is altered by a change in its initial value $\mathbf{y}_j$. Obviously the corresponding question is important to numerical methods, too. Let us consider the simple scalar problem.

$$u' = \lambda u, \qquad u(x_j) = y_j \tag{1.2}$$

to see what can happen. The true solution is $u(x_{j+1}) = \exp(h\lambda)\, y_j$. The forward Euler method has $y_{j+1} = (1 + h\lambda)\, y_j$. Because the problem is linear and scalar, we can solve the algebraic equation of the backward Euler formula easily to get

$$y_{j+1} = \left(\frac{1}{1 - h\lambda}\right) y_j$$

When $|h\lambda|$ is small, the factors in these expressions are

$$\begin{aligned}
\exp(h\lambda) &= 1 + h\lambda + \frac{(h\lambda)^2}{2} + \mathbb{O}((h\lambda)^3),\\
1 + h\lambda &= 1 + h\lambda,\\
(1 - h\lambda)^{-1} &= 1 + h\lambda + (h\lambda)^2 + \mathbb{O}((h\lambda)^3).
\end{aligned}$$

Comparison of the approximations to the true solution at the end of the step shows that both the first order methods approximate the local solution there to $\mathbb{O}(h^2)$, as expected.

Suppose we change y_j to $y_j + \delta$. This results in a change in the true solution at the end of the step, $u(x_{j+1})$, of $\delta \exp(h\lambda)$. Similarly, a change of δ in y_j results in a change in y_{j+1} of $\delta(1 + h\lambda)$ in the one case and $\delta(1 - h\lambda)^{-1}$ in the other. Clearly the stability of the differential equation is imitated by both numerical methods when $|h\lambda|$ is small. However, the situation is quite different when $|h\lambda|$ is not small. The differential equation is stable for all $\lambda \leq 0$. Here this means that the change of $\delta \exp(h\lambda)$ at the end of the step is no greater in magnitude than the change δ at the beginning of the step. By stability of the numerical method we mean that it imitates the behavior of the equation to the extent that a change in y_j is not amplified in y_{j+1}. The implicit method is stable for any h when $\lambda \leq 0$ because

$$\left|\frac{1}{1 - h\lambda}\right| \leq 1.$$

(For convenience in theoretical discussions, we always assume that the problem is redefined if necessary to make h positive.) The explicit method is stable only when

$$|1 + h\lambda| \leq 1, \tag{1.3}$$

hence only when

$$h \leq 2/|\lambda|.$$

Stability restricts the step size that can be used with the forward Euler method. When $\lambda \ll -1$, the restriction is so severe that the backward Euler method becomes interesting despite the complication and expense due to it being implicit.

Generally the step size must be restricted both to get the accuracy desired and to keep the computation stable. The interesting implicit methods are much more stable than the popular explicit methods. It is not necessarily the case that an implicit method is more expensive to evaluate and even when it is, if stability restricts sufficiently the step size that might be used by an explicit method, it will be more efficient to use the implicit method. In the case of *stiff* problems—stable problems with large Lipschitz constants and solutions easy to approximate—the stability constraint is so severe that even very expensive steps with a highly stable method can be a bargain.

It is the presence of derivatives that is troublesome with the Taylor series approach because the problem is specified by **F** alone and it may be expensive, inconvenient, or even impossible to obtain its derivatives. We look now for procedures that require only the evaluation of **F**. A broad and important class of methods is based on integration. In the special case

$$u' = f(x), \qquad u(x_j) = y_j,$$

we have

$$u(x_j + h) = y_j + \int_{x_j}^{x_j+h} f(t)\, dt.$$

A *numerical quadrature formula* of order μ is a set of constants w_k, θ_k for $k = 1, \ldots, \nu$ such that for any sufficiently smooth function $f(t)$,

$$\int_{x_j}^{x_j+h} f(t)\, dt = h \sum_{k=1}^{\nu} w_k f(x_j + \theta_k h) + \mathcal{O}(h^{\mu+1}).$$

It is conventional in the context to shorten the term numerical quadrature to *quadrature* and this will be done from now on.

For later use let us recall how an *interpolatory quadrature* formula is constructed. Choose distinct θ_k, $k = 1, \ldots, \nu$, and form the interpolating polynomial $P(x)$ of degree at most $\nu - 1$ such that

$$P(x_j + \theta_k h) = f(x_j + \theta_k h) \qquad k = 1, \ldots, \nu.$$

In terms of the fundamental Lagrangian interpolation polynomials $L_k(x)$,

$$L_k(x) = \prod_{\substack{m=1 \\ m \neq k}}^{\nu} \left(\frac{x - (x_j + \theta_m h)}{(x_j + \theta_k h) - (x_j + \theta_m h)} \right), \qquad k = 1, \ldots, \nu,$$

this polynomial is

$$P(x) = \sum_{k=1}^{\nu} L_k(x)\, f(x_j + \theta_k h).$$

We then approximate

$$\int_{x_j}^{x_j+h} f(t)\, dt \approx \int_{x_j}^{x_j+h} P(t)\, dt = h \sum_{k=1}^{\nu} w_k f(x_j + \theta_k h).$$

Here

$$h\, w_k = \int_{x_j}^{x_j+h} L_k(t)\, dt,$$

so that

$$w_k = \int_0^1 \prod_{\substack{m=1 \\ m \neq k}}^{\nu} \left(\frac{\eta - \theta_m}{\theta_k - \theta_m} \right) d\eta.$$

From interpolation theory (see the appendix), for each t there is a point ξ such that

$$f(t) = P(t) + \frac{f^{(\nu)}(\xi)}{\nu!} \prod_{m=1}^{\nu} (t - (x_j + \theta_m h)).$$

This implies that

$$\left| \int_{x_j}^{x_j+h} f(t)\, dt - \int_{x_j}^{x_j+h} P(t)\, dt \right| \leq C\, h^{\nu+1}.$$

Here the constant C depends on the maximum of $|f^{(\nu)}|$ over the smallest interval that includes both $\{x_j, x_j + h\}$ and the *nodes* $\{x_j + \theta_1 h, \ldots, x_j + \theta_\nu h\}$. Clearly the procedure allows us to construct quadrature formulas of any order that we wish.

The local solution $\mathbf{u}(x)$ satisfies

$$\mathbf{u}(x_j + h) = \mathbf{y}_j + \int_{x_j}^{x_j+h} \mathbf{F}(t, \mathbf{u}(t))\, dt.$$

If we replace the integral here with a quadrature formula of order μ, we have

$$\mathbf{u}(x_j + h) = \mathbf{y}_j + h \sum_{k=1}^{\nu} w_k\, \mathbf{F}(x_j + \theta_k h, \mathbf{u}(x_j + \theta_k h)) + \mathcal{O}(h^{\mu+1})$$

This suggests a way to approximate $\mathbf{u}(x_j + h)$, but it is not immediately useful because it involves a set of unknown solution values $\mathbf{u}(x_j + \theta_k h)$. To get a formula, we must approximate these values by computable quantities

$$\mathbf{Y}_k \approx \mathbf{u}(x_j + \theta_k h).$$

The representation

$$\mathbf{u}(x_j + \theta_k h) = \mathbf{y}_j + \int_{x_j}^{x_j+\theta_k h} \mathbf{F}(t, \mathbf{u}(t))\, dt$$

suggests that we obtain the $\mathbf{Y}_k$ from quadrature formulas, too. We seem to be compounding our difficulties, but it all works out because these auxiliary values do not have to be as accurate as the approximation $\mathbf{y}_{j+1}$ that we seek. We are supposing that the quadrature formula is of order

μ. Suppose now that we have at our disposal a collection of formulas $\mathbf{\Phi}_k$ (not necessarily different) each being of order at least $\mu - 1$. For each $k = 1, \ldots, \nu$, we define

$$\mathbf{Y}_k = \mathbf{y}_j + \theta_k h \, \mathbf{\Phi}_k(x_j, \mathbf{y}_j, \mathbf{F}, \theta_k h),$$

that is, we take a step of size $\theta_k h$ from x_j to approximate $\mathbf{u}(x_j + \theta_k h)$ by $\mathbf{Y}_k$. Then, by our assumption about the accuracy of the formula $\mathbf{\Phi}_k$,

$$\mathbf{Y}_k = \mathbf{u}(x_j + \theta_k h) + \mathcal{O}(h^\mu),$$

or

$$\|\mathbf{u}(x_j + \theta_k h) - \mathbf{Y}_k\| \leq C_k \, h^\mu, \tag{1.4}$$

for a suitable constant C_k. Let us now define a recipe for $\mathbf{y}_{j+1}$ by replacing the $\mathbf{u}(x_j + \theta_k h)$ in the quadrature formula by the approximations $\mathbf{Y}_k$:

$$\mathbf{y}_{j+1} = \mathbf{y}_j + h \sum_{k=1}^{\nu} w_k \, \mathbf{F}(x_j + \theta_k h, \mathbf{Y}_k).$$

To see that these lower order intermediate solution approximations do not destroy the accuracy of the quadrature formula, we first observe that

$$\|\mathbf{u}(x_j + h) - \mathbf{y}_{j+1}\| = \left\| h \sum_{k=1}^{\nu} w_k [\mathbf{F}(x_j + \theta_k h, \mathbf{u}(x_j + \theta_k h)) - \mathbf{F}(x_j + \theta_k h, \mathbf{Y}_k)] + \mathcal{O}(h^{\mu+1}) \right\|,$$

and then using the Lipschitz condition on $\mathbf{F}$ and the triangle inequality for the norm,

$$\|\mathbf{u}(x_j + h) - \mathbf{y}_{j+1}\| \leq h \sum_{k=1}^{\nu} |w_k| \, L \|\mathbf{u}(x_j + \theta_k h) - \mathbf{Y}_k\| + \mathcal{O}(h^{\mu+1}).$$

On using the bounds (1.4), we see that the factor of h multiplying the sum causes the right hand side to be of order $\mu + 1$, hence that the formula yielding $\mathbf{y}_{j+1}$ is of order μ.

This construction shows how to obtain an explicit formula of order μ whenever we have available an explicit formula of order $\mu - 1$ and a quadrature formula of order μ. The construction for interpolatory quadrature formulas provides formulas of any order we want. The forward Euler method provides us with an explicit one-step method of order 1. By a "bootstrap" process, the construction can be used to generate explicit one-step methods of any order. To illustrate the process, let us consider the trapezoidal rule:

$$\int_{x_j}^{x_j+h} f(t)\, dt = \frac{h}{2}\,[f(x_j) + f(x_j + h)] + \mathcal{O}(h^3)$$

With this quadrature formula, the method has the form

$$\mathbf{y}_{j+1} = \mathbf{y}_j + h\left[\frac{1}{2}\,\mathbf{F}(x_j, \mathbf{y}_j) + \frac{1}{2}\,\mathbf{F}(x_j + h, \mathbf{Y}_1)\right].$$

The intermediate value $\mathbf{y}_1$ is an approximation to $\mathbf{y}(x_j + h)$. If we use $\mathbf{y}_{j+1}$ for this value, we get an implicit formula of order two called the *trapezoidal rule*

$$\mathbf{y}_{j+1} = \mathbf{y}_j + h\left[\frac{1}{2}\,\mathbf{F}(x_j, \mathbf{y}_j) + \frac{1}{2}\,\mathbf{F}(x_j + h, \mathbf{y}_{j+1})\right].$$

An explicit formula is obtained by computing $\mathbf{Y}_1$ with the forward Euler method,

$$\mathbf{Y}_1 = \mathbf{y}_j + h\,\mathbf{F}(x_j, \mathbf{y}_j).$$

According to the construction, combining the quadrature formula of order two with intermediate results computed by a method of order one results in a formula of order two. Notice that the formula requires two evaluations of $\mathbf{F}$ per step. The formula is known both as the *improved Euler formula* and as *Heun's method*. It is used, for example, in the PHASER package of Koçak [1989] and in the solution of partial differential equations by semi-discretization.

A more complicated example will make another point. *Simpson's rule* is another familiar quadrature formula:

$$\int_{x_j}^{x_j+h} f(t)\, dt = \frac{h}{6}\left[f(x_j) + 4f\left(x_j + \frac{h}{2}\right) + f(x_j + h)\right] + \mathcal{O}(h^5)$$

The construction says that the formula

$$\mathbf{y}_{j+1} = \mathbf{y}_j + h\left[\frac{1}{6}\,\mathbf{F}(x_j, \mathbf{y}_j) + \frac{4}{6}\,\mathbf{F}\left(x_j + \frac{h}{2}, \mathbf{Y}_1\right) + \frac{1}{6}\,\mathbf{F}(x_j + h, \mathbf{Y}_2)\right]$$

will be of order four if we compute the intermediate approximations $\mathbf{Y}_1$, $\mathbf{Y}_2$ with formulas of order three. If we compute the intermediate approximations with formulas of order two, the resulting formula is just of order three. It is straightforward to use Heun's method to obtain the intermediate approximations if only we remember that $\mathbf{Y}_1$ is an approximation to $\mathbf{u}(x_j +$

$h/2$) so that the method must be used with a step size of $h/2$. Now

$$\mathbf{U}_1 = \mathbf{y}_j + \frac{h}{2}\,\mathbf{F}(x_j, \mathbf{y}_j),$$

$$\mathbf{Y}_1 = \mathbf{y}_j + \frac{h}{2}\left[\frac{1}{2}\,\mathbf{F}(x_j, \mathbf{y}_j) + \frac{1}{2}\,\mathbf{F}\left(x_j + \frac{h}{2}, \mathbf{U}_1\right)\right],$$

$$\mathbf{W}_1 = \mathbf{y}_j + h\,\mathbf{F}(x_j, \mathbf{y}_j),$$

$$\mathbf{Y}_2 = \mathbf{y}_j + h\left[\frac{1}{2}\,\mathbf{F}(x_j, \mathbf{y}_j) + \frac{1}{2}\,\mathbf{F}(x_j + h, \mathbf{W}_1)\right].$$

Naturally this is to be programmed so that $\mathbf{F}(x_j, \mathbf{y}_j)$ is evaluated only once. Reusing function evaluations like this is quite common in Runge–Kutta formulas. If we wished, this new formula of order three could be used in conjunction with Simpson's rule to generate a formula of order four.

EXERCISE 2. The midpoint rule has

$$\int_{x_j}^{x_j+h} f(t)\,dt = h\,f\left(x_j + \frac{1}{2}h\right) + \mathcal{O}(h^3).$$

Derive an explicit second-order formula using this quadrature rule and Euler's method that involves two evaluations of $\mathbf{F}$ per step. The formula is called the *midpoint*, or *modified, Euler formula.* It is an option in the MacMath package. Like the improved Euler method it is sometimes used to solve partial differential equations by semi-discretization.

EXERCISE 3. A two point Radau quadrature rule has

$$\int_{x_j}^{x_j+h} f(t)\,dt = \frac{h}{4}\left[f(x_j) + 3\,f\left(x_j + \frac{2}{3}\,h\right)\right] + \mathcal{O}(h^4).$$

Using an explicit second-order formula and this quadrature rule, derive an explicit formula of order three that involves only three evaluations of $\mathbf{F}$ per step.

The general form of formulas constructed in this way is

$$X_0 = x_j, \qquad \mathbf{Y}_0 = \mathbf{y}_j, \qquad \mathbf{F}_0 = \mathbf{F}(X_0, \mathbf{Y}_0)$$

and for $k = 1, \ldots, s$

$$X_k = x_j + \alpha_k h$$

$$\mathbf{Y}_k = \mathbf{y}_j + h \sum_{m=0}^{s} \beta_{k,m} \mathbf{F}_m \tag{1.5}$$

$$\mathbf{F}_k = \mathbf{F}(X_k, \mathbf{Y}_k).$$

The solution at the end of the step is

$$\mathbf{y}_{j+1} = \mathbf{y}_j + h \sum_{k=0}^{s} \gamma_k \mathbf{F}_k$$

The coefficients α_k, $\beta_{k,m}$, and γ_k define the formula. The $\mathbf{F}_k$ are called the *stages* of the formula, and a procedure of this kind is called a *Runge–Kutta formula* of $s + 1$ stages.

It is easier to derive Runge–Kutta formulas for equations in autonomous form. If the problem arises in the non-autonomous form (1.1), it can be converted to autonomous form for a larger system in a new independent variable t with $dx/dt = 1$, $x(a) = a$. Applying the formula to this system results in

$$\begin{pmatrix} \mathbf{Y}_k \\ X_k \end{pmatrix} = \begin{pmatrix} \mathbf{y}_j \\ x_j \end{pmatrix} + h \sum_{m=0}^{s} \beta_{k,m} \begin{pmatrix} \mathbf{F}(X_m, \mathbf{Y}_m) \\ 1 \end{pmatrix}.$$

Separating out the old independent variable leads to

$$X_k = x_j + h \sum_{m=0}^{s} \beta_{k,m}$$

and

$$\mathbf{Y}_k = \mathbf{y}_j + h \sum_{m=0}^{s} \beta_{k,m} \mathbf{F}(X_m, \mathbf{Y}_m).$$

Evidently, if we assume that

$$\alpha_k = \sum_{m=0}^{s} \beta_{k,m} \quad \text{for } k = 1, 2, \ldots, s,$$

we obtain the form (1.5) of the formula that applies to an equation in the general form (1.1). The assumption allows us to do the theoretical work with the autonomous form and just write down the formula needed for the typical code that accepts problems in non-autonomous form. All the practical formulas assume this and we do the same in future without further comment.

The $\mathbf{Y}_k$ are only defined implicitly by (1.5), so it is not clear that the procedure is even well defined. In §3 we shall see that for all sufficiently small h, it is. Everything *is* clear when (1.5) has the special form

$$\mathbf{Y}_k = \mathbf{y}_j + h \sum_{m=0}^{k-1} \beta_{k,m} \mathbf{F}_m,$$

because then $\mathbf{F}_k$ depends only on previously computed stages and this is an explicit recipe for $\mathbf{y}_{j+1}$. *Explicit Runge–Kutta formulas* of $s + 1$ stages have the form (1.5) with $\beta_{k,m} = 0$ for all $m \geq k$.

A standard way to present a Runge–Kutta formula is by means of an array called a *Butcher array* after John Butcher. For an explicit formula this array has the form

$$\begin{array}{c|ccccc}
\alpha_0 & & & & & \\
\alpha_1 & \beta_{1,0} & & & & \\
\vdots & \vdots & & & & \\
\alpha_s & \beta_{s,0} & \beta_{s,1} & \cdots & \beta_{s,s-1} & \\
\hline
 & \gamma_0 & \gamma_1 & \cdots & \gamma_{s-1} & \gamma_s
\end{array}$$

For an implicit formula the $\beta_{k,m}$ form a square array. By convention the $\beta_{k,m}$ are not displayed when $m \geq k$ for explicit formulas because they are all 0.

The next two exercises show that explicit Runge–Kutta formulas can be combined to obtain other explicit Runge–Kutta formulas. The results will be important later when we take up the estimation of error. The manipulations are a little easier for $\mathbf{F}$ in autonomous form.

EXERCISE 4. Suppose you have an explicit Runge–Kutta formula that yields a result $\mathbf{y}_{j+1}$ after a step of h from $(x_j, \mathbf{y}_j)$. Consider the scheme that consists of using this formula to form $\tilde{\mathbf{y}}_{j+1/2}$ as the result of a step of $h/2$ from $(x_j, \mathbf{y}_j)$ and then taking another step of $h/2$ from $(x_j + h/2, \tilde{\mathbf{y}}_{j+1/2})$ to obtain $\tilde{\mathbf{y}}_{j+1}$ as an approximation to the solution $\mathbf{y}(x_j + h)$. This new scheme is also an explicit Runge–Kutta formula. Work out the details for the improved Euler formula and in particular, write down the Butcher array definition of the resulting formula of four stages. If you are keen, show the general result.

EXERCISE 5. Suppose you have a formula of order p that yields a result $\mathbf{y}_{j+1}$ after a step of h from $(x_j, \mathbf{y}_j)$ and you also have a formula of order q that yields a result $\mathbf{y}^*_{j+1}$ after a step of h from $(x_j, \mathbf{y}_j)$. Let ζ be a constant that is not equal to 0 nor to 1. What is the order of the scheme with result

$$\tilde{\mathbf{y}}_{j+1} = \zeta\, \mathbf{y}_{j+1} + (1 - \zeta)\mathbf{y}^*_{j+1}?$$

Suppose now that both formulas are explicit Runge–Kutta formulas. The new scheme is also an explicit Runge–Kutta formula. Work out the details when the two formulas are the forward Euler method and the improved Euler method. In particular, write out the Butcher array. If you are keen, show the general result.

As mentioned earlier in connection with Taylor series methods, at present a great deal of attention is being devoted to providing an interpolation capability for Runge–Kutta methods, see, e.g., Horn [1983], Shampine [1985b], Enright et al. [1986], Dormand and Prince [1986]. Let us briefly consider how this is accomplished. As a rule the intermediate approximations $\mathbf{Y}_k$ formed in the evaluation of (1.5) are low-order approximations to $\mathbf{u}(x)$ at $X_k = x_j + \alpha_k h$, but they can be of the same order as $\mathbf{y}_{j+1}$. Implicit in the quadrature approach to Runge–Kutta formulas is the fact that $\mathbf{F}_k$ approximates $\mathbf{u}'(X_k)$ to the same order of accuracy that $\mathbf{Y}_k$ approximates $\mathbf{u}(X_k)$. Let us now make this explicit. Suppose that $\mathbf{Y}_k$ agrees with the local solution to order μ, i.e., (1.4) holds. Then

$$\|\mathbf{u}'(X_k) - \mathbf{F}_k\| = \|\mathbf{F}(X_k, \mathbf{u}(X_k)) - \mathbf{F}(X_k, \mathbf{Y}_k)\| \leq L\|\mathbf{u}(X_k) - \mathbf{Y}_k\| \leq L\, C_k\, h^{\mu}.$$

If there are enough $\mathbf{Y}_k$ and $\mathbf{F}_k$ of sufficiently high order of accuracy, an interpolant will provide approximations to $\mathbf{u}(x)$ of the desired accuracy throughout the step. The interpolation is done using data for the current step only, so the character of the method is not changed. Approximations to the value and slope of the local solution are always available at x_j and the method produces an approximation $\mathbf{y}_{j+1}$ to the value of the local solution at x_{j+1}. It is natural to interpolate at both ends of the step. Doing so results in a piecewise polynomial function that is continuous on the whole interval of integration—for each j the value at x_{j+1} of the interpolant for the interval $[x_j, x_{j+1}]$ is the same as the value of the interpolant for the interval $[x_{j+1}, x_{j+2}]$, namely $\mathbf{y}_{j+1}$. The (forward) Euler method furnishes a simple example. Of course, it is a Taylor series method so that the interpolant

$$\mathbf{y}_{j+\sigma} = \mathbf{y}_j + \sigma h\, \mathbf{F}_j$$

for a value approximating $\mathbf{u}(x_j + \sigma h)$ might be described as arising from interpolation to the value and slope at the beginning of the step. However, in the present context we would write it as linear interpolation to $\mathbf{y}_j$ and $\mathbf{y}_{j+1}$, namely

$$\mathbf{y}_{j+\sigma} = \mathbf{y}_j + \sigma(\mathbf{y}_{j+1} - \mathbf{y}_j).$$

Because we are concerned with solutions that have at least a continuous derivative, it is natural to ask that the piecewise polynomial interpolating function also have a continuous derivative. The interpolants of Taylor series methods do not have this property. Indeed, it is not possible to create such interpolants with the data we have been assuming; we cannot interpolate the slope $\mathbf{F}(x_{j+1}, \mathbf{y}_{j+1})$ until we compute it. In practice interpolation is done only after we have decided that the step to x_{j+1} is acceptable. When using an explicit Runge–Kutta method, the first stage of the next step, $\mathbf{F}_0$, is precisely the slope of the (new) local solution at x_{j+1}. Thus if we form $\mathbf{F}(x_{j+1}, \mathbf{y}_{j+1})$ in the current step in order to obtain a smoother interpolant, this value will be used in the next step, hence may be fairly regarded as "free." This slope not only allows us to obtain a smoother interpolant, the extra information can be used to obtain a more accurate interpolant.

The simplest example of an interpolant with a continuous derivative arises from interpolating value and slope at the two ends of the step. This Hermite interpolating polynomial of degree 3 might be used for any formula. The question, then, is whether it yields intermediate approximations that are as accurate as $\mathbf{y}_{j+1}$ is. There are two sources of error. One is from the interpolation process itself, because even if it were supplied the true values of $\mathbf{u}(x)$ and its first derivative, it would not reproduce the local solution exactly. The other arises from the fact that the interpolant is not provided the true values of $\mathbf{u}(x)$ and its derivative, rather approximations to them. Let $\mathcal{P}(x)$ be the cubic Hermite interpolant to the true values of $\mathbf{u}(x)$:

$$\mathcal{P}(x_m) = \mathbf{u}(x_m), \qquad \mathcal{P}'(x_m) = \mathbf{u}'(x_m), \qquad m = j, j + 1$$

It is useful in some contexts to know that derivatives of the interpolant approximate derivatives of the local solution. For example, this might be used to find where a component has a maximum. Because of this, a standard result for the error of this interpolant will be stated that includes approximation of derivatives. We are interested in x for which $x_j \leq x \leq x_j + h \leq x_j + H$. If $\mathbf{u}$ has four continuous derivatives on the fixed interval $[x_j, x_j + H]$, then for $0 \leq K < 4$,

$$\|\mathbf{u}^{(K)}(x) - \mathcal{P}^{(K)}(x)\| \leq \max_{[x_j, x_j + H]} \|\mathbf{u}^{(4)}\| \frac{h^{4-K}}{(4 - K)!}.$$

This tells us that if we had exact data, the interpolant would approximate the local solution to $\mathcal{O}(h^4)$. Although there are circumstances in which an interpolant less accurate than the formula is useful, generally we want intermediate values to be of accuracy comparable to the result $\mathbf{y}_{j+1}$ at the end of the step. Accordingly, this interpolant is not appropriate for methods of orders four and up. The question we must now address is the effect of interpolating inaccurate data.

We shall need an explicit form for the Hermite interpolant. It is convenient to write $\mathbf{y}'_j$ for $\mathbf{F}(x_j, \mathbf{y}_j)$ and $\mathbf{y}'_{j+1}$ for $\mathbf{F}(x_{j+1}, \mathbf{y}_{j+1})$. It is also convenient to introduce the scaled variable σ defined by $x = x_j + \sigma h$. For this variable, the value $\sigma = 0$ corresponds to the beginning of the step and $\sigma = 1$ corresponds to the end, so we are interested in the range $0 \le \sigma \le 1$. The interpolating polynomial is

$$\mathbf{P}(x) = A_1(\sigma)\mathbf{y}_j + A_2(\sigma)\mathbf{y}_{j+1} + B_1(\sigma)\mathbf{y}'_j + B_2(\sigma)\mathbf{y}'_{j+1}, \tag{1.6}$$

where

$$A_1(\sigma) = (\sigma - 1)^2(1 + 2\sigma), \qquad A_2(\sigma) = (3 - 2\,\sigma)\sigma^2,$$
$$B_1(\sigma) = h\,\sigma(\sigma - 1)^2, \qquad B_2(\sigma) = h\,\sigma^2(\sigma - 1).$$

EXERCISE 6. Verify that the $\mathbf{P}(x)$ of (1.6) *is* the interpolating polynomial, that is, verify that $\mathbf{P}(x_j) = \mathbf{y}_j$, $\mathbf{P}(x_{j+1}) = \mathbf{y}_{j+1}$, $\mathbf{P}'(x_j) = \mathbf{y}'_j$, and $\mathbf{P}'(x_{j+1}) = \mathbf{y}'_{j+1}$. *This is conveniently done in terms of the variable σ, but you must then take into account that*

$$\frac{d}{dx} = \frac{1}{h}\frac{d}{d\sigma}.$$

The error of the interpolant satisfies

$$\begin{aligned}\|\mathbf{u}(x) - \mathbf{P}(x)\| &= \|\mathbf{u}(x) - \mathcal{P}(x) + \mathcal{P}(x) - \mathbf{P}(x)\|,\\ &\le \|\mathbf{u}(x) - \mathcal{P}(x)\| + \|\mathcal{P}(x) - \mathbf{P}(x)\|,\end{aligned}$$

where $\mathcal{P}(x)$ is the interpolant to the exact data. We just learned that the first term on the right here is $\mathcal{O}(h^4)$, so we need only investigate the second term. From the explicit form of the interpolant we have first that

$$\mathcal{P}(x) = A_1(\sigma)\mathbf{y}_j + A_2(\sigma)\mathbf{u}(x_{j+1}) + B_1(\sigma)\mathbf{y}'_j + B_2(\sigma)\mathbf{u}'(x_{j+1}),$$

and then after an obvious manipulation that

$$\|\mathcal{P}(x) - \mathbf{P}(x)\| \le |A_2(\sigma)|\,\|\mathbf{u}(x_{j+1}) - \mathbf{y}_{j+1}\| + |B_2(\sigma)|\,\|\mathbf{u}'(x_{j+1}) - \mathbf{y}'_{j+1}\|.$$

For $0 \le \sigma \le 1$, the coefficient $A_2(\sigma)$ is clearly uniformly bounded. There is a factor of h appearing in the expression for $B_2(\sigma)$ and it is clearly $\mathcal{O}(h)$. If the formula is of order p, $\mathbf{y}_{j+1}$ agrees with $\mathbf{u}(x_{j+1})$ to $\mathcal{O}(h^{p+1})$ and we have seen that $\mathbf{y}'_{j+1}$ agrees with $\mathbf{u}'(x_{j+1})$ to the same order. Putting all these observations together, we find that $\mathbf{P}(x)$ approximates $\mathbf{u}(x)$ to order $\min(p + 1, 4)$. Thus this simple procedure can be used with any formula of order three and lower to get intermediate solution values that have the same order of accuracy as $\mathbf{y}_{j+1}$. Naturally it would be interesting to know if the constants in the order statements are of about the same size. There are techniques in the research literature for investigating this. It is worth mentioning that the question has a very nice answer in this particular case for methods of order two and lower: To leading order, the error within the step arises from interpolation rather than from inaccurate data. Because of this the distribution of the error is independent of the formula and the problem. As σ increases from 0 to 1, the local error increases smoothly and monotonely from 0 at the beginning of the step to the local error of $\mathbf{y}_{j+1}$ at the end of the step. We could scarcely hope to find procedures with more satisfactory behavior. The reader should not be misled by the success of interpolation at low orders. Interpolation at higher orders is much more difficult and expensive because there are not enough $\mathbf{Y}_k$ and $\mathbf{F}_k$ of the necessary accuracy available from the computation of $\mathbf{y}_{j+1}$.

EXERCISE 7. In this way of obtaining approximations to the local solution at intermediate points it is easy to show that accurate approximations to derivatives of the local solution are also obtained. Following the line of proof for the order of approximation of the local solution by the $\mathbf{P}(x)$ of (1.6), show that $\mathbf{P}^{(K)}(x)$ approximates $\mathbf{u}^{(K)}(x)$ to order $(\min(p + 1, 4) - K)$. In the course of doing this you will need to show that $A_2^{(K)}(\sigma)$ is order $(-K)$ and $B_2^{(K)}(\sigma)$ is order $(1 - K)$.

There is a way of investigating interpolants that appears to be quite different. We have already seen an example in the case of the Euler method. In the form suggested by its alias as a Taylor series method, it is seen that the interpolant amounts to a formula for stepping from x_j to $x_j + \sigma h$. A Runge–Kutta formula (1.5) provides an answer at $x_j + h$ in the form

$$\mathbf{y}_{j+1} = \mathbf{y}_j + h \sum_{k=0}^{s} \gamma_k \mathbf{F}_k.$$

The idea is to seek an answer at $x_j + \sigma h$ in the form

$$\mathbf{y}_{j+\sigma} = \mathbf{y}_j + (\sigma h) \sum_{k=0}^{s^*} \gamma_k(\sigma) \mathbf{F}_k. \tag{1.7}$$

What we would like is to take $s^* = s$ and take the stages here, the $\mathbf{F}_k$, to be the stages evaluated for the computation of $\mathbf{y}_{j+1}$ so that forming $\mathbf{y}_{j+\sigma}$ would have a negligible cost. A formula of the form (1.7) is a Runge–Kutta formula for each σ and its accuracy can be studied just like that of the formula for $\mathbf{y}_{j+1}$. What we would like is to determine coefficients $\gamma_k(\sigma)$ such that for all $0 \leq \sigma \leq 1$, the result is of the same order of accuracy as $\mathbf{y}_{j+1}$. As long as we are making up a "wish list," we would like rather more. We would like for $\mathbf{y}_{j+\sigma}$ obtained in this way to approach $\mathbf{y}_j$ when $\sigma \to 0$ and to approach $\mathbf{y}_{j+1}$ when $\sigma \to 1$ so that we have a continuous interpolant. Indeed, as with the interpolation approach, we would like a continuous first derivative. Only for the lowest order formulas is all this possible. Generally it is necessary to add some stages if we are to be able to interpolate, i.e., take $s^* > s$. Earlier it was pointed out that to get an interpolant with continuous derivative, it is necessary to evaluate $\mathbf{F}(x_{j+1}, \mathbf{y}_{j+1})$. Examination of the form (1.5) shows that we can regard $\mathbf{y}_{j+1}$ as another intermediate approximation, say $\mathbf{Y}_{s+1}$, and this function evaluation is just another stage formed from this "intermediate" approximation. A virtue of this approach to interpolation is that we create a family of Runge–Kutta formulas depending on a parameter σ and for any given σ, we can study the formula in the same manner that we study the formula for $\mathbf{y}_{j+1}$. In this view of interpolation, the formula (1.7) is often described as a *continuous extension* of (1.5). It is not immediately obvious, but in most cases interpolants obtained with one approach can be written in the form required by the other approach.

EXERCISE 8. Suppose that a Runge–Kutta formula of $s + 1$ stages (1.5) is interpolated by the cubic Hermite interpolant $\mathbf{P}(x)$ of (1.6). Introducing an additional stage, $\mathbf{F}_{s+1}$, show that this approximation to $\mathbf{u}(x_j + \sigma h)$ can be written in the form (1.7) with $s^* = s + 1$.

§1.2 Choosing a Runge–Kutta Formula—Accuracy

The Runge–Kutta formulas derived in the manner described in the last subsection are not efficient unless great care is taken. The trick is to reuse stages $\mathbf{F}_m$ from the construction of some $\mathbf{Y}_k$ when constructing other $\mathbf{Y}_q$. The quadrature approach to Runge–Kutta formulas can be used to derive efficient formulas, but the most efficient formulas have usually been derived in a different way. The idea is to start with the general form of a Runge–Kutta formula, expand its error in a Taylor series in the step size h, and search for coefficients that make the error of high order. This abstraction is important to the derivation, and especially to the understanding of the quality, of a formula, but without a familiarity with the quadrature approach, it makes Runge–Kutta formulas seem rather mysterious.

To see what is involved in this other approach, let us examine explicit Runge–Kutta formulas of two stages. In an example and an exercise we have already seen two such formulas that will appear here as special cases. For $\mathbf{F}$ in autonomous form, such a formula has the form

$$\mathbf{F}_0 = \mathbf{F}(\mathbf{y}_j),$$

$$\mathbf{Y}_1 = \mathbf{y}_j + h\,\beta_{1,0}\mathbf{F}_0, \qquad \mathbf{F}_1 = \mathbf{F}(\mathbf{Y}_1),$$

$$\mathbf{y}_{j+1} = \mathbf{y}_j + h[\gamma_0\,\mathbf{F}_0 + \gamma_1\,\mathbf{F}_1].$$

We seek coefficients $\beta_{1,0}$, γ_0, γ_1 such that $\mathbf{y}_{j+1}$ agrees as well as possible with the local solution. We saw earlier that component i of this solution has the expansion

$$u_i(x_j + h) = u_i(x_j) + h\, u_i'(x_j) + \frac{h^2}{2!} u_i''(x_j) + \frac{h^3}{3!} u_i'''(x_j) + \ldots .$$

In the compact notation for the derivatives, this is

$$(\mathbf{u}(x_j + h))^i = (\mathbf{y}_j)^i + h(\mathbf{D}_1^1)^i + \frac{h^2}{2}(\mathbf{D}_1^2)^i + \frac{h^3}{6}[(\mathbf{D}_1^3)^i + (\mathbf{D}_2^3)^i] + \ldots$$

$$= y_j^i + h\, F^i + \frac{h^2}{2} F^i_{,j} F^j + \frac{h^3}{6}[F^i_{,j,k}\, F^j\, F^k + F^i_{,j}\, F^j_{,k}\, F^k] + \ldots .$$

Here all functions are evaluated at $(x_j, \mathbf{y}_j)$. We must expand component i of $\mathbf{y}_{j+1}$ and compare the two expressions. To do this we must expand $\mathbf{F}_1$. In general, Taylor expansion of component i of a function $\mathbf{F}(\mathbf{v} + h\boldsymbol{\delta})$ has the form

$$F^i(\mathbf{v} + h\boldsymbol{\delta}) = F^i + h\, F^i_{,j}\delta^j + \frac{h^2}{2!} F^i_{,j,k}\delta^j\delta^k + \frac{h^3}{3!} F^i_{,j,k,m}\delta^j\delta^k\delta^m + \ldots,$$

where all functions are evaluated at $\mathbf{v}$, δ^k is component k of $\boldsymbol{\delta}$, etc. Applying this to the expansion of $\mathbf{F}_1$ with $\mathbf{v} = \mathbf{y}_j$ and $\boldsymbol{\delta} = \beta_{1,0}\, \mathbf{F}_0$ results in

$$F_1^i = F^i + h\, F^i_{,j}(\beta_{1,0}\, F^j) + \frac{h^2}{2} F^i_{,j,k}(\beta_{1,0}\, F^j)(\beta_{1,0}\, F^k) + \ldots .$$

Combining this with the other terms, we find that the expansion of component i of the result of a step is

$$(\mathbf{y}_{j+1})^i = (\mathbf{y}_j)^i + h[\gamma_0 + \gamma_1]\, F^i + h^2\, \gamma_1\, \beta_{1,0}\, F^i_{,j}\, F^j + \frac{h^3}{2}\gamma_1\, \beta_{1,0}^2\, F^i_{,j,k}\, F^j\, F^k + \mathcal{O}(h^4).$$

Comparison of the two expressions shows that we get agreement to $\mathcal{O}(h^2)$, a formula of order one, if we have $\gamma_0 + \gamma_1 = 1$. If the agreement is to hold to this order for *all* smooth $\mathbf{F}$, this condition is also necessary. We get a formula of order two if in addition, $\gamma_1\, \beta_{1,0} = 1/2$. Notice that it is not possible to get an explicit Runge–Kutta formula of order three using only two stages because there is an elementary differential appearing in the expansion of the true solution that does not appear in the expansion of the result of the formula.

From this analysis it is easy to determine all two-stage, second-order explicit Runge–Kutta formulas. Obviously it must be the case that $\gamma_1 \neq 0$ if we are to get order 2. Taking γ_1 to be a parameter ζ, the other two coefficients are fully determined: $\gamma_0 = 1 - \zeta$, $\gamma_1 = \zeta \neq 0$, and $\beta_{1,0} =$

$1/(2\,\zeta)$. It is typical that Runge–Kutta formulas are derived as families of formulas. Indeed, there are often so many free parameters that it is an embarrassment. The purpose of all this analysis is to solve initial value problems and to solve a problem, we must choose a set of parameters for our formula. The question is, how? Historically parameters were chosen to result in "simple" coefficients. This is a reason for the popular choices of ζ equal to 1 and 1/2 that we saw earlier—the improved Euler and midpoint Euler formulas. It was later suggested that the parameters be chosen so that the formula is particularly accurate. The local error is

$$h^3\left[\left(\frac{1}{6} - \frac{1}{2}\gamma_1\,\beta_{1,0}^2\right)\mathbf{D}_1^3 + \frac{1}{6}\,\mathbf{D}_2^3\right] + \mathcal{O}(h^4). \tag{1.8}$$

Since it is not possible to make the leading term vanish for all problems, it seems reasonable to choose the parameter to make it "small" for a "typical" problem. The two elementary differentials here depend on the problem and have no relation to one another (they are independent), so we do not know how they will combine in the local error. Still, it seems reasonable here to make the coefficient of $\mathbf{D}_1^3$ vanish by taking $\zeta = 3/4$. We must make a choice and this one seems likely to result in relatively small errors. This line is pursued further in present practice. To be sure, there are many issues of quality that we have not yet taken up that must be taken into account, but the idea is to exploit the choice of parameters to select a formula of high quality. The coefficients are often not the simple ones of the classical literature, but a computer does not "care" about this.

For a number of the numerical methods taken up in this chapter, the error made in a single step depends only on the local solution $\mathbf{u}(x)$. Because of this it is worth emphasizing that in general the error of a Runge–Kutta method depends on $\mathbf{F}$, too. This can be seen in the expression (1.8) for the error of two-stage, second-order formulas. The exercise that follows considers a concrete problem. To get a little practice with the autonomous form and the error expressions, we begin this exercise as an example.

EXAMPLE 1. Suppose we wish to solve the initial value problem

$$y' = 2x, \qquad y(1) = 1,$$

which has the obvious solution $y = x^2$. To use the error expression derived earlier, we need to convert the problem to autonomous form:

$$\frac{d}{dt}\mathbf{Y} = \frac{d}{dt}\begin{pmatrix} y \\ x \end{pmatrix} = \begin{pmatrix} 2x \\ 1 \end{pmatrix} = \mathbf{F}(\mathbf{Y}).$$

We have then that $F^1 = 2x$ and $F^2 = 1$. The first component of the elementary differential $\mathbf{D}_2^3$ is

$$(\mathbf{D}_2^3)^1 = F^1_{,j} F^j_{,k} F^k = F^1_{,1} F^1_{,1} F^1 + F^1_{,1} F^1_{,2} F^2 + F^1_{,2} F^2_{,1} F^1 + F^1_{,2} F^2_{,2} F^2.$$

(A compact notion is convenient for the theory, but when it is necessary actually to evaluate an elementary differential, it is found that a lot of terms were swept under the carpet.) The first component of $\mathbf{Y}$ is y and the second is x, so

$$F^1_{,1} = 0, \qquad F^1_{,2} = 2, \qquad F^2_{,1} = 0, \qquad F^2_{,2} = 0,$$

and this implies that the first component of $\mathbf{D}_2^3$ is 0. The second component is

$$(\mathbf{D}_2^3)^2 = F^2_{,j} F^j_{,k} F^k.$$

We have just seen that both partial derivatives of F^2 appearing here vanish, so this component must also vanish. The other elementary differential

$$(\mathbf{D}_1^3)^i = F^i_{,j,k} F^j F^k$$

is even easier to evaluate. Because all the second partial derivatives vanish, this differential vanishes. We see, then, that for this particular problem, the error introduced in one step is of higher order than usual (at least three) because the usual leading term in the error expansion vanishes.

The initial value problem

$$y' = 2y/x, \qquad y(1) = 1,$$

also has $y = x^2$ as its solution. Going to autonomous form in the same way, we now have $F^1 = 2y/x$ and $F^2 = 1$. The first partial derivatives are

$$F^1_{,1} = 2/x, \qquad F^1_{,2} = -2y/x^2, \qquad F^2_{,1} = 0, \qquad F^2_{,2} = 0.$$

Now we have

$$(\mathbf{D}_2^3)^1 = F^1_{,j} F^j_{,k} F^k = 4y/x^3,$$

which is generally not zero. However, just as with the other equation, $(\mathbf{D}_2^3)^2$ is zero. This problem has the same solution as the other one, but in this case the leading term in the error expansion does not vanish identically and the method is therefore of order two.

EXERCISE 9. Both of the problems

$$y' = 2x, \qquad y(1) = 1,$$
$$y' = 2y/x, \qquad y(1) = 1,$$

have the solution $y = x^2$. Solve them with one of the two-stage, second-order Runge–Kutta methods and a constant step size. On comparing the numerical solutions to the true solution, you will observe that one problem is solved more accurately than the other. More specifically, for $h = 1/N$ and $N = 10, 10^2, 10^3, 10^4$, calculate $y_N \approx y(2)$ and compare the errors $y(2) - y_N$ for the two equations. (You will probably notice roundoff effects when solving one equation when h is small because the formula is so accurate then.) Compute the ratio $(y(2) - y_N)/h^m$ for $m = 1, 2, 3$ to ascertain the order of the method. For example, if the method is convergent of order two, then as $h \to 0$, the ratio will tend to 0 for $m = 1$, tend to a non-zero value for $m = 2$, and become unbounded for $m = 3$.

As with the related expansions of Taylor series formulas, a considerable amount of technique has been developed for expansion of the error of a Runge–Kutta method in powers of the step size. We have no need to develop this technique, so refer the reader to texts such as Butcher [1987] and Hairer, Nørsett and Wanner [1987] for the derivations and explicit expressions for the coefficients. (There are other techniques for expanding the error, such as those of Cooper and Verner [1972] and Albrecht [1987], that offer some advantages.) We do need to become acquainted with the form of the result, so we state here the expressions for the lowest order terms. As usual, it is convenient to take the differential equation to be in autonomous form. For smooth $\mathbf{F}$, the difference between the local solution and the result of the Runge–Kutta formula (1.5) can be expanded as

$$\mathbf{u}(x_j + h) - \mathbf{y}_{j+1} = \sum_{i=1}^{\infty} h^i \left(\sum_{n=1}^{\lambda_i} T_n^i \, \mathbf{D}_n^i \right). \tag{1.9}$$

Here the $\mathbf{D}_n^i$ are *elementary differentials*, sums of products of partial derivatives of components of $\mathbf{F}$, that depend only on the problem through $\mathbf{F}$ and its argument $\mathbf{y}_j$. The *truncation error coefficients* T_n^i depend only on the formula as specified by its coefficients. A short *table of truncation error coefficients* follows.

$\lambda_1 = 1$	$T_1^1 = \left(1 - \sum_i \gamma_i\right)$
$\lambda_2 = 1$	$T_1^2 = \left(\frac{1}{2} - \sum_i \gamma_i \alpha_i\right)$

$\lambda_3 = 2$	$T_1^3 = \frac{1}{2} \times \left(\frac{1}{3} - \sum_i \gamma_i \, \alpha_i^2\right)$
	$T_2^3 = \left(\frac{1}{6} - \sum_{ij} \gamma_i \, \beta_{ij} \, \alpha_j\right)$
$\lambda_4 = 4$	$T_1^4 = \frac{1}{6} \times \left(\frac{1}{4} - \sum_i \gamma_i \, \alpha_i^3\right)$
	$T_2^4 = \left(\frac{1}{8} - \sum_{ij} \gamma_i \, \alpha_i \, \beta_{ij} \, \alpha_j\right)$
	$T_3^4 = \frac{1}{2} \times \left(\frac{1}{12} - \sum_{ij} \gamma_i \, \beta_{ij} \, \alpha_j^2\right)$
	$T_4^4 = \left(\frac{1}{24} - \sum_{ijk} \gamma_i \, \beta_{ij} \, \beta_{jk} \, \alpha_k\right)$
$\lambda_5 = 9$	$T_1^5 = \frac{1}{24} \times \left(\frac{1}{5} - \sum_i \gamma_i \, \alpha_i^4\right)$
	$T_2^5 = \frac{1}{2} \times \left(\frac{1}{10} - \sum_{ij} \gamma_i \, \alpha_i^2 \, \beta_{ij} \, \alpha_j\right)$
	$T_3^5 = \frac{1}{2} \times \left(\frac{1}{20} - \sum_{ijk} \gamma_i \, \beta_{ij} \, \alpha_j \, \beta_{ik} \, \alpha_k\right)$
	$T_4^5 = \frac{1}{2} \times \left(\frac{1}{15} - \sum_{ij} \gamma_i \, \alpha_i \, \beta_{ij} \, \alpha_j^2\right)$
	$T_5^5 = \frac{1}{6} \times \left(\frac{1}{20} - \sum_{ij} \gamma_i \, \beta_{ij} \, \alpha_j^3\right)$
	$T_6^5 = \left(\frac{1}{30} - \sum_{ijk} \gamma_i \, \alpha_i \, \beta_{ij} \, \beta_{jk} \, \alpha_k\right)$
	$T_7^5 = \left(\frac{1}{40} - \sum_{ijk} \gamma_i \, \beta_{ij} \, \alpha_j \, \beta_{jk} \, \alpha_k\right)$
	$T_8^5 = \frac{1}{2} \times \left(\frac{1}{60} - \sum_{ijk} \gamma_i \, \beta_{ij} \, \beta_{jk} \, \alpha_k^2\right)$
	$T_9^5 = \left(\frac{1}{120} - \sum_{ijkm} \gamma_i \, \beta_{ij} \, \beta_{jk} \, \beta_{km} \, \alpha_m\right)$

The expressions are valid for both implicit and explicit formulas. The sums are over all indices that are relevant. For example, the sum in T_2^3 does not have a term with $i = 0$ because there are no coefficients $\beta_{0,j}$. Further, if the formula is explicit, $\beta_{i,j} = 0$ for $j \geq i$ and the sum involves only those j with $j < i$.

If all the truncation error coefficients $T_n^i = 0$ for all $i \leq p$, then according to (1.9),

$$\mathbf{u}(x_j + h) - \mathbf{y}_{j+1} = h^{p+1}\left(\sum_{n=1}^{\lambda_{p+1}} T_n^{p+1}\mathbf{D}_n^{p+1}\right) + \mathcal{O}(h^{p+2}), \qquad (1.10)$$

which says that the formula is of order p. Similarly, if the formula is to be of order p for all sufficiently smooth $\mathbf{F}$, that is, for all possible elementary differentials $\mathbf{D}_n^i$ with $i \leq p$, then all the truncation error coefficients T_n^i must vanish for all $i \leq p$. These equations for a formula to be of order p are called the *equations of condition* or *order conditions*. They are usually derived and written so that it is a multiple of the truncation error coefficient that is set to zero. Each expression given above for a truncation error coefficient contains in parentheses the factor that is usually equated to zero to form the equation of condition. With explicit expressions for the truncation error coefficients, it is straightforward to determine the order of a given Runge–Kutta formula. On the other hand, to derive a formula of a given order p, it is necessary to find coefficients that make all these T_n^i vanish, i.e., to solve a complicated system of nonlinear equations for the parameters of the formula.

In the early derivation of formulas the coefficients were chosen to be "simple." A formula of this kind that has received a lot of attention is the *classic four-stage, fourth-order formula* with the array

$$\begin{array}{c|cccc} 0 & & & & \\ 1/2 & 1/2 & & & \\ 1/2 & 0 & 1/2 & & \\ 1 & 0 & 0 & 1 & \\ \hline & 1/6 & 1/3 & 1/3 & 1/6 \end{array}$$

EXERCISE 10. Verify that this formula is of order 4 by computing its truncation error coefficients.

EXERCISE 11. After the advantages of estimating the local error were recognized, researchers considered how to estimate the error of the classic fourth-order formula. As we shall see in Chapter 7, there are some general principles that can be used for this purpose, but researchers hoped to find less expensive ways to estimate the error. One discovered by Zonneveld [1964] adds a fifth stage to the step,

$$\mathbf{Y}_4 = \mathbf{y}_j + h\left(\frac{5}{32}\mathbf{F}_0 + \frac{7}{32}\mathbf{F}_1 + \frac{13}{32}\mathbf{F}_2 - \frac{1}{32}\mathbf{F}_3\right), \qquad \mathbf{F}_4 = \mathbf{F}\left(x_j + \frac{3}{4}h, \mathbf{Y}_4\right),$$

and then computes

$$\mathbf{d} = h\left(-\frac{2}{3}\mathbf{F}_0 + 2\mathbf{F}_1 + 2\mathbf{F}_2 + 2\mathbf{F}_3 - \frac{16}{3}\mathbf{F}_4\right)$$

as a measure of the local error. In this way an assessment of the error is obtained at the cost of one extra function evaluation per step. We must defer to Chapter 7 a proper interpretation of this measure of the error, but the key is that when **d** is added (Or subtracted? You are to find out.) to $\mathbf{y}_{j+1}$, a result $\mathbf{y}^*_{j+1}$ is obtained that is of order 3. Derive this formula by computing the truncation error coefficients for the two possibilities of adding or subtracting. Write out the third order formula directly in terms of the stages (rather than **d**).

EXERCISE 12. The following formula of Merson [1957] has been used in a number of popular codes, e.g., D02BAF in the NAG library and the code of Christiansen [1970]:

0					
1/3	1/3				
1/3	1/6	1/6			
1/2	0	1/8	3/8		
1	1/2	0	−3/2	2	
	1/6	0	0	2/3	1/6

Verify that this five-stage formula is of order 4. In addition, verify that the intermediate quantity $\mathbf{Y}_4$ is an approximation to $\mathbf{y}(x_{j+1})$ of order 3; this is the reason for the extra line in the Butcher array. The way that this is used to estimate the local error is explained in Example 4 of Chapter 7.

EXERCISE 13. One of the ways taken up in Chapter 7 to estimate the error of the classic fourth-order formula uses the formula to take a step of size $h/2$ to get $\mathbf{y}_{j+1/2}$ and then another step of size $h/2$ to get $\mathbf{y}_{j+1}$. In Exercise 4 it was seen that this $\mathbf{y}_{j+1}$ can be regarded as the result of an explicit Runge–Kutta formula. One way tried to estimate the error in this result is based on Simpson's rule. As in the quadrature approach to Runge–

Kutta formulas, these intermediate results and the rule lead to a formula

$$\mathbf{y}^s_{j+1} = \mathbf{y}_j + \frac{h}{6}\left[\mathbf{F}(x_j, \mathbf{y}_j) + 4\mathbf{F}\left(x_j + \frac{h}{2}, \mathbf{y}_{j+1/2}\right) + \mathbf{F}(x_j + h, \mathbf{y}_{j+1})\right].$$

The theory developed for the approach guarantees that $\mathbf{y}^s_{j+1}$ is a formula of order at least four. Verify that it is *not* of order five.

The quantity $\|\mathbf{y}^s_{j+1} - \mathbf{y}_{j+1}\|$ is used as an estimate of the local error of $\mathbf{y}_{j+1}$ in a number of codes, e.g., DSL in Shah [1976] and CSMP in Speckhard and Green [1976]. This estimate is "free" and the thought apparently was that the results of the two half steps would be improved when used with Simpson's rule. It is when the basic formula is of order three, but not when the basic formula is of order four. Still, the scheme has not performed badly in practice, so we ought to ask, why? An answer is given in Shampine and Baca [1985] that you are to work out for yourself now. Calculate T^5_i, $i = 1, 2, \ldots, 9$, the truncation error coefficients of $\mathbf{y}_{j+1}$, and $T^{5,s}_i$, the truncation error coefficients of $\mathbf{y}^s_{j+1}$. Show that in a *very rough* way,

$$T^5_i \approx \frac{1}{8}\left(T^5_i - T^{5,s}_i\right),$$

for each i and then argue that

$$\|\mathbf{y}^s_{j+1} - \mathbf{y}_{j+1}\| \approx 8\,\|\mathbf{y}(x_{j+1}) - \mathbf{y}_{j+1}\|.$$

We see, then, that controlling $\|\mathbf{y}^s_{j+1} - \mathbf{y}_{j+1}\|$ does provide a rough control of the local error in $\mathbf{y}_{j+1}$. The fact that the estimate is "usually" about an order of magnitude too big helps compensate for the crudeness of the control.

When a Runge–Kutta formula of a given order p is sought, it is usual that a family of such formulas is found and there is then the question of how to choose the "best" formula in the family. We have seen that in general the error of a Runge–Kutta formula depends on the problem. Because of this, no one formula in a family can be the most accurate for all problems. It is plausible that if one formula has truncation error coefficients that are uniformly smaller than those of another, the first formula will "usually" be more accurate. This can be quantified (Shampine [1985b]) by working with a bound on the error. If the role of the problem is quantified by defining

$$\Gamma_{p+1} = \max_{1 \le n \le \lambda_{p+1}} \|\mathbf{D}^{p+1}_n\|,$$

the error (1.10) can be bounded by

$$\|\mathbf{u}(x_j + h) - \mathbf{y}_{j+1}\| \le h^{p+1}\,\Gamma_{p+1}\left(\sum_{n=1}^{\lambda_{p+1}} |T_n^{p+1}|\right) + \mathcal{O}(h^{p+2}).$$

If the T_n^{p+1} are thought of as components in a vector $\mathbf{T}^{p+1}$ of λ_{p+1} elements, this bound can be written in tidy fashion by using a norm of the vector, namely the one norm:

$$\|\mathbf{u}(x_j + h) - \mathbf{y}_{j+1}\| \le h^{p+1}\,\Gamma_{p+1}\,\|\mathbf{T}^{p+1}\|_1 + \mathcal{O}(h^{p+2}).$$

Alternatively, different measures of the size of the truncation error coefficients can be employed by simple manipulation of the basic inequality:

$$\|\mathbf{u}(x_j + h) - \mathbf{y}_{j+1}\| \le h^{p+1}\,\Gamma_{p+1}\,\lambda_{p+1}\,\|\mathbf{T}^{p+1}\|_\infty + \mathcal{O}(h^{p+2}),$$

$$\|\mathbf{u}(x_j + h) - \mathbf{y}_{j+1}\| \le h^{p+1}\,\Gamma_{p+1}\,\sqrt{\lambda_{p+1}}\,\|\mathbf{T}^{p+1}\|_2 + \mathcal{O}(h^{p+2}).$$

Regardless of the particular measure of size, the idea is to choose the parameters defining the formula so as to make the size of the truncation error coefficients about as small as possible within the family of possible formulas. It is desirable that a formula have a relatively uniform behavior. For this reason present practice is to make the T_n^{p+1} of similar size, insofar as this is possible. For reasons to be taken up in another chapter, it is important to good software that none of these coefficients vanish. Because of these considerations, it does not matter a great deal which measure of size is used, but the Euclidean (two) norm is the most popular choice. A choice of parameters must be made. It is realized that no choice is best for all problems, but a reasonable way to make a choice is to make a bound on the error small.

Like the example of two-stage, second-order Runge–Kutta formulas, there is a general solution for three-stage, third-order formulas, see, e.g., Ralston [1965]. There are two parameters and we would like to choose them so as to have an "optimal" formula. However, also like the case of the second order formulas, one of the truncation error coefficients does not depend on the parameters. The best that can be done is to make the remaining truncation error coefficients small. It is possible to make some of the coefficients zero. This has the consequence that the formula will be of order 4 for some classes of problems. Ralston [1962] chose parameter values that make two of the truncation error coefficients equal to zero and make the Euclidean norm of the truncation error coefficients about as small as possible. His choice satisfies a couple of minor criteria in the construction of formulas. It is useful that the α_k are all distinct because in the step from x_j, the formula is sampling the behavior of the solution at $x_j + \alpha_k h$. Other things being equal, the more samples, the better because the additional samples make it more likely that unexpected changes in the problem, such as discontinuities or the step size being out of scale, will be revealed. Ralston did choose "nice" coefficients and this is a minor advantage in communicating the formula. These characteristics of his formula are verified in the exercise that follows. The formula is the basis for a pair developed by Bogacki and Shampine [1989] that will be taken up in Chapter 7. The pair is used in the TI-85 engineering graphics calculator. Moreover, it is one of the pairs available in the nice teaching package Differential Systems of Gollwitzer [1991] and one of the pairs available in RKSUITE.

EXERCISE 14. Verify that the following three-stage formula of Ralston [1962] is of order three:

$$\begin{array}{c|ccc} 0 & & & \\ \frac{1}{2} & \frac{1}{2} & & \\ \frac{3}{4} & 0 & \frac{3}{4} & \\ \hline & \frac{2}{9} & \frac{1}{3} & \frac{4}{9} \end{array}$$

In the course of this verification, you are to show that two of the truncation error coefficients T_n^4 are zero.

There are subtleties in the choice of formula that cannot be fully explained at this point, but some of the issues can be understood easily enough. Although early work focussed on achieving a given order in as few stages as possible, there is no need to do this and using more stages than necessary increases the possibilities for deriving an accurate formula of high quality. A reasonable measure of the cost of an explicit Runge–Kutta formula is the number of stages. Let us consider two formulas of order p and distinguish them by attaching asterisks to all quantities associated with one of the formulas. If one formula has s stages and the other $s^* > s$ stages, it is not necessarily true that the formula with the greater number of stages is less efficient. If it is more accurate, it might be able to use a step size h^* that is enough bigger than h to make the formula a bargain. The possibility is clear, but it is difficult to quantify efficiency because it depends on the problem. Again we resort to bounds to help us make a choice. Suppose that the step size h is to yield an accuracy of ε according to the bound on the error,

$$\varepsilon \approx h^{p+1}\, \Gamma_{p+1}\sqrt{\lambda_{p+1}}\, \|\mathbf{T}^{p+1}\|_2,$$

and h^* is to yield the same error with the other formula,

$$\varepsilon \approx (h^*)^{p+1} \Gamma_{p+1} \sqrt{\lambda_{p+1}} \, \|(\mathbf{T}^*)^{p+1}\|_2.$$

A natural measure of efficiency of an explicit Runge–Kutta formula is how far the integration is advanced in a step of length h divided by the cost s of the step. The matter of most interest now is the relative efficiency of two formulas:

$$\left(\frac{h}{s}\right) \Big/ \left(\frac{h^*}{s^*}\right) = \left(\frac{s^*}{s}\right) \left(\frac{\|(\mathbf{T}^*)^{p+1}\|_2}{\|\mathbf{T}^{p+1}\|_2}\right)^{1/(p+1)}.$$

With explicit expressions for the truncation error coefficients, this measure of relative efficiency can be used to find the most efficient formula in a family or to compare formulas from different families. Because of the dependence of the error of a Runge–Kutta formula on the problem, a formula that is more efficient than another by this measure might actually be less efficient for a given problem. This is especially true because only bounds on the error are being considered. We have here an example of the art of numerical analysis. The situation cannot be fully analyzed (well, no one has managed it yet!), but a choice must be made. Working with bounds is a reasonable way to proceed that has become accepted by workers in the field.

One subtlety in the preceding analysis is that it cannot be taken too far. Suppose for example that we wish to derive a formula of order two that involves three stages. It is possible to achieve order three with three stages. Because of this, if we minimize $\|\mathbf{T}^3\|_2$ in a suitable family of formulas, we get a minimum value of 0 and a formula of order 3. We need an additional measure of quality that prevents this degeneration. Several measures of quality being used at present accomplish this. One that can be appreciated at this point is that in the analysis presented, it is supposed that the leading term in the error expansion dominates so that higher order terms can be neglected. To be specific, in the case of a second order formula the leading term involves h^3. Unless its coefficient "accidentally" vanishes for the particular $\mathbf{F}$ and $\mathbf{y}_j$, this term will dominate successive terms in the expansion for all "sufficiently small" step sizes h. If its coefficient is small compared to the coefficient of the term involving h^4, the step size might have to be very small before the assumption that the leading term dominates is valid. This is not hypothetical. E. Fehlberg derived many very successful formulas. In Fehlberg [1970] he presents a second-order formula that has exactly this difficulty. Only for rather small step sizes does it "look" like a second-order formula. This formula is impractical because second-order formulas are simply not used when small step sizes are required. (One reason will be taken up shortly.) The size of the truncation error coefficients of the leading term in the error expansion must be compared to those of the next term in the expansion and some authors go so far as to examine the coefficients of several succeeding terms. In Chapter 7 we return to this matter and for now leave it with an appreciation that the coefficients of the leading term should not be greatly smaller than those of succeeding terms.

EXERCISE 15. The approach fails when we attempt to compare the efficiency of formulas of different orders p and p^*. How? It can be used to prove that as $\varepsilon \to 0$, the higher order formula becomes the more efficient, no matter what its relative cost. Do so. This result corroborates a more general result proven in §4.

There are several things about Runge–Kutta formulas that should concern us. At even a moderate order, a Runge–Kutta formula requires more stages than some of its competitors. This is frequently cited as a reason for preferring a competitor, but it is superficial. A formula that uses $2E$ evaluations per step may be more efficient than one that uses E evaluations because it can take a step more than twice as large. Obviously a proper comparison involves accuracy as well as cost. A more serious matter that might concern us is that it is not clear how we can estimate the local error. We shall see in a later chapter that this is easy enough to do; the trick is to do it efficiently. It is also not clear how to approximate $\mathbf{u}(x)$ between x_j and x_{j+1}. Current research has made considerable progress with this question. How this is done has been outlined and the reader should turn to the literature cited for the details of the formulas currently thought to be the "best."

§2 Methods with Memory

The methods presented in this section exploit the fact that when we reach a point x_j in the integration, we have at our disposal previously computed solution values $\mathbf{y}_j, \mathbf{y}_{j-1}, \mathbf{y}_{j-2}, \ldots$, and ordinarily approximate derivative values $\mathbf{F}(x_j, \mathbf{y}_j), \mathbf{F}(x_{j-1}, \mathbf{y}_{j-1}), \mathbf{F}(x_{j-2}, \mathbf{y}_{j-2}), \ldots$, as well. These values are reused to achieve a high order of accuracy with just a few additional evaluations of $\mathbf{F}$. The most important formulas of this kind are the Adams methods and the backward differentiation formulas, BDFs. They are members of a general class of formulas called *linear multistep methods*, LMMs, and some attention will be given to the whole class. There is a closely related class of methods called *predictor–corrector methods*. A brief discussion of how the implicit Adams formulas are evaluated in practice is used to explain how predictor–corrector methods are related to LMMs.

A more complete discussion of how implicit formulas are evaluated is deferred to §3.

§2.1 Adams Methods

A variation on the theme of quadrature provides very inexpensive formulas. Recall that in the last section we had

$$\mathbf{u}(x_j + h) = \mathbf{y}_j + h \sum_{k=1}^{\nu} w_k \mathbf{F}(x_j + \theta_k h, \mathbf{u}(x_j + \theta_k h)) + \mathcal{O}(h^{\mu+1}),$$

and the difficulty was that we did not know the $\mathbf{u}(x_j + \theta_k h)$. Earlier these auxiliary quantities were computed independently with lower order one-step formulas. A different tack is to use previously computed solution values. To do this we rely on the fact that we can derive interpolatory quadrature formulas that interpolate at any set of nodes $\{x_j + \theta_k h\}$ we like. The idea of the *Adams formulas* is to take the nodes to be the points where we have previously computed the solution so that we can use these solution values as the auxiliary quantities. There are a number of possibilities. The *Adams–Bashforth family* is based on the interpolatory quadrature formula arising from the polynomial $\mathbf{P}_\nu(x)$ defined by the interpolation conditions

$$\mathbf{P}_\nu(x_{j+1-k}) = \mathbf{F}(x_{j+1-k}, \mathbf{y}_{j+1-k}) = \mathbf{F}_{j+1-k} \qquad k = 1, 2, \ldots, \nu.$$

The vector-valued polynomial interpolates independently the components of $\mathbf{F}$. In terms of the *fundamental Lagrangian polynomials*

$$L_k(x) = \prod_{\substack{m=1 \\ m \neq k}}^{\nu} \left(\frac{x - x_{j+1-m}}{x_{j+1-k} - x_{j+1-m}} \right) \qquad k = 1, 2, \ldots, \nu,$$

the interpolant is

$$\mathbf{P}_\nu(x) = \sum_{k=1}^{\nu} \mathbf{F}_{j+1-k} L_k(x).$$

The formula for $\mathbf{y}_{j+1}$ is

$$\mathbf{y}_{j+1} = \mathbf{y}_j + \int_{x_j}^{x_j+h_j} \mathbf{P}_\nu(t)\, dt.$$

hence

$$\mathbf{y}_{j+1} = \mathbf{y}_j + h_j \sum_{k=1}^{\nu} \beta^*_{\nu,k} \mathbf{F}_{j+1-k} \tag{2.1}$$

where

$$\beta^*_{\nu,k} = \frac{1}{h_j} \int_{x_j}^{x_j+h_j} \prod_{\substack{m=1 \\ m \neq k}}^{\nu} \left(\frac{t - x_{j+1-m}}{x_{j+1-k} - x_{j+1-m}} \right) dt.$$

Notice that because the coefficients $\beta^*_{\nu,k}$ depend on the relative spacing of the mesh points x_{j+1-k}, they must be computed at each step. Techniques have been developed to do this efficiently and the cost is kept down in the codes by changing the step size only when it appears to be worth the cost of computing a new set of coefficients. Traditionally integrations were done with a constant step size. One reason for this is that the $\beta^*_{\nu,k}$ are constant then because it is the *relative* spacing that affects the coefficients. Specifically, with a constant step size h,

$$x_{j+1-k} = x_{j+1} - kh, \qquad k = 1, 2, \ldots, \nu,$$

and

$$\beta^*_{\nu,k} = \int_0^1 \prod_{\substack{m=1 \\ m \neq k}}^{\nu} \left(\frac{m - s}{m - k} \right) ds.$$

An Adams–Bashforth formula requires only one evaluation of $\mathbf{F}$ per step: Suppose we have reached x_j in the integration and have at our disposal $\mathbf{y}_j$, the preceding mesh points $x_{j-1}, x_{j-2}, \ldots$, and the preceding function values $\mathbf{F}_{j-1}, \mathbf{F}_{j-2}, \ldots$. To advance a step, we evaluate $\mathbf{F}_j = \mathbf{F}(x_j, \mathbf{y}_j)$, calculate, if necessary, the coefficients of the formula (2.1), and finally form $\mathbf{y}_{j+1}$ according to (2.1). The arguments made about the accuracy of formulas generated from a quadrature formula of order ν apply here as well. They say that if the auxiliary approximations, here the values $\mathbf{y}_{j-1}$, $\mathbf{y}_{j-2}, \ldots$, are accurate of order at least $\nu - 1$, then $\mathbf{y}_{j+1}$ is of order ν. Just as before, we can obtain explicit formulas of whatever order we wish, but now at the cost of only one stage, or as is said in this context, of one function evaluation per step.

The simplest example of an Adams–Bashforth formula of order ν, ABν, is the case $\nu = 1$, the case of no "memorized" function values. From the origin of the formula it is clear that the AB1 formula is just the forward Euler method. With one memorized value we get the second-order AB2 formula. In this case

$$\beta^*_{2,2} = \frac{1}{h_j} \int_{x_j}^{x_j+h_j} \left(\frac{t - x_j}{x_{j-1} - x_j} \right) dt = -\frac{1}{2} \frac{h_j}{h_{j-1}},$$

$$\beta^*_{2,1} = \frac{1}{h_j} \int_{x_j}^{x_j+h_j} \left(\frac{t - x_{j-1}}{x_j - x_{j-1}} \right) dt = 1 + \frac{1}{2} \frac{h_j}{h_{j-1}},$$

where $h_j = x_{j+1} - x_j$ and $h_{j-1} = x_j - x_{j-1}$. If the step is a constant h, the AB2 formula reduces to

$$\mathbf{y}_{j+1} = \mathbf{y}_j + h\left[\frac{3}{2}\mathbf{F}_j - \frac{1}{2}\mathbf{F}_{j-1}\right].$$

EXERCISE 16. Verify that AB3, the Adams–Bashforth formula of order 3, is

$$\mathbf{y}_{j+1} = \mathbf{y}_j + h\left[\frac{23}{12}\mathbf{F}_j - \frac{16}{12}\mathbf{F}_{j-1} + \frac{5}{12}\mathbf{F}_{j-2}\right]$$

when the step size is a constant h.

For numerical experimentation it is convenient to have a *table of Adams–Bashforth formulas*. In the form (2.1) with constant step size h, they are

ν	$\beta^*_{\nu,1}$	$\beta^*_{\nu,2}$	$\beta^*_{\nu,3}$	$\beta^*_{\nu,4}$	$\beta^*_{\nu,5}$	$\beta^*_{\nu,6}$
1	1					
2	$\frac{3}{2}$	$-\frac{1}{2}$				
3	$\frac{23}{12}$	$-\frac{16}{12}$	$\frac{5}{12}$			
4	$\frac{55}{24}$	$-\frac{59}{24}$	$\frac{37}{24}$	$-\frac{9}{24}$		
5	$\frac{1901}{720}$	$-\frac{2774}{720}$	$\frac{2616}{720}$	$-\frac{1274}{720}$	$\frac{251}{720}$	
6	$\frac{4277}{1440}$	$-\frac{7923}{1440}$	$\frac{9982}{1440}$	$-\frac{7298}{1440}$	$\frac{2877}{1440}$	$-\frac{475}{1440}$

The form assumed by the Adams–Bashforth formula is a consequence of the form chosen for the interpolating polynomial. The Lagrangian form we have used is convenient because the role of the quantities interpolated, the $\mathbf{F}_{j+1-k}$, is clear. Other forms of the interpolating polynomial are more convenient for some tasks in interpolation and the same is true of the corresponding form of the Adams–Bashforth formula. When the step size is a constant h, the Newton backward difference form of the interpolating polynomial is particularly convenient for estimating the error of the interpolation. It is also very convenient for selecting and changing the order of interpolation. As stated in the appendix, this form of the interpolating polynomial is

$$\mathbf{P}_\nu(x) = \mathbf{F}_j + \frac{(x - x_j)}{h} \nabla \mathbf{F}_j + \dots$$

$$+ \frac{(x - x_j)(x - x_{j-1}) \dots (x - x_{j+2-\nu})}{h^{\nu-1}(\nu - 1)!} \nabla^{\nu-1} \mathbf{F}_j,$$

where the backward difference operator has its usual meaning that

$$\nabla^0 \mathbf{F}_j = \mathbf{F}_j, \quad \nabla \mathbf{F}_j = \mathbf{F}_j - \mathbf{F}_{j-1}, \quad \text{and in general,} \quad \nabla^{k+1} \mathbf{F}_j = \nabla(\nabla^k \mathbf{F}_j).$$

This form of the interpolating polynomial is analogous to a Taylor series and correspondingly the Adams–Bashforth method in this form is analogous to a Taylor series method. The Adams–Bashforth formula is the result of integrating the interpolating polynomial, so

$$\mathbf{y}_{j-1} = \mathbf{y}_j + h \sum_{k=1}^{\nu} \gamma_{k-1} \nabla^{k-1} \mathbf{F}_j$$

where

$$\gamma_0 = 1 \quad \text{and for} \quad k \geq 1,$$

$$\gamma_k = \int_{x_j}^{x_j+h} \frac{(t - x_j) \cdots (t - x_{j+1-k})}{h^k \, k!} \, dt$$

$$= \frac{1}{k!} \int_0^1 s(s + 1) \cdots (s + k - 1) \, ds.$$

Notice that the γ_k do not depend on the order like the $\beta^*_{\nu,k}$ do. This makes changing order very easy. Examination of the recipe for $\mathbf{y}_{j+1}$ shows that if $\mathbf{y}^\nu_{j+1}$ is the result of the formula of order ν and $\mathbf{y}^{\nu+1}_{j+1}$ the result of the formula of order $\nu + 1$, then

$$\mathbf{y}^{\nu+1}_{j+1} = \mathbf{y}^\nu_{j+1} + h \, \gamma_\nu \, \nabla^\nu \mathbf{F}_j.$$

In this form it is clear that increasing the number of terms in the sum of the recipe for $\mathbf{y}_{j+1}$ increases the order of the formula, just as with a Taylor series expansion. In the days of hand computation, a table of backward differences was carried along with the solution. Inspection of this table suggested how many terms were appropriate in the formula. This is all very similar to what might be done with Taylor series methods. Like Taylor series, it turns out that the error can be approximated by the first term omitted in the expansion. Before going into this important matter, let us first note that it is straightforward to compute the constants γ_k for small k. By using the method of generating functions Henrici [1962, p. 192 ff.] it is possible to derive a simple recurrence for these constants:

$$\text{Let } \gamma_0 = 1 \text{ and then}$$

$$\gamma_k + \frac{1}{2}\gamma_{k-1} + \frac{1}{3}\gamma_{k-2} + \ldots + \frac{1}{k+1}\gamma_0 = 1, \qquad k = 1, 2, \ldots .$$

EXERCISE 17. Calculate the first few γ_k directly from their definitions as integrals and verify that they satisfy the recurrence stated. Using the recurrence verify that

k	1	2	3	4	5
γ_k	$\frac{1}{2}$	$\frac{5}{12}$	$\frac{3}{8}$	$\frac{251}{720}$	$\frac{95}{288}$

There are some important conceptual difficulties with methods with memory that are almost always glossed over. When we said earlier that $\mathbf{y}_{j+1}$ is of order p, we relied upon the general argument made for quadrature formulas. This argument requires the values $\mathbf{y}_j, \mathbf{y}_{j-1}, \ldots$ to be accurate of at least order $p - 1$. Methods with memory are qualitatively different from one-step methods. In the case of a one-step method, we speak of the error made in a step from x_j to $x_j + h$—this is a local matter. In the case of a method with memory, the error depends on the accuracy of previously computed values, and their accuracy depends on the accuracy of values computed still earlier, and so forth—this is a global matter. It is not even clear how to model practical computation. In practice we integrate to a point x_j, try a step of length h, and if the result is not acceptable, we reduce h and try again. This is modeled directly for one-step methods, but what about methods with memory? We certainly do not alter previously computed values and in particular, we do not alter the mesh spacing just prior to x_j. The analysis of convergence presented in the next chapter considers the integration as a whole and convergence is established as the maximum step size tends to zero. This is needed to justify the hypothesis that the previously computed values are accurate. When a constant step size h is reduced at x_j in this model of the computation, it is supposed that the whole integration is done with the new step size. To reconcile this with the way codes work, we can pretend that there is a regular distribution of step sizes that is to be used and what we are doing at x_j is finding out what the distribution is. The integration actually made is regarded as one member of a sequence of integrations done with step sizes taken from the regular distribution with the maximum step size tending to zero. We shall return to this issue later when we can be more specific.

In the convergence proofs, a key issue is how well the difference scheme imitates the behavior of the differential equation. This is measured by how well the solution of the differential equation satisfies the difference scheme, the (local) truncation error. Let us now examine this for Adams–Bashforth formulas. To see how well $\mathbf{y}(x)$, or rather its values at mesh points, satisfies the difference equation (2.1), we rely upon interpolation theory. We want to determine $\boldsymbol{\delta}_j$ in

$$\mathbf{y}(x_{j+1}) = \mathbf{y}(x_j) + h_j \sum_{k=1}^{\nu} \beta^*_{\nu,k}\, \mathbf{F}(x_{j+1-k}, \mathbf{y}(x_{j+1-k})) + \boldsymbol{\delta}_j.$$

In this case the underlying polynomial $\mathbf{P}_\nu$ interpolates the first derivative of the solution:

$$\mathbf{P}_\nu(x_{j+1-k}) = \mathbf{F}(x_{j+1-k}, \mathbf{y}(x_{j+1-k})) = \mathbf{y}'(x_{j+1-k}) \qquad k = 1, 2, \ldots, \nu.$$

From interpolation theory, for each t and each component i, there is an argument ξ such that

$$(\mathbf{y}'(t))^i = (\mathbf{P}_\nu(t))^i + \frac{(\mathbf{y}^{(\nu+1)}(\xi))^i}{\nu!} \prod_{k=1}^{\nu} (t - x_{j+1-k}).$$

As with Taylor expansions earlier, this expression and similar ones that follow will be written in the simpler notation

$$\mathbf{y}'(t) = \mathbf{P}_\nu(t) + \frac{\mathbf{y}^{(\nu+1)}(*)}{\nu!} \prod_{k=1}^{\nu} (t - x_{j+1-k}).$$

Then

$$\mathbf{y}(x_{j+1}) = \mathbf{y}(x_j) + \int_{x_j}^{x_j+h_j} \mathbf{P}_\nu(t)\, dt + \int_{x_j}^{x_j+h_j} \frac{\mathbf{y}^{(\nu+1)}(*)}{\nu!} \prod_{k=1}^{\nu} (t - x_{j+1-k})\, dt.$$

In this form it is clear that the discrepancy $\boldsymbol{\delta}_j$ is the second integral. The product term in this integral is of one sign, so a mean value theorem leads to

$$\boldsymbol{\delta}_j = \frac{\mathbf{y}^{(\nu+1)}(*)}{\nu!} \int_{x_j}^{x_j+h_j} \prod_{k=1}^{\nu} (t - x_{j+1-k})\, dt$$

where the components of the derivative are evaluated at points in the span of the nodes used in the formula. If H is a bound for the step sizes in this span, it is easy to see that the local truncation error

$$\|\boldsymbol{\delta}_j\| \leq \|\mathbf{y}^{\nu+1)}\| \, H^{\nu+1},$$

hence that the formula is of order ν. When the step size is a constant h, the truncation error simplifies to

$$\boldsymbol{\delta}_j = \mathbf{y}^{(\nu+1)}(*)h^{\nu+1} \frac{1}{\nu!} \int_0^1 \prod_{m=0}^{\nu-1} (s + m) \, ds = \gamma_\nu \, \mathbf{y}^{(\nu+1)}(*)h^{\nu+1}.$$

This analysis suggests a way to estimate the truncation error. If one more node is used so that $\mathbf{P}_{\nu+1}(x)$ can be formed, the error is

$$\mathbf{y}(x_{j+1}) = \mathbf{y}(x_j) + \int_{x_j}^{x_j+h_j} \mathbf{P}_{\nu+1}(t) \, dt + \mathcal{O}(H^{\nu+2}).$$

Subtracting this expression from the one satisfied by $\mathbf{P}_\nu(x)$ and a little manipulation leads to

$$\boldsymbol{\delta}_j = \int_{x_j}^{x_j+h_j} \mathbf{P}_{\nu+1}(t) \, dt - \int_{x_j}^{x_j+h_j} \mathbf{P}_\nu(t) \, dt + \mathcal{O}(H^{\nu+2}).$$

The expressions here are easily evaluated when the step size is a constant h and the interpolating polynomials are in backward difference form:

$$\boldsymbol{\delta}_j = h \, \gamma_\nu \nabla^\nu \, \mathbf{F}_j + \mathcal{O}(h^{\nu+2}).$$

The estimate of the truncation error is simply the first term omitted in the sum for the formula. This is just like using a Taylor series expansion.

Obviously the backward difference form of the interpolating polynomial is very convenient for estimating how well the formula imitates the differential equation and for changing the order of the formula used. These advantages are not limited to constant step size. The theory of interpolation resorts to divided differences to generalize backward differences to a general mesh and naturally the representation can be used to derive the Adams–Bashforth formulas for a general mesh. Of course the integration coefficients, the equivalent of the γ_k, now depend on the relative spacing of the mesh points $x_j, x_{j-1}, \ldots$, hence must be calculated at each step. A certain amount of technique has been developed for this purpose and the algorithms in the codes try to minimize this cost by changing the step size only when it is necessary or appears to be worth the cost. The treatment of variable step size is different in detail, but not in kind, so the details are omitted. A full treatment can be found in Shampine and Gordon [1975].

The *Adams–Moulton family* of formulas is a variation on the theme. In this case the polynomial $\mathbf{P}_\nu(x)$ of degree at most $\nu - 1$ is defined by

$$\mathbf{P}_\nu(x_{j+1-k}) = \mathbf{F}_{j+1-k}, \qquad k = 0, 1, \ldots, \nu - 1.$$

These formulas are different in nature from the Adams–Bashforth formulas because interpolating at $x_{j+1}(k = 0)$ makes the formula implicit:

$$\mathbf{y}_{j+1} = \mathbf{P}_\nu(x_{j+1}) = \mathbf{F}(x_{j+1}, \mathbf{y}_{j+1}).$$

Accepting for now the implicit definition, integration of the polynomial leads to a formula exactly as in the case of the Adams–Bashforth formulas. Specifically,

$$\mathbf{y}_{j+1} = \mathbf{y}_j + h_j \sum_{k=0}^{\nu-1} \beta_{\nu,k}\, \mathbf{F}_{j+1-k}$$

where

$$\beta_{\nu,k} = \frac{1}{h_j} \int_{x_j}^{x_j+h_j} \prod_{\substack{m=0 \\ m \neq k}}^{\nu-1} \left(\frac{t - x_{j+1-m}}{x_{j+1-k} - x_{j+1-m}} \right) dt.$$

For the evaluation of the formula it is useful to express more clearly the terms involving the new approximate solution:

$$\mathbf{y}_{j+1} = \mathbf{y}_j + h_j\, \beta_{\nu,0}\, \mathbf{F}(x_{j+1}, \mathbf{y}_{j+1}) + h_j \sum_{k=1}^{\nu-1} \beta_{\nu,k}\, \mathbf{F}_{j+1-k}. \tag{2.2}$$

Again the simplest case of $\nu = 1$ degenerates to a one-step method that is obviously the backward Euler method. However, the case of $\nu = 2$ also degenerates to an implicit one-step formula that is called the trapezoidal rule. We leave it and the non-degenerate case of $\nu = 3$ as exercises.

EXERCISE 18. Show that when the step size is a constant h, the Adams–Moulton formula of order 2, AM2, is

$$\mathbf{y}_{j+1} = \mathbf{y}_j + h\left[\frac{1}{2}\,\mathbf{F}(x_{j+1}, \mathbf{y}_{j+1}) + \frac{1}{2}\,\mathbf{F}_j\right].$$

EXERCISE 19. Show that when the step size is a constant h, the Adams–Moulton formula of order 3, AM3, is

$$\mathbf{y}_{j+1} = \mathbf{y}_j + h\left[\frac{5}{12}\,\mathbf{F}(x_{j+1}, \mathbf{y}_{j+1}) + \frac{8}{12}\,\mathbf{F}_j - \frac{1}{12}\,\mathbf{F}_{j-1}\right].$$

For numerical experimentation it is convenient to have a *table of Adams–Moulton formulas*. In the form (2.2) with constant step size h, they are

ν	$\beta_{\nu,0}$	$\beta_{\nu,1}$	$\beta_{\nu,2}$	$\beta_{\nu,3}$	$\beta_{\nu,4}$	$\beta_{\nu,5}$
1	1					
2	$\frac{1}{2}$	$\frac{1}{2}$				
3	$\frac{5}{12}$	$\frac{8}{12}$	$-\frac{1}{12}$			
4	$\frac{9}{24}$	$\frac{19}{24}$	$-\frac{5}{24}$	$\frac{1}{24}$		
5	$\frac{251}{720}$	$\frac{646}{720}$	$-\frac{264}{720}$	$\frac{106}{720}$	$-\frac{19}{720}$	
6	$\frac{475}{1440}$	$\frac{1427}{1440}$	$-\frac{798}{1440}$	$\frac{482}{1440}$	$-\frac{173}{1440}$	$\frac{27}{1440}$

As with the Adams–Bashforth formulas, when the step size is a constant h, the backward difference form of the interpolating polynomial leads to a formula convenient for error estimation and change of order:

$$\mathbf{y}_{j+1} = \mathbf{y}_j + h \sum_{k=1}^{\nu} \gamma_{k-1}^* \nabla^{k-1} \mathbf{F}_{j+1}$$

where

$$\gamma_0^* = 1 \quad \text{and for} \quad k \geq 1,$$

$$\gamma_k^* = \frac{1}{k!} \int_0^1 (s - 1)(s) \cdots (s + k - 2)\, ds.$$

Again, using the method of generating functions Henrici [1962, p. 194 ff.] it is possible to derive a simple recurrence for these constants:

Let $\gamma_0^* = 1$ and then

$$\gamma_k^* + \frac{1}{2} \gamma_{k-1}^* + \frac{1}{3} \gamma_{k-2}^* + \cdots + \frac{1}{k+1} \gamma_0^* = 0, \qquad k = 1, 2, \ldots .$$

EXERCISE 20. Calculate the first few γ_k^* directly from their definitions as integrals and verify that they satisfy the recurrence stated. Using the recurrence verify that

k	1	2	3	4	5
γ_k^*	$-\frac{1}{2}$	$-\frac{1}{12}$	$-\frac{1}{24}$	$-\frac{19}{720}$	$-\frac{3}{160}$

EXERCISE 21. Using the recurrences prove that $\gamma_k^* = \gamma_k - \gamma_{k-1}$ for $k \geq 1$.

Just as with the Adams–Bashforth family, interpolation theory can be used to find out how well the formula imitates the differential equation. We want to determine $\boldsymbol{\delta}_j$ in

$$\mathbf{y}(x_{j+1}) = \mathbf{y}(x_j) + h_j\beta_{\nu,0}\,\mathbf{F}(x_{j+1}, \mathbf{y}(x_{j+1})) + h_j \sum_{k=1}^{\nu-1} \beta_{\nu,k}\,\mathbf{F}(x_{j+1-k}, \mathbf{y}(x_{j+1-k})) + \boldsymbol{\delta}_j.$$

Again the product term in the integral is of one sign so that a mean value theorem gives

$$\boldsymbol{\delta}_j = \frac{\mathbf{y}^{(\nu+1)}(*)}{\nu!} \int_{x_j}^{x_j+h_j} \prod_{k=0}^{\nu-1} (t - x_{j+1-k})\,dt$$

where the components of the derivative are evaluated at points in the span of the nodes. If H is a bound for the step sizes in this span, it follows that

$$\|\boldsymbol{\delta}_j\| \leq \|\mathbf{y}^{(\nu+1)}\|\, H^{\nu+1},$$

hence that the formula is of order ν. When the step size is a constant h,

$$\boldsymbol{\delta}_j = \mathbf{y}^{(\nu+1)}(*)h^{\nu+1} \frac{1}{\nu!} \int_0^1 \prod_{k=0}^{\nu-1} (s + k - 1)\,ds = \gamma_\nu^*\, \mathbf{y}^{(\nu+1)}(*)h^{\nu+1}.$$

As with the Adams–Bashforth formulas, the formulation in terms of differences provides a convenient estimate of the truncation error:

$$\boldsymbol{\delta}_j = h\,\gamma_\nu^*\,\nabla^\nu\,\mathbf{F}_{j+1} + \mathcal{O}(h^{\nu+2}).$$

We observed earlier that the forward Euler and backward Euler methods have essentially the same accuracy. In their present guise as AB1 and AM1, the expressions just given quantify this observation. This order is the exception. Comparison of γ_ν to γ_ν^* shows that the Adams–Moulton formula of even moderately large order ν is considerably more accurate than the Adams–Moulton formula of the same order (γ_ν^* is considerably smaller than γ_ν).

§2.2 Backward Differentiation Formulas (BDFs)

We have exploited numerical integration to derive formulas and it is natural to ask if numerical differentiation might also lead to interesting formulas. The backward Euler method can be derived by means of a backward difference approximation to the derivative as in

$$\mathbf{y}'(x_{j+1}) = \mathbf{F}(x_{j+1}, \mathbf{y}(x_{j+1})) \approx \frac{\mathbf{y}(x_{j+1}) - \mathbf{y}(x_j)}{h}.$$

Similarly, the forward Euler method can be derived from the forward difference

$$\mathbf{y}'(x_j) = \mathbf{F}(x_j, \mathbf{y}(x_j)) \approx \frac{\mathbf{y}(x_{j+1}) - \mathbf{y}(x_j)}{h}.$$

In this section we pursue the argument leading to the backward Euler method in a systematic fashion to derive the family of *backward differentiation formulas*, BDFs.

Suppose we have previously computed values $\mathbf{y}_j, \mathbf{y}_{j-1}, \ldots$ and wish to determine $\mathbf{y}_{j+1}$. Once again we approximate the solution by an interpolating polynomial $\mathbf{P}_\nu(x)$ of degree at most ν. Now, however, we interpolate the solution itself:

$$\mathbf{P}_\nu(x_{j+1-k}) = \mathbf{y}_{j+1-k}, \qquad k = 0, 1, \ldots, \nu.$$

A way to approximate the derivative of a function is to differentiate an interpolating polynomial. This is done at $x = x_{j+1}$. Somehow the information that $\mathbf{y}$ is a solution of the differential equation must be taken into account. This is done by relating $\mathbf{y}'(x_{j+1})$ to $\mathbf{y}(x_{j+1})$ by means of the differential equation. Thus it is required that

$$\mathbf{P}'_\nu(x_{j+1}) = \mathbf{F}(x_{j+1}, \mathbf{P}_\nu(x_{j+1})) = \mathbf{F}(x_{j+1}, \mathbf{y}_{j+1}).$$

This represents a set of algebraic equations for $\mathbf{y}_{j+1}$, hence is an implicit formula. To see just what kind of formula has been obtained, the fundamental Lagrangian polynomials

$$L_k(x) = \prod_{\substack{m=0 \\ m \neq k}}^{\nu} \left(\frac{x - x_{j+1-m}}{x_{j+1-k} - x_{j+1-m}} \right), \qquad k = 0, 1, \ldots, \nu$$

can be used to write

$$\mathbf{P}_\nu(x) = \sum_{k=0}^{\nu} L_k(x)\, \mathbf{y}_{j+1-k}.$$

and then

$$\mathbf{P}'(x_{j+1}) = \sum_{k=0}^{\nu} L'_k(x_{j+1})\, \mathbf{y}_{j+1-k} = \mathbf{F}(x_{j+1}, \mathbf{y}_{j+1}). \tag{2.3}$$

This implicit formula has some resemblance to an Adams–Moulton formula and when the step size is a constant h, the coefficients simplify in the same way. Let η be defined by $x = x_j + \eta h$. Then

$$L_k(\eta) = \prod_{\substack{m=0\\m\neq k}}^{\nu} \left(\frac{\eta + m - 1}{m - k}\right)$$

and

$$\frac{d}{dx} L_k(x)\bigg|_{x_{j+1}} = \frac{1}{h}\frac{d}{d\eta} L_k(\eta)\bigg|_1 = \frac{1}{h}\alpha_k.$$

With this definition of α_k and multiplication by h, (2.3) becomes

$$\alpha_0\, \mathbf{y}_{j+1} + \alpha_1\, \mathbf{y}_j + \ldots + \alpha_\nu\, \mathbf{y}_{j+1-\nu} = \text{h}\, \mathbf{F}(x_{j+1}, \mathbf{y}_{j+1}). \qquad (2.4)$$

The simplest example of $\nu = 1$ has $\mathbf{P}_1(x)$ as the linear polynomial that interpolates $\mathbf{y}_{j+1}$ and $\mathbf{y}_j$. Its derivative is simply the constant slope of this straight line so that (2.3) is

$$\mathbf{P}_1'(\text{x}_{j+1}) = \frac{1}{h}(\mathbf{y}_{j+1} - \mathbf{y}_j) = \mathbf{F}(x_{j+1}, \mathbf{y}_{j+1}).$$

After an obvious rearrangement, this is the familiar backward Euler method. In this particular case the Adams–Moulton family and the BDF overlap. They are generally quite different. In a way, they are both extreme cases among the LMMs because the one uses the minimum number of **y** values and the other uses the minimum number of **F** values.

EXERCISE 22. Show that for constant step size h, the backward differentiation formula for $\nu = 2$ is

$$\frac{3}{2}\mathbf{y}_{j+1} - 2\mathbf{y}_j + \frac{1}{2}\mathbf{y}_{j-1} = h\, \mathbf{F}(x_{j+1}, \mathbf{y}_{j+1}).$$

For later reference we give a *table of the BDFs* in the form (2.4):

ν	α_0	α_1	α_2	α_3	α_4	α_5	α_6
1	1	-1					
2	$\frac{3}{2}$	-2	$\frac{1}{2}$				
3	$\frac{11}{6}$	-3	$\frac{3}{2}$	$-\frac{1}{3}$			
4	$\frac{25}{12}$	-4	3	$-\frac{4}{3}$	$\frac{1}{4}$		

ν	α_0	α_1	α_2	α_3	α_4	α_5	α_6
5	$\frac{137}{60}$	-5	5	$-\frac{10}{3}$	$\frac{5}{4}$	$-\frac{1}{5}$	
6	$\frac{147}{60}$	-6	$\frac{15}{2}$	$-\frac{20}{3}$	$\frac{15}{4}$	$-\frac{6}{5}$	$\frac{1}{6}$

The family of formulas has a particularly attractive form when written in terms of backward differences. This may account for the name backward differentiation formulas or, as we shall write, the BDFs. C.W. Gear popularized the use of these formulas for the solution of stiff problems. Although they had been studied (Henrici [1962]) and their use advocated for stiff problems (Curtiss and Hirschfelder [1952]) much earlier, the well-deserved popularity of the code DIFSUB (Gear [1971a,b,c]) led many to refer to the BDFs as *Gear's methods.*

Expressed in terms of backward differences, the interpolating polynomial is

$$\mathbf{P}_\nu(x) = \mathbf{y}_{j+1} + \frac{(x - x_{j+1})}{h} \nabla \mathbf{y}_{j+1} + \ldots + \frac{(x - x_{j+1}) \cdots (x - x_{j+2-\nu})}{h^\nu \, \nu!} \nabla^\nu \mathbf{y}_{j+1}.$$

The BDF arises from differentiating this expression and evaluating it at x_{j+1}. The coefficient of the term involving $\nabla^{k+1} \mathbf{y}_{j+1}$ is then

$$\frac{1}{h^{k+1}(k+1)!} \frac{d}{dx} \prod_{m=0}^{k} (x - x_{j+1-m}) \Bigg|_{x_{j+1}}$$

$$= \frac{1}{h^{k+1}(k+1)!} \sum_{m=0}^{k} \prod_{\substack{r=0 \\ r \neq m}}^{k} (x - x_{j+1-r}) \Bigg|_{x_{j+1}}$$

$$= \frac{1}{h^{k+1}(k+1)!} \prod_{r=1}^{k} (x_{j+1} - x_{j+1-r}) = \frac{h^k k!}{h^{k+1}(k+1)!} = \frac{1}{h(k+1)}.$$

The BDF is then

$$\mathbf{F}(x_{j+1}, \mathbf{y}_{j+1}) = \mathbf{P}'_\nu(x_{j+1}, \mathbf{y}_{j+1}) = \frac{1}{h} \nabla \mathbf{y}_{j+1} + \cdots + \frac{1}{h\nu} \nabla^\nu \mathbf{y}_{j+1}.$$

Scaling by h as with the Lagrangian form leads to

$$h\,\mathbf{F}(x_{j+1}, \mathbf{y}_{j+1}) = \sum_{k=1}^{\nu} \frac{1}{k} \nabla^k \mathbf{y}_{j+1}.$$

This is an attractively simple form of the BDF, but it is not so convenient computationally because in solving iteratively this algebraic equation for $\mathbf{y}_{j+1}$, the right hand side requires the repeated formation of differences to take account of the successive approximations to $\mathbf{y}_{j+1}$. A more convenient form can be derived after we consider how to predict a starting value for the iterative process. This is done in all the codes by interpolating previously computed solution values at x_{j+1-k} for $k = 1, 2, \ldots, \nu + 1$ and evaluating this interpolant at x_{j+1}. The interpolating polynomial in backward difference form is

$$\mathbf{Q}_\nu(x) = \mathbf{y}_j + \frac{(x - x_j)}{h} \nabla \mathbf{y}_j + \ldots + \frac{(x - x_j) \cdots (x - x_{j+1-\nu})}{h^\nu \, \nu!} \nabla^\nu \mathbf{y}_j,$$

so the predicted value is

$$\mathbf{y}_{j+1}^* = \mathbf{Q}_\nu(x_{j+1}) = \mathbf{y}_j + \frac{(h)}{h} \nabla \mathbf{y}_j + \ldots + \frac{(h) \cdots (\nu h)}{h^\nu \nu!} \nabla^\nu \mathbf{y}_j = \sum_{k=0}^{\nu} \nabla^k \mathbf{y}_j.$$

This is a very convenient predictor given that the differences here are retained from the last step. For later use we relate this predicted value to $\mathbf{y}_{j+1}$:

$$\begin{aligned}
\mathbf{y}_{j+1} - \mathbf{y}_{j+1}^* &= (\mathbf{y}_{j+1} - \mathbf{y}_j) - \nabla \mathbf{y}_j - \nabla^2 \mathbf{y}_j - \ldots - \nabla^\nu \mathbf{y}_j, \\
&= (\nabla \mathbf{y}_{j+1} - \nabla \mathbf{y}_j) - \nabla^2 \mathbf{y}_j - \ldots - \nabla^\nu \mathbf{y}_j, \\
&= (\nabla^2 \mathbf{y}_{j+1} - \nabla^2 \mathbf{y}_j) - \ldots - \nabla^\nu \mathbf{y}_j = \ldots = \nabla^{\nu+1} \mathbf{y}_{j+1}.
\end{aligned}$$

It is also useful to predict a value for the derivative of the solution by differentiating $\mathbf{Q}_\nu(x)$ and evaluating it at x_{j+1}. The manipulations are very much like those leading to the formula itself and result in

$$\mathbf{d}_{j+1} = \mathbf{Q}_\nu'(x_{j+1}) = \frac{1}{h} \sum_{k=1}^{\nu} \delta_k^* \, \nabla^k \mathbf{y}_j,$$

where now the constants

$$\delta_k^* = \sum_{m=1}^{k} \frac{1}{m}.$$

Notice that

$$\delta_1^* = 1 \quad \text{and} \quad \delta_k^* - \delta_{k-1}^* = \frac{1}{k} \quad \text{for} \quad k > 1.$$

Because of this relation,

$$\begin{aligned}\mathbf{F}(x_{j+1}, \mathbf{y}_{j+1}) &= \frac{1}{h}\sum_{k=1}^{\nu}\frac{1}{k}\,\nabla^k\mathbf{y}_{j+1}\\ &= \frac{1}{h}\,\nabla\mathbf{y}_{j+1} + \frac{1}{h}\sum_{k=2}^{\nu}(\delta_k^* - \delta_{k-1}^*)\nabla^k\mathbf{y}_{j+1}\\ &= \frac{1}{h}\sum_{k=1}^{\nu}\delta_k^*\nabla^k\mathbf{y}_{j+1} - \frac{1}{h}\sum_{k=2}^{\nu}\delta_{k-1}^*(\nabla^{k-1}\mathbf{y}_{j+1} - \nabla^{k-1}\,\mathbf{y}_j)\\ &= \frac{1}{h}\,\delta_\nu^*\nabla^\nu\,\mathbf{y}_{j+1} + \frac{1}{h}\sum_{m=1}^{\nu-1}\delta_m^*\nabla^m\,\mathbf{y}_j.\end{aligned}$$

This result and previous ones lead to

$$\begin{aligned}\mathbf{F}(x_{j+1}, \mathbf{y}_{j+1}) - \mathbf{d}_{j+1} &= \frac{1}{h}\,\delta_\nu^*\nabla^\nu\,\mathbf{y}_{j+1} - \frac{1}{h}\,\delta_\nu^*\nabla^\nu\,\mathbf{y}_j,\\ &= \frac{1}{h}\,\delta_\nu^*\nabla^{\nu+1}\,\mathbf{y}_{j+1} = \frac{1}{h}\,\delta_\nu^*\,(\mathbf{y}_{j+1} - \mathbf{y}_{j+1}^*).\end{aligned}$$

Finally, the algebraic equation can be written as

$$(\mathbf{F}(x_{j+1}, \mathbf{y}_{j+1}) - \mathbf{d}_{j+1}) - \frac{1}{h}\,\delta_\nu^*(\mathbf{y}_{j+1} - \mathbf{y}_{j+1}^*) = 0.$$

In this form there is less arithmetic in an iteration because the two terms depending on the solution are handy. This is also a good way of writing the equation because we can compute the solution $\mathbf{y}_{j+1}$ as a correction to the predicted value $\mathbf{y}_{j+1}^*$ and similarly for the derivative.

The Adams formulas are based on interpolation to the approximations to the first derivative $\mathbf{y}'_{j-k}$ ($= \mathbf{F}_{j-k}$). The BDFs are based on interpolation to the solution values $\mathbf{y}_{j-k}$ themselves. Just as with the Adams methods, we can use an error result for polynomial interpolation to get the order and the local truncation error of the formula. Suppose that $\mathbf{P}_\nu(x)$ interpolates $\mathbf{y}(x)$:

$$\mathbf{P}_\nu(x_{j+1-k}) = \mathbf{y}(x_{j+1-k}), \qquad k = 0, 1, \ldots, \nu.$$

A standard result for the error in the approximation of the derivative has a simple form at a node:

$$\mathbf{y}'(x_{j+1}) = \mathbf{P}'_\nu(x_{j+1}) + \frac{1}{(\nu + 1)!}\,\mathbf{y}^{(\nu+1)}(*)\prod_{k=1}^{\nu}(x_{j+1} - x_{j+1-k}).$$

We can bound this error in terms of a bound for $\|\mathbf{y}^{(\nu+1)}\|$ and a bound H on the step sizes in the span of the data to deal with general meshes and from it conclude that the method is of order ν. Rather than work out such a bound in detail, let us work out the case of a constant step size h because it is more important for this family of formulas. For constant step size h it is easy to see that

$$\mathbf{y}'(x_{j+1}) = \mathbf{F}(x_{j+1}, \mathbf{y}(x_{j+1})) = \mathbf{P}'_\nu(x_{j+1}) + \frac{1}{\nu + 1}\mathbf{y}^{(\nu+1)}(*)h^\nu.$$

Now

$$\alpha_0\,\mathbf{y}(x_{j+1}) + \alpha_1\,\mathbf{y}(x_j) + \cdots + \alpha_k\,\mathbf{y}(x_{j+1-k}) - h\,\mathbf{F}(x_{j+1}, \mathbf{y}(x_{j+1}))$$

$$= \boldsymbol{\delta}_j = h(\mathbf{P}'_\nu(x_{j+1}) - \mathbf{F}(x_{j+1}, \mathbf{y}(x_{j+1}))) = -\frac{1}{\nu + 1}\mathbf{y}^{(\nu+1)}(*)h^{\nu+1}.$$

This says that the formula is of order ν.

Although this construction produces formulas of any order, the high-order formulas cannot be used in practice. Specifically, if the step size is constant, the formulas of orders 7 and up are useless computationally because they are not stable. Nothing we have done so far would suggest that the BDF of order 7 would be any different in character from the BDF of order 6, or for that matter, from the Adams–Moulton formula of order 7. The reason we do not see any difference is that so far we have been looking only at what happens in a single step. The next chapter deals with stability and convergence, the cumulative effect of errors. It turns out that a BDF is not nearly as accurate as an Adams–Moulton formula of the same order and only relatively low order BDFs are stable. Nevertheless, the stable BDFs are so much more stable than the Adams formulas of the same order that they are essential for the solution of those problems for which stability is a severe constraint on the step size, the stiff problems.

§2.3 Linear Multistep and Predictor–Corrector Methods

A good question at this point is, why do we bother with implicit formulas? Earlier we considered the matter briefly with the forward and backward Euler methods, methods that in this context we call AB1 and AM1, respectively. As with the first order formulas, the Adams–Moulton formulas of moderately high orders are much more stable than the Adams–Bashforth formulas of the same order. Unlike the first order formulas, the Adams–Moulton formulas of moderately high order are considerably more accurate than the Adams–Bashforth formulas of the same order. Because of these facts, an Adams–Moulton formula of moderate to high order permits a considerably bigger step size than the Adams–Bashforth formula of the same order. The main question is whether this bigger step size will compensate for the higher cost of evaluating an implicit formula. In the popular codes this cost varies from step to step, but is generally in the range of 1.5 to 2 evaluations of **F** per step. This is sufficiently inexpensive that the Adams–Moulton formulas are a bargain. In §3 we take up the evaluation of the formulas, but a brief description is needed here to lead into another class of formulas.

EXERCISE 23. To study the assertion made about the cost of a step in an Adams–Moulton code, solve some problems using one of the Adams–Moulton codes like DDRIV2, DIFSUB, EPISODE, LSODE, VODE, or their variants found in popular software libraries. Count the number of steps and the cost in evaluations of **F** required to solve the problem to determine an average cost per step. The codes cited make this data available to you, though in different ways. This easy computation is somewhat superficial because it does not account for the extra costs associated with starting and failed steps, but you will be able to get a fair idea of the average cost. Two examples given in Chapter 8 illustrate what you might find. Example 1 is the solution of an unstable problem with LSODE for which the average cost is about 1.4 evaluations per step. Example 7 is the solution of a stiff problem with EPISODE for which the average cost is about 1.7 evaluations per step.

There is a broad class of formulas called *linear multistep methods*, LMMs, that includes the Adams formulas and the BDFs. These methods were first defined and investigated at a time when it was assumed that the step size would be a constant h. It is still useful to investigate this special case. For some kinds of computations, in particular those associated with the solution of partial differential equations by semi-discretization, it is still common to integrate with a constant step size. General-purpose integrators based on methods with memory tend to work with a constant step size for a variety of practical reasons. It is hardly surprising that we can understand better in a theoretical way the behavior of an integrator when the step size is constant, so that investigating this special case provides valuable insight.

When the step size is a constant h, the coefficients defining the formulas derived so far are constants, so the general form of a LMM is taken to be

$$\sum_{i=0}^{k} \alpha_i \mathbf{y}_{j+1-i} = h \sum_{i=0}^{k} \beta_i \mathbf{F}(x_{j+1-i}, \mathbf{y}_{j+1-i}). \tag{2.5}$$

The coefficients α_i and β_i define the formula. This is a recipe for $\mathbf{y}_{j+1}$ that is to be valid for all sufficiently small h, so we must require that $\alpha_0 \neq 0$. When $\beta_0 = 0$, this is an explicit recipe for $\mathbf{y}_{j+1}$ in terms of the previously computed values $\mathbf{y}_{j+1-i}$ and $\mathbf{F}_{j+1-i} = \mathbf{F}(x_{j+1-i}, \mathbf{y}_{j+1-i})$ for $i = 1, \ldots, k$. When $\beta_0 \neq 0$, the new value $\mathbf{y}_{j+1}$ is defined only implicitly by the recipe, but it is shown in §3 that for all sufficiently small h, there is a unique $\mathbf{y}_{j+1}$ that satisfies (2.5). It is usually assumed that $|\alpha_k| + |\beta_k| \neq 0$ so that the formula is genuinely k-step. Notice that the results of such a formula would not be altered if all the coefficients were divided by any positive number. Depending on the context, different normalizations of this kind are preferred. It will turn out that a polynomial defined in terms of the α_i will play a key role in the analysis of such formulas. This (*first*) *characteristic polynomial* $\rho(\theta)$ is

$$\rho(\theta) = \alpha_0\theta^k + \alpha_1\theta^{k-1} + \ldots + \alpha_k.$$

When it is necessary to work with a *normalized formula*, it will be assumed that $\rho'(1) = 1$. Notice that as the Adams formulas were defined earlier, $\rho(\theta) = \theta^k - \theta^{k-1}$ and the formulas are normalized. Later we shall see that as derived earlier, the BDFs are normalized, too.

If $\beta_0 \neq 0$, the formula is implicit. Let us write the formula in a way that makes the role of $\mathbf{y}_{j+1}$ clearer:

$$\mathbf{y}_{j+1} = h\frac{\beta_0}{\alpha_0}\mathbf{F}(x_{j+1}, \mathbf{y}_{j+1}) - \sum_{i=1}^{k}\frac{\alpha_i}{\alpha_0}\mathbf{y}_{j+1-i} + h\sum_{i=1}^{k}\frac{\beta_i}{\alpha_0}\mathbf{F}_{j+1-i}.$$

Implicit formulas in this class are normally evaluated by an iterative procedure called *simple iteration*, or *functional iteration*. An explicit formula is used to generate an initial approximation $\mathbf{y}_{j+1}^{(0)}$:

$$\mathbf{y}_{j+1}^{(0)} = -\sum_{i=1}^{k^*}\frac{\alpha_i^*}{\alpha_0^*}\mathbf{y}_{j+1-i} + h\sum_{i=1}^{k^*}\frac{\beta_i^*}{\alpha_0^*}\mathbf{F}_{j+1-i}.$$

Iterate number $r + 1$ of simple iteration is given by the (explicit) recipe

$$\mathbf{y}_{j+1}^{(r+1)} = h\frac{\beta_0}{\alpha_0}\mathbf{F}(x_{j+1}, \mathbf{y}_{j+1}^{(r)}) - \sum_{i=1}^{k}\frac{\alpha_i}{\alpha_0}\mathbf{y}_{j+1-i} + h\sum_{i=1}^{k}\frac{\beta_i}{\alpha_0}\mathbf{F}_{j+1-i}. \tag{2.6}$$

Obviously each iteration costs one evaluation of $\mathbf{F}$. In §3 it is shown that if hL is small enough (recall that L is a Lipschitz constant for $\mathbf{F}$), there is a unique solution $\mathbf{y}_{j+1}$ of the algebraic equation (2.5), and the iterates (2.6) converge quickly to this value. The essence of the matter is easily seen. Assuming that the equation (2.5) has a solution, we use equation (2.6) to get an expression for the error of iterate number $r + 1$:

$$\mathbf{y}_{j+1} - \mathbf{y}_{j+1}^{(r+1)} = h\frac{\beta_0}{\alpha_0}[\mathbf{F}(x_{j+1}, \mathbf{y}_{j+1}) - \mathbf{F}(x_{j+1}, \mathbf{y}_{j+1}^{(r)})]$$

Taking norms and using the Lipschitz condition, we find that

$$\|\mathbf{y}_{j+1} - \mathbf{y}_{j+1}^{(r+1)}\| \le h \left|\frac{\beta_0}{\alpha_0}\right| L \, \|\mathbf{y}_{j+1} - \mathbf{y}_{j+1}^{(r)}\|.$$

The error of each iterate decreases by at least a factor of $hL\,|\beta_0/\alpha_0|$, hence the scheme converges for all sufficiently small h.

In the classical theory a fixed step size is used and it is assumed that implicit formulas are evaluated exactly. This is *not* the way that implicit LMMs are implemented in practice. The typical code limits the number of evaluations of $\mathbf{F}$ per step, a common limit being three. If $\mathbf{y}_{j+1}$ is not approximated to a prescribed accuracy after this many iterations, the step is rejected and tried again with a smaller h. We shall see that the practical process has essentially the same properties as the ideal one. The distinction here is in how accurately the formula is evaluated. Evaluating the formula exactly is called *iteration to completion*. In practice the formula is evaluated to a fixed accuracy.

A *predictor–corrector method* is a combination of an explicit *predictor* formula and an implicit *corrector* formula. The idea is to do a fixed number of iterations with simple iteration. Such formulas are described as a prediction P, followed by an evaluation E of $\mathbf{F}$, a correction C consisting of substitution in the corrector formula, and so forth. By far the most common procedure is *PECE*. To exemplify the approach, let us consider the combination of the forward Euler method (AB1) and the trapezoidal rule (AM2):

$$\begin{array}{lll}
\text{P} & \text{AB1} & \mathbf{y}_{j+1}^{*} = \mathbf{y}_j + h\,\mathbf{F}_j \\
\text{E} & & \mathbf{F}_{j+1}^{*} = \mathbf{F}(x_{j+1}, \mathbf{y}_{j+1}^{*}) \\
\text{C} & \text{AM2} & \mathbf{y}_{j+1} = \mathbf{y}_j + \dfrac{h}{2}[\mathbf{F}_j + \mathbf{F}_{j-1}^{*}] \\
\text{E} & & \mathbf{F}_{j+1} = \mathbf{F}(x_{j+1}, \mathbf{y}_{j+1})
\end{array}$$

This particular example arose earlier in a different context—it is the two-stage, second-order explicit Runge–Kutta formula called the improved Euler formula. Of course, predictor–corrector methods are generally not one-step formulas, but they are always explicit formulas.

EXERCISE 24. Write out the PECE algorithm based on AB2 and AM2 with constant step size h. Though of the same order as the AB1–AM2 pair, the behavior of this pair differs in detail.

An Adams–Bashforth formula is usually used as a predictor for an Adams–Moulton formula because of some convenient relations between the formulas when expressed in difference form. Let us write the result of the Adams–Bashforth formula of order ν as $\mathbf{y}^*_{j+1}$ and recall that

$$\mathbf{y}^*_{j+1} = \mathbf{y}_j + h \sum_{k=1}^{\nu} \gamma_{k-1} \nabla^{k-1} \mathbf{F}_j.$$

The Adams–Moulton formula of order ν is

$$\mathbf{y}_{j+1} = \mathbf{y}_j + h \sum_{k=1}^{\nu} \gamma^*_{k-1} \nabla^{k-1} \mathbf{F}_{j+1}.$$

The differences in the Adams–Moulton formula involve

$$\mathbf{F}_{j+1} = \mathbf{F}(x_{j+1}, \mathbf{y}_{j+1}),$$

which is replaced in a predictor–corrector pair in PECE mode by

$$\mathbf{F}^*_{j+1} = \mathbf{F}(x_{j+1}, \mathbf{y}^*_{j+1}),$$

so that

$$\mathbf{y}_{j+1} = \mathbf{y}_j + h \sum_{k=1}^{\nu} \gamma^*_{k-1} \nabla^{k-1} \mathbf{F}^*_{j+1}.$$

The differences here are easy to compute successively from the differences $\nabla^m \mathbf{F}_j$ stored from the computation of the preceding step:

$$\begin{aligned}
\nabla^0 \mathbf{F}^*_{j+1} &= \mathbf{F}^*_{j+1}, \\
\nabla^1 \mathbf{F}^*_{j+1} &= \mathbf{F}^*_{j+1} - \mathbf{F}_j = \nabla^0 \mathbf{F}^*_{j+1} - \nabla^0 \mathbf{F}_j, \\
&\vdots \\
\nabla^{m+1} \mathbf{F}^*_{j+1} &= \nabla^m \mathbf{F}^*_{j+1} - \nabla^m \mathbf{F}_j, \\
&\vdots
\end{aligned}$$

These differences are used only in the formation of y_{j+1} and in error estimation, so do not need to be stored. After forming $\mathbf{y}_{j+1}$, the value $\mathbf{F}_{j+1}$ is formed, and the differences are updated to correspond to x_{j+1} by

$$\begin{aligned}
\nabla^0 \mathbf{F}_{j+1} &= \mathbf{F}_{j+1}, \\
\nabla^1 \mathbf{F}_{j+1} &= \mathbf{F}_{j+1} - \mathbf{F}_j = \nabla^0 \mathbf{F}_{j+1} - \nabla^0 \mathbf{F}_j, \\
&\vdots \\
\nabla^{m+1} \mathbf{F}_{j+1} &= \nabla^m \mathbf{F}_{j+1} - \nabla^m \mathbf{F}_j, \\
&\vdots
\end{aligned}$$

The new differences may be stored over the old ones as they are formed, so storage management is also convenient. There is an identity that makes prediction with the Adams–Bashforth formula especially easy. For the

ABν–AMν pair,

$$\mathbf{y}_{j+1} = \mathbf{y}^*_{j+1} + h\,\gamma_{\nu-1}\nabla^\nu\,\mathbf{F}^*_{j+1}.$$

To see this, subtract the formula for the predictor from that for the corrector to get

$$\mathbf{y}_{j+1} - \mathbf{y}^*_{j+1} = h\sum_{k=1}^{\nu}\gamma^*_{k-1}\,\nabla^{k-1}\,\mathbf{F}^*_{j+1} - h\sum_{k=1}^{\nu}\gamma_{k-1}\,\nabla^{k-1}\,\mathbf{F}_j.$$

The first term in each sum combines to form

$$h\,\gamma^*_0\nabla^0\,\mathbf{F}^*_{j+1} - h\,\gamma_0\,\nabla^0\,\mathbf{F}_j = h\,\gamma_0\,\nabla^1\,\mathbf{F}^*_{j+1},$$

on using the fact that $\gamma^*_0 = \gamma_0$. When the second term in each sum is added to this, the result simplifies as follows:

$$h\,\gamma_0\,\nabla^1\,\mathbf{F}^*_{j+1} + h\,\gamma^*_1\,\nabla^1\,\mathbf{F}^*_{j+1} - h\,\gamma_1\,\nabla^1\,\mathbf{F}_j$$

$$= h(\gamma_0 + \gamma^*_1)\,\nabla^1\,\mathbf{F}^*_{j+1} - h\,\gamma_1\,\nabla^1\,\mathbf{F}_j$$

$$= h\,\gamma_1(\nabla^1\,\mathbf{F}^*_{j+1} - \nabla^1\,\mathbf{F}_j) = h\,\gamma_1\,\nabla^2\,\mathbf{F}^*_{j+1}.$$

Here the identity $\gamma^*_m = \gamma_m - \gamma_{m-1}$ of Exercise 21 is used. Repetition of this argument leads to the result stated about the relation of the predictor to the corrector.

The Adams–Moulton formula of order ν does not use the value $\mathbf{F}_{j+1-\nu}$ that the Adams–Bashforth formula of order ν uses. This suggests that the Adams–Moulton formula of order $\nu + 1$ be used as a corrector for the Adams–Bashforth predictor of order ν because it takes advantage of this value. It is a little surprising that this predictor is related to the corrector in almost the same way as the other predictor, namely

$$\mathbf{y}_{j+1} = \mathbf{y}^*_{j+1} + h\,\gamma_\nu\,\nabla^\nu\,\mathbf{F}^*_{j+1}.$$

Clearly it is just as easy to use the higher order corrector as not and doing so is quite popular.

Predictor–corrector methods are similar to linear multistep methods, but they are not identical in theory nor in practice. A clear difference is that the predictor affects the efficiency of a code based on an implicit LMM, but not the results themselves. In contrast, after a fixed number of iterations, the predictor still affects the results of a predictor–corrector code. Let us apply AB1–AM1 in PECE implementation to the scalar problem (1.2). A little calculation leads to

$$y_{j+1} = (1 + h\lambda + (h\lambda)^2)y_j.$$

Comparison of this expression with the true solution (1.2) shows that for small $h\lambda$, it is just as accurate as AM1 when it is evaluated exactly as in (1.3). It is usual to employ a predictor of the same order as the corrector, or at most one order lower. We shall see that if this is done, a single correction already provides a formula of the same order of accuracy as the corrector when used by itself as an implicit formula. The role of the iterations in the evaluation of an implicit formula is widely misunderstood. Iterating to completion (or a fixed accuracy) does not provide a more accurate result; it provides a result with the stability of the implicit formula. To see the distinction here, recall (see (1.3) and the discussion there) that if $\lambda \leq 0$, the backward Euler rule AM1 is stable for all step sizes h. The AB1–AM1 PECE implementation is stable only when

$$|1 + h\lambda + (h\lambda)^2| \leq 1.$$

Because this polynomial in $h\lambda$ is unbounded as $|h\lambda| \to \infty$, the predictor–corrector pair cannot be stable for all step sizes like the implicit formula is. This shows that in some respects the behavior of a predictor–corrector pair is qualitatively different from that of the implicit formula used as corrector.

Predictor–corrector pairs can be thought of as a variation on linear multistep methods. Though based on an implicit LMM method, they have the advantage in simplicity of being explicit. A proper match of predictor and corrector results in a formula with essentially the same accuracy as the underlying implicit LMM. On the other hand, the stability of a predictor–corrector pair is normally very different from that of the underlying implicit LMM. Later we explore the matter in more detail and come to the conclusion that for many kinds of problems, there is little difference in practice between careful implementations of the two approaches.

Our derivation of the Adams formulas supplied us with the order of the formula, that is, the order of the discrepancy when the solution $\mathbf{y}(x)$, or rather the sequence $\{\mathbf{y}(x_j)\}$, is substituted into the formula. For a given LMM (2.5) we can test how well the formula imitates the behavior of the differential equation over one step by expansion in Taylor series. Let us calculate the local truncation error $\boldsymbol{\delta}_j$ in

$$\sum_{i=0}^{k} \alpha_i \, \mathbf{y}(x_{j+1-i}) = h \sum_{i=0}^{k} \beta_i \, \mathbf{F}(x_{j+1-i}, \mathbf{y}(x_{j+1-i})) + \boldsymbol{\delta}_j$$

or, equivalently,

$$\boldsymbol{\delta}_j = \sum_{i=0}^{k} [\alpha_i \, \mathbf{y}(x_{j+1-i}) - h \, \beta_i \, \mathbf{y}'(x_{j+1-i})].$$

We think of x_{j+1} as being fixed and expand all functions about this point. For the solution itself this results in

$$\begin{aligned}\mathbf{y}(x_{j+1-i}) &= \mathbf{y}(x_{j+1}) - ih\,\mathbf{y}'(x_{j+1}) + \frac{(-ih)^2}{2!}\mathbf{y}''(x_{j+1}) + \dots \\ &= \mathbf{y}(x_{j+1}) + \sum_{s=1}^{p+1}\frac{(-ih)^s}{s!}\mathbf{y}^{(s)}(x_{j+1}) + \mathcal{O}(h^{p+2}).\end{aligned}$$

Similarly

$$\begin{aligned}\mathbf{y}'(x_{j+1-i}) &= \mathbf{y}'(x_{j+1}) - ih\,\mathbf{y}''(x_{j+1}) + \dots \\ &= \sum_{s=1}^{p+1}\frac{(-ih)^{s-1}}{(s-1)!}\mathbf{y}^{(s)}(x_{j+1}) + \mathcal{O}(h^{p+2}).\end{aligned}$$

Substitution results in

$$\begin{aligned}\boldsymbol{\delta}_j &= \mathbf{y}(x_{j+1})\sum_{i=0}^{k}\alpha_i \\ &+ \sum_{s=1}^{p+1} h^s\,\mathbf{y}^{(s)}(x_{j+1})\left((-1)^s\sum_{i=0}^{k}\left[\frac{i^s\,\alpha_i}{s!} + \frac{i^{s-1}\,\beta_i}{(s-1)!}\right]\right) + \mathcal{O}(h^{p+2}).\end{aligned}$$

We see that if the formula is to imitate the differential equation at all, it must be the case that

$$\sum_{i=0}^{k}\alpha_i = 0.$$

In terms of the characteristic polynomial, this condition is $\rho(1) = 0$, so we have found that for any reasonable LMM, the quantity 1 will be a root of the characteristic polynomial. We are interested only in methods that are more accurate than this. If

$$C_s = (-1)^s\sum_{i=0}^{k}\left[\frac{i^s\,\alpha_i}{s!} + \frac{i^{s-1}\,\beta_i}{(s-1)!}\right] = 0$$

for $s = 1, \dots, p$ and $C_{p+1} \neq 0$, then

$$\boldsymbol{\delta}_j = C_{p+1}\,h^{p+1}\,\mathbf{y}^{(p+1)}(x_{j+1}) + \mathcal{O}(h^{p+2}), \tag{2.7}$$

where

$$C_{p+1} = (-1)^{p+1}\sum_{i=0}^{k}\left[\frac{i^{p+1}\,\alpha_i}{(p+1)!} + \frac{i^p\,\beta_i}{p!}\right].$$

Note that in (2.7) we could to the same degree of approximation evaluate $\mathbf{y}^{(p+1)}$ at x_j or any other point in the span of the memory. The constant

C_{p+1} is called the *error constant* of the method. It depends on the normalization of the coefficients, so when comparing the accuracy of formulas, we must be careful that they are normalized in the same way. Earlier the error constant was called γ_p when we derived the Adams–Bashforth formula of order p and γ_p^* when we derived the Adams–Moulton formula of order p. In the case of the BDF of order p, the error constant was found to be $-1/(p+1)$.

The equations $C_s = 0$ for $s \leq p$ are the equivalent of the equations of condition for a Runge–Kutta method of order p. They are comparatively simple for LMMs because these formulas are linear in the solution and its first derivative. In the terminology used earlier, (2.7) states that the LMM is of order p. What we did earlier was to use interpolation theory to obtain the local truncation error directly. Then we had equality rather than equality only up to terms of order $p + 2$ because the components of $\mathbf{y}^{(p+1)}$ were evaluated at unknown points within the span of the memory.

The *second characteristic polynomial* of a LMM is defined to be

$$\sigma(\theta) = \beta_0\theta^k + \beta_1\theta^{k-1} + \ldots + \beta_k.$$

We are interested only in LMM for which the order $p \geq 1$, hence only in methods for which

$$C_1 = (-1)\sum_{i=0}^{k} [i\alpha_i + \beta_i] = 0.$$

This condition states that

$$\sigma(1) = \sum_{i=0}^{k} \beta_i = -\sum_{i=0}^{k} i\alpha_i.$$

Now

$$\rho'(\theta) = \sum_{i=0}^{k} (k - i)\alpha_i\theta^{k-i-1},$$

hence

$$\rho'(1) = k\sum_{i=0}^{k} \alpha_i - \sum_{i=0}^{k} i\alpha_i = k\rho(1) + \sigma(1) = \sigma(1).$$

The two conditions on the characteristic polynomials, $\rho(1) = 0$ and $\rho'(1) = \sigma(1)$, are easily seen to be equivalent to the requirement that the LMM have order $p \geq 1$. This result can be applied to show that the BDFs in the form derived earlier are normalized. Reference to equation (2.4) shows that $\sigma(\theta) = \theta^k$ for these formulas. Because the formula has order $p \geq 1$, we have $1 = \sigma(1) = \rho'(1)$, hence the formula is normalized.

EXAMPLE 2. As an example of the calculation of the order of a LMM, let us work out the order of AB2 directly. The formula is

$$\mathbf{y}_{j+1} - \mathbf{y}_j = h\left[\frac{3}{2}\mathbf{F}(x_j, \mathbf{y}_j) - \frac{1}{2}\mathbf{F}(x_{j-1}, \mathbf{y}_{j-1})\right].$$

We identify from this the coefficients $\alpha_0 = 1, \alpha_1 = -1, \alpha_2 = 0$. Obviously

$$\sum_{i=0}^{2} \alpha_i = 0,$$

so the minimal requirement is satisfied. We also identify $\beta_0 = 0$, $\beta_1 = 3/2$, $\beta_2 = -1/2$. Then

$$C_1 = (-1)\sum_{i=0}^{2} [i\alpha_i + \beta_i]$$
$$= -\left[(0 \times 1 + 0) + \left(1 \times (-1) + \frac{3}{2}\right) + \left(2 \times 0 + \left(-\frac{1}{2}\right)\right)\right] = 0.$$

This implies that the formula is of order at least one. Continuing on,

$$C_2 = (-1)^2 \sum_{i=0}^{k} \left[\frac{i^2\alpha_i}{2} + i\beta_i\right] = \left[(0+0) + \left(-\frac{1}{2} + \frac{3}{2}\right) + (0 - 1)\right] = 0,$$

so the formula is of order at least two. In the same way, we find that

$$C_3 = (-1)^3 \sum_{i=0}^{k} \left[\frac{i^3\,\alpha_i}{6} + \frac{i^2\,\beta_i}{2}\right]$$
$$= -\left[(0 + 0) + \left(-\frac{1}{6} + \frac{3}{4}\right) + (0 - 1)\right] = \frac{5}{12}.$$

Because $C_3 \neq 0$ and the formula is normalized, C_3 is the error constant. This tells us that the formula is of order 2 and

$$\boldsymbol{\delta}_j = \frac{5}{12} h^3 \mathbf{y}^{(3)}(x_j) + \mathcal{O}(h^4).$$

EXERCISE 25. The midpoint quadrature rule was used earlier to derive a Runge–Kutta formula. It can also be used to derive a method with memory that is rather like AB2. The trick is to use a different integrated form of the differential equation, namely

$$\mathbf{y}(x_j + h) = \mathbf{y}(x_j - h) + \int_{x_j - h}^{x_j + h} \mathbf{F}(t, \mathbf{y}(t))\, dt.$$

Approximating the integral by the midpoint quadrature rule leads a formula called the midpoint rule,

$$\mathbf{y}_{j+1} - \mathbf{y}_{j-1} = 2h\,\mathbf{F}(x_j, \mathbf{y}_j).$$

Normalize the formula and find its error constant.

EXERCISE 26. Verify that the formula

$$\mathbf{y}_{j+1} + 4\mathbf{y}_j - 5\mathbf{y}_{j-1} = h[4\mathbf{F}(x_j, \mathbf{y}_j) + 2\mathbf{F}(x_{j-1}, \mathbf{y}_{j-1})]$$

is of order 3. Normalize it and find its error constant.

Now let us take up the error of a predictor–corrector pair. The predictor is an explicit LMM with coefficients that we distinguish with asterisks. The tentative value at x_{j+1} will also be denoted with an asterisk. The formula has the form

$$\mathbf{y}^*_{j+1} = -\sum_{i=1}^{k}\left(\frac{\alpha_i^*}{\alpha_0^*}\mathbf{y}_{j+1-i} - h\frac{\beta_i^*}{\alpha_0^*}\mathbf{F}_{j+1-i}\right).$$

The true solution satisfies this difference equation with a discrepancy (local truncation error) $\boldsymbol{\delta}_j^*$ that we have already investigated. After an obvious manipulation, we have

$$\mathbf{y}(x_{j+1}) - \frac{1}{\alpha_0^*}\boldsymbol{\delta}_j^* = -\sum_{i=1}^{k}\left(\frac{\alpha_i^*}{\alpha_0^*}\mathbf{y}(x_{j+1-i}) - h\frac{\beta_i^*}{\alpha_0^*}\mathbf{F}(x_{j+1-i}, \mathbf{y}(x_{j+1-i}))\right)$$

The implicit LMM is made explicit in a predictor–corrector pair by using the tentative value $\mathbf{y}^*_{j+1}$ for $\mathbf{y}_{j+1}$ in the function evaluation:

$$\alpha_0\mathbf{y}_{j+1} + \sum_{i=1}^{k}\alpha_i\mathbf{y}_{j+1-i} = h\beta_0\mathbf{F}(x_{j+1}, \mathbf{y}^*_{j+1}) + h\sum_{i=1}^{k}\beta_i\mathbf{F}_{j+1-i}.$$

We would like to measure how well $\mathbf{y}(x)$ satisfies this formula, that is, determine the local truncation error. This is a little confusing because of the form. If we spell out the predicted value in this formula, we have

$$\mathbf{F}(x_{j+1}, \mathbf{y}^*_{j+1}) = \mathbf{F}\left(x_{j+1}, -\sum_{i=1}^{k}\left(\frac{\alpha_i^*}{\alpha_0^*}\mathbf{y}_{j+1-i} - h\frac{\beta_i^*}{\alpha_0^*}\mathbf{F}_{j+1-i}\right)\right),$$

from which it is clearer that substituting values of the true solution at the mesh points leads to

$$\mathbf{F}(x_{j+1}, \mathbf{y}(x_{j+1}) - \frac{1}{\alpha_0^*}\boldsymbol{\delta}_j^*).$$

With this observation, substituting values of the true solution into the equation leads to a discrepancy

$$\sum_{i=0}^{k} \alpha_i \, \mathbf{y}(x_{j+1-i}) - h\, \beta_0 \, \mathbf{F}\left(x_{j+1}, \mathbf{y}(x_{j+1}) - \frac{1}{\alpha_0^*} \boldsymbol{\delta}_j^* \right) - h \sum_{i=1}^{k} \beta_i \, \mathbf{F}(x_{j+1-i}, \mathbf{y}(x_{j+1-i})).$$

This can be related to the discrepancy $\boldsymbol{\delta}_j$ by which $\mathbf{y}(x)$ satisfies the implicit LMM, the local truncation error of this formula,

$$\sum_{i=0}^{k} \alpha_i \, \mathbf{y}(x_{j+1-i}) - h \sum_{i=0}^{k} \beta_i \, \mathbf{F}(x_{j+1-i}, \mathbf{y}(x_{j+1-i})) = \boldsymbol{\delta}_j,$$

in a simple way. Adding and subtracting a term, we find the discrepancy for the pair is

$$\boldsymbol{\delta}_j + h\, \beta_0 \left\{ \mathbf{F}(x_{j+1}, \mathbf{y}(x_{j+1})) - \mathbf{F}\left(x_{j+1}, \mathbf{y}(x_{j+1}) - \frac{1}{\alpha_0^*} \boldsymbol{\delta}_j^* \right) \right\}.$$

Expansion about $\mathbf{y}(x_{j+1})$ here leads to the local truncation error for the predictor–corrector pair

$$\boldsymbol{\delta}_j + h \frac{\beta_0}{\alpha_0^*} \frac{\partial \mathbf{F}}{\partial \mathbf{y}} (x_{j+1}, \mathbf{y}(x_{j+1}))\boldsymbol{\delta}_j^* + \mathcal{O}(\|\boldsymbol{\delta}_j^*\|^2). \tag{2.8}$$

Let us now try to understand what the expression (2.8) means. If the predictor formula is of order p^*, then by definition its truncation error $\boldsymbol{\delta}_j^*$ is of order $p^* + 1$ and if the corrector formula is of order p, its truncation error $\boldsymbol{\delta}_j$ is of order $p + 1$. The expression (2.8) tells us that the truncation error of the predictor–corrector pair is of order $\min(p + 1, p^* + 2)$, hence that the formula is of order $\min(p, p^* + 1)$. This is why we were interested earlier in predicting with a formula of the same order as the corrector, or perhaps one lower. Notice that if the predictor is of the same order as the corrector, the truncation error of the pair agrees to leading order with the truncation error of the implicit formula used as the corrector—to leading order even one correction provides all the accuracy of the implicit formula. If the predictor is of order one lower than the corrector, the order of the pair is the same as that of the implicit formula used as the corrector, but the truncation error itself is different. This has all been done for PECE because it is the most common implementation. It is easy enough to see that each correction reduces the effect of the predictor by a factor of h.

§2.4 General Remarks About Memory

So far we have been discussing what linear multistep and predictor–corrector methods are. In view of what we now know about one-step methods, let us consider the advantages and disadvantages of methods with memory in general. One big advantage of methods with memory is that we can get a high order of accuracy with just a few evaluations of **F**. However, this is achieved by reusing previously computed solution values. Error in these values feeds back into the computation of the current approximate solution. We should be concerned that these errors might be amplified to the extent that the formula is useless. This concern is well founded; many linear multistep methods that look attractive because of their accuracy are useless for just this reason. An advantage of the Adams formulas and the BDFs is that they are based on polynomial interpolants that provide an obvious way to approximate the solution of the differential equation between mesh point. The trouble is that the accuracy of these interpolants depends on the accuracy of previously computed solution values. We must defer consideration of these interpolants until we have established convergence and so have at our disposal results about the accuracy of previously computed values. This contrasts with the situation for interpolation of Runge–Kutta methods for which the analysis is confined to what happens in a single step.

Methods with memory make use of previously computed values. Where do the "previously computed" values come from when the integration is started? The classical theory spends little time on this important practical matter. Also, how to change the step size is scarcely mentioned in the classical theory because it is regarded as equivalent to restarting. The usual suggestion was to start by computing the necessary values with a method having no memory, such as an explicit Runge–Kutta formula. This is not a very good idea, at least when done in the obvious way. One reason is that for even moderately high order formulas, it is expensive. The expense would be acceptable if incurred only once, but the step size might be changed many times in the course of an integration. In the classical theory changing the step size is ignored, but in practice it cannot be ignored. For one thing, it may be necessary to reduce the step size to obtain an acceptable rate of convergence of the procedure for evaluating an implicit formula. For another, the step size must be reduced when the approximate solution is not sufficiently accurate. Though not actually necessary, it is important in practice to increase the step size when possible because this makes the integration more efficient. Indeed, it is not practical to solve stiff problems without taking advantage of this. An important reason for varying the step size to control the local error is that it has the side effect of producing a step size for which the integration is stable.

It is convenient to discuss first how to change the step size. There are two approaches seen in popular codes. One takes a rather classical view of the matter: The step size is changed only when necessary—when the step is a failure or the convergence of the iteration for evaluating an implicit formula is unacceptably slow—and when a substantial increase makes the expense of a change worthwhile. In this view, we change from an integration with a constant step size h to an integration with a different constant step size H. This is done in practice by interpolating the memorized values computed with the step size h to generate a set of fictitious previously computed values corresponding to the new step size H. These values are used to continue the integration with the step size H. The approach is inexpensive and with some restrictions on how it is done, it can be proven to "work." For stiff problems and the backward differentiation formulas, there is a very important practical reason for preferring this quasi-constant step size implementation that arises in the way the implicit formulas must be evaluated. The other approach takes advantage of the fact that popular formulas are defined on any mesh. A change of step size requires only the computation of the coefficients defining the formula. Although in principle any mesh might be used, to keep down the cost of evaluating coefficients, the fully variable step size implementations work with a constant step size much of the time. As one might guess, interpolation can amplify the errors in previously computed values and experience shows that a fully variable step size implementation can be rather more stable than a quasi-constant step size implementation. For the Adams methods there are theoretical results to this effect. Nevertheless, the interpolatory approach is quite satisfactory when implemented carefully.

The popular multistep formulas such as Adams and BDF come in families of orders 1, 2, 3, ... They are commonly implemented so as to vary the order. The lowest order formula is a one-step formula. Starting with this formula and then quickly building up the order to an efficient value is the usual way to get going in these codes. Of course the point of higher orders method is to permit bigger step sizes, so the step size is increased rapidly as the order is increased. In a mechanical sense all this is easy to do and it seems to work well enough. Despite its importance—it is used by all the popular generalpurpose codes based on LMMs and predictor–corrector methods—it is only very recently (Shampine and Zhang [1990]) that any theoretical justification of the idea was presented. Brankin et al. [1988] have proposed that such codes be started with Runge–Kutta formulas that have an interpolation capability. The idea is to take a step with an explicit Runge–Kutta formula of moderate order that has an interpolant. Using the fact that derivatives of the interpolant approximate derivatives of the local solution, it is possible to estimate what step sizes might be used at several orders. After the most efficient order is recognized in this way, the necessary starting values are obtained by evaluating the interpolant. There are practical difficulties, but the approach is efficient and has the virtue of theoretical support.

§3 Implicit Methods

We have seen a number of implicit methods in this chapter and deferred consideration of how to evaluate them. In this section we go into how this is done. We begin with linear multistep methods. A little manipulation exposes the value $\mathbf{y}_{j+1}$ that is to be computed:

$$\mathbf{y}_{j+1} = h \frac{\beta_0}{\alpha_0} \mathbf{F}(x_{j+1}, \mathbf{y}_{j+1}) + h \sum_{i=1}^{k} \frac{\beta_i}{\alpha_0} \mathbf{F}_{j+1-i} - \sum_{i=1}^{k} \frac{\alpha_i}{\alpha_0} \mathbf{y}_{j+1-i}.$$

It is the form of the algebraic equation that matters here. We abstract it as

$$\mathbf{v} = h\gamma\, \mathbf{F}(\mathbf{v}) + h\,\boldsymbol{\delta} + \boldsymbol{\mu} \tag{3.1}$$

for constants γ, $\boldsymbol{\delta}$, $\boldsymbol{\mu}$. Because the independent variable is held fixed in $\mathbf{F}$, it is suppressed in the notation here. The key to the analysis is the assumption that $\mathbf{F}$ satisfies a Lipschitz condition with constant L. First we observe that for all sufficiently small h, there is at most one solution of (3.1). Suppose to the contrary that $\mathbf{v}$ and $\mathbf{w}$ are two different solutions of (3.1). Then

$$\mathbf{w} = h\gamma\, \mathbf{F}(\mathbf{w}) + h\,\boldsymbol{\delta} + \boldsymbol{\mu},$$

and

$$\|\mathbf{w} - \mathbf{v}\| = \|h\gamma\, \mathbf{F}(\mathbf{w}) - h\gamma\, \mathbf{F}(\mathbf{v})\| \leq |h\gamma| L \|\mathbf{w} - \mathbf{v}\|,$$

on using the Lipschitz condition. If h is small enough that

$$|h\gamma| L = \rho < 1, \tag{3.2}$$

then

$$\|\mathbf{w} - \mathbf{v}\| < \|\mathbf{w} - \mathbf{v}\|,$$

which is impossible.

The iteration suggested earlier for evaluating an implicit formula is in the present notation

$$\mathbf{u}_{m+1} = h\gamma\, \mathbf{F}(\mathbf{u}_m) + h\,\boldsymbol{\delta} + \boldsymbol{\mu},$$

where the iterates are now indicated by subscripts. A starting value $\mathbf{u}_0$ is obtained somehow. In practice it is computed using an explicit formula to predict $\mathbf{y}_{j+1}$. If a solution $\mathbf{v}$ of (3.1) exists, condition (3.2) implies that the sequence $\{\mathbf{u}_{m+1}\}$ converges to $\mathbf{v}$. To see this, we observe from the definitions that

$$\|\mathbf{v} - \mathbf{u}_{m+1}\| = \|h\gamma\, \mathbf{F}(\mathbf{v}) - h\gamma\, \mathbf{F}(\mathbf{u}_m)\|$$

and then that

$$\|\mathbf{v} - \mathbf{u}_{m+1}\| \leq |h\gamma| L \|\mathbf{v} - \mathbf{u}_m\| = \rho\|\mathbf{v} - \mathbf{u}_m\|.$$

Notice that the convergence rate ρ is $\mathbb{O}(h)$. To explore the implications of this, suppose that the initial guess $\mathbf{u}_0$ is obtained from a formula of order q so that

$$\mathbf{u}_0 = \mathbf{u}(x_{j+1}) + \mathbb{O}(h^{q+1}).$$

If the LMM we wish to evaluate is of order $p \geq q$, we have

$$\mathbf{v} = \mathbf{y}_{j+1} = \mathbf{u}(x_{j+1}) + \mathbb{O}(h^{p+1}).$$

Together these imply that the error of the initial guess, $\|\mathbf{v} - \mathbf{u}_0\|$, is $\mathbb{O}(h^{q+1})$. Because ρ is $\mathbb{O}(h)$, each iterate approximates $\mathbf{v} = \mathbf{y}_{j+1}$ to one higher order of accuracy in h. With a predictor of order comparable to the corrector and "small" h, the iteration converges very quickly to a solution to the algebraic equations. It is usual to require h to be small enough that a few iterations will suffice. For example, some codes insist that the estimated ρ be no more than 0.1 and others that three iterations be enough to achieve convergence.

This process is an example of a *contraction mapping*. It is not difficult to show the existence of a solution to (3.1) in addition to its uniqueness and convergence of the iteration. Often the function $\mathbf{F}(x, \mathbf{y})$ does not satisfy a Lipschitz condition for all $\mathbf{y}$. An extension of the contraction mapping argument provides existence, uniqueness, and convergence of the iteration when $\mathbf{F}$ satisfies a Lipschitz condition only for arguments $\mathbf{y}$ near the solution $\mathbf{y}(x)$ of the given initial value problem. We do not pursue this matter, merely affirm that the extension will deal with nearly all practical problems.

Implicit Runge–Kutta formulas are handled similarly except that at each step we need to solve a system of equations of the form

$$\begin{aligned}
\mathbf{Y}_1 &= \mathbf{y}_j + h \sum_{m=1}^{s} \beta_{1,m}\, \mathbf{F}(x_j + \theta_m h, \mathbf{Y}_m) \\
&\vdots \\
\mathbf{Y}_s &= \mathbf{y}_j + h \sum_{m=1}^{s} \beta_{s,m}\, \mathbf{F}(x_j + \theta_m h, \mathbf{Y}_m)
\end{aligned}$$

Simple iteration is

$$\mathbf{Y}_l^{(r+1)} = \mathbf{y}_j + h \sum_{m=1}^{s} \beta_{1,m} \mathbf{F}(x_j + \theta_m h, \mathbf{Y}_m^{(r)})$$

$$\vdots$$

$$\mathbf{Y}_s^{(r+1)} = \mathbf{y}_j + h \sum_{m=1}^{s} \beta_{s,m} \mathbf{F}(x_j + \theta_m h, \mathbf{Y}_m^{(r)})$$

The only difference in the analysis is that we combine vectors in R^n to form longer vectors in $R^{n \times s}$ such as

$$\mathcal{Y}^{(r)} = \begin{pmatrix} \mathbf{Y}_1^{(r)} \\ \mathbf{Y}_2^{(r)} \\ \vdots \\ \mathbf{Y}_s^{(r)} \end{pmatrix}.$$

In this space of longer vectors the analysis and conclusions are exactly the same as for LMMs.

It is important to appreciate that simple iteration is not all that we might want. Obviously we are concerned that an implicit formula be well defined. Now we can understand why all the kinds of implicit formulas that we have considered are well defined for all sufficiently small h when $\mathbf{F}$ satisfies a Lipschitz condition. When the step size is small enough, simple iteration provides a high rate of convergence and a suitable predictor formula provides a good starting value. The snag is that we may not want to work with step sizes this small. It is better to defer details until we take up stiffness in Chapter 8, but we have already examined the issue of stability enough to see the difficulty. In §1 we looked at the solution of the scalar equation (1.2) by the backward Euler method. (Later we found that this method is a member of the Adams–Moulton family with the alias AM1 and of the backward differentiation formulas with the alias BDF1.) The formula itself is stable for all h when $\lambda \le 0$. We saw, however, that a predictor–corrector pair based on the formula, specifically AB1–AM1 implemented as PECE, is not stable for all h. For this reason we want to evaluate the implicit formula. Earlier we developed a bound on the rate of convergence of simple iteration in terms of h and the Lipschitz constant L. A problem as simple as (1.2) allows us to see more clearly what is happening. We wish to solve

$$y_{j+1} = y_j + h\lambda\, y_{j+1},$$

which we know has a solution for all sufficiently small h. In point of fact, for $\lambda \le 0$, there is a solution for all $h > 0$. Simple iteration improves an approximation $y_{j+1}^{(r)}$ by

$$y_{j+1}^{(r+1)} = y_j + h\lambda\, y_{j+1}^{(r)}.$$

Subtracting these two expressions, we find immediately that

$$|y_{j+1}^{(r+1)} - y_{j+1}| = |h\lambda|\, |y_{j+1}^{(r)} - y_{j+1}|.$$

Clearly the process does not converge unless $|h\lambda| < 1$. The point of this simple example is that the process for evaluating an implicit formula may itself impose a restriction on the step size. This restriction can be just as severe as the stability restriction imposed by implementing the formula as a predictor–corrector pair, an explicit formula. To exploit the superb stability properties of the backward Euler formula, and the other BDFs for that matter, we must resort to a different way of evaluating implicit formulas. A variant of Newton's method is used for this purpose. Compared to simple iteration, such procedures are complicated and expensive. They are justified only when they make possible a very much larger step size. Stiff problems are problems that cannot be solved in a reasonable time without resorting to this kind of approach.

§4 General Remarks About Order

In deriving formulas some emphasis has been placed on high order formulas. This is because in a certain sense, if you ask for enough accuracy, raising the order of the formula used increases the efficiency of the integration. This is true on very general grounds. Suppose that $\mathbf{y}_{j+1}$ is the result of a step to $x_j + h$ with a formula of order p that requires s evaluations of $\mathbf{F}$ per step. The local error of the step has the form

$$\mathbf{u}(x_j + h) - \mathbf{y}_{j+1} = \mathbf{d}\, h^{p+1} + \mathcal{O}(h^{p+2}),$$

where generally $\mathbf{d} \neq \mathbf{0}$. Similarly, if $\mathbf{Y}_{j+1}$ is the result of a step to $x_j + H$ with a formula of order $P > p$ that requires $S > s$ evaluations of $\mathbf{F}$ per step, then

$$\mathbf{u}(x_j + h) - \mathbf{Y}_{j+1} = \mathbf{D}\, H^{P+1} + \mathcal{O}(H^{P+2}),$$

where generally $\mathbf{D} \neq \mathbf{0}$. We have to account for the fact that the formulas do not cost the same per step. The distance advanced divided by the number of evaluations of $\mathbf{F}$ will be the same if we take $H = h(S/s)$. The question then is which formula provides the more accurate answer. Rewriting the expression for the error of the higher order formula to account for the difference in cost, we have

$$\mathbf{u}(x_j + h) - \mathbf{Y}_{j+1} = \mathbf{D}\left(\frac{S}{s}\right)h^{P+1} + \mathbb{O}(h^{P+2}).$$

Comparing the two expressions, we see that for all sufficiently small h, the higher order formula provides the more accurate approximation. Put differently, a given accuracy is obtained more efficiently by the higher order formula for all sufficiently high accuracies, regardless of the relative cost per step. This argument does *not* imply that a low order formula is to be preferred for low accuracies, just that a high order one is to be preferred for high accuracies.

Having observed that high order can be advantageous at high accuracies, we must hasten to qualify the observation. First, the method must actually be of high order. By this we mean that our statements about the order of a formula are based on the assumption that the problem is smooth. The order of a formula is reduced when it is not. Some get the impression that high order formulas cannot be used for such problems. This is incorrect. They can be used, but the integration does not converge as fast as when the problem is smooth. The matter is taken up in the chapter devoted to convergence. Second, the higher order method is more efficient only when it is permitted to use a larger step size. Reference has been made to constraints on the step size due to output and there are other constraints in practice. Third, when comparing formulas that are both of rather high order or when the higher is very much more expensive per step, the result might be true in theory but misleading in practice because the accuracies for which the higher order formula is the more efficient are not attainable in the precision available.

Another aspect of the order of a formula is that it is not generally possible to decide whether a problem will be hard or easy just from the appearance of the coefficients in the differential equation or an expectation about the behavior of the solution. There are many reasons for this, but one is directly related to the order of a formula. The accuracy of a formula depends on a relatively high order derivative and the size of this derivative may not be apparent—coefficients that look smooth may correspond to a solution with a large derivative or conversely, a solution that changes rapidly might have some derivative that is small. A contrived example will help make the point.

EXAMPLE 3. The function $u(x) = \exp(-x)$ cannot be distinguished visually from the function $v(x) = \exp(-x) + 10^{-6}\sin(10^3 x)$ on the interval

[0, 2]. The first is the solution of the initial value problem

$$y' = -y, \qquad 0 \le x \le 2, \qquad y(0) = 1,$$

and the second is the solution of

$$y' = -y + 10^{-3}\cos(10^3 x) + 10^{-6}\sin(10^3 x),$$
$$0 \le x \le 2, \qquad y(0) = 1.$$

The coefficients of the two problems can scarcely be distinguished visually. With this contrived example, the origin of the difficulty is easily seen because the change to the coefficient of the first problem to get the second is small in magnitude but of high frequency so that derivatives of the solution of the second problem are not small. When the first problem was solved with the Runge–Kutta code RKF45 with a pure relative error tolerance of 10^{-8}, only 151 evaluations of the equation were necessary to obtain a result at $x = 2$ with an error of 7.8×10^{-9}. In contrast, when the second problem was solved, it cost 3968 evaluations to obtain a result at $x = 2$ with an error of 1.3×10^{-7}. The step size in this code is chosen to be appropriate to a formula of order 4 and still larger differences are seen with methods of higher order.

EXERCISE 27. Try this yourself with an integrator based on a formula, or formulas, of moderate to high order.

EXERCISE 28. Suppose that you wish to integrate from 0 to 1 a problem with solution $y(x) = 1/(1.3 - x)^2$. On this interval, the solution looks fine. Investigate the truncation error at $x = 1$ of a step of size h for a method of relatively high order. The form of the solution makes it easy to evaluate the truncation error for formulas with errors that depend only on a derivative of the solution like a Taylor series method, an Adams method, or a BDF. Although the solution itself appears innocuous, the derivatives grow pretty rapidly and a high order method is less accurate than you might expect.

EXERCISE 29. Suppose you wish to solve $\mathbf{y}' = \mathbf{F}(x, \mathbf{y})$, $\mathbf{y}(a) = \mathbf{A}$. If there is a constant vector $\mathbf{v}$ such that $\mathbf{v}^T \mathbf{F}(x, \mathbf{y}) \equiv \mathbf{0}$, then any solution $\mathbf{y}(x)$ of the initial value problem satisfies the linear conservation law $\mathbf{v}^T \mathbf{y}(x) = \mathbf{v}^T\mathbf{A}$. Recall that this was proved in Example 3 of Chapter 1 by observing that $\mathbf{v}^T \mathbf{y}'(x) = \mathbf{v}^T \mathbf{F}(x, \mathbf{y}(x)) = \mathbf{0}$ and integrating. All the standard methods preserve all linear conservation laws. To see how this goes, prove that any Runge–Kutta method started with the correct initial value, $\mathbf{y}_0 = \mathbf{A}$, will produce $\mathbf{y}_j \approx \mathbf{y}(x_j)$ that also satisfy the conservation law, i.e., $\mathbf{v}^T\mathbf{y}_j = \mathbf{v}^T\mathbf{A}$.

Further prove that any LMM,

$$\sum_{i=0}^{k} \alpha_i \, \mathbf{y}_{j+1-i} = h \sum_{i=0}^{k} \beta_i \, \mathbf{F}(x_{j+1-i}, \mathbf{y}_{j+1-i}),$$

started with initial values $\mathbf{y}_r$, $r = 0, \ldots, k-1$, that satisfy the conservation law will produce $\mathbf{y}_j \approx \mathbf{y}(x_j)$ that also satisfy the conservation law.

5

Convergence and Stability

In the last chapter we asked how well the numerical methods approximate the solution of a differential equation over a single step. In this chapter we ask how well they approximate the solution over an interval $[a, b]$. In the classical theory the step size is a constant h and numerical methods are investigated for "small" h by letting h tend to 0. For one-step methods it turns out to be as easy to analyze variable step size as constant. Adams methods and others of similar form can be analyzed in a way rather like that used for one-step methods. In particular, it is possible to account for variation of the step size. Unfortunately, the important BDFs do not have this form and if we are to study them in a relatively simple manner, we have to make the classical assumption that the step size is constant. There are other reasons for taking up the special case of constant step size. For one, it is possible to understand better the behavior of LMMs and predictor–corrector methods. This will show us why some methods are of no practical value despite an origin and appearance that are quite similar to those of methods important in practice. For another, some of the tools developed for this analysis are needed for the study of the stability of methods when the step size is not "small" which is taken up in the next chapter. Variation of the step size is of great practical importance, so it is necessary to consider how to extend the classical approach of constant step size to accommodate it. The discrete variable methods we study produce approximate solutions on a mesh in the interval $[a, b]$. Taylor series methods produce approximate solutions everywhere and the same is true of the Adams methods and the BDFs. We shall see that the approximations between mesh points have the same order of accuracy as the approximations at the mesh points themselves.

The solution $\mathbf{y}(x)$ of

$$\mathbf{y}' = \mathbf{F}(x, \mathbf{y}), \qquad a \leq x \leq b, \qquad \mathbf{y}(a) = \mathbf{A}, \tag{1}$$

is approximated on a mesh $a = x_0 < x_1 < \cdots$ by a sequence $\{\mathbf{y}_j\}$. To study the error of this approximation, like quantities must be compared. The usual approach is to compare the sequence $\{\mathbf{y}_j\}$ to the sequence $\{\mathbf{y}(x_j)\}$. A less obvious approach is to construct a piecewise smooth approximating function $\mathbf{v}(x)$ that interpolates the discrete values $\{\mathbf{y}_j\}$ as in §2 of Chapter 2 and then compare $\mathbf{v}(x)$ to $\mathbf{y}(x)$. The idea is to show that the numerical method results in an approximating function $\mathbf{v}(x)$ that satisfies the problem (1) with a small discrepancy. Stability of the differential equation then implies that $\mathbf{v}(x)$ is close to $\mathbf{y}(x)$. The approach is a valuable one, but the first approach is by far the more common and it illuminates other aspects of the methods, so it is the one developed in this chapter. The basic idea remains the same. The distinction is that now the stability of the approximating difference equation plays the role of the stability of the differential equation. The stability of the differential equation is fundamental. The most we can expect of a numerical method is that its stability imitate that of the differential equation and it is reasonable to suppose that this is possible only for sufficiently small step sizes. The key to the analysis of convergence is to establish such a stability result for the difference equation. If the sequence $\{\mathbf{y}(x_j)\}$ satisfies the difference equation with a small discrepancy, stability then implies that the members of this sequence are close to the members of the sequence $\{\mathbf{y}_j\}$.

§1 One-Step Methods

To see how convergence proofs work, it is best to start with simple methods and work up to more complicated ones. For this reason, we start with explicit one-step methods. As defined in Chapter 4, such methods have the form

$$\mathbf{y}_{j+1} = \mathbf{y}_j + h_j\,\mathbf{\Phi}(x_j, \mathbf{y}_j, \mathbf{F}, h_j) \tag{1.1}$$

when advancing the solution of

$$\mathbf{u}' = \mathbf{F}(x, \mathbf{u}), \qquad \mathbf{u}(x_j) = \mathbf{y}_j$$

one step of size h_j from x_j to $x_{j+1} = x_j + h_j$. It is possible to prove convergence when $\mathbf{y}_{j+1}$ scarcely approximates $\mathbf{u}(x_{j+1})$. Taylor series expansion of the local solution tells us that

$$\mathbf{u}(x_j + h_j) = \mathbf{y}_j + h_j\,\mathbf{F}(x_j, \mathbf{y}_j) + \mathcal{O}(h_j^2).$$

Comparison to (1.1) shows that about the least that might be asked of $\mathbf{y}_{j+1}$ is that

$$\lim_{h_j \to 0} \mathbf{\Phi}(x_j, \mathbf{y}_j, \mathbf{F}, h_j) = \mathbf{\Phi}(x_j, \mathbf{y}_j, \mathbf{F}, 0) = \mathbf{F}(x_j, \mathbf{y}_j).$$

A method that satisfies this condition is said to be *consistent* with the differential equation. We are concerned only with methods that are more accurate than this. Specifically, we always suppose that $\mathbf{y}_{j+1}$ is an approximation to $\mathbf{u}(x_{j+1})$ of order $p \geq 1$, meaning that for all $(x_j, \mathbf{y}_j)$,

$$\mathbf{y}_{j+1} = \mathbf{u}(x_j + h_j) + \mathcal{O}(h_j^{p+1}),$$

or, put differently,

$$\mathbf{u}(x_j + h_j) = \mathbf{y}_j + h_j \mathbf{\Phi}(x_j, \mathbf{y}_j, \mathbf{F}, h_j) + \mathcal{O}(h_j^{p+1}).$$

Here and throughout this chapter, it is supposed that $\mathbf{F}$ and $\mathbf{\Phi}$ have as many continuous derivatives as are required. Formal statements of precise conditions for the results are distracting and do not provide new insight, so they are omitted throughout.

In Chapter 1 the effect of changes of initial value and of $\mathbf{F}$ on the solution $\mathbf{y}(x)$ of the initial value problem (1) was considered. Here the corresponding question must be studied for the numerical method. Suppose that the values $\{\mathbf{z}_j\}$ arise from an initial value $\mathbf{z}_0 = \mathbf{y}_0 + \Delta_0$ that is only a little different from $\mathbf{y}_0$ and a recipe that is only a little different from the recipe producing $\mathbf{y}_j$:

$$\mathbf{z}_{j+1} = \mathbf{z}_j + h_j \mathbf{\Phi}(x_j, \mathbf{z}_j, \mathbf{F}, h_j) + h_j \boldsymbol{\delta}_j. \tag{1.2}$$

The one-step method is said to be *stable* if there exists a constant S such that

$$\max_j \|\mathbf{z}_j - \mathbf{y}_j\| \leq S \max(\|\Delta_0\|, \max_J \|\boldsymbol{\delta}_J\|).$$

This is analogous to what was meant by stability of the differential equation itself. Obviously a numerical method must be stable with respect to computational errors if it is to be of practical value, but stability is useful in the theory, too. Indeed, stability leads quickly to the result we want: A stable one-step method of order p is convergent of order p. To prove this, it is first necessary to state clearly what is meant by convergence.

In practice the initial value $\mathbf{y}_0$ is taken to be the given value of the solution $\mathbf{y}(a)$, but for theoretical purposes, it is useful to permit $\mathbf{y}_0$ to be different. After the starting value has been specified somehow, a code will use the formula (1.1) to step from a to b producing results $\mathbf{y}_{j+1}$ successively at points x_{j+1}. The step size $h_j = x_{j+1} - x_j$ is determined automatically in most codes so as to yield an accurate solution as efficiently as possible. In the course of Example 7 of Chapter 2, the variation of the step size when solving a two-body problem with the Runge–Kutta code in the NAG library was presented as Figure 2.7. It shows that a suitable step size might vary considerably in the course of an integration. The range seen in the figure is not unusual. The range of 16 orders of magnitude required for the solution of the problem taken up in Example 4 of Chapter 3 is remarkable.

For the analysis of convergence, it is supposed only that a sequence of step sizes $\{h_j\}$ spanning the interval has been specified somehow. By spanning the interval is meant

$$h_0 + h_1 + \cdots + h_{N-1} = (x_1 - a) + (x_2 - x_1) + \cdots + (b - x_{N-1}) = b - a.$$

Let H be the largest of these step sizes:

$$H = \max_j h_j.$$

Our model of computation is that a sequence of integrations is done with the maximum step size H tending to zero. Early treatments of convergence look at what happens at a particular value of x, let us say b, and ask whether $\mathbf{y}_N$ tends to $\mathbf{y}(b)$ as $N \to \infty$. Instead we define *convergence* as

$$\max_{0 \leq j \leq N} \|\mathbf{y}(x_j) - \mathbf{y}_j\| \to 0 \text{ as } H \to 0. \tag{1.3}$$

The distinction is that in the one case, as H tends to 0, the meshes must be refined in such a way that the point of interest is always a mesh point and in the other, the meshes need not be related at all. It turns out that there is little difference for the problems and methods investigated here. The uniform convergence of (1.3) shows more clearly that the function $\mathbf{y}(x)$ is approximated throughout the interval of interest. Convergence is said to be of order p if

$$\max_{0 \leq j \leq N} \|\mathbf{y}(x_j) - \mathbf{y}_j\| \quad \text{is} \quad \mathcal{O}(H^p).$$

The local truncation error $h_j \tau_j$ satisfies

$$\mathbf{y}(x_j + h_j) = \mathbf{y}(x_j) + h_j \mathbf{\Phi}(x_j, \mathbf{y}(x_j), \mathbf{F}, h_j) + h_j \boldsymbol{\tau}_j.$$

Written in this way, it is clear that the $\{\mathbf{y}(x_j)\}$ satisfy a perturbation of the recipe that defines the numerical solution $\{\mathbf{y}_j\}$. Let us now apply the definition of stability to the sequence $\{\mathbf{y}(x_j)\}$. The sequence satisfies a difference equation like the one satisfied by the sequence $\{\mathbf{y}_j\}$, but the starting value is perturbed by a starting error and the recipe is perturbed by local truncation errors. For a stable method

$$\max_j \|\mathbf{y}(x_j) - \mathbf{y}_j\| \le S \max(\|\mathbf{y}(a) - \mathbf{y}_0\|, \max_J \|\boldsymbol{\tau}_J\|).$$

For a method of order p, the $\|\boldsymbol{\tau}_J\|$ are all $\mathbb{O}(H^p)$ by definition, so if the initial value is correct to order q, that is,

$$\mathbf{y}_0 = \mathbf{y}(a) + \mathbb{O}(H^q),$$

this inequality says that the method is convergent of order $\min(p, q)$. When actually solving a problem with a one-step method, the starting value is taken to be the given initial value and with this choice, convergence is of order p. This is why the method is said to be of order p. The bound does, however, suggest that it is possible to spoil the accuracy of the method by starting inaccurately. This is not merely a possibility—starting errors *do* persist. Though not troublesome for one-step methods, starting accurately methods with memory is a matter that requires attention in practical computation.

It is important to understand what has been done. The solution values $\mathbf{y}(x_j)$ satisfy the difference equation with a discrepancy called the local truncation error. There also may be a discrepancy in the initial value. Stability of the numerical scheme says that these small changes to the recipe lead to small changes in the values produced, i.e., the $\mathbf{y}(x_j)$ are uniformly close to the $\mathbf{y}_j$. In Chapter 4 we saw how to construct formulas that are very accurate, i.e., are of high order p. The task that remains is to understand the stability of these formulas.

For the differential equation itself, a Lipschitz condition is sufficient to guarantee stability on a finite interval. It proves sufficient for one-step methods as well. It will be assumed that

$$\|\boldsymbol{\Phi}(x, \mathbf{v}, \mathbf{F}, h) - \boldsymbol{\Phi}(x, \mathbf{w}, \mathbf{F}, h)\| \le \mathcal{L}\|\mathbf{v} - \mathbf{w}\|$$

for all x in $[a, b]$, all $\mathbf{v}$, $\mathbf{w}$, and all $0 < h \le h^*$. In the case of Runge–Kutta methods this follows from the fact that $\mathbf{F}$ satisfies a Lipschitz condition, but in the case of Taylor series, this places additional conditions on the smoothness of $\mathbf{F}$. To see this in the case of explicit Runge–Kutta methods, recall from Chapter 4 that

$$\boldsymbol{\Phi}(x, \mathbf{v}, \mathbf{F}, h) = \sum_{i=0}^{s} \gamma_i \mathbf{F}_i(\mathbf{v})$$

where

$$\mathbf{F}_0(\mathbf{v}) = \mathbf{F}(x_j, \mathbf{v}),$$

$$\mathbf{F}_i(\mathbf{v}) = \mathbf{F}\left(x_j + \alpha_i h, \mathbf{v} + h \sum_{m=0}^{i-1} \beta_{i,m} \mathbf{F}_m(\mathbf{v})\right), \qquad i = 1, \ldots, s.$$

The idea is to prove successively that each of the stages $\mathbf{F}_i$ satisfies a Lipschitz condition. The first stage is trivial because this is a statement about $\mathbf{F}$ itself:

$$\|\mathbf{F}_0(\mathbf{v}) - \mathbf{F}_0(\mathbf{w})\| = \|\mathbf{F}(x_j, \mathbf{v}) - \mathbf{F}(x_j, \mathbf{w})\| \le L \, \|\mathbf{v} - \mathbf{w}\|.$$

The next stage shows how it all goes:

$$\begin{aligned}\|\mathbf{F}_1(\mathbf{v}) - \mathbf{F}_1(\mathbf{w})\| &= \|\mathbf{F}(x_j + \alpha_1 h, \mathbf{v} + h\beta_{1,0}\mathbf{F}_0(\mathbf{v})) \\ &\quad - \mathbf{F}(x_j + \alpha_1 h, \mathbf{w} + h\beta_{1,0}\,\mathbf{F}_0(\mathbf{w}))\| \\ &\le L\,\|(\mathbf{v} + h\beta_{1,0}\,\mathbf{F}_0(\mathbf{v})) - (\mathbf{w} + h\beta_{1,0}\,\mathbf{F}_0(\mathbf{w}))\| \\ &\le L\,\|\mathbf{v} - \mathbf{w}\| + Lh|\beta_{1,0}|\;\|\mathbf{F}_0(\mathbf{v}) - \mathbf{F}_0(\mathbf{w})\| \\ &\le (L + L^2 h^*|\beta_{1,0}|)\;\|\mathbf{v} - \mathbf{w}\|.\end{aligned}$$

Notice here that the Lipschitz constant for the stage depends on the maximum step size permitted, h^*. It is easy now to repeat the argument for all the stages and finally for the sum in $\boldsymbol{\Phi}$ to see that the increment function $\boldsymbol{\Phi}$ satisfies a Lipschitz condition.

To establish stability suppose now that the sequence $\mathbf{y}_{j+1}$ is generated by the formula (1.1) and the sequence $\mathbf{z}_{j+1}$ by the perturbed formula (1.2). Subtracting the two formulas results in

$$\mathbf{z}_{j+1} - \mathbf{y}_{j+1} = \mathbf{z}_j - \mathbf{y}_j + h_j[\boldsymbol{\Phi}(x_j, \mathbf{z}_j, \mathbf{F}, h_j) - \boldsymbol{\Phi}(x_j, y_j, \mathbf{F}, h_j)] + h_j\boldsymbol{\delta}_j.$$

The Lipschitz condition on the increment function then implies that

$$\|\mathbf{z}_{j+1} - \mathbf{y}_{j+1}\| \le (1 + h_j\mathcal{L})\;\|\mathbf{z}_j - \mathbf{y}_j\| + h_j\|\boldsymbol{\delta}_j\|. \tag{1.4}$$

To deal with more complicated situations that arise later, this will be interpreted by means of a lemma.

Lemma 1. *Let the sequence $\{w_m\}$ be the solution of the difference equation*

$$w_{m+1} = (1 + h_m\mathcal{L})w_m + h_m\delta_m \qquad m = 0, 1, \ldots \tag{1.5}$$

for given non-negative w_0, $\{h_m\}$, $\{\delta_m\}$, and $\mathcal{L}$. Then for all j

$$w_j \le e^{\mathcal{L}(x_j - x_0)}w_0 + \sum_{m=0}^{j-1} e^{\mathcal{L}(x_j - x_{m+1})}h_m\delta_m.$$

Proof. The inequalities $1 \le 1 + x \le \exp(x)$, valid for $x \ge 0$, follow from Taylor series expansion of $\exp(x)$. Using them, the result follows easily by induction: The case $j = 0$ is trivial. The case $j = 1$ is

$$w_1 = (1 + h_1\mathcal{L})w_0 + h_0\delta_0 \le e^{h_1\mathcal{L}}w_0 + e^0 h_0\delta_0.$$

Supposing that the inequality holds for the case j, the difference equation for $m = j$ leads to

$$\begin{aligned} w_{j+1} &= (1 + h_j\mathcal{L})w_j + h_j\delta_j \\ &\le e^{h_j\mathcal{L}}\left(e^{\mathcal{L}(x_j - x_0)}\, w_0 + \sum_{m=0}^{j-1} e^{\mathcal{L}(x_j - x_{m+1})} h_m\delta_m \right) + h_j\delta_j \\ &= e^{\mathcal{L}(x_{j+1} - x_0)} w_0 + \sum_{m=0}^{j} e^{\mathcal{L}(x_{j+1} - x_{m+1})} h_m\delta_m \end{aligned}$$

because $h_j + (x_j - x_r) = x_{j+1} - x_j + x_j - x_r = x_{j+1} - x_r$. By induction, the bound holds for all j.

To apply this bound to the present task, we define $w_0 = \|\mathbf{z}_0 - \mathbf{y}_0\|$ and the sequence $\{w_j\}$ by (1.5) with $\delta_j = \|\boldsymbol{\delta}_j\|$ for each j. With this definition it is easy to see that for each j,

$$\|\mathbf{z}_j - \mathbf{y}_j\| \le w_j.$$

This is obviously true for $j = 0$. Assuming that it is true for $j = m$, (1.4) leads to

$$\|\mathbf{z}_{m+1} - \mathbf{y}_{m+1}\| \le (1 + h_m\mathcal{L})w_m + h_m\delta_m = w_{m+1},$$

hence the inequality is true for all j. Using the bound developed for the sequence $\{w_m\}$ of (1.5), this last inequality leads to

$$\begin{aligned} \|\mathbf{z}_j - \mathbf{y}_j\| &\le e^{\mathcal{L}(x_j - x_0)}\|\mathbf{z}_0 - \mathbf{y}_0\| + \sum_{m=0}^{j-1} e^{\mathcal{L}(x_j - x_{m+1})} h_m \|\boldsymbol{\delta}_m\|, \\ &\le e^{\mathcal{L}(x_j - x_0)}\|\mathbf{z}_0 - \mathbf{y}_0\| + e^{\mathcal{L}(x_j - x_0)} \max_{m<j} \|\boldsymbol{\delta}_m\| \sum_{m=0}^{j-1} h_m, \\ &\le e^{\mathcal{L}(x_j - x_0)}[\,\|\mathbf{z}_0 - \mathbf{y}_0\| + (x_j - x_0) \max_{m<j} \|\boldsymbol{\delta}_m\|\,]. \end{aligned}$$

Finally,

$$\|\mathbf{z}_j - \mathbf{y}_j\| \le e^{\mathcal{L}(b-a)} 2\max(b - a, 1)\max(\|\mathbf{z}_0 - \mathbf{y}_0\|, \max_J \|\boldsymbol{\delta}_J\|).$$

This states that the one-step method is stable with constant

$$S = e^{\mathcal{L}(b-a)} 2\max(b - a, 1).$$

Implicit one-step methods introduce few complications to the analysis. These methods have the form

$$\mathbf{y}_{j+1} = \mathbf{y}_j + h_j\, \boldsymbol{\Phi}(x_j, \mathbf{y}_{j+1}, \mathbf{y}_j, \mathbf{F}, h_j). \tag{1.6}$$

Because the method is implicit, we must assume that $\mathbf{y}_{j+1}$ is well defined for all sufficiently small h_j, say $h_j \leq h^*$. The Lipschitz condition now has the form

$$\|\mathbf{\Phi}(x, \mathbf{v}_{j+1}, \mathbf{v}_j, \mathbf{F}, h) - \mathbf{\Phi}(x, \mathbf{w}_{j+1}, \mathbf{w}_j, \mathbf{F}, h)\| \leq \mathcal{L} \max_{0 \leq i \leq 1} \|\mathbf{v}_{j+1-i} - \mathbf{w}_{j+1-i}\|. \quad (1.7)$$

Suppose that $\mathbf{y}_{j+1}$ is the result of a step from $(x_j, \mathbf{y}_j)$ of size h_j and $\mathbf{z}_{j+1}$ is the result of a step from $(x_j, \mathbf{z}_j)$ of the same size. Subtracting the recipes satisfied by these values and using the Lipschitz condition leads now to

$$\begin{aligned}\|\mathbf{y}_{j+1} - \mathbf{z}_{j+1}\| &\leq \|\mathbf{y}_j - \mathbf{z}_j + h_j[\mathbf{\Phi}(x_j, \mathbf{y}_{j+1}, \mathbf{y}_j, \mathbf{F}, h_j) - \mathbf{\Phi}(x_j, \mathbf{z}_{j+1}, \mathbf{z}_j, \mathbf{F}, h_j)]\|, \\ &\leq (1 + h_j\mathcal{L}) \|\mathbf{y}_j - \mathbf{z}_j\| + h_j\mathcal{L} \|\mathbf{y}_{j+1} - \mathbf{z}_{j+1}\|.\end{aligned}$$

The thing that is different with the implicit formula is that the change in $\mathbf{y}_{j+1}$ appears on both sides of the inequality. By making the upper bound h^* on h_j smaller if necessary, it can be assumed that $h_j\mathcal{L} \leq 1/2$ with the consequence that

$$\frac{1}{1 - h_j\mathcal{L}} \leq 2,$$

and

$$\frac{(1 + h_j\mathcal{L})}{(1 - h_j\mathcal{L})} \leq 1 + 4\, h_j\mathcal{L}.$$

A little manipulation of the basic inequality then leads to

$$\|\mathbf{y}_{j+1} - \mathbf{z}_{j+1}\| \leq \frac{(1 + h_j\mathcal{L})}{(1 - h_j\mathcal{L})} \|\mathbf{y}_j - \mathbf{z}_j\| \leq (1 + 4\, h_j\mathcal{L}) \|\mathbf{y}_j - \mathbf{z}_j\|.$$

This differs from the case of an explicit one-step method only in that $\mathcal{L}$ has been increased to $4\mathcal{L}$. In the same way, if $\mathbf{y}_{j+1}$ is the result of the implicit formula and $\mathbf{z}_{j+1}$ is the result arising from perturbations $h_j\boldsymbol{\delta}_j$, then

$$\begin{aligned}\|\mathbf{z}_{j+1} - \mathbf{y}_{j+1}\| &\leq \frac{(1 + h_j\mathcal{L})}{(1 - h_j\mathcal{L})} \|\mathbf{z}_j - \mathbf{y}_j\| + \frac{h_j}{(1 - h_j\mathcal{L})} \|\boldsymbol{\delta}_j\|, \\ &\leq (1 + 4h_j\mathcal{L}) \|\mathbf{z}_j - \mathbf{y}_j\| + 2h_j \|\boldsymbol{\delta}_j\|.\end{aligned}$$

Again, this is just like the inequality for the explicit formula except that $\mathcal{L}$ and $\|\boldsymbol{\delta}_j\|$ have been increased by constant factors. Increasing these constants does not alter any of the arguments establishing stability and the result now is

$$||\mathbf{z}_j - \mathbf{y}_j|| \le e^{4\mathcal{L}(b-a)} 4\max(b - a, 1)\max(||\mathbf{z}_0 - \mathbf{y}_0||, \max_J ||\boldsymbol{\delta}_J||).$$

Convergence is established just as for an explicit one-step method with, however, the proviso that H be small enough that the method is well defined and the analysis applicable.

As we have seen, these stability results lead immediately to convergence results in our model of the computation. In stating formally such results, we do not repeat the assumption we make throughout that the function **F** in (1) is continuous on $[a, b] \times R^n$, satisfies there a Lipschitz condition with constant L, and is as smooth as necessary for the stated order of the formula used. Collecting the results developed, we have

Theorem 1. *A one-step method (1.6) is used to approximate the solution of (1). Suppose that the method satisfies a Lipschitz condition (1.7) and that it is well defined for all sufficiently small step sizes. It follows that the method is stable for all sufficiently small step sizes. Further, if the method is of order* p *and the starting value* $\mathbf{y}_0$ *is accurate of order* q*, then the numerical solution* $\{y_j\}$ *is convergent of order min*(p, q).

Corollary Runge–Kutta methods satisfy the hypotheses of the theorem.

EXAMPLE 1. We return now to an interesting example of Hubbard and West [1990, 1992]. In their numerical solution of $y' = f(x, y) = y^2 - x$ with the fixed step size midpoint Euler formula of MacMath, a spurious solution is seen. This numerical solution appears to be stable. We saw in Example 1 of Chapter 2 that the same is true of the classic four-stage, fourth-order Runge–Kutta formula. Working with the simpler formula, we now reconcile these experiments and the convergence theory of this section. To imitate in the discrete case the isocline analysis of the continuous case, we write the midpoint Euler formula in the form of a difference quotient:

$$Dy_j = \frac{y_{j+1} - y_j}{h} = h\, f(x_j + 0.5h,\, y_j + 0.5h\, f(x_j, y_j)).$$

For the equation at hand, this is

$$Dy_j = \frac{y_{j+1} - y_j}{h} = (y_j + 0.5h(y_j^2 - x_j))^2 - x_j - 0.5h.$$

A little calculation shows that the solutions of $Dy_j = 0$, the null "isoclines" are given by

$$y_j = -\frac{1}{h} \pm \sqrt{-\frac{1}{h^2} + x_j \pm \frac{2}{h}\sqrt{x_j + \frac{h}{2}}}.$$

These "isoclines" are plotted in Figure 5.1a for $h = 0.3$, the step size used in Example 21.2a of Hubbard and West [1992]. The signs of the difference quotients indicated in the figure help us to understand the stability of the discrete problem, just like the signs of the derivatives helped us in Chapter 2 to understand the stability of the differential equation. The unstable null "isocline" at the top of the figure approximates well the unstable isocline $\sqrt{x}$ of the differential equation. The behavior of the "isocline" that approximates the stable isocline $-\sqrt{x}$ is quite interesting. It is stable and a reasonably accurate approximation up to the point $h^{-2} - \sqrt{2/h}$ where it

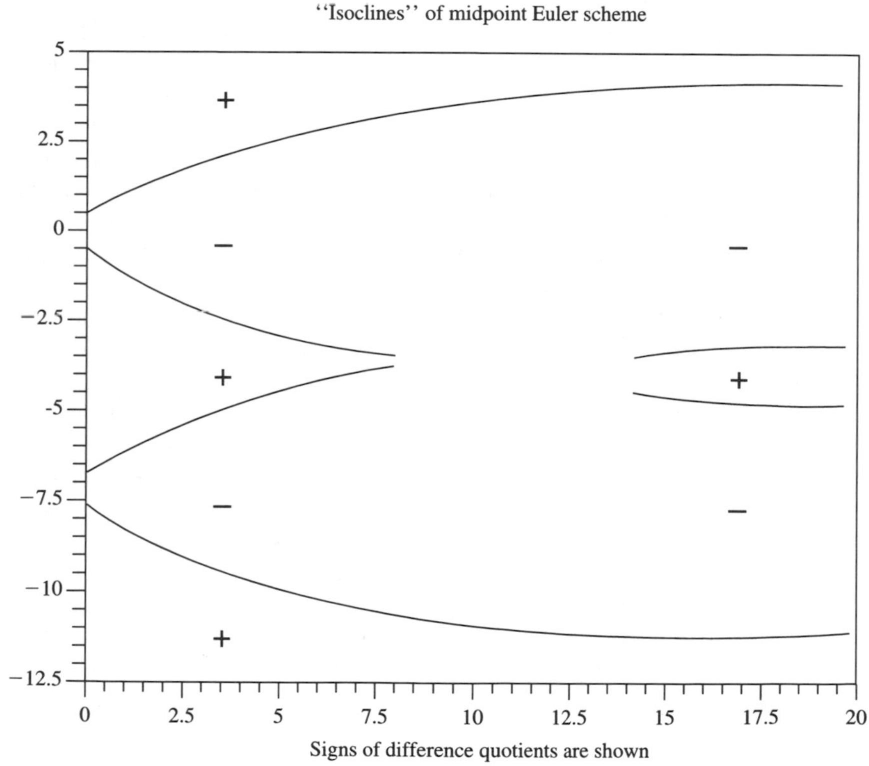

Figure 5.1a

merges with an unstable "isocline" and comes to an end. The two "isoclines" reappear at the point $h^{-2} + \sqrt{2/h}$. There is an exchange of stability at this point—one "isocline" continues to be a good approximation of the stable asymptotic solution $-\sqrt{x}$, but it is not stable. Although there is a gap in the "isoclines," it is easily verified that along the curve $y(x) = -\sqrt{x}$, the difference quotient Dy_j has the small, constant value $-h/2$. Indeed, it is relatively small near this curve throughout the gap between "isoclines." As a consequence, solution curves starting near the stable null isocline of the differential equation approach the stable "isocline" of the discrete problem until it ceases to exist. They are then rather flat until a new stable "isocline" emerges and they approach it. This last curve is the spurious solution of the differential equation remarked by Hubbard and West.

Often one is inclined to think that a step size small enough to resolve a solution will be small enough for the method to produce a reasonable approximation to the solution. That is not the case here. Convergence results guarantee an accurate solution only when the step size is sufficiently small. Just how small the step size must be to get a reasonable numerical solution depends on the problem. Here the behavior of the numerical "isocline" approximates that of the stable isocline of the differential equation up to the point $h^{-2} - \sqrt{2/h}$ where two "isoclines" merge and disappear. The figure shows the "isoclines" when $h = 0.3$. If we had taken $h = 0.1$, this point would be about 95.5, hence the numerical solution would be in qualitative agreement with the solution of the differential equation over the entire interval of the plot. This is precisely what happens with the fourth-order formula in Example 21.2a of Hubbard and West [1992]—the interval they use is short enough that only a hint of a spurious solution appears. There is another interesting phenomenon for the midpoint Euler formula only hinted at in their plot. The "isocline" approximating the stable isocline of the differential equation becomes unstable as the integration progresses and curves starting just below it move towards the bottom of the plot. These curves are moving toward a stable "isocline" identified in our analysis. This additional stable "isocline" is another qualitative difference between the discrete and continuous problems. As with the spurious solution, it moves out of the range of interest as $h \to 0$.

Figure 5.1b shows the results of integrations with step sizes 0.3 and 0.15. The great difference in these two solutions from the middle of the interval on makes it clear that a constant step size of 0.3 is too big for the convergence theory to be applicable. Notice that the solutions differ substantially at the beginning of the interval, too—reintegration with a smaller tolerance or step size is a valuable way to gain some confidence in a numerical solution, but this example serves as a warning that consistent

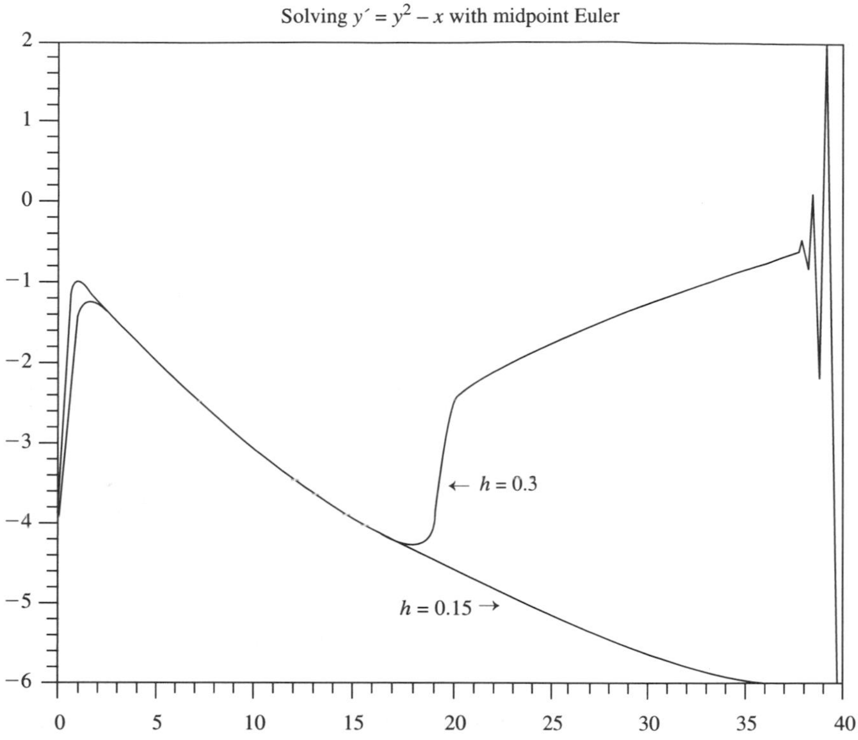

Figure 5.1b

results at a point, or even throughout an interval as in this example, does not imply that the integration is accurate at preceding points.

§2 Some Methods with Memory

Some methods with memory can be analyzed in a way quite similar to that used for one-step methods in the last section. A crucial difference is that the coefficients of the formulas depend on the relative spacing of mesh points. In keeping with our usual way of proceeding, let us begin with the relatively simple case of an explicit formula. Although the argument will be presented in terms of the Adams–Bashforth formulas, it should be appreciated that it is the form of the method that makes the simple proof possible. One reason for working with the Adams–Bashforth formulas is that we have seen how to define them for an arbitrary mesh.

Recall from Chapter 4 that the Adams–Bashforth formula of order ν, ABν, has the form

$$\mathbf{y}_{j+1} = \mathbf{y}_j + h_j \sum_{k=1}^{\nu} \beta^*_{\nu,k} \mathbf{F}_{j+1-k}$$

where

$$\beta^*_{\nu,k} = \frac{1}{h_j} \int_{x_j}^{x_j+h_j} \prod_{\substack{m=1 \\ m \neq k}}^{\nu} \left(\frac{t - x_{j+1-m}}{x_{j+1-k} - x_{j+1-m}} \right) dt.$$

To establish convergence, the solution values $\{\mathbf{y}(x_j)\}$ must be shown to satisfy a perturbed version of this recipe. That was done in Chapter 4 where it was found that

$$\mathbf{y}(x_{j+1}) = \mathbf{y}(x_j) + h_j \sum_{k=1}^{\nu} \beta^*_{\nu,k} \mathbf{F}(x_{j+1-k}, \mathbf{y}(x_{j+1-k})) + h_j \boldsymbol{\tau}_j.$$

Here the local truncation error has the form

$$h_j \boldsymbol{\tau}_j = \frac{\mathbf{y}^{(\nu+1)}(*)}{\nu!} \int_{x_j}^{x_j+h_j} \prod_{k=1}^{\nu} (t - x_{j+1-k})\, dt$$

where the components of the derivative are evaluated at points in the span of the nodes. For H a bound for the step sizes in this span, the local truncation error is of order $\nu + 1$, specifically,

$$\|h_j \boldsymbol{\tau}_j\| \leq \|\mathbf{y}^{(\nu+1)}\|\, H^{\nu+1}$$

Again let us note a way to interpret the local truncation error. If the previously computed approximate solutions $\mathbf{y}_j, \ldots, \mathbf{y}_{j+1-\nu}$ are exact, i.e., $\mathbf{y}_j = \mathbf{y}(x_j), \ldots, \mathbf{y}_{j+1-\nu} = \mathbf{y}(x_{j+1-\nu})$, then the local truncation error is the difference between $\mathbf{y}(x_{j+1})$ and $\mathbf{y}_{j+1}$.

In addition to the recipe for calculating $\mathbf{y}_{j+1}$, there must be a collection of procedures for calculating the initial values, $\mathbf{y}_r = \mathbf{s}_r(H)$ for $0 \leq r < \nu$. Only the accuracy of the starting procedures matters for convergence, so we assume that there is a constant C for which

$$\|\mathbf{y}(x_r) - \mathbf{y}_r\| \leq C\, H^q, \qquad 0 \leq r < \nu.$$

In the next section we return to the matter of starting the integration and comment here only that one way to calculate the $\mathbf{y}_r$ is to use a one-step method of order $q - 1$ with step size $x_r - x_0$.

In the last section we established stability and then deduced convergence. For the sake of variety and to make clear that it is possible, this time we establish convergence directly. For each j, let the true error of the numerical solution be denoted by

$$\mathbf{e}_j = \mathbf{y}(x_j) - \mathbf{y}_j.$$

Subtracting the relation for $\mathbf{y}_{j+1}$ from that for $\mathbf{y}(x_j)$ and taking norms leads to

$$\|\mathbf{e}_{j+1}\| \leq \|\mathbf{e}_j\| + h_j L \sum_{k=1}^{\nu} |\beta^*_{\nu,k}| \, \|\mathbf{e}_{j+1-k}\| + h_j \|\boldsymbol{\tau}_j\|, \qquad j \geq \nu - 1.$$

In contrast to the situation with one-step methods, the coefficients $\beta^*_{\nu,k}$ here depend on the mesh. To bound the growth of errors, we have to assume that there is a constant β such that

$$\sum_{k-1}^{\nu} |\beta^*_{\nu,k}| \leq \beta \tag{2.1}$$

for all thc cocfficicnts $\beta^*_{\nu,k}$ that might arisc in thc intcgration. For onc-stcp methods we defined a sequence $\{w_m\}$ such that w_j is an upper bound for the norm of the error at x_j. A difficulty now is the presence of errors at mesh points prior to x_j. The trick for dealing with this is to take w_j to be an upper bound for not only $\|\mathbf{e}_j\|$, but for all previous errors. Thus, we aim to define a sequence $\{w_m\}$ such that

$$w_m \geq \|\mathbf{e}_r\| \quad \text{for} \quad r = 0, 1, \ldots, m. \tag{2.2}$$

To accomplish this, first let w_0 be a bound on the starting errors,

$$w_0 = \max_{0 \leq r < \nu} \|\mathbf{e}_r\|.$$

Let $\mathcal{L} = L\beta$ and for $j \geq 0$, let $\delta_j = \|\boldsymbol{\tau}_j\|$ and

$$w_{j+1} = (1 + h_j \mathcal{L}) w_j + h_j \delta_j.$$

The sequence is obviously non-decreasing so that $w_m \geq w_0 \geq \|\mathbf{e}_r\|$ for $r = 0, \ldots, \nu - 1$. This says that (2.2) holds for all $m \leq \nu - 1$. Suppose now that it holds for $m = j$. The definitions and the bound on $\|\mathbf{e}_{j+1}\|$ then lead to

$$\|\mathbf{e}_{j+1}\| \leq w_j + h_j L \beta w_j + h_j \delta_j = w_{j+1}.$$

Because $w_j \geq \|\mathbf{e}_r\|$ for all $r = 0, \ldots, j$ by assumption and w_{j+1} is at least as big as w_j, we find that $w_{j+1} \geq \|\mathbf{e}_r\|$ for all $r = 0, \ldots, j + 1$, which is to say that (2.2) holds for $m = j + 1$. By induction (2.2) holds for all m. The difference equation for the w_m is the same one that arose in bounding the error for one-step methods, namely (1.5), and the rest of the proof of convergence is also the same. The result is

$$\|\mathbf{y}(x_j) - \mathbf{y}_j\|$$

$$\leq e^{\mathcal{L}(b-a)} 2\max(b - a, 1)\max\left(\max_{0\leq r<\nu} \|\mathbf{y}(x_r) - \mathbf{y}_r\|, \max_{J} \|\tau_J\|\right).$$

For a method of order ν and starting values that are accurate to $\mathcal{O}(H^q)$, this states that there is convergence of order $\min(\nu, q)$. As remarked in connection with one-step methods, the order really can be reduced if q is less than ν, so it is important to start off with accurate values. Note, however, that if a formula of order 1 is used for one step, it produces a value accurate to $\mathcal{O}(H^2)$, hence is accurate enough to start a formula of order 2. In particular, AB2 might be started with Euler's method, AB1, without reducing the overall order of convergence.

The bound (2.1) is crucial. When a constant step size h is used, the bound is trivial and the convergence proof for Adams–Bashforth formulas is complete. To complete the proof for variable step size, we must guarantee somehow that the coefficients in the formulas are uniformly bounded for all the step sizes that might be used. Fortunately there is a very simple condition, easily realized in practice, that guarantees this for Adams–Bashforth formulas. Before going into the matter, let us first make some observations about constant step size integrations. Because the convergence proof depends on the form of the formula rather than anything particular to Adams–Bashforth formulas, a somewhat broader class of formulas can be accommodated by a little modification of the proof:

Theorem 2. *A LMM of the form*

$$\mathbf{y}_{j+1} = \mathbf{y}_{j-m} + h \sum_{k=1}^{\nu} \beta_k \mathbf{F}_{j+1-k}$$

is used to approximate the solution of (1). Here $m \geq 0$ and the integration is done with a constant step size h. If the method is of order p and if the starting values $\mathbf{y}_r$ are accurate of order q, then the numerical solution $\{\mathbf{y}_j\}$ is convergent of order $\min(p, q)$.

Remarks *(1) Adams–Bashforth formulas satisfy the hypotheses of the theorem.*

(2) The midpoint rule,

$$\mathbf{y}_{j+1} = \mathbf{y}_{j-1} + 2h\, \mathbf{F}(x_j, \mathbf{y}_j),$$

satisfies the hypotheses of the theorem.

EXERCISE 1. Prove Theorem 2.

EXERCISE 2. This way of establishing convergence cannot be used for the formula

$$\mathbf{y}_{j+1} + 4\mathbf{y}_j - 5\mathbf{y}_{j-1} = h[4\mathbf{F}(x_j, \mathbf{y}_j) + 2\mathbf{F}(x_j, \mathbf{y}_j)]$$

shown earlier to be of order 3. Why? What goes wrong in the proof?

We return now to the matter of the bound (2.1) for a variable step implementation of Adams–Bashforth methods. In Chapter 4, the coefficients for AB2 were worked out as an example. In this case

$$\sum_{k=1}^{2} |\beta^*_{2,k}| = \left(1 + \frac{1}{2}\frac{h_j}{h_{j-1}}\right) + \left(\frac{1}{2}\frac{h_j}{h_{j-1}}\right).$$

If there is to be a bound of the form (2.1), it is clearly necessary that the ratio of successive step sizes be uniformly bounded above. All the codes that select step sizes automatically put bounds of this kind on the rate of increase of step size for practical reasons—the asymptotic arguments used to justify the selection of an appropriate change of step size are not credible when "large" changes are contemplated. A bound of 2 is common, but bounds of 5 and 10 are not unusual in popular codes. In the case of Adams–Bashforth formulas, we shall now prove that any upper bound will do and no other restriction on the mesh is necessary.

Suppose there is a constant $\Gamma \geq 1$ such that $h_{j+1} \leq \Gamma\, h_j$ for all j and all integrations in the model of convergence. It is claimed that this implies a uniform bound on the magnitude of

$$\beta^*_{\nu,k} = \frac{1}{h_j}\int_{x_j}^{x_j+h_j} \prod_{\substack{m=1\\ m\neq k}}^{\nu} \left(\frac{t - x_{j+1-m}}{x_{j+1-k} - x_{j+1-m}}\right) dt.$$

The integral here is over an interval of length h_j and $m \geq 1$, so

$$|\beta^*_{\nu,k}| \leq \prod_{\substack{m=1\\ m\neq k}}^{\nu} \left|\frac{x_{j+1} - x_{j+1-m}}{x_{j+1-k} - x_{j+1-m}}\right|$$

because the product is a bound on the integrand valid for all t in the interval. To bound the individual factors in this product, it is convenient to consider two cases:

Case 1: When $k > m$, the denominator

$$|x_{j+1-k} - x_{j+1-m}| = x_{j+1-m} - x_{j+1-k} \geq h_{j+1-k},$$

and the numerator

$$|x_{j+1} - x_{j+1-m}| = h_{j+1-m} + h_{j+2-m} + \cdots + h_j$$
$$\leq h_{j+1-m}[1 + \Gamma + \Gamma^2 + \cdots + \Gamma^{m-1}] \leq h_{j+1-m}[m\Gamma^m].$$

Now

$$h_{j+1-k+1} \leq \Gamma\, h_{j+1-k}, \qquad h_{j+1-k+2} \leq \Gamma^2\, h_{j+1-k}, \ldots,$$
$$h_{j+1-k+(k-m)} = h_{j+1-m} \leq \Gamma^{k-m}\, h_{j+1-k},$$

hence

$$|x_{j+1} - x_{j+1-m}| \leq h_{j+1-m}[m\,\Gamma^m] \leq h_{j+1-k}[m\,\Gamma^k].$$

Then

$$\frac{|x_{j+1} - x_{j+1-m}|}{|x_{j+1-k} - x_{j+1-m}|} \leq \frac{h_{j+1-k}[m\,\Gamma^k]}{h_{j+1-k}} \leq \nu\,\Gamma^\nu.$$

Case 2: When $k < m$, the denominator

$$|x_{j+1-k} - x_{j+1-m}| = x_{j+1-k} - x_{j+1-m} \geq h_{j+1-m},$$

and as in the other case, the numerator

$$|x_{j+1} - x_{j+1-m}| \leq h_{j+1-m}[m\,\Gamma^m] \leq h_{j+1-m}[\nu\,\Gamma^\nu],$$

so that

$$\frac{|x_{j+1} - x_{j+1-m}|}{|x_{j+1-k} - x_{j+1-m}|} \leq \frac{h_{j+1-m}[\nu\,\Gamma^\nu]}{h_{j+1-m}} = \nu\,\Gamma^\nu.$$

These bounds for the $\nu - 1$ factors are valid for each of the ν values of k, so they lead to a bound

$$\sum_{k=1}^{\nu} |\beta^*_{\nu,k}| \leq \nu(\nu\,\Gamma^\nu)^{\nu-1} = \beta.$$

This is a crude bound, but for the theory, any bound will do.

As with one-step methods, implicit methods do not introduce much complication to the analysis. They are exemplified by the Adams–Moulton formulas. Recall from Chapter 4 that these formulas have the form

$$\mathbf{y}_{j+1} = \mathbf{y}_j + h_j\beta_{\nu,0}\,\mathbf{F}(x_{j+1}, \mathbf{y}_{j+1}) + h_j \sum_{k=1}^{\nu-1} \beta_{\nu,k}\,\mathbf{F}_{j+1-k},$$

where

$$\beta_{\nu,k} = \frac{1}{h_j}\int_{x_j}^{x_j+h} \prod_{\substack{m=0\\ m\neq k}}^{\nu-1} \left(\frac{t - x_{j+1-m}}{x_{j+1-k} - x_{j+1-m}}\right) dt.$$

The formula is well defined for all sufficiently small step sizes. As was shown in Chapter 4, a sufficient condition for $\mathbf{y}_{j+1}$ to be uniquely defined is

$$h_j|\beta_{\nu,0}|L \leq \frac{1}{2}. \tag{2.3}$$

This suggests that a uniform bound on the magnitude of the coefficient $\beta_{\nu,0}$ is needed for a discussion of convergence. However, if the proof for Adams–Bashforth methods is to be followed, a bound is needed on all the coefficients:

$$\sum_{k=0}^{\nu-1} |\beta_{\nu,k}| \leq \beta \tag{2.4}$$

With a bound on the rate of growth of the step size, the existence of a suitable constant β can be demonstrated in the same way as the corresponding bound for the Adams–Bashforth formulas.

EXERCISE 3. Suppose that for some $\Gamma \geq 1$, $h_{j+1} \leq \Gamma\, h_j$ for all j and all integrations in the convergence model. Show the existence of a β for which (2.4) is true for AMν.

Recall from Chapter 4 that for AMν,

$$\mathbf{y}(x_{j+1}) = \mathbf{y}(x_j) + h_j\beta_{\nu,0}\,\mathbf{F}(x_{j+1}, \mathbf{y}(x_{j+1})) + h_j \sum_{k=1}^{\nu-1} \beta_{\nu,k}\,\mathbf{F}(x_{j+1-k}, \mathbf{y}(x_{j+1-k})) + h_j\boldsymbol{\tau}_j,$$

and

$$\|\boldsymbol{\tau}_j\| \leq \|\mathbf{y}^{(\nu+1)}\|\,H^{\nu}.$$

Arguing as we did with ABν, we see that the error satisfies the inequality

$$\|\mathbf{e}_{j+1}\| \leq \|\mathbf{e}_j\| + h_jL \sum_{k=0}^{\nu-1} |\beta_{\nu,k}|\;\|\mathbf{e}_{j+1-k}\| + h_j\|\boldsymbol{\tau}_j\|.$$

As with implicit one-step methods, the complication here is that $\|\mathbf{e}_{j+1}\|$ appears on both sides of the inequality. The way out is the same. The argument is simplified a little if it is assumed that all the step sizes satisfy

$$h_jL\beta \leq \frac{1}{2},$$

which is more stringent than (2.3). With this assumption

$$\begin{aligned} \|\mathbf{e}_{j+1}\| &\le \frac{1}{(1 - h_j L\beta)} \left[\|\mathbf{e}_j\| + h_j L \sum_{k=1}^{\nu-1} |\beta_{\nu,k}| \, \|\mathbf{e}_{j+1-k}\| + h_j \|\boldsymbol{\tau}_j\| \right], \\ &\le \frac{(1 + h_j L\beta)}{(1 - h_j L\beta)} \max_{1 \le k \le \nu - 1} \|\mathbf{e}_{j+1-k}\| + \frac{h_j}{(1 - h_j L\beta)} \|\boldsymbol{\tau}_j\|, \\ &\le (1 + 4h_j L\beta) \max_{1 \le k \le \nu - 1} \|\mathbf{e}_{j+1-k}\| + 2h_j \|\boldsymbol{\tau}_j\|. \end{aligned}$$

The key inequality has the same form as in the case of the explicit formula ABν. The analysis is virtually identical and the same conclusion is reached: When the starting values are all accurate to $\mathbb{O}(H^q)$, there is convergence of order $\min(\nu, q)$. Both implicit and explicit formulas require a bound on the rate of increase of the step size, but any bound will do. In the case of the implicit Adams–Moulton formulas, the step sizes must be sufficiently small both to guarantee that the formula is well defined and to make the proof given here valid.

THEOREM 3. *An Adams–Bashforth method of order ν is used to approximate the solution of (1). Suppose that the ratio of successive step sizes is bounded above by a constant $\Gamma \ge 1$. Then if the starting values $\mathbf{y}_r$ are accurate of order q, the numerical solution $\{\mathbf{y}_j\}$ is convergent of order $\min(\nu, q)$. With the same assumptions, the Adams–Moulton method of order ν is well defined for all sufficiently small maximum step sizes H and is convergent of order $\min(\nu, q)$.*

Let us now take up the convergence of Adams–Bashforth–Moulton predictor–corrector (PECE) pairs. Suppose that an Adams–Bashforth formula of order ν^* is used as a predictor. Using asterisks to refer to all quantities associated with the predictor, the formula is

$$\mathbf{y}_{j+1}^* = \mathbf{y}_j + h_j \sum_{k=1}^{\nu^*} \beta_{\nu,k}^* \, \mathbf{F}_{j+1-k}$$

and

$$\mathbf{y}(x_{j+1}) = \mathbf{y}(x_j) + h_j \sum_{k=1}^{\nu^*} \beta_{\nu,k}^* \, \mathbf{F}(x_{j+1-k}, \mathbf{y}(x_{j+1-k})) + h_j \boldsymbol{\tau}_j^*.$$

In a manner very similar to the proof of convergence for ABν^*, it is seen that the error of the predicted value, $\mathbf{e}_m^* = \mathbf{y}(x_m) - \mathbf{y}_m^*$, satisfies

$$\begin{aligned} \|\mathbf{e}_{j+1}^*\| &\le \|\mathbf{e}_j\| + h_j L \sum_{k=1}^{\nu^*} |\beta_{\nu,k}^*| \, \|\mathbf{e}_{j+1-k}\| + h_j \|\boldsymbol{\tau}_j^*\|, \\ &\le (1 + h_j L\beta^*) w_j + h_j \|\boldsymbol{\tau}_j^*\|, \end{aligned}$$

where

$$\sum_{k=1}^{\nu^*} |\beta^*_{\nu,k}| \le \beta^*,$$

and

$$w_j \ge \|\mathbf{e}_m\| \quad \text{for} \quad m = 0, 1, \ldots, j.$$

The corrector formula of order ν, AMν, is

$$\mathbf{y}_{j+1} = \mathbf{y}_j + h_j\beta_{\nu,0}\,\mathbf{F}(x_{j+1}, \mathbf{y}^*_{j+1}) + h_j \sum_{k=1}^{\nu-1} \beta_{\nu,k}\,\mathbf{F}(x_{j+1-k}, \mathbf{y}_{j+1-k})$$

and

$$\mathbf{y}(x_{j+1}) = \mathbf{y}(x_j) + h_j\beta_{\nu,0}\,\mathbf{F}(x_{j+1}, \mathbf{y}(x_{j+1})) + h_j \sum_{k=1}^{\nu-1} \beta_{\nu,k}\,\mathbf{F}(x_{j+1-k}, \mathbf{y}(x_{j+1-k})) + h_j\boldsymbol{\tau}_j.$$

From these definitions,

$$\begin{aligned}\|\mathbf{e}_{j+1}\| &\le \|\mathbf{e}_j\| + h_jL|\beta_{\nu,0}|\,\|\mathbf{e}^*_{j+1}\| \\ &\quad + h_jL \sum_{k=1}^{\nu-1} |\beta_{\nu,k}|\,\|\mathbf{e}_{j+1-k}\| + h_j\|\boldsymbol{\tau}_j\|, \\ &\le (1 + h_jL\beta)w_j + h_jL\beta\,\|\mathbf{e}^*_{j+1}\| + h_j\|\boldsymbol{\tau}_j\|,\end{aligned}$$

where

$$\sum_{k=0}^{\nu-1} |\beta_{\nu,k}| \le \beta.$$

Taking account of the bound on the error of the predicted value, this leads to

$$\|\mathbf{e}_{j+1}\| \le (1 + h_jL\beta(2 + h_jL\beta^*))w_j + h_j(h_jL\beta\,\|\boldsymbol{\tau}^*_j\| + \|\boldsymbol{\tau}_j\|).$$

Once again w_{j+1} is defined by the difference equation

$$w_{j+1} = (1 + h_j\mathscr{L})w_j + h_j\delta_j$$

with obvious definitions of $\mathscr{L}$ and δ_j. Because of the predictor formula, both these quantities are increased from the values of the earlier proof. Notice that the order of

$$\delta_j = h_jL\beta\,\|\boldsymbol{\tau}^*_j\| + \|\boldsymbol{\tau}_j\|$$

is $\min(\nu^* + 1, \nu)$ because of the predictor. The rest of the proof of convergence is the same and it is found that when the starting values are accurate to $\mathbb{O}(H^q)$, convergence is at the rate $\min(q, \nu, \nu^* + 1)$. In practice starting procedures and a predictor are chosen to be sufficiently accurate that the order of convergence is that of the corrector formula, i.e., $\nu^* + 1 \geq \nu$ and $q \geq \nu$. In the case of Adams–Bashforth–Moulton pairs, both $\nu^* = \nu$ and $\nu^* = \nu - 1$ are seen in the codes.

THEOREM 4. *An Adams–Bashforth–Moulton ABν^*–AMν predictor–corrector (PECE) pair is used to approximate the solution of (1). Suppose that the ratio of successive step sizes is bounded above by a constant $\Gamma \geq 1$. Then if the starting values $\mathbf{y}_r$ are accurate of order q, the numerical solution $\{\mathbf{y}_j\}$ is convergent of order $\min(\nu, \nu^* + 1, q)$.*

When the step size is constant, it is easy to accommodate more general pairs that have a similar form. For example, when the step size is constant, we can use the midpoint rule as a predictor for the trapezoidal rule, AM2, and obtain a predictor–corrector pair that converges at the same rate as the AB2–AM2 pair. This might be attractive because the midpoint–trapezoidal pair has a shorter memory than the AB2–AM2 pair. The proof of the following theorem is left as an exercise. Notice that the predictor can be any explicit LMM. The role of the predictor is subordinate to that of the corrector in a pair and the form of the corrector alone is enough to make a simple convergence proof possible. Recall from Chapter 4 that the order of a predictor–corrector (PECE) pair formed from an explicit LMM of order s^* and an implicit LMM of order s is $p = \min(s, s^* + 1)$.

THEOREM 5. *A predictor–corrector (PECE) pair of order p having the form*

$$\mathbf{y}_{j+1}^* = -\sum_{k=1}^{\nu^*} \alpha_k^* \mathbf{y}_{j+1-k} + h \sum_{k=1}^{\nu^*} \beta_k^* \mathbf{F}_{j+1-k}$$

$$\mathbf{y}_{j+1} = \mathbf{y}_{j-m} + h\beta_0 \mathbf{F}(x_{j+1}, \mathbf{y}_{j+1}^*) + h \sum_{k=1}^{\nu} \beta_k \mathbf{F}_{j+1-k}$$

is used to approximate the solution of (1). Here $m \geq 0$ and the integration is done with a constant step size h. If the starting values $\mathbf{y}_r$ are accurate of order q, then the numerical solution $\{\mathbf{y}_j\}$ is convergent of order $\min(p, q)$.

EXERCISE 4. Prove Theorem 5.

§3 Starting Methods with Memory

If the method has a memory, where do the starting values come from? A natural way to construct them, the one seen in the early codes, is to use a one-step method. What is not generally realized is that when done in the usual manner, a little analysis is required to demonstrate that this works. It is true that if we use a one-step method of order q with a step size of $x_r - a$ to compute *independently* $\mathbf{y}_r$, this step size will be no larger than rH and

$$\mathbf{y}_r = \mathbf{s}_r(H) = \mathbf{y}(x_r) + \mathcal{O}((rH)^{q+1}) = \mathbf{y}(x_r) + \mathcal{O}(H^{q+1}).$$

Clearly any method with memory can be started in this way. However, a more natural way to proceed, the one invariably used in practice, is to take $k - 1$ *successive* steps. In this case we must show that $\mathbf{y}_r$ actually approximates $\mathbf{y}(x_r)$ to the desired accuracy because it depends on how well $\mathbf{y}_{r-1}$ approximates $\mathbf{y}(x_{r-1})$, and so forth. We start with the given value $\mathbf{y}(a)$ for $\mathbf{y}_0$, that is, $\mathbf{s}_0(H) = \mathbf{y}(x_0)$. We then use the one-step method of order q with step size h_0 to produce $\mathbf{y}_1$ for which

$$\mathbf{y}(x_1) = \mathbf{y}_1 + \mathcal{O}(h_0^{q+1}) = \mathbf{y}_1 + \mathcal{O}(H^{q+1}),$$

that is, $\mathbf{s}_1(H) = \mathbf{y}_1$. It is at the next step that some analysis is required. We step from x_1 to $x_2 = x_1 + h_1$ to get the result $\mathbf{y}_2$. In this step the formula approximates the local solution $\mathbf{u}(x)$ defined by

$$\mathbf{u}' = \mathbf{F}(x, \mathbf{u}), \qquad \mathbf{u}(x_1) = \mathbf{y}_1,$$

rather than $\mathbf{y}(x)$. A formula of order q yields a result such that

$$\mathbf{u}(x_2) = \mathbf{y}_2 + \mathcal{O}(h_1^{q+1}).$$

Stability of the differential equation implies that

$$\|\mathbf{u}(x_2) - \mathbf{y}(x_2)\| \leq \exp(Lh_1)\|\mathbf{u}(x_1) - \mathbf{y}(x_1)\| = \exp(Lh_1)\|\mathbf{y}_1 - \mathbf{y}(x_1)\|.$$

From this we conclude that the local solution $\mathbf{u}(x)$ and the global solution $\mathbf{y}(x)$ differ by $\mathcal{O}(H^{q+1})$ at x_2 because they differ by $\mathcal{O}(H^{q+1})$ at x_1 and the difference is amplified in a distance of h_1 by no more than $\exp(Lh_1) = 1 + \mathcal{O}(h_1) = 1 + \mathcal{O}(H)$. This result then leads to

$$\mathbf{y}(x_2) = \mathbf{u}(x_2) - (\mathbf{u}(x_2) - \mathbf{y}(x_2)) = (\mathbf{y}_2 + \mathcal{O}(H^{q+1})) + \mathcal{O}(H^{q+1}).$$

Consequently, if we define $\mathbf{s}_2(H) = \mathbf{y}_2$, we have

$$\mathbf{y}(x_2) = \mathbf{s}_2(H) + \mathcal{O}(H^{q+1}),$$

as we need. It is clear that we can repeat this argument for the rest of the fixed, finite number of starting values.

If a method is of order q, the error made in a single step is $\mathcal{O}(H^{q+1})$. It is the effect of taking $\mathcal{O}(H^{-1})$ steps that results in convergence of order q, whence the term. This may be loosely described as saying that for a convergent method, the worst that can happen is that the errors add up. When we consider what happens over a bounded number of steps of length $\mathcal{O}(H)$ as in the start, the final error is still $\mathcal{O}(H^{q+1})$. Because of this we can start a method of order p with a one-step method of order $p - 1$ without reducing the order of convergence. We shall see that if we start with a method of order p or higher, the behavior of the error is essentially the same as if we had started with exact values and if we start with a method of order $p - 1$, the behavior is different, although of the same order.

EXAMPLE 2. The Adams–Bashforth formula of order two requires starting values $\mathbf{y}_0$ and $\mathbf{y}_1$. To get starting values that lead to convergence of order two, we might take $\mathbf{y}_0$ to be the given initial value and then form $\mathbf{y}_1$ with Euler's method of order 1 or with one of the two-stage, second-order Runge–Kutta formulas. The problem $y' = -y$, $y(0) = 1$ was integrated over [0, 1] with a range of constant step sizes and AB2. The quantities

$$E(h) = \max_{0 \le j \le N} |y(x_j) - y_j| \quad \text{and} \quad \text{"ratio"} = E(h)/h^2$$

show the behavior of the error. With the Euler start it was found that

h	10^{-1}	10^{-2}	10^{-3}	10^{-4}
ratio	0.4837	0.4983	0.4998	0.49998

Evidently the ratio is approaching a non-zero limit so that the maximum error is $\mathcal{O}(h^2)$. Because the starting values were included in the computation of the maximum error, we confirm that the first-order Euler's method does yield an accuracy of $\mathcal{O}(h^2)$ in the start. When the start was done with the second-order Runge–Kutta formula based on the trapezoidal rule (improved Euler method), some of the results were

h	10^{-2}	10^{-4}
ratio	0.15327	0.15328

Evidently the maximum error is again $\mathcal{O}(h^2)$, but the behavior of the error is different from that of the integration started with the lower order formula. For this simple example the solution is known so that the integrations with AB2 could be started with the true value $y(x_1)$. When this was done, the ratios turned out to be the same as those computed with the improved Euler start to the number of digits reported here.

EXERCISE 5. You should do some experimentation yourself along the lines of this example. For a simple problem with a known solution, solve the problem for a sequence of constant step sizes and measure the rate of convergence. Suppose that the order of the starting formula is q and the order of the formula used for the integration is p. It has been shown that if $q \geq p - 1$, the rate of convergence is p. It has been claimed that the behavior of the error might be different in the two cases $q = p - 1$ and $q > p - 1$. The convergence proof says that convergence is of order $q + 1$ when $q < p - 1$. This statement does not preclude convergence at a faster rate, but in general the rate *is* $q + 1$, hence an inaccurate starting procedure "spoils" the accuracy of the integration. It is not necessary to use the same one-step formula to get all the starting values. Your experiments should be designed to confirm or disprove some or all of these statements.

EXERCISE 6. Solve the problem $y' = y^2$, $y(0) = 1/2$ on $0 \leq x \leq 1$ for its solution $y(x) = 1/(2 - x)$. Integrate with AB3 and a constant step size. You are to study three procedures for starting. In all cases take $y_0 = y(0) = 1/2$.

(a) Compute y_1, y_2 using AB1 (the forward Euler method) successively. Show that the convergence is $\mathcal{O}(h^2)$.

(b) Compute y_1, y_2 using the improved Euler method successively. Show that the convergence is $\mathcal{O}(h^3)$.

(c) Use exact values $y_1 = y(h)$, $y_2 = y(2h)$. Show that the convergence is $\mathcal{O}(h^3)$ as in case (b), but the error is different.

§4 Convergence with Constant Step Size

In this section we study the convergence of LMMs and predictor–corrector (PECE) pairs when the step size is constant. We have two goals. One is to establish convergence for the backward differentiation formulas. The other is to gain more insight about the effects of memory on the convergence of discrete variable methods. In particular, we would like to understand why methods that look little different from successful methods cannot be used in practice. First we develop some simple necessary conditions for stability. This involves the solution of some constant coefficient difference equations. A study of this matter will pay dividends in the next chapter when we discuss the stability of formulas for "large" step sizes. After writing LMMs and predictor–corrector pairs in a way much like the one-step methods we have studied, we generalize the convergence proof for one-step methods to deal with these formulas. Finally we show how to apply the results so as to model better the way computations are done in practice.

§4.1 Necessary Conditions for Stability of LMMs and Predictor–Corrector Methods

Linear multistep methods, LMMs, have the form

$$\sum_{i=0}^{k} \alpha_i \mathbf{y}_{j+1-i} = h \sum_{i=0}^{k} \beta_i \, \mathbf{F}(x_{j+1-i}, \mathbf{y}_{j+1-i}) \tag{4.1}$$

where $\alpha_0 \neq 0$. Associated with the method is its (first) characteristic polynomial,

$$\rho(\theta) = \sum_{i=0}^{k} \alpha_i \, \theta^{k-i}.$$

We assume that the formula (4.1) is normalized so that $\rho'(1) = 1$. It was seen in Chapter 4 that as they were derived there, the Adams and the backward differentiation formulas are normalized in this way. It was also seen that if the formula is of order $p \geq 1$, then $\rho(1) = 0$. We suppose that there is a collection of starting procedures

$$\mathbf{y}_r = \mathbf{s}_r(h), \qquad 0 \leq r < k \tag{4.2}$$

that supply the initial values. There is no need to be specific about the procedures themselves, but, of course, they must actually approximate the solution.

The idea of stability is that small changes of the problem lead to small changes in the solution. Consider $\{\mathbf{z}_j\}$ defined by

$$\mathbf{z}_r = \mathbf{s}_r(h) + \boldsymbol{\Delta}_r \qquad 0 \leq r < k,$$

$$\sum_{i=0}^{k} \alpha_i \, \mathbf{z}_{j+1-i} = h \sum_{i=0}^{k} \beta_i \, \mathbf{F}(x_{j+1-i}, \mathbf{z}_{j+1-i}) + h \, \boldsymbol{\delta}_j \quad \text{for} \quad j \geq k-1$$

for given perturbations $\{\boldsymbol{\Delta}_j\}$ to the starting values (4.2) and given perturbations $\{\boldsymbol{\delta}_j\}$ to the formula (4.1). If there is a constant S such that for all sufficiently small step sizes

$$\max_j \|\mathbf{z}_j - \mathbf{y}_j\| \leq S \max\left(\max_{0 \leq r < k} \|\boldsymbol{\Delta}_r\|, \max_{k-1 \leq J} \|\boldsymbol{\delta}_J\| \right),$$

the method is said to be *zero-stable* or *D-stable* (after Germund Dahlquist) or simply *stable*. (This generalizes the definition of §1 to account for more starting values.)

The situation with predictor–corrector pairs implemented in PECE form is not much different. The predictor is an explicit LMM with coefficients that we distinguish with asterisks and we also distinguish the tentative value at x_{j+1} with an asterisk:

$$\mathbf{y}_{j+1}^{*} = -\sum_{i=1}^{k^*}\left(\frac{\alpha_i^*}{\alpha_0^*}\,\mathbf{y}_{j+1-i} - h\,\frac{\beta_i^*}{\alpha_0^*}\,\mathbf{F}_{j+1-i}\right).$$

An implicit LMM is made explicit on replacing one appearance of $\mathbf{y}_{j+1}$ in the recipe by the predicted value $\mathbf{y}_{j+1}^{*}$:

$$\alpha_0\,\mathbf{y}_{j+1} + \sum_{i=1}^{k}\alpha_i\,\mathbf{y}_{j+1-i} = h\,\beta_0\,\mathbf{F}(x_{j+1}, \mathbf{y}_{j+1}^{*}) + h\sum_{i=1}^{k}\beta_i\,\mathbf{F}_{j+1-i}.$$

It is convenient here to introduce zero coefficients as necessary so that $k = k^*$. After doing this the characteristic polynomial is again defined as

$$\rho(\theta) = \sum_{i=0}^{k}\alpha_i\theta^{k-i}.$$

When k is increased because of the use of a predictor, this is no longer the characteristic polynomial of the implicit LMM. An example will make the effect clear. The explicit formula

$$\mathbf{y}_{j+1} + 4\mathbf{y}_j - 5\mathbf{y}_{j-1} = h[4\mathbf{F}(x_j, \mathbf{y}_j) + 2\mathbf{F}(x_{j-1}, \mathbf{y}_{j-1})] \qquad (4.3)$$

was seen in Exercise 26 of Chapter 4 to be of order 3. When used as a predictor for AM2, the pair is

$$\mathbf{y}_{j+1}^{*} = -4\mathbf{y}_j + 5\mathbf{y}_{j-1} + h[4\mathbf{F}(x_j, \mathbf{y}_j) + 2\mathbf{F}(x_{j-1}, \mathbf{y}_{j-1})], \qquad (4.4a)$$

$$\mathbf{y}_{j+1} = \mathbf{y}_j + h\left[\frac{1}{2}\,\mathbf{F}(x_j, \mathbf{y}_j) + \frac{1}{2}\,\mathbf{F}(x_{j+1}, \mathbf{y}_{j+1}^{*})\right]. \qquad (4.4b)$$

Because of the prediction, $\mathbf{y}_{j+1}$ depends on previously computed values as far back as $\mathbf{y}_{j-1}$, so we write this last equation as

$$\mathbf{y}_{j+1} - \mathbf{y}_j + 0\mathbf{y}_{j-1} = h\left[\frac{1}{2}\,\mathbf{F}(x_j, \mathbf{y}_j) + \frac{1}{2}\,\mathbf{F}(x_{j+1}, \mathbf{y}_{j+1}^{*})\right],$$

with characteristic polynomial $\rho(\theta) = \theta^2 - \theta = \theta(\theta - 1)$. We see that the characteristic polynomial of AM2 is multiplied by θ to obtain the characteristic polynomial of this predictor–corrector pair.

EXAMPLE 3. An example will show that there is reason to be concerned about stability when the method has a memory. We applied the explicit LMM (4.3) to the solution of $y' = -y$, $y(0) = 1$ with constant step size $h = 0.1$ and two sets of starting values. One set consisted of the true solution values at x_0 and x_1: $y_0 = 1$, $y_1 = \exp(-h)$. The other differed in that one value was perturbed by a small amount: $z_0 = 1$, $z_1 = y_1 + 0.001$.

In both computations, the LMM was evaluated exactly at all steps. The results were:

x_j	y_j	z_j
0.0	1.00000	1.00000
0.1	0.904837	0.905837
⋮	⋮	⋮
0.8	0.197151	−100.797
0.9	1.74427	537.503
1.0	−6.72848	−2848.84

The approximations y_j were started with exact values of the solution $\exp(-x)$ and the formula is accurate, yet the results after only a few steps are terrible approximations to the solution—after all, the solution $\exp(-x)$ is positive and monotonely decreasing! Clearly, imitating well the behavior of the equation over one step, a high order of accuracy, is not enough to guarantee that a numerical method is accurate after even a few steps. Comparison of the y_j to the z_j shows that a small change in y_1 is amplified enormously after just a few steps, suggesting that the method is not zero-stable.

EXERCISE 7. Repeat this numerical experiment with several formulas:

(a) Use the unstable formula (4.3) by itself to reproduce the computations reported and see instability for yourself.
(b) Use the stable AM3 by itself. For this particular problem it is easy to evaluate the implicit formula exactly—do so.
(c) Use the unstable formula (4.3) as a predictor and the stable AM3 as a corrector in a PECE implementation. You will observe that the computation is stable.

It is easy to establish some necessary conditions for stability. We certainly want a numerical method to be stable for the solution of the trivial problem $y' = 0$, $y(0) = 0$. Linear multistep methods and predictor–corrector methods have a very special form for such problems. If we start with exact values, $s_r(h) = y(rh) \doteq 0$ for $0 \leq r < k$, the numerical solution (4.2) is also exact, $y_j = 0$ for all j. Suppose the starting values are perturbed to Δ_r for $0 \leq r < k$, but the formula is evaluated exactly so that $\delta_j = 0$ for $j \geq k - 1$. This results in a numerical solution z_j such that

$$z_r = \Delta_r \qquad 0 \leq r < k, \tag{4.5a}$$

$$\sum_{i=0}^{k} \alpha_i \, z_{j+1-i} = 0 \qquad j \geq k - 1. \tag{4.5b}$$

This is a constant coefficient difference equation for $\{z_j\}$. The solution of such equations is closely analogous to the better known solution of constant coefficient differential equations. The idea is to look for a solution in the form $z_j = \gamma\zeta^j$ for suitable non-zero constants γ, ζ. If there is to be a solution of this form, we must have

$$\sum_{i=0}^{k} \alpha_i \gamma\zeta^{j+1-i} = \gamma \zeta^{j+1-k} \sum_{i=0}^{k} \alpha_i \zeta^{k-i} = \gamma \zeta^{j+1-k}\rho(\zeta) = 0.$$

The key to solving this equation is the characteristic polynomial because it is obvious that any nonzero root ζ of $\rho(\theta)$ provides a solution of the difference equation for any constant $\gamma \neq 0$.

We are investigating necessary conditions for stability, so we are free to specify any (small) perturbations we like. Let us take the initial perturbations to be

$$\Delta_r = \gamma\zeta^r \qquad \text{for } 0 \leq r < k,$$

so that $z_j = \gamma\zeta^j$ satisfies the initial conditions and the difference equation. In the recipe for $\{z_j\}$ all the perturbations δ_J are zero, so

$$\max\left(\max_{0\leq r<k} |\Delta_r|, \max_{k-1\leq J} |\delta_J|\right) = \max_{0\leq r<k} |\gamma\zeta^r|,$$

By choosing γ properly, the perturbations can be made arbitrarily small. For the formula to be stable, there would have to be a constant S such that

$$\max_{0\leq j\leq N} |z_j - y_j| \leq S \max_{0\leq r<k} |\gamma\zeta^r|.$$

The right hand side of this inequality is constant, so if the formula is to be stable, the difference between z_j and y_j would have to be uniformly bounded. For the perturbations considered, we see that if $\rho(\zeta)$ has a root ζ with $|\zeta| > 1$, then

$$\max_{0\leq j\leq N} |y_j - z_j| = |\gamma|\ |\zeta|^N,$$

which is not bounded as $N \to \infty$. This says that such a formula cannot be stable.

This argument provides a necessary condition for zero-stability. It shows by example that if the characteristic equation has a root of magnitude greater than 1, there are small perturbations that are greatly amplified, so much so that they are not bounded as the step size tends to zero. The formula of Example 3 is $\mathbf{y}_{j+1} + 4\mathbf{y}_j - 5\mathbf{y}_{j-1} = \cdots$. For this method $\rho(\theta) = \theta^2 + 4\theta - 5 = (\theta + 5)(\theta - 1)$ with roots $\zeta = -5, 1$. The root $\zeta = -5$

shows that the formula is not zero-stable. Indeed, the root is so large in magnitude that perturbations are amplified very quickly, just as we saw in the example.

EXERCISE 8. Find the most accurate LMM possible of the form

$$\alpha_0\, \mathbf{y}_{j+1} + \alpha_1\, \mathbf{y}_j + \alpha_2\, \mathbf{y}_{j-1} + \alpha_3\, \mathbf{y}_{j-2} = h\, \mathbf{F}(x_j, \mathbf{y}_j).$$

What is its order? What is its error constant? (Remember to normalize the formula.) Prove that this formula is not zero-stable. Hint: Use the fact that 1 is a root of $\rho(\theta)$.

The argument also serves to demonstrate an important fact about convergence—the order of convergence is generally not faster than the order of accuracy of the starting values. When solving $y' = 0$, $y(0) = 0$, if we take as starting values $z_r = s_r(h) = \gamma = h^q$ for $0 \leq r < k$, then the numerical solution is $z_j = \gamma = h^q$ for all j because 1 is a root of $\rho(\theta) = 0$. Then

$$\max_j |y(x_j) - z_j| = \max_j |z_j| = h^q,$$

and the order of convergence is precisely the order of accuracy of the starting values.

There is a technical matter that was glossed over. The coefficients of the characteristic polynomial are real numbers, but the roots need not be real. The analysis just presented would for a complex root ζ involve starting perturbations and a solution that are complex numbers. This situation will recur in our study of the numerical solution of the initial value problem. It seems odd to introduce complex numbers when we are concerned only with real problems, but doing so simplifies the analysis. If we wanted to, we could work entirely with real numbers. To illustrate this, suppose that the root $\zeta = r \exp(i\phi)$ of the characteristic equation is a non-zero complex number of magnitude r and phase ϕ. We have seen that

$$z_j = \zeta^j = r^j(\sin(j\phi) + i \cos(j\phi))$$

is a solution of the difference equation (4.5). Because the coefficients α_i of $\rho(\theta)$ are real, the conjugate of ζ is also a root of $\rho(\theta)$, hence

$$z_j = \bar{\zeta}^j = r^j(\sin(j\phi) - i \cos(j\phi))$$

is a solution of (4.5), too. Because the equation is linear and homogeneous, the real sequence

$$z_j = \zeta^j + \bar{\zeta}^j = 2r^j \sin(j\phi)$$

is also a solution of (4.5). The analysis is now just as before, but it is applied to this real sequence. Clearly $r = |\zeta| > 1$ is incompatible with stability.

A root that has magnitude 1 is a marginal case. We might suspect that sometimes such methods are stable, and sometimes they are not. This turns out to be so. Unfortunately, we cannot just ignore this marginal case because any LMM of order $p \geq 1$ has $\zeta = 1$ as a root of its characteristic polynomial. We shall now modify the counterexample to stability for roots ζ of magnitude greater than 1 to show that if $|\zeta| = 1$ and ζ is a multiple root, the method cannot be stable. First we note that if ζ is a multiple root, then in addition to $\rho(\zeta) = 0$, it must be the case that $\rho'(\zeta) = 0$. Supposing that ζ is a multiple root, let us now look for a solution of the difference equation (4.5) of the form

$$z_j = \gamma(j + 1)\zeta^j.$$

(This is mysterious when stated baldly, but it can be derived in a natural way as a limit of a pair of solution sequences corresponding to a pair of simple roots as the roots coalesce to form a double root. It is analogous to the way a constant coefficient differential equation is solved when its characteristic equation has a multiple root.) For this sequence to be a solution of the difference equation, it must be the case that for $j \geq k - 1$,

$$\sum_{i=0}^{k} \alpha_i \, \gamma(j + 2 - i)\zeta^{j+1-i} = 0.$$

That this is true follows from the identity

$$\begin{aligned} \frac{d}{d\theta}\left(\theta^{j+2-k}\,\rho(\theta)\right) &= \sum_{i=0}^{k} \alpha_i(j + 2 - i)\theta^{j+1-i} \\ &= (j + 2 - k)\theta^{j+1-k}\,\rho(\theta) + \theta^{j+2-k}\,\rho'(\theta) \end{aligned}$$

after evaluation at $\theta = \zeta$. Proceeding as before, we take the initial values to be

$$z_r = \Delta_r = \gamma(r + 1)\zeta^r \qquad 0 \leq r < k,$$

and take all the perturbations δ_j to be zero. Now

$$\max_{0 \leq r < k} |\Delta_r| = \max_{0 \leq r < k} |\gamma(r + 1)\zeta^r| = |\gamma| k$$

because ζ has magnitude 1. Because

$$\max_{0 \leq j \leq N} |z_j - y_j| = \max_{0 \leq j \leq N} |\gamma(j + 1)\zeta^j| = |\gamma|(N + 1)$$

is not bounded as $N \to \infty$, the method is not stable. Notice that the instability is not nearly so strong as for a root of magnitude greater than 1, reflecting the fact that this is a marginal case.

These simple computations provide important *necessary conditions for stability*. They are usually stated in terms of ζ being outside the unit circle in the complex plane rather than $|\zeta| > 1$ and the like. In summary:

THEOREM 6. *If the (first) characteristic polynomial of a LMM or predictor–corrector (PECE) pair has a root outside the unit circle in the complex plane, or a multiple root on the unit circle, the method is not zero-stable.*

Because of the importance of these results there is some associated terminology:

Definitions A method is said to satisfy the *root condition* if the roots of the characteristic polynomial $\rho(\theta)$ all lie within or on the unit circle, those on the unit circle being simple. The *strong root condition* requires in addition that the only root on the unit circle be $+1$. A method with even minimal accuracy, a *consistent* method, always has $\zeta = 1$ as a root. It is called the *principal root*. Other roots ζ_m with $|\zeta_m| = 1$ are called *essential roots*. Roots with $|\zeta_m| < 1$ are *non-essential roots* that are also called *spurious*, *extraneous*, or *parasitic roots*.

The Adams methods have the form $\mathbf{y}_{j+1} - \mathbf{y}_j = \ldots$, this being so whether we consider the explicit Adams–Bashforth formulas, the implicit Adams–Moulton formulas, or predictor–corrector pairs with an Adams–Moulton formula as corrector. At first glance one might think that the characteristic polynomial is always $\theta - 1$. However, because values as far back as x_{j+1-k} appear in the recipe for $\mathbf{y}_{j+1}$, the characteristic polynomial is $\theta^k - \theta^{k-1} = (\theta - 1)\theta^{k-1}$. Obviously these methods all satisfy the strong root condition. A method with memory must have spurious roots, but in a sense, these methods are as close as we can get to a method without memory because all the spurious roots are 0—suggesting that they will have a minimal effect on amplification of perturbations.

It is not obvious that the BDFs are zero-stable and the fact is that some are and some are not. For any specific BDF, we need only compute numerically the roots of the characteristic polynomial to decide. This suffices in practice because only a relatively small range of orders is interesting. However, it is possible to analyze the stability of the family theoretically. One way or the other, it is found that the BDFs of orders 6 and lower satisfy the strong root condition and the higher order BDFs are not even zero-stable because they do not satisfy the root condition. A fact we use later is that for the stable BDFs, all the roots are distinct. The BDF1 (alias AM1 and backward Euler method) is

$$\mathbf{y}_{j+1} - \mathbf{y}_j = h\,\mathbf{F}(x_{j+1}, \mathbf{y}_{j+1}).$$

The characteristic polynomial is $\theta - 1$ and the roots obviously have the properties claimed. The normalized BDF2 is

$$\frac{3}{2}\,\mathbf{y}_{j+1} - 2\mathbf{y}_j + \frac{1}{2}\,\mathbf{y}_{j-1} = h\,\mathbf{F}(x_{j+1}, \mathbf{y}_{j+1})$$

and its characteristic polynomial is

$$\rho(\theta) = \frac{3}{2}\,\theta^2 - 2\theta + \frac{1}{2} = \frac{3}{2}\,(\theta - 1)\left(\theta - \frac{1}{3}\right).$$

With the two roots of 1 and 1/3, it is seen that the formula satisfies the strong root condition and the roots are distinct.

EXERCISE 9. The normalized BDF3 is

$$\frac{11}{6}\,\mathbf{y}_{j+1} - 3\mathbf{y}_j + \frac{3}{2}\,\mathbf{y}_{j-1} - \frac{1}{3}\,\mathbf{y}_{j-2} = h\,\mathbf{F}(x_{j+1}, \mathbf{y}_{j+1}).$$

Verify analytically that the characteristic polynomial of this formula satisfies the strong root condition and that its roots are distinct. Hint: Take into account the principal root.

EXERCISE 10. Compute the roots of the characteristic polynomial for each of the BDFs of orders 1–6 and verify that the formula satisfies the strong root condition and that the roots are all distinct. (In addition to calculating the roots, you will need to calculate their magnitude.) With a tool like MATLAB this can be done in a few minutes, making the point that numerical verification of the stability of a given LMM is an easy matter. The roots already given for the lowest order formulas provide checks, but as an additional check, you should find in the case of BDF6 that there are two real roots and two pairs of complex roots and further, of the non-essential roots, the largest have a magnitude of about 0.8634.

Because we make use of these properties of the BDFs later, let us state them clearly:

Properties of the BDFs When the step size is constant, the backward differentiation formulas of orders 6 and lower satisfy the strong root condition and all roots of the characteristic equation are distinct. The backward differentiation formulas of orders 7 and higher do not satisfy the root condition.

The midpoint rule

$$\mathbf{y}_{j+1} - \mathbf{y}_{j-1} = 2h\,\mathbf{F}(x_j, \mathbf{y}_j)$$

has $\rho(\theta) = \theta^2 - 1 = (\theta - 1)(\theta + 1)$, hence has an essential root -1 in addition to the principal root $+1$. This is the usual form of the formula, but notice that $\rho'(1) = 2$, so the formula has not been normalized like the others we have defined. Normalized in the way we are supposing, it is

$$\frac{1}{2}\,\mathbf{y}_{j+1} - \frac{1}{2}\,\mathbf{y}_{j-1} = h\,\mathbf{F}(x_j, \mathbf{y}_j).$$

We cannot use a formula by itself that is not zero-stable, but that does not mean that such a formula cannot be used as a predictor in a predictor–corrector pair. This is clear from the convergence result of Theorem 5. Zero-stability constrains the accuracy that can be achieved with a given number of values from the past, so we might want to use an unstable predictor because it is especially accurate. Whether a predictor–corrector combination is zero-stable depends only on the corrector. In particular, the formula (4.3) has been shown to be unstable when used by itself. However, when it is used as a predictor for AM2 as in (4.4), the pair is zero-stable. This concept of stability is concerned with the limit $h \to 0$. When we take up stability for "large" step sizes in the next chapter, we shall find that the behavior then depends on both members of the pair. The situation is entirely different when an implicit LMM is considered. The only role the predictor plays for an implicit formula is to provide a good start to an iterative scheme for evaluating the formula.

§4.2 *Formulation as One-Step Methods*

When the step size is a constant h, LMMs and predictor–corrector methods can be written as one-step methods of the form

$$\mathbf{y}_{j+1} = Q\,\mathbf{v}_j + h\,\Phi(x_j, \mathbf{v}_{j+1}, \mathbf{v}_j, \mathbf{F}, h). \tag{4.6}$$

This form differs from that considered in §1 by the presence of the matrix Q. For simplicity only the scalar case $y' = F(x, y)$ will be treated in detail. Some examples will show how it all goes. The explicit midpoint rule is

$$y_{j+1} = y_{j-1} + 2h\,F(x_j, y_j).$$

Introducing vectors, this is obviously equivalent to

$$\begin{pmatrix} y_{j+1} \\ y_j \end{pmatrix} = \begin{pmatrix} 0 & 1 \\ 1 & 0 \end{pmatrix}\begin{pmatrix} y_j \\ y_{j-1} \end{pmatrix} + h\begin{pmatrix} 2F(x_j, y_j) \\ 0 \end{pmatrix}.$$

Similarly, the implicit AM2,

$$y_{j+1} = y_j + h\left[\frac{1}{2}\,F(x_j, y_j) + \frac{1}{2}\,F(x_{j+1}, y_{j+1})\right],$$

is equivalent to

$$\begin{pmatrix} y_{j+1} \\ y_j \end{pmatrix} = \begin{pmatrix} 1 & 0 \\ 1 & 0 \end{pmatrix} \begin{pmatrix} y_j \\ y_{j-1} \end{pmatrix} + h \begin{pmatrix} \frac{1}{2} F(x_j, y_j) + \frac{1}{2} F(x_{j+1}, y_{j+1}) \\ 0 \end{pmatrix}.$$

The predictor–corrector (PECE) pair of the midpoint rule and AM2 is equivalent to

$$\begin{pmatrix} y_{j+1} \\ y_j \end{pmatrix} = \begin{pmatrix} 1 & 0 \\ 1 & 0 \end{pmatrix} \begin{pmatrix} y_j \\ y_{j-1} \end{pmatrix} + h \begin{pmatrix} \frac{1}{2} F(x_j, y_j) + \frac{1}{2} F(x_{j+1}, y^*_{j+1}) \\ 0 \end{pmatrix}$$

where

$$y^*_{j+1} = y_{j-1} + 2h\, F(x_j, y_j)$$

is a function of the vector $(y_j, y_{j-1})^T$.

With these examples to guide us it is now easy to see how to write LMMs and predictor–corrector (PECE) methods in the form (4.6). For each j the vector $\mathbf{v}_j$ is taken to be $(y_j, y_{j-1}, \ldots, y_{j+1-k})^T$. The matrix Q is

$$Q = \begin{bmatrix} -\frac{\alpha_1}{\alpha_0} & -\frac{\alpha_2}{\alpha_0} & -\frac{\alpha_3}{\alpha_0} & & -\frac{\alpha_k}{\alpha_0} \\ 1 & 0 & 0 & \cdots & 0 \\ 0 & 1 & 0 & \cdots & 0 \\ \vdots & & \ddots & \ddots & \vdots \\ \vdots & & & \ddots \quad \ddots & \vdots \\ 0 & & & 0 \quad 1 & 0 \end{bmatrix}. \tag{4.7}$$

In the case of a LMM the increment function $\mathbf{\Phi}(x_j, \mathbf{v}_{j+1}, \mathbf{v}_j, F, h)$ is

$$\left(\sum_{i=0}^{k} \frac{\beta_i}{\alpha_0} F(x_{j+1-i}, y_{j+1-i}), 0, \ldots, 0\right)^T.$$

The only component of $\mathbf{v}_{j+1}$ that does not appear in $\mathbf{v}_j$ is y_{j+1}. Accordingly the increment function depends on $\mathbf{v}_{j+1}$ if, and only if, the LMM is implicit. In the case of a predictor–corrector pair, the increment function is

$$\left(\frac{\beta_0}{\alpha_0} F(x_{j+1}, y^*_{j+1}) + \sum_{i=1}^{k} \frac{\beta_i}{\alpha_0} F(x_{j+1-i}, y_{j+1-i}), 0, \ldots, 0\right)^T,$$

where

$$y^*_{j+1} = -\sum_{i=1}^{k} \left(\frac{\alpha^*_i}{\alpha^*_0} y_{j+1-i} - h \frac{\beta^*_i}{\alpha^*_0} F(x_{j+1-i}, y_{j+1-i})\right).$$

Predictor–corrector pairs are explicit and their increment function does not depend on $\mathbf{v}_{j+1}$.

EXERCISE 11. Write the predictor–corrector pair (4.4) in the form (4.6).

For both LMMs and predictor–corrector pairs it is obvious that the increment function is identically zero when $F(x, y)$ is. It is also easy to show that the increment function satisfies a Lipschitz condition because F does. As an example of this, we consider the case of an explicit LMM. Because only one component of the increment function is non-zero,

$$\|\mathbf{\Phi}(x_j, \mathbf{z}, F, h) - \mathbf{\Phi}(x_j, \boldsymbol{\zeta}, F, h)\|_\infty =$$

$$= \left| \sum_{i=1}^{k} \frac{\beta_i}{\alpha_0} [F(x_{j+1-i}, z_{j+1-i}) - F(x_{j+1-i}, \zeta_{j+1-i})] \right|$$

$$\leq \sum_{i=1}^{k} \left| \frac{\beta_i}{\alpha_0} \right| L |z_{j+1-i} - \zeta_{j+1-i}| \leq \mathcal{L} \|\mathbf{z} - \boldsymbol{\zeta}\|_\infty$$

where

$$\mathcal{L} = L \sum_{i=1}^{k} \left| \frac{\beta_i}{\alpha_0} \right|.$$

The way that the indices have been defined, the starting values are $\mathbf{v}_{k-1} = (y_{k-1}, y_{k-2}, \ldots, y_1, y_0)^T$. It is an annoyance in the analysis to have the integration start with index $k - 1$, but it is convenient to have y_j associated with $\mathbf{v}_j$ and the lemma that played so fundamental a role in the analysis of one-step methods can be applied to a sequence starting at index $k - 1$ just like a sequence starting at index 0.

In the case of a LMM, the solution $y(x)$ of the differential equation satisfies the difference equation with a discrepancy $h\,\tau_j$ (the local truncation error):

$$\sum_{i=0}^{k} \alpha_i\, y(x_{j+1-i}) = h \sum_{i=0}^{k} \beta_i F(x_{j+1-i}, y(x_{j+1-i})) + h\tau_j.$$

For each j let $\mathbf{u}_j = (y(x_j), y(x_{j-1}), \ldots, y(x_{j+1-k}))^T$ and $\boldsymbol{\mu}_j = (\tau_j, 0, \ldots, 0)^T$. The vector of values of the true solution then satisfies (4.6) with a discrepancy $h\boldsymbol{\mu}_j$:

$$\mathbf{u}_{j+1} = Q\mathbf{u}_j + h\,\mathbf{\Phi}(x_j, \mathbf{u}_{j+1}, \mathbf{u}_j, F, h) + h\boldsymbol{\mu}_j. \tag{4.8}$$

The same is true of predictor–corrector pairs with an obvious change in the definition of $\boldsymbol{\mu}_j$. Notice that the order of the norm of the vector $\boldsymbol{\mu}_j$ is the same as the order of the scalar τ_j, namely the order of the method.

§4.3 Convergence

As usual, we look first at the convergence of explicit methods. Subtracting the recipe (4.6) for the numerical solution from the perturbed version (4.8) satisfied by the true solution, taking norms, and using the Lipschitz condition satisfied by the increment function leads in (by now) familiar fashion to

$$\|\mathbf{e}_{j+1}\| \le \|Q\| \; \|\mathbf{e}_j\| + h\mathcal{L}\|\mathbf{e}_j\| + h\|\boldsymbol{\mu}_j\|,$$

where for each j the error $\mathbf{e}_j = \mathbf{u}_j - \mathbf{v}_j$. If we could bound $\|Q\|$ by 1, this would be the same inequality studied in §1 and convergence would follow in the same way. With some background in the theory of matrices and norms, it is not hard to show that if no eigenvalue of the matrix Q has magnitude greater than 1 and if all eigenvalues of magnitude 1 are simple, i.e., the eigenvalues of Q satisfy the root condition, then there is a norm for which $\|Q\| \le 1$. Convergence in this particular norm is established in the way outlined and then use is made of the fact that in a finite dimensional vector space, convergence in one norm implies covergence in all norms. We proceed in an equivalent way that requires less background in matrix theory.

The matrices Q that concern us have the form (4.7). Such a matrix is a *companion matrix* so that its characteristic equation is the polynomial $\rho(\theta)$ and the eigenvalues of Q are the roots of $\rho(\theta) = 0$. This is a standard result in matrix theory, but it is easy to see how it goes. Suppose that $\mathbf{w} = (w_{k-1}, w_{k-2}, \ldots, w_0)^T$ is an eigenvector of Q associated with an eigenvalue λ. In components, the equation $\lambda\mathbf{w} = Q\mathbf{w}$ is $\lambda w_0 = w_1$, $\lambda w_1 = w_2$, $\ldots$, $\lambda w_{k-2} = w_{k-1}$, and

$$\lambda w_{k-1} = -\frac{\alpha_1}{\alpha_0} w_{k-1} - \frac{\alpha_2}{\alpha_0} w_{k-2} - \cdots - \frac{\alpha_k}{\alpha_0} w_0.$$

We have here $w_1 = \lambda w_0$, $w_2 = \lambda w_1 = \lambda^2 w_0$, etc., which on substitution in the last equation results in

$$\lambda^k w_0 = -\frac{\alpha_1}{\alpha_0} \lambda^{k-1} w_0 - \frac{\alpha_2}{\alpha_0} \lambda^{k-2} w_0 - \cdots - \frac{\alpha_k}{\alpha_0} w_0.$$

This is $\rho(\lambda) w_0 = 0$, which states that the eigenvalue λ is a root of $\rho(\theta)$.

Returning to the proof of convergence, let us again subtract (4.6) from (4.8) to get

$$\mathbf{u}_{j+1} - \mathbf{v}_{j+1} = Q(\mathbf{u}_j - \mathbf{v}_j) + h[\boldsymbol{\Phi}(x_j, \mathbf{u}_j) - \boldsymbol{\Phi}(x_j, \mathbf{v}_j)] + h\boldsymbol{\mu}_j.$$

We now introduce a new variable $\boldsymbol{\zeta}_j$ and bound it in the max norm. For a non-singular matrix T, let $\boldsymbol{\zeta}_j = T^{-1}\mathbf{e}_j$. Multiplying the equation by T^{-1} and restating it in terms of ζ_j leads to

$$\zeta_{j+1} = T^{-1}QT\zeta_j + hT^{-1}[\Phi(x_j, \mathbf{u}_j) - \Phi(x_j, \mathbf{v}_j)] + hT^{-1}\boldsymbol{\mu}_j.$$

Taking norms and using the Lipschitz condition on the increment function leads to

$$||\zeta_{j+1}|| \leq ||T^{-1}QT||\ ||\zeta_j|| + h\mathcal{L}||T^{-1}||\ ||\mathbf{u}_j - \mathbf{v}_j|| + h||T^{-1}||\ ||\boldsymbol{\mu}_j||.$$

Proceeding in this way, we need to demonstrate that if ρ satisfies the root condition, then we can find a T such that $||T^{-1}QT||_\infty \leq 1$. This technical issue will be deferred for the moment. With a bound $||T^{-1}QT|| \leq 1$, we have

$$||\zeta_{j-1}|| \leq ||\zeta_j|| + h\mathcal{L}||T^{-1}||\ ||\mathbf{u}_j - \mathbf{v}_j|| + h||T^{-1}||\ ||\mathbf{u}_j||.$$

The factor $||\mathbf{u}_j - \mathbf{v}_j||$ arising from the Lipschitz condition can be bounded in terms of ζ_j.

$$||\mathbf{u}_j - \mathbf{v}_j|| = ||\mathbf{e}_j|| = ||T\zeta_j|| \leq ||T||\ ||\zeta_j||,$$

and then the fundamental inequality becomes

$$||\zeta_{j+1}|| \leq (1 + h\mathcal{L}\kappa)\ ||\zeta_j|| + h||T^{-1}||\ ||\boldsymbol{\mu}_j||$$

where $\kappa = ||T^{-1}||\ ||T||$. This is exactly the kind of inequality that we studied in §1 and the lemma developed there leads to a bound for $j \geq k - 1$,

$$||\zeta_j|| \leq e^{\mathcal{L}\kappa(b-a)}2\max(b - a, 1)\max(||\zeta_{k-1}||, ||T^{-1}|| \max_{k-1\leq J} ||\boldsymbol{\mu}_J||).$$

Now we return to the original variables. The starting error

$$\begin{aligned} ||\zeta_{k-1}||_\infty = ||T^{-1}\mathbf{e}_{k-1}||_\infty &\leq ||T^{-1}||_\infty\ ||\mathbf{e}_{k-1}||_\infty \\ &= ||T^{-1}||_\infty \max_{0\leq r<k} |y(x_r) - y_r|. \end{aligned}$$

As noted earlier, the norm of the truncation error term is just $||\boldsymbol{\mu}_J||_\infty = |\tau_j|$. Furthermore,

$$||\mathbf{e}_j|| = ||T\zeta_j|| \leq ||T||\ ||\zeta_j||,$$

and the norm of $\mathbf{e}_j$ is related to the quantity of interest by

$$||\mathbf{e}_j||_\infty = \max_{1\leq i<k} |y(x_{j+1-i}) - y_{j+1-i}| \geq |y(x_j) - y_j|.$$

Putting all these observations together, we finally obtain

$$|y(x_j) - y_j| \le S \max\left(\max_{0 \le r < k} |y(x_r) - y_r|, \max_{k-1 \le J} |\tau_J| \right)$$

where

$$S = \kappa\, e^{\mathcal{L}\kappa(b-a)} 2\max(b - a, 1).$$

Let us return now to the technical matter of establishing the existence of a non-singular matrix T such that $||T^{-1}QT||_\infty \le 1$. The matter is especially simple when the $k \times k$ matrix Q has k eigenvectors that are linearly independent because if we take the columns of T to be the eigenvectors of Q, the assumption about linear independence implies that the inverse of T exists and that $T^{-1}QT$ is a diagonal matrix D with the eigenvalues of Q on the diagonal. In this case

$$||T^{-1}QT||_\infty = ||D||_\infty = \max_i \sum_j |D_{i,j}| = \max_i |\lambda_i| \le 1$$

because of the root condition. If the eigenvalues of Q are distinct, that is, if the roots of $\rho(\theta)$ are distinct, then there *are* k linearly independent eigenvectors. Earlier we stated that when the step size is constant, the BDFs of orders 6 and lower have the properties we require now. Namely, they satisfy the strong root condition and have distinct roots. Also, the BDFs of orders 7 and higher do not satisfy the root condition, which according to Theorem 6 implies that they are not stable. In summary,

THEOREM 7. *A BDF of order ν is used to approximate the solution of (1). Suppose that the step size is constant in each integration and that the starting values $\mathbf{y}_r$ are accurate of order q. The method is well defined for all sufficiently small step sizes. For orders $\nu \le 6$, the method is convergent of order min(ν, q). For orders $\nu \ge 7$, the method is not stable.*

The reader who does not wish to go into the details that remain for formulas with multiple roots of the characteristic equation can skip the next three paragraphs.

It is not always possible to find a matrix T for which $T^{-1}QT$ is a diagonal matrix D. In general the best that can be done is to put Q into the Jordan canonical form. This result says that there is a matrix M for which $M^{-1}QM$ is a block diagonal matrix D with the eigenvalues of Q on the diagonal. The blocks have two forms. All simple eigenvalues appear in 1×1 blocks. A multiple eigenvalue λ can appear in a 1×1 block, but it can also appear in an $r \times r$ block of the form

$$\begin{pmatrix} \lambda & 1 & & 0 \\ 0 & \ddots & \ddots & \\ \vdots & \ddots & \lambda & 1 \\ 0 & \cdots & 0 & \lambda \end{pmatrix}.$$

(A multiple eigenvalue λ can appear in more than one block of this kind and the blocks can be of different size.) The Jordan canonical form is often cited, but not so often proven. For a discussion and a nice proof, one might turn to Strang [1988].

The maximum norm of a matrix is the largest of the sums of the magnitudes of the elements in a row. For a Jordan block of the kind displayed, one such sum is $|\lambda| + 1$. If we are to get a maximum norm that is less than 1, we must somehow reduce this sum. For a number $e > 0$, let the matrix $E = \text{diag}\{e, e^2, \ldots, e^k\}$. The inverse of E is $E^{-1} = \text{diag}\{e^{-1}, e^{-2}, \ldots, e^{-k}\}$. The effect of multiplying a matrix on the left with a diagonal matrix is extremely simple, all elements in a row being multiplied by the element in that row of the diagonal matrix. If the block displayed has its first row as row m in $M^{-1}QM$, then in $EM^{-1}QM$ the block becomes

$$\begin{pmatrix} \lambda e^m & e^m & & 0 \\ 0 & \ddots & \ddots & \\ \vdots & \ddots & \lambda e^{m+r-2} & e^{m+r-2} \\ 0 & \cdots & 0 & \lambda e^{m+r-1} \end{pmatrix}.$$

Multiplication on the right is equally simple with all elements in a column being multiplied by the element in that column of the diagonal matrix. Thus, in the matrix $EM^{-1}QME^{-1}$ the block displayed becomes

$$\begin{pmatrix} \lambda & e^{-1} & & 0 \\ 0 & \ddots & \ddots & \\ \vdots & \ddots & \lambda & e^{-1} \\ 0 & \cdots & 0 & \lambda \end{pmatrix}.$$

Let $T = ME^{-1}$ so that we are discussing the matrix $T^{-1}QT$. The maximum norm of this matrix is the largest of the sum of the magnitudes of the elements in a row. There are two situations. One corresponds to 1×1 blocks and to the last row in $r \times r$ blocks. In this situation the only non-zero element in the row is the eigenvalue λ on the diagonal. Because of the root condition, the magnitude of this element is no larger than 1, hence the sum of the magnitudes of the elements in the row is no larger than 1. The root condition requires all eigenvalues of magnitude 1 to be simple and because all simple eigenvalues appear in 1×1 blocks, we have now dealt with all the eigenvalues of magnitude 1. In the other situation the row sum has the form $|\lambda| + e^{-1}$ and because of the root condition, $|\lambda| < 1$. If we take the parameter e sufficiently large, this row sum, indeed, all such row sums, will be no greater than 1. In this way we find that for any matrix Q with eigenvalues that satisfy the root condition, there is a nonsingular matrix T such that $||T^{-1}QT||_\infty \leq 1$.

The BDFs are a special case in this analysis because there is a matrix T that diagonalizes Q. The Adams methods are another special case. They have only the essential root 1 and a multiple root 0. The Jordan canonical form is not a diagonal matrix for Adams methods. However, it is easy to see that the max norm of Q is 1 for such methods.

EXERCISE 12. What is the $k \times k$ matrix Q for an Adams method? Show that $||Q||_\infty = 1$. The eigenvalues of Q are obvious—why? Show that the rank of $Q - \lambda I$ is $k - 1$ for both eigenvalues λ, hence that Q has only two linearly independent eigenvectors.

As in our previous convergence results, it is not much more complicated to deal with implicit methods. This proof has been for a single equation. When systems are treated, the vectors that arise are longer because each component y_j in $\mathbf{v}_j$ approximates the solution of the scalar problem at x_j and this must be replaced in the general case by a vector $\mathbf{y}_j$. Correspondingly, each entry in Q is expanded to a block. Some knowledge about the eigenvalues and eigenvectors of block matrices is needed, but there is no essential difference in the proof for systems. Because of this and for the sake of simplicity, the details are omitted.

THEOREM 8. *A predictor–corrector (PECE) pair of order p is used to approximate the solution of (1). Suppose that the step size is constant in each integration and that the starting values $\mathbf{y}_r$ are accurate of order q. If the method satisfies the root condition, then it is convergent of order min(p, q). With the same assumptions, the same conclusion is true of a LMM of order p.*

These convergence results resemble those that we established earlier for Runge–Kutta methods and Adams methods. The important distinction is in the step sizes that are permitted. In the case of Runge–Kutta methods we asked only that the step sizes span the interval of integration and that the maximum step size tend to zero. In the case of the Adams methods, there was the additional requirement that the rate of increase of the step size be uniformly bounded. In the present context the step size is constant. Not all methods with memory converge as the step size tends to zero. Necessary conditions for convergence were developed that can be checked by calculating the roots of the characteristic polynomial. We have just seen that these necessary conditions are also sufficient for convergence of LMMs and predictor–corrector methods when the step size is constant. Let us state this important conclusion clearly:

Remark When the step size is constant, the root condition is both necessary and sufficient for the convergence of LMMs and predictor–corrector (PECE) pairs. When the step size is constant, "stable" is a synonym for "satisfies the root condition."

§4.4 Quasi-Constant Step Size

It is very important in practice to adapt the step size to the solution. The convergence results we have established for Runge–Kutta and Adams methods place mild restrictions on the step size used, so implementations that exploit this are called *fully variable step size* implementations. In this section we extend the convergence results for constant step size to a convergence model involving a *quasi-constant step size*. By this we mean that when integrating an equation from a to b, the step size is permitted to change at a number of fixed points $a < c_1 < c_2 < \cdots < c_s < b$ and nowhere else. Because fully variable step size codes try to work with a constant step size, fully variable step size implementation might be viewed with some justice as a way to change from one constant step size to another.

The classical way of working with a quasi-constant step size is easily justified. We consider a LMM or a predictor–corrector pair of order p and for simplicity suppose that the starting procedures produce values that are accurate of order p. On reaching c_1 after an integration with the constant step size h_1, a step size h_2 is chosen and the integration is continued from c_1 to c_2 with this new constant step size. The classical way of obtaining starting values for the second integration is to restart from the solution $\mathbf{y}_N$ approximating $\mathbf{y}(c_1)$. Our convergence theorems applied to the interval $[a, c_1]$ tell us that $\mathbf{y}_N$ agrees with $\mathbf{y}(c_1)$ to $\mathcal{O}(h_1^p)$. This approximate solution will be the starting value $\mathbf{y}_0$ for the next integration. We assume now that $h_2 \leq \Gamma h_1$ for some constant $\Gamma \geq 1$. This implies that h_2 is $\mathcal{O}(h_1)$, hence that $\mathbf{y}_N$ agrees with $\mathbf{y}(c_1)$ to $\mathcal{O}(h_2^p)$. The remaining starting values at c_1 are to be computed just like those at a. We can now apply our convergence theorems to the interval $[c_1, c_2]$ and so conclude that the errors of this second integration tend to zero at the rate $\mathbb{O}(h_2^p)$. Because this rate is $\mathbb{O}(h_1^p)$, we have the errors tending to zero at the rate $\mathbb{O}(h_1^p)$ on all of $[a, c_2]$. Repetition of this argument shows that with a change of step size at a finite number of fixed points in the interval $[a, b]$, there is convergence on all of $[a, b]$ at the rate $\mathbb{O}(H^p)$. As usual, H is the maximum step size here. This is stated formally as a theorem.

THEOREM 9. *A stable predictor–corrector (PECE) pair of order p is used to approximate the solution of (1). Suppose that the starting procedures produce values that are accurate of order at least p. The step size is permitted to change only at a finite number of fixed points in the interval and the ratio of successive step sizes is bounded above by a constant $\Gamma \geq 1$. When the step size is changed by means of a restart, the method is convergent of order p. With the same assumptions, the same conclusion is true of a stable LMM of order p.*

This result is much like that proven for fully variable step size formulations. The main difference is in how the step size is changed. Changing the step size in a fully variable step size implementation of a method with memory is by no means "free," but it is considerably less expensive than a complete restart. Fortunately, there is a way to change the step size in a quasi-constant step size model that is quite competitive. In this approach to a change of step size, the starting values for the integration from c_1 to c_2 come from points prior to the initial point c_1. A review of the convergence proofs given earlier shows that they are easily modified to accommodate this. In more detail, the integration from c_1 to c_2 is started with

$$\mathbf{y}_{-r} = \mathbf{s}_{-r}(h_2) \qquad 0 \leq r < k,$$

and a recipe of the generic form

$$\sum_{i=0}^{k} \alpha_i \, \mathbf{y}_{j+1-i} = h_2 \, \mathbf{\Phi}(x_j, \mathbf{y}_{j+1}, \ldots, \mathbf{y}_{j+1-k}, \mathbf{F}, h_2)$$

is applied for $j = 0, 1, \ldots$. In the present context, it is clear that $\mathbf{F}$ is defined prior to $x = c_1$, but for a proof that such a start is possible in general, it is obviously necessary to assume that $\mathbf{F}(x, \mathbf{y})$ is defined for $x < c_1$, say for $c_1 - \varepsilon \leq x \leq c_2$ and some $\varepsilon > 0$. With this minor qualification, the proofs of stability are almost identical to those given earlier and stability implies that the error on $[c_1, c_2]$ is bounded by a multiple of the larger of the truncation error and the starting error. For the matter at hand, the only question is whether the starting values for the integration from $[c_1, c_2]$ are obtained in a stable and accurate way from the results of the integration from $[a, c_1]$. Granted this, the theory will deal with integrations that involve changes from one constant step size to another at a finite number of points specified in advance provided only that the ratio of successive step sizes is bounded.

A very natural way to get starting values at the new step size is to interpolate. On integrating from a to c_1 with step size h_1, we have available

$$\mathbf{y}_{N-r} \approx \mathbf{y}(x_{N-r}), \qquad r = 0, 1, \ldots, m,$$

where $x_{N-r} = c_1 - rh_1$. To start the integration from c_1 to c_2 with step size h_2, we need values $\mathbf{Y}_{N-r} \approx \mathbf{y}(c_1 - rh_2)$ for $r = 0, 1, \ldots, k - 1$. Of course, it is natural to take $\mathbf{Y}_N$ to be the last result of the first integration, $\mathbf{y}_N$, but it is not necessary to distinguish this starting value from the others. Note that the number $m + 1$ of values interpolated has not yet been specified—an appropriate number will be identified by the analysis that follows. Let us define a polynomial $\mathbf{P}(x)$ of degree at most m by the interpolation conditions $\mathbf{P}(b - rh_1) = \mathbf{y}_{N-r}$, $r = 0, 1, \ldots, m$. In terms of the fundamental Lagrangian polynomials

$$L_r(x) = \prod_{\substack{q=0 \\ q \neq r}}^{m} \left(\frac{\eta - q}{r - q} \right) \qquad r = 0, 1, \ldots, m,$$

where the scaled variable η is defined by $x = c_1 - \eta h_1$, this polynomial is

$$\mathbf{P}(x) = \sum_{r=0}^{m} \mathbf{y}_{N-r} L_r(x).$$

The values $\mathbf{Y}_{N-j} = \mathbf{P}(b - jh_2)$, $j = 0, 1, \ldots, k - 1$, provide the "fictitious" values on a mesh with constant spacing h_2 that are needed to start the second integration. How well does $\mathbf{Y}_{N-j}$ approximate $\mathbf{y}(c_1 - jh_2)$? There are two kinds of error. One arises in the interpolation process itself and the other in the fact that the values interpolated are themselves in error. To separate the effects, we define the interpolant to the true solution

$$\mathcal{P}(x) = \sum_{r=0}^{m} \mathbf{y}(c_1 - rh_1) L_r(x).$$

Then

$$\begin{aligned} &\|\mathbf{Y}_{N-j} - \mathbf{y}(c_1 - jh_2)\| \\ &\qquad \leq \|\mathbf{P}(c_1 - jh_2) - \mathcal{P}(c_1 - jh_2)\| + \|\mathcal{P}(c_1 - jh_2) - \mathbf{y}(c_1 - jh_2)\|. \end{aligned}$$

A standard result stated in the appendix says that for each x,

$$\mathbf{y}(x) = \mathcal{P}(x) + \frac{\mathbf{y}^{(m+1)}(*)}{(m+1)!} \prod_{q=0}^{m} (x - (c_1 - qh_1)).$$

For smooth $\mathbf{y}$, this implies that the interpolation error $\|\mathcal{P}(c_1 - jh_2) - \mathbf{y}(c_1 - jh_2)\|$ is $\mathcal{O}(h_1^{m+1})$. The effect of the data errors is measured by

$$\mathcal{P}(c_1 - jh_2) - \mathbf{P}(c_1 - jh_2) = \sum_{r=0}^{m} [\mathbf{y}(c_1 - rh_1) - \mathbf{y}_{N-r}]L_r(c_1 - jh_2).$$

If the method is convergent of order p, then the $\|\mathbf{y}(c_1 - jh_1) - \mathbf{y}_{N-j}\|$ are $\mathcal{O}(h_1^p)$. Notice that this is a statement about global error rather than local error—it is the error at mesh points near $x = c_1$ after $\mathcal{O}(h_1^{-1})$ steps. If the $L_r(c_1 - jh_2)$ are uniformly bounded in magnitude, then the differences $\|\mathcal{P}(c_1 - jh_2) - \mathbf{P}(c_1 - jh_2)\|$ are $\mathcal{O}(h_1^p)$. To see that these quantities are uniformly bounded, notice first that when $x = c_1 - jh_2$, we have $\eta = jh_2/h_1$, hence

$$L_r(c_1 - jh_2) = \sum_{\substack{q=0\\q\neq r}}^{m} \left(\frac{jh_2/h_1 - q}{r - q}\right).$$

Assuming that the ratio of the two step sizes is bounded above by a constant, $h_2/h_1 \leq \Gamma$, it is clear that $|L_r(c_1 - jh_2)|$ is bounded as $h_1 \to 0$. Putting together all these bounds,

$$\|\mathbf{Y}_{N-j} - \mathbf{y}(c_1 - jh_2)\| = \mathcal{O}(h_1^{m+1}) + \mathcal{O}(h_1^p).$$

The $\mathbf{Y}_{N-j}$ are the starting values for the second integration. Any value $m \geq p - 1$ will preserve the order of the integration, but m greater than $p - 1$ does not increase the accuracy and requires more previously computed values to be retained. When phrased in terms of the number of points interpolated, $m + 1$, this leads to

THEOREM 10. *A stable predictor–corrector (PECE) pair of order p is used to approximate the solution of (1). Suppose that the starting procedures produce values that are accurate of order at least p. The step size is permitted to change only at a finite number of fixed points in the interval and the ratio of successive step sizes is bounded above by a constant $\Gamma \geq 1$. When the step size is changed by means of interpolating at least p consecutive solution values, the method is convergent of order p. With the same assumptions, the same conclusion is true of a stable LMM of order p.*

Early codes change the step size by restarting, but nowadays all the popular codes based on constant step size formulas use interpolation because it is much more efficient and often more convenient. For this reason, whenever we refer to a quasi-constant step size implementation, we mean one that changes step size by interpolation. Such an implementation is comparable in efficiency to a fully variable step size implementation.

A few additional remarks should be made about interpolatory step size changing. As stated, the interpolation scheme is related to the formula only through the number of solution values that must be interpolated. When the formula is actually based on an interpolating polynomial, it is often very convenient to use this polynomial for changing the step size. The BDFs are based on a polynomial that interpolates solution values and our analysis applies immediately. The Adams methods are based on a polynomial that interpolates previously computed approximate derivative values and just one approximate solution value. It is natural and very convenient to evaluate this polynomial to change the step size, but our analysis must be extended to justify this. This is easily done and the same conclusions reached. The very influential early code DIFSUB of Gear is a quasi-constant step size code that changes step size using these natural interpolants. The organization of the code and the Nordsieck representation used make changing the step size easy.

There is an approach to changing the step size that is intermediate to fully variable and quasi-constant step size formulations called a *fixed leading coefficient* formulation (Jackson and Sacks-Davis [1980]). Suppose that a code uses a constant step size h_1 up to a point c_1 where it changes to a new constant step size h_2. A fully variable step size formulation can be used with any choice of step sizes so, of course, it can be used with this one. It is the same as the constant step size formula up to the point c_1. Thereafter, as long as the memory of the formula involves mesh points prior to c_1, the coefficients of the formula will differ from the values they have when the step size is constant. When the memory no longer involves points prior to c_1 so that the mesh spacing is constant, the formula has the same form as before the change of step size. This means that the effect of changing from one constant step size to another is spread over a number of steps, the number depending on how long a memory the formula has. In contrast, the quasi-constant step size formulation gets the change over all at once. On reaching c_1, a set of fictitious computed solution values on a mesh of spacing h_2 is obtained by interpolation and the integration continues with the usual constant step size formula. The fixed leading coefficient approach also uses the constant step size formula all the time. However, new fictitious computed solution values on a mesh of spacing h_2 are computed at each step by interpolation on the actual computational mesh. That is, on passing c_1, the values interpolated involve points prior to c_1 at a mesh spacing of h_1 and points after c_1 at a mesh spacing of h_2.

This is the same data used by the fully variable formulation, but it is used in such a way as to work with a constant step size. When written out as linear combinations of the approximate solution and derivative values it is found that the coefficient of $\mathbf{y}_{j+1}$ remains constant through the change of step size. It might be expected that in practice the stability of this approach would be intermediate between the quasi-constant step size and fully variable step size implementations and this seems to be the case. Practical considerations when solving stiff problems with BDFs strongly favor a quasi-constant step size. For this reason the fully variable step size implementation of EPISODE performs better than its quasi-constant step size predecessor GEAR only in circumstances making special demands on the stability of step changing. The fixed leading coefficient approach seen in the recent code VODE appears to be a good compromise.

§5 Interpolation

In a natural way the BDFs and the variants of the Adams methods produce approximate solutions between mesh points. Recall, for example, how the BDFs were defined for a general mesh spacing. A polynomial $\mathbf{P}_\nu(x)$ of degree at most ν is defined by the interpolation conditions

$$\mathbf{P}_\nu(x_{j+1-k}) = \mathbf{y}_{j+1-k} \qquad k = 0, 1, \ldots, \nu. \tag{5.1}$$

The new solution value $\mathbf{y}_{j+1}$ is determined by the algebraic equation

$$\mathbf{P}'_\nu(x_{j+1}) = \mathbf{F}(x_{j+1}, \mathbf{P}_\nu(x_{j+1})) = \mathbf{F}(x_{j+1}, \mathbf{y}_{j+1}).$$

After a successful step to x_{j+1}, it is natural to approximate $\mathbf{y}(x)$ at any x in (x_j, x_{j+1}) by the value $\mathbf{P}_\nu(x)$. This is inexpensive because it amounts merely to evaluating a known polynomial. Furthermore, this permits the codes to select the largest step size that meets the specified accuracy requirement and then obtain answers at specific points by interpolation. Proceeding in this way, "frequent" output has little impact on the cost of the integration. All quality general-purpose BDF codes exploit this possibility.

An interpolating polynomial can be constructed by (5.1) to get intermediate solution values for any discrete variable method. There are two sources of error—that of the interpolation process itself and that due to interpolating values $\mathbf{y}_{j+1-k}$ that are not exactly equal to the true values $\mathbf{y}(x_{j+1-k})$. When the step size is constant, justification of interpolation is almost identical to the analysis of §4.4, so there is no point in working through the details. In the present circumstances we are interested in the value at a point between x_{j+1} and x_j, which corresponds in the earlier analysis to an $h_2 < h_1$. We showed that in such a situation the interpolation coefficients are uniformly bounded and the interpolant agrees with the true solution to order $\min(m + 1, p)$. If we simply interpolate at enough preceding points, the value of the interpolant at any point in (x_j, x_{j+1}) has the same order of accuracy as the results at the mesh points. In particular, if the natural interpolant of a BDF is used, the order of the method is preserved.

This approach provides a piecewise polynomial approximation to $\mathbf{y}(x)$ on all of $[a, b]$. Because it is defined separately for the various subintervals (x_r, x_{r+1}), it is important to ask how the pieces fit together. Let $\mathbf{P}_j(x)$ be the interpolant for (x_{j-1}, x_j) and let $\mathbf{P}_{j+1}(x)$ be the interpolant for (x_j, x_{j+1}). Because both polynomials interpolate to $\mathbf{y}_j$ at x_j, the approximation to $\mathbf{y}(x)$ formed by these two pieces is continuous at x_j On the other hand, there is no reason to think that the two polynomials will have the same derivative at x_j and in general they will not. For this reason the piecewise polynomial approximations provided by the typical BDF code have jumps in the first and higher derivatives at mesh points, a fact rarely appreciated by the users of popular codes. Of course, by going to an interpolant of higher degree it is possible to interpolate not just the solution values at x_j and x_{j+1} for each j, but also the approximate first derivative at x_j and x_{j+1}. It is easy enough to see that this does not affect the order of convergence and when these polynomials are assembled, the result is not only continuous but also has a continuous derivative throughout $[a, b]$. For some purposes the smoother interpolant is important, so some of the newer codes like the D02E?? codes in the NAG library provide a C^1 interpolant.

EXERCISE 13. Experiment with one of the BDF codes available to you to see if its interpolant is C^1. To do this easily you will have to work with a code, or a level in a code, that returns after taking a step and allows you to evaluate an approximation to the derivative of the solution wherever you wish. Evaluate the solution and the approximation to its derivative at a sequence of points that approach the mesh point. Do the interpolated values appear to approach the values at the mesh point?

A variable step size complicates matters only in that constraints on the step size are necessary to keep the coefficients of the interpolating polynomial bounded. Essentially the same matter was taken up for the Adams methods in §2. One thing that is different about the natural interpolants for the Adams methods is that they interpolate approximations to $\mathbf{y}'(x)$ rather than approximations to $\mathbf{y}(x)$. Suppose that we have just taken a successful step to x_{j+1} with an Adams–Moulton formula or Adams–Moulton–Bashforth predictor–corrector pair. For the prediction of the result at x_{j+2}, a polynomial $P_\nu(x)$ is defined by the interpolation conditions

$$\mathbf{P}_\nu(x_{j+2-k}) = \mathbf{F}(x_{j+2-k}, \mathbf{y}_{j+2-k}) = \mathbf{F}_{j+2-k} \qquad k = 1, 2, \ldots, \nu.$$

The Adams–Bashforth formula for $\mathbf{y}_{j+2}$ is

$$\mathbf{y}_{j+2} = \mathbf{y}_{j+1} + \int_{x_{j+1}}^{x_{j+1}+h_{j+1}} \mathbf{P}_\nu(t)\, dt = \mathbf{y}_{j+1} + h_{j+1} \sum_{k=1}^{\nu} \beta^*_{\nu,k}\, \mathbf{F}_{j+2-k},$$

where

$$\beta^*_{\nu,k} = \frac{1}{h_{j+1}} \int_{x_{j+1}}^{x_{j+1}+h_{j+1}} \prod_{\substack{m=1\\ m\neq k}}^{\nu} \left(\frac{t - x_{j+2-m}}{x_{j+2-k} - x_{j+2-m}} \right) dt.$$

Thc polynomial

$$\mathbf{Q}(x) = \mathbf{y}_{j+1} + \int_{x_{j+1}}^{x} \mathbf{P}_\nu(t)\, dt$$

provides a natural way to approximate $\mathbf{y}(x)$ for x between x_j and x_{j+1}. It interpolates $\mathbf{y}_{j+1}$ at x_{j+1} and its derivative interpolates the approximate derivatives $\mathbf{y}'_{j+2-k} = \mathbf{F}_{j+2-k}$ at x_{j+2-k} for $k = 1, 2, \ldots, \nu$.

Evaluation of $\mathbf{Q}(x)$ is not as straightforward as that of the polynomial used with the BDF. We want an approximation to $\mathbf{y}$ at $x = x_j + \theta h_j$ for $0 < \theta < 1$. The value of the interpolant is

$$\begin{aligned}\mathbf{y}_{j+\theta} &= \mathbf{Q}(x_j + \theta h_j) = \mathbf{y}_{j+1} + \int_{x_{j+1}}^{x_j+\theta h_j} \mathbf{P}_\nu(t)\, dt,\\ &= \mathbf{y}_{j+1} + (\theta - 1)h_j \sum_{k=1}^{\nu} \beta^*_{\nu,k}(\theta)\mathbf{F}_{j+1-k},\end{aligned}$$

where

$$\beta^*_{\nu,k}(\theta) = \frac{1}{(\theta - 1)h_j} \int_{x_{j+1}}^{x_{j+1}+(\theta-1)h_j} \prod_{\substack{m=1\\ m\neq k}}^{\nu} \left(\frac{t - x_{j+2-m}}{x_{j+2-k} - x_{j+2-m}} \right) dt.$$

For each point where we want an intermediate solution approximation, it is necessary to compute the coefficients $\beta^*_{\nu,k}(\theta)$. A code set up to integrate on a general mesh will necessarily have a module for this computation so that interpolation can be accomplished in a straightforward way. As written, interpolation is equivalent to constructing a family of formulas depending on a parameter θ. The member corresponding to θ provides an approximate solution at $x_j + \theta h_j$. The key point is that no evaluations of $\mathbf{F}$ are needed other than those made in advancing to x_{j+1}.

Justification of these interpolants is like that for interpolation of the solution values. Again the interpolation coefficients, the $\beta^*_{\nu,k}(\theta)$ in this case, must be uniformly bounded in magnitude. This has already been investigated in proving convergence for the Adams methods and no additional constraint on the step sizes is needed for interpolation. It is perhaps worth pointing out that this kind of interpolant could be used for any formula, just like interpolation to solution values alone, but it is hardly worth the extra complication of evaluating the polynomial. What seems odd about this interpolant is that because it is based on interpolating the first derivative, $\mathbf{Q}'(x)$ is continuous when moving from the subinterval (x_{j-1}, x_j) to (x_j, x_{j+1}), but $\mathbf{Q}(x)$ generally is not. Thus in the popular suite of Adams codes ODE/STEP, INTRP there are jumps in the approximate solution values at mesh points, but not in the approximate first derivative values. For some purposes this is inappropriate. By going to a higher degree interpolant that interpolates the solution at x_j as well as x_{j+1}, a continuous piecewise polynomial approximation with continuous first derivative is obtained. This was done in the successor to ODE called DEABM.

EXERCISE 14. Experiment with one of the Adams codes available to you to see if its interpolant is continuous as well as having a continuous derivative. To do this easily you will have to work with a code, or a level in a code, that returns after taking a step and allows you to evaluate an approximation to the derivative of the solution wherever you wish. Evaluate the solution and the approximation to its derivative at a sequence of points that approach the mesh point. Do the interpolated values appear to approach the values at the mesh point?

These arguments about interpolating the approximate solution values do not depend on how the values were computed, so it appears that one could interpolate results obtained with Runge–Kutta formulas. In principle, this is correct, but there are serious difficulties in practice. One of the attractions of Runge–Kutta methods is that they have no memory. Retaining previously computed values for the purpose of interpolation introduces a number of unwelcome complications into the codes. For example, what is to be done in the start when there are not enough values to interpolate? For the interpolation to be stable, it is necessary to impose a restriction on the rate of change of the step size that is not needed for the method itself. In a code in the NAG library, Gladwell [1979] does this for the fourth-order Merson Runge–Kutta formula. At this relatively low order, he could minimize the number of previously computed values retained by resorting to Hermite interpolation, i.e., interpolation at each x_j to not only $\mathbf{y}_j$ but also $\mathbf{y}'_j$. In some contexts it is not unusual to make computations with constant step size and Runge–Kutta formulas of rather low order. Interpolation is quite practical then. The situation is different for moderate to high-order Runge–Kutta formulas. These formulas do a considerable amount of work in the course of a single step. For them to be competitive with, say, Adams methods, and they are, it is necessary that they take a relatively large step. By this it is meant that the solution might change quite a lot in the course of a single step. The difficulty is that this step size might be too large for interpolation to be accurate. The theory says that as the maximum step size tends to 0, the interpolation process will yield an accurate result. However, in practice step sizes appropriate to a formula of many stages are often too big to obtain interpolated values of comparable accuracy. On the other hand, it is not always the case that the intermediate results need to be of the same accuracy as results at mesh points. For example, a stringent tolerance might be necessary to achieve even a modest accuracy after a "long" integration. In such a situation an interpolant of modest accuracy might be quite acceptable. Current research into interpolation for Runge–Kutta methods takes advantage of the stages formed in the course of a step, plus others formed solely for the interpolant, to obtain intermediate results that are as accurate as the formula itself.

§6 Computational Errors

Generally we try to ignore the effects of roundoff, but for "small" step sizes this is not possible. As a consequence of finite precision arithmetic, the computed $\mathbf{y}_{j+1}$ do not satisfy the formula exactly. Because of the variety of formulas considered, it is awkward to discuss them all at once. As a generic form, the formula is written here as

$$\sum_{i=0}^{k} \alpha_i \, \mathbf{y}_{j+1-i} = h \, \mathbf{\Phi}(x_j, \mathbf{y}_{j+1}, \ldots, \mathbf{y}_{j+1-k}, \mathbf{F}, h)$$

and it should be appreciated that the form is a little different for one-step methods and for the Adams methods with a fully variable step size. With this notation, the computed $\mathbf{y}_{j+1}$ satisfy

$$\sum_{i=0}^{k} \alpha_i \, \mathbf{y}_{j+1-i} = h \, \mathbf{\Phi}(x_j, \mathbf{y}_{j+1}, \ldots, \mathbf{y}_{j+1-k}, \mathbf{F}, h) + \mathbf{r}_j,$$

that is, when substituted into the recipe there is a *residual* or *defect* $\mathbf{r}_j$. In the various analyses of convergence given earlier, this situation is accommodated easily by simply adding these residuals to the local truncation error. It is important to understand how these residuals might arise in practice and what their effects might be. Roundoff errors in evaluating $\mathbf{\Phi}$

get multiplied by h, so have the form $h\boldsymbol{\mu}_j$. The $\boldsymbol{\mu}_j$ look like additional local truncation errors in the convergence proof. Generally these roundoff errors do not decrease as $h \to 0$. With such a term added to the discretization error in the convergence result, the global error is bounded by a term going to zero like h^p (from the discretization error of the formula) plus a term due to roundoff that does not go to zero at all. According to the bounds developed on the global error, there is then a limit on the accuracy that can be attained. It is easy to demonstrate that this happens in practice. Another source of error is in calculating $\mathbf{y}_{j+1}$ from

$$\mathbf{y}_{j+1} = -\frac{1}{\alpha_0} \sum_{i=1}^{k} \alpha_i \, \mathbf{y}_{j+1-i} + \cdots .$$

Because these errors are not multiplied by a factor of h, they can add up in $\mathbb{O}(h^{-1})$ steps to cause an actual growth in error as $h \to 0$. Fortunately this part of the computation is for many formulas trivial—think of Runge–Kutta and Adams methods—and generates little error. Formally these residuals $\mathbf{r}_j$ are accommodated in the convergence proofs by writing $\mathbf{r}_j = h\boldsymbol{\mu}_j$ where $\boldsymbol{\mu}_j = (\mathbf{r}_j/h)$ grows like $\mathbb{O}(h^{-1})$.

EXAMPLE 4. To illustrate the effects of precision, we solved $y' = y^2$, $y(0) = 1/2$ on $0 \le x \le 1$ for its solution $y(x) = 1/(2 - x)$. Using the improved Euler method (alias AB1–AM2 PECE method) and a constant step size $h = 10^{-M}$ for $M = 1, 2, \ldots, 5$, we integrated from 0 to 1 and recorded the maximum absolute error $E(h)$. In single precision on an IBM 3081 this resulted in

h	10^{-1}	10^{-2}	10^{-3}	10^{-4}	10^{-5}
$E(h)$	2.3×10^{-3}	3.2×10^{-5}	5.7×10^{-5}	2.9×10^{-4}	2.6×10^{-3}
$E(h)/h^2$	2.3×10^{-1}	3.2×10^{-1}	5.7×10^{1}	2.9×10^{4}	2.6×10^{7}

In theory the error tends to 0 like h^2 and this is seen for the larger step sizes, but finite precision arithmetic ruins the accuracy when enough steps are taken. Figure 5.2 presents results obtained with a microcomputer. The errors are similar but somewhat smaller because of the superior IEEE arithmetic. For each h the maximum error was scaled by a factor of 5×10^5 and its logarithm plotted against the logarithm of $1/h$. The theoretical h^2 behavior appears as a straight line because of the logarithmic scales.

EXERCISE 15. Try this yourself using AB2 with exact starting values. Do the computation in single precision so that the errors will be more visible.

Generally we can ignore roundoff errors except at very stringent tolerances, but there is another source of computational error that we cannot ignore. This is the error in the evaluation of an implicit formula. We do

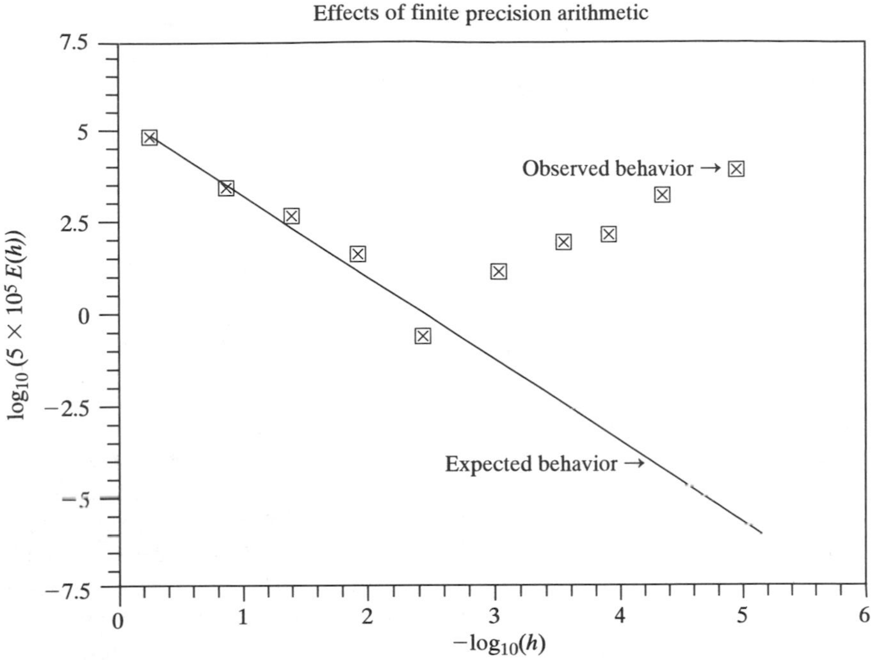

Figure 5.2

not evaluate $\mathbf{y}_{j+1}$ exactly, so the computed value satisfies the recipe with a residual. Of course in exact arithmetic we could make this residual as small as we wish, but it is not practical to make it very small. As described in Chapter 4, an implicit formula is evaluated by simple iteration, namely by first using an explicit formula to calculate an initial approximation $\mathbf{y}_{j+1}^{(0)}$ (predict) and then iterating (correcting) by

$$\mathbf{y}_{j+1}^{(m+1)} = -\frac{1}{\alpha_0}\sum_{i=1}^{k} \alpha_i\, \mathbf{y}_{j+1-i} + \frac{h}{\alpha_0}\, \mathbf{\Phi}(x_j, \mathbf{y}_{j+1}^{(m)}, \ldots, \mathbf{y}_{j+1-k}, \mathbf{F}, h).$$

This can be rewritten in terms of the residual of the iterate $\mathbf{y}_{j+1}^{(m)}$:

$$\alpha_0\, \mathbf{y}_{j+1}^{(m)} + \sum_{i=1}^{k} \alpha_i\, \mathbf{y}_{j+1-i} = h\, \mathbf{\Phi}(x_j, \mathbf{y}_{j+1}^{(m)}, \ldots, \mathbf{y}_{j+1-k}, \mathbf{F}, h) + h\, \boldsymbol{\mu}_j^{(m)}.$$

Suppose the current iterate is $\mathbf{y}_{j+1}^{(m)}$. There are a couple of ways of proceeding, depending on how one decides when to quit iterating, but either will make the point here. In one way of proceeding, the residual $\boldsymbol{\mu}_j^{(m)}$ of the current iterate is first evaluated. If it is smaller in norm than a prescribed accuracy, the iterate is accepted as the computed $\mathbf{y}_{j+1}$ and the integration continued. If it is not accurate enough, the next iterate is calculated from

$$\mathbf{y}_{j+1}^{(m+1)} = \mathbf{y}_{j+1}^{(m)} - \frac{h}{\alpha_0} \boldsymbol{\mu}_j^{(m)},$$

and so forth.

EXERCISE 16. Verify this recipe for the next iterate in terms of the residual of the current iterate.

As with the roundoff errors, the effect of iteration errors that are kept smaller than a prescribed tolerance is to add this tolerance to the term bounding the effect of the truncation error in the bound on the error of the entire integration. If we set the tolerance to a value somewhat smaller than the local accuracy we want from the formula itself, it is now seen that it is not important that we do not satisfy the algebraic equations exactly. The process described is computationally convenient and it differs from practice mainly in that no provision is made in this analysis for reducing the step size to ensure that an acceptable iterate is found in just a few iterations. Reducing the residual to, say, one tenth of the truncation error that we are willing to accept is a natural approach for non-stiff problems and it works fine, but it is not appropriate for stiff problems. The typical code for stiff problems computes approximations that agree with $\mathbf{y}_{j+1}$ to a given tolerance rather than approximations with a residual smaller than a given tolerance. They do this because the computation is stable if the residuals are merely bounded and for stiff problems, the distance of the iterate $\mathbf{y}_{j+1}^{(r)}$ from $\mathbf{y}_{j+1}$ is typically much smaller than the norm of the residual—the residual does not have to be anything like as small as the tolerance on the local error to obtain satisfactory results.

EXERCISE 17. Solve the problem $y' = F(x, y) = -y$, $y(0) = 1$ on $[0, 1]$ with AM2 and a constant step size h. Iterate as described above until the residual is smaller than a tolerance τ. To make the effect of iterating more visible, use AB1 as a predictor. For a range of step sizes measure the maximum error seen in the course of the integration. For each step size, try several different τ to explore the effect of changing τ on the maximum number of iterations to meet the tolerance, on the cost of the whole integration as measured by the number of evaluations of F, and on the maximum error of the numerical solution.

§7 Variation of Order

Modern Adams and BDF codes vary the order of the formula used in the course of the integration. There is an enormous amount of computational experience that shows the advantages of doing this, but there is little theory to justify it. Starting these methods with memory is a special task. It was observed quite early that if the code could vary its order as well as its step size, the first step could be taken with the lowest order member of the family, a one-step method, and thereafter the usual adaptation of order and step size would cause the order and step size to increase quickly to values appropriate to the solution and the accuracy specified. It is reasonable to describe this as a rather complex starting procedure for a formula that is to be used for the remainder of the integration or at least for some time. Only recently (Shampine and Zhang [1990]) has starting such codes by variation of order been justified theoretically.

The effects of varying the order after the start are much harder to analyze. It is not clear how to apply arguments based on "small" step sizes to the selection of the most efficient formula. Although it is found in practice that it is important to vary the order, the theory says that for all sufficiently small step sizes, the most efficient formula is the highest order formula available. Just where a change of order takes place is not fixed—as more accuracy is requested, a higher order becomes more efficient earlier in the integration. A few details about why the obvious model does not describe practical computation will help in understanding this. Suppose that the integration proceeds from a to c_1 with a formula of one order and continues on from c_1 to c_2 with a formula of a different order. With a change of order at a fixed point c_1, it is easy to analyze the rate of convergence. When the order is lowered, say from p to $p - 1$, the convergence on $[a, c_1]$ is of order p and the convergence on $[c_1, c_2]$ is of order $p - 1$. This is fine as far as it goes, but in the limit as the step size tends to zero, there is no advantage to reducing the order—it makes the results less accurate! The flaws of the model are even more evident when increasing the order from p to $p + 1$. According to the bound on the error, starting errors may persist. More concretely, we have shown by example that the order of convergence is generally no higher than the order of accuracy of the starting values. Accordingly, in this model the starting errors prevent the order of convergence from being raised to order $p + 1$ as intended.

Some numerical results were presented in Figure 2.7 to illustrate the variation of step size when a fixed order Runge–Kutta code from the NAG library was used to integrate a two body problem. The same problem was solved with the Adams code in this library to illustrate the variation of order. In Figure 5.3 is shown the order used at each step of the integration

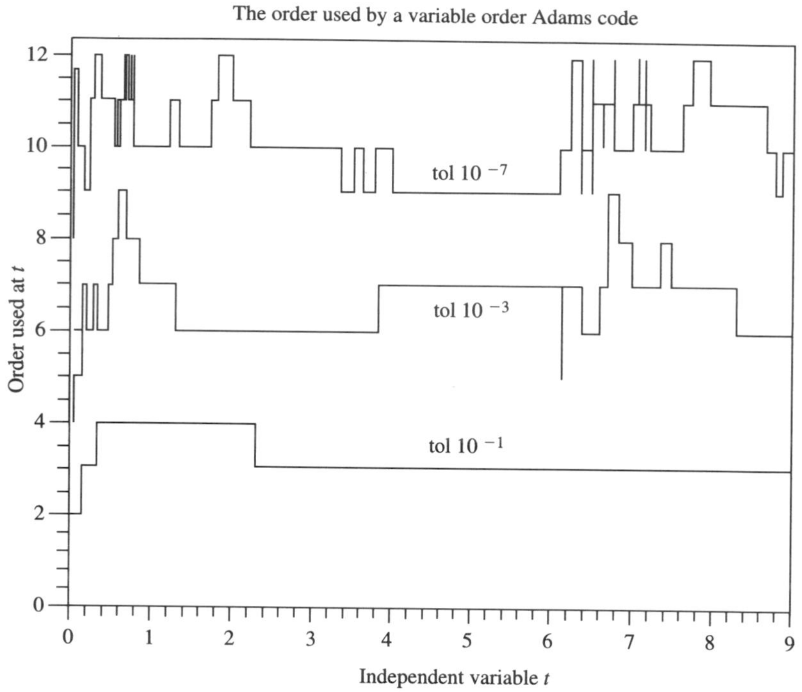

Figure 5.3

for several tolerances. For this particular problem and particular code, a more-or-less constant order is appropriate with the order depending in the expected way on the accuracy demanded. The code starts with the minimum order of 2 and quickly raises the order to a value more appropriate to the tolerance. As it happens, the starting point is a point of closest approach, a difficult point in the integration. There is a considerable variation of the order then as is seen later in the run when the two bodies are again close. What is happening is that the code is manipulating the order in an attempt to avoid reducing the step size. For theoretical purposes it is reasonable to assume that a variable order code will hold the order constant for some while, but this example makes it clear that this is not always a good assumption.

Variable order Adams and BDF codes are exceedingly important in practical computation. In the author's opinion, an understanding of the variation of order is the most important theoretical question open in this area. There have been important advances, notably stability results and an understanding of the start, but much remains to be done. Our purpose in this section is to identify the problem and to point out that it is not obvious how one should model computation with such codes.

§8 Problems That Are Not Smooth

The convergence results we have developed in this chapter assume that **F** is as smooth as necessary, but as we observed in Chapters 1 and 2, problems do arise that are not smooth. To be specific, let us discuss the model problem

$$y' = 2|x|y, \qquad -1 \leq x \leq 1, \qquad y(-1) = \exp(-1). \tag{8.1}$$

It is easily verified that $y(x) = \exp(-x^2)$ for $x \leq 0$ and $y(x) = \exp(x^2)$ for $x > 0$. The solution is smooth except at $x = 0$ where the second derivative has a jump. We have seen that by breaking up the initial value problem into pieces we can establish existence, uniqueness, and stability of problems such as this one that have discontinuities in the independent variable. Problems with discontinuities in a dependent variable present additional difficulties. Breaking up the problem can be useful in practice as well as in theory. In the case of (8.1), suppose that we integrate from -1 to 0 and then restart at 0 for an integration to 1. Theorem 1 in the case of one-step methods, or Theorem 9 in the case of methods with memory, shows that the whole integration is then convergent at the rate expected for **F** that are smooth on the whole interval. Unfortunately, there can be difficulties with the approach in practice. If there are a great many places where the solution is not smooth, the restarts become expensive. In extreme cases one-step methods of relatively low order are to be preferred. This is because restarting a LMM is comparatively expensive and the higher the order, the more so because of the longer memory. Restarting a one-step method is trivial—it just amounts to making sure that the "bad" point is a mesh point. High order Runge–Kutta methods involve a relatively large number of function evaluations per step, so they are advantageous only when they are free to take the longer step that the high order makes possible. If there are a great many places where the solution is not smooth, the step size might be restricted to the point that a low-order method of fewer stages is more efficient. All this presumes that we know where the bad places are. When this depends on the solution itself, the matter becomes much more difficult because it is necessary to locate the bad points in order to restart there. By way of an example, the problem

$$\begin{aligned} y'' &= -y - \operatorname{sgn}(y) - 3\sin(2x),\ 0 \leq x \leq 8\pi, \\ y(0) &= 0,\ y'(0) = 3, \end{aligned} \tag{8.2}$$

involves a discontinuous function

$$\operatorname{sgn}(y) = \begin{cases} +1 & \text{if } y \geq 0, \\ -1 & \text{if } y < 0, \end{cases}$$

that causes the second derivative of $y(x)$ to have a jump discontinuity when $y(x)$ changes sign. This is a simplified model of a system with a relay. It has been used (Shampine et al. [1976]) to test how well codes cope with a lack of smoothness. Generally a proper treatment of such a problem would require integrating until y changes sign, locating carefully where this happened, and restarting there with the new equation corresponding to y of the new sign. With the particular initial conditions of (8.2), the solution is periodic and there is a change of sign at multiples of $\pi/2$.

Through ignorance or sloth, users often pay no attention to the presence of bad points in the interval where the solution is not smooth enough for the method to have its usual order. The analysis of this chapter provides insight. To be specific, we discuss the solution of (8.1) with constant step size. The integration of the smooth equation from -1 to 0 is routine. Suppose first that a one-step method is used. If the origin happens to be a mesh point, the step from the origin can be regarded as the first step of an integration from 0 to 1. The initial error is of the order of the global error of the routine integration, so the integration from 0 to 1 is also routine. If, on the other hand, we step over the origin, the local error made in this step is of lower order than usual because $\mathbf{F}$ is not smooth there. Because an inaccurate start normally spoils the accuracy of the whole integration, we must expect the order of accuracy of the integration from 0 to 1 to be the same as that of the error made in stepping over the origin. In a sequence of integrations it is possible to observe an erratic behavior of the error. For example, suppose we take $h = 2/N$ for $N = 10, 11, 12, \ldots$. When N is even, the origin is a mesh point and the order of the method is preserved, but when N is odd, the order of the accuracy of the step over the origin is less than the order of the method and the accuracy of the integration from 0 to 1 is of lower order than expected. An integration using a method with memory will have a similar behavior except that the accuracy is always reduced after passing the origin because the local error is of reduced order for the several steps taken while the origin is within the span of the memory.

EXAMPLE 5. The initial value problem (8.1) was integrated with AB6 using a constant step size of $h = 2/N$ for $N = 12, 24, 48, 96, 192$. Even N were chosen so that an answer would be produced at $x = 0$. The absolute value of the error was measured there and at $x = 1$. The integrations were started with exact initial values.

h	error at $x = 0$	error at $x = 1$
$1.7D - 1$	$3.7D - 4$	$2.2D - 2$
$8.3D - 2$	$1.9D - 5$	$6.9D - 3$
$4.2D - 2$	$3.5D - 7$	$1.6D - 3$
$2.1D - 2$	$5.6D - 9$	$3.9D - 4$
$1.0D - 2$	$8.7D - 11$	$9.8D - 5$

At $x = 0$ we see that the error 5.6×10^{-9} obtained with step size $h = 2/96$ is approximately 64 times the error 8.7×10^{-11} obtained with step size $h/2$. This corresponds to a factor of 2^6 as we expect for a method of order 6 applied to a smooth problem. At $x = 1$, the error 3.9×10^{-4} is only 4 times the error 9.8×10^{-5}, corresponding to a factor of 2^2 characteristic of a method of order 2. The jump in the second derivative at the origin means that the formula provides approximations that are accurate there only to $O(h^2)$. This amounts to a starting error for the integration on (0, 1], so we cannot expect convergence on this interval that is any faster than order 2, and this is what we see in the computations. A high-order method can be used to solve a problem that is not smooth; it just does not converge as fast as it would for a smooth problem. More details about the analysis of situations like this one are to be found in Shampine and Zhang [1988].

EXERCISE 18. Verify that the solution of the initial value problem (8.2) is

$$\begin{aligned} y(x) &= \cos(x) + \sin(x) - 1 + \sin(2x) \qquad 0 \leq x \leq \pi/2, \\ &= \cos(x) - \sin(x) + 1 + \sin(2x) \qquad \pi/2 < x \leq \pi, \end{aligned}$$

and periodic with period π thereafter.

EXERCISE 19. Integrate either problem (8.1) or (8.2) with a fixed step LMM and measure the global error where the solution is smooth and after passing a bad point where the solution has reduced smoothness.

EXERCISE 20. Integrate either problem (8.1) or (8.2) with a fixed step Runge–Kutta method. Make runs that (effectively) restart at the bad points where the solution has reduced smoothness by including them in the mesh. Make other runs that step over the bad points. Compare the behavior of the error in the two sets of runs before and after the first bad point.

General purpose codes rely upon automatic selection of the step size. Just how this is done will be taken up in a later chapter, but it is useful here to appreciate what the codes try to do and some of the consequences when the problem is not smooth. If the local error estimator recognizes a "large" error because the problem is not smooth in the course of the current step, the step will be rejected and tried again with a smaller step size. All the algorithms are derived for smooth problems, so it can happen that a code does not "notice" that the problem is not as smooth as it is supposed to be. Still, the local error estimators do pretty well at this and the authors of quality codes do make some provision for the possibility that the problem is not smooth—an instance of the art of numerical analysis. What is likely to happen with one of the better codes is that it, in effect, locates the bad points in the integration and uses small step sizes near these points to get an accurate solution there. It is much better practice to inform a code of the location of bad points, both on grounds of reliability and efficiency, but often it is possible to ignore the matter and get away with it. Example 1 of Chapter 3 discusses the solution of a problem with a forcing function consisting of square pulses. Though the emphasis there was on the effect of output, it is an example of a code failing to locate bad points automatically. The example discusses how to inform the code of these bad points. If there are a great many places where the solution is not smooth, a general purpose code is even more expensive than the method of restarting presented earlier because the code must not only restart at bad points, it must first find them. When interpolating experimental data, users often create problems with a great many bad points without realizing it. For example, fitting data with cubic splines is an effective and popular way to represent the data. Although a cubic spline appears smooth to the eye, it is not at all smooth when used as a coefficient in a differential equation because it has a jump in its second derivative at all breakpoints. This can be devastating to a general purpose code of high order because it must locate and restart at all the breakpoints. An appropriate remedy is to use an approximating function that is smoother—not more accurate, just smoother. A simple example will make the point.

EXAMPLE 6. The solution of $y' = -y + \exp(-x)$, $y(0) = 0$ is $y(x) = x\exp(-x)$. Even for stringent accuracy requests, it is inexpensive to integrate this problem from $x = 0$ to 1. The (7, 8) Runge–Kutta pair in RKSUITE requires only 53 evaluations of the equation to solve the problem when given a relative error tolerance of 10^{-9} and a threshold of 10^{-20}. Comparing the computed solution at $x = 1$ to the true solution shows it to have a relative error of about 4×10^{-12}.

To exemplify the approximation of a coefficient, we solve the problem $y' = -y + S(x)$, $y(0) = 0$ where $S(x)$ is the linear spline interpolating $\exp(-x)$ at $x_i = i \times 10^{-3}$ for $i = 0, 1, \ldots, 10^3$:

$$S(x) = \left(\frac{x - x_i}{x_{i+1} - x_i}\right)e^{-x_{i+1}} + \left(\frac{x_{i+1} - x}{x_{i+1} - x_i}\right)e^{-x_i}, \qquad x_i \le x \le x_{i+1}.$$

A standard expression for the error of linear interpolation shows that $S(x)$ differs from $\exp(-x)$ on $[0, 1]$ by no more than 5×10^{-7}. The solutions of the two problems differ at $x = 1$ by less than 10^{-7}. The two problems look almost the same to us, but they look very different to codes—RKSUITE required 6933 evaluations to solve the approximating problem! This great increase in cost is due in part to the cost of locating the points where the solution is not smooth. More important here is the restriction of the step size to 10^{-3} imposed by the lack of smoothness of the approximating problem; this is far smaller than the step size that this high-order method can use when integrating the smooth problem.

EXERCISE 21. Try this yourself. Solve both problems numerically with your favorite code and a stringent accuracy request. Do you observe a similar difference in cost? The example makes use of a formula that does a comparatively large amount of work per step. How expensive is each step with the method you use? Does the cost per step play an important role?

6

Stability for Large Step Sizes

Because the stability results of Chapter 5 were proven only for all "sufficiently small" step sizes, it is not clear whether an integration with a particular choice of step sizes will be stable. In this chapter we investigate stability in the context of a single integration. Specifically, if the solution sequence $\{\mathbf{y}_j\}$ is the result of exact computation and $\{\mathbf{z}_j\}$ is the result after some $\mathbf{y}_m$ is perturbed slightly to $\mathbf{z}_m$, we ask whether the difference between $\mathbf{y}_j$ and $\mathbf{z}_j$ grows as j increases. Stability in this sense is obviously very important in practice, but it is difficult to analyze. The same is true of the differential equation. To study the behavior of solutions with respect to a change in the initial values, we had to look at small changes. This restriction is not so bad, but we found that the behavior is specified in terms of the solution of a differential equation that itself depends on the solution of the original problem and as a consequence it is difficult to get any insight about stability. By restricting our attention to linear problems we could escape the limitation of small changes. By further restricting ourselves to quite special linear problems, the effects of changes could be worked out and useful guidelines about stability obtained. In this chapter we travel the same path for difference equations. There is an additional complication with difference equations due to the mesh. In the case of one-step methods, it is possible to work with a general mesh. In the case of methods with memory, we restrict ourselves to constant step size so as to simplify the task to the point that guidelines can be obtained. These guidelines are useful in practice because codes that select the step size automatically tend to work with a constant step size and it is often the case that there are sequences of steps of constant size to which the model applies. The results derived in this chapter are useful for selecting methods and when applied with care, can be useful in describing practical computation, but they are not all that one might hope for.

§1 Stability with Respect to Small Perturbations

Because we restrict our attention here to constant step size when discussing methods with memory, the generic form of the methods is

$$\mathbf{y}_r \quad \text{given for} \quad 0 \le r < k,$$

$$\sum_{i=0}^{k} \alpha_i \, \mathbf{y}_{j+1-i} = h \, \Phi(x_j, \mathbf{y}_{j+1}, \ldots, \mathbf{y}_{j+1-k}, \mathbf{F}, h), \qquad j \ge k - 1. \quad (1.1)$$

This form is valid for one-step methods, too, but for such methods we may permit the step size to vary. For numerical purposes it is important that small changes in the initial values $\mathbf{y}_r$ lead to small changes in the $\mathbf{y}_j$ as the integration proceeds. This is clear enough, but in a single integration we do not compute two sequences $\{\mathbf{y}_j\}$ and $\{\mathbf{z}_j\}$, so how does instability show up in the computation? If the equation is autonomous and the step size is constant, there is a way to get some insight.

The methods we investigate have increment functions that do not depend on x when $\mathbf{F}$ does not depend on x (the system is autonomous). For a computed solution sequence $\{\mathbf{y}_j\}$, let us define another sequence $\{\mathbf{z}_j\}$ with starting values

$$\mathbf{z}_r = \mathbf{y}_r + (\mathbf{y}_{r+1} - \mathbf{y}_r) = \mathbf{y}_{r+1}, \qquad 0 \le r < k.$$

It follows that $\mathbf{z}_m = \mathbf{y}_{m+1}$ for all m. This is true by definition for $m < k$. To see that it is true in general, suppose that it is true for $m \le j - 1$. By definition

$$\sum_{i=0}^{k} \alpha_i \, \mathbf{y}_{j+1-i} = \alpha_0 \, \mathbf{y}_{j+1} + \sum_{i=1}^{k} \alpha_i \, \mathbf{y}_{j+1-i}$$

$$= h \, \Phi(\mathbf{y}_{j+1}, \mathbf{y}_j, \ldots, \mathbf{y}_{j+1-k}, \mathbf{F}, h),$$

where the fact that the increment function does not depend on x has been taken into account in the notation. On using the induction hypothesis this is

$$\alpha_0 \, \mathbf{y}_{j+1} + \sum_{i=1}^{k} \alpha_i \, \mathbf{z}_{j-i} = h \, \Phi(\mathbf{y}_{j+1}, \mathbf{z}_{j-1}, \ldots, \mathbf{z}_{j-k}, \mathbf{F}, h),$$

from which it is clear that $\mathbf{z}_j = \mathbf{y}_{j+1}$. The sequence $\mathbf{z}_j$ arises from a particular set of perturbations $\boldsymbol{\delta}_r$ (namely, $\boldsymbol{\delta}_r = \mathbf{y}_{r+1} - \mathbf{y}_r$) to the initial values of $\mathbf{y}_j$. If the computation with this set of initial perturbations is unstable, the difference $\|\mathbf{z}_j - \mathbf{y}_j\| = \|\mathbf{y}_{j+1} - \mathbf{y}_j\|$ will grow rapidly as the integration progresses. Furthermore, any block of k successive approximations $\mathbf{y}_m$ can be regarded as initial values for the rest of the integration, so if perturbations of this kind on any block of k successive solution values lead to an unstable integration, this will be observed by a rapid growth in the difference of successive solution values. In principle it can happen that an integration will not exhibit instability for some perturbations even when it is very unstable for others. In practice we expect that if a class of perturbations leads to instability, then the effects of finite precision arithmetic will eventually introduce perturbations in the class and instability will always be observed. Later we shall pursue this matter with further assumptions about the problem to get more insight.

It is only reasonable to ask for stability of the numerical method when the differential equation is itself stable. To investigate these questions, we simplify the task by restricting our attention to small perturbations. First let us recall what happens with the differential equation. If $\mathbf{y}(x)$ is the solution of

$$\mathbf{y}' = F(x, \mathbf{y}), \qquad \mathbf{y}(a) = \mathbf{A},$$

and $\mathbf{u}(x, \varepsilon)$ the solution of

$$\mathbf{u}' = \mathbf{F}(x, \mathbf{u}), \qquad \mathbf{u}(a, \varepsilon) = \mathbf{A} + \varepsilon\, \mathbf{e}(a) + \mathcal{O}(\varepsilon^2),$$

then

$$\mathbf{u}(x, \varepsilon) = \mathbf{y}(x) + \varepsilon\, \mathbf{e}(x) + \mathcal{O}(\varepsilon^2),$$

where the function $\mathbf{e}(x)$ is the solution of

$$\mathbf{e}'(x) = \frac{\partial \mathbf{F}}{\partial \mathbf{y}}(x, \mathbf{y}(x))\, \mathbf{e}(x), \qquad \mathbf{e}(a) \text{ given.}$$

(This result is taken up in the appendix.) Let us proceed in an analogous, formal way by defining $\{\mathbf{z}(\varepsilon)_j\}$ as the solution of

$$\mathbf{z}(\varepsilon)_r = \mathbf{y}_r + \varepsilon\, \mathbf{e}_r + \mathcal{O}(\varepsilon^2) \qquad 0 \le r < k,$$

$$\sum_{i=0}^{k} \alpha_i\, \mathbf{z}(\varepsilon)_{j+1-i} = h\, \Phi(x_j, \mathbf{z}(\varepsilon)_{j+1}, \ldots, \mathbf{z}(\varepsilon)_{j+1-k}, \mathbf{F}, h).$$

We ask if it is possible to write

$$\mathbf{z}(\varepsilon)_j = \mathbf{y}_j + \varepsilon\, \mathbf{e}_j + \mathcal{O}(\varepsilon^2)$$

for all j. The form is valid for $0 \le j < k$ with given $\mathbf{e}_j$. A formal expansion of the difference equation leads to an equation for the $\mathbf{e}_j$ for $j \ge k$:

$$\sum_{i=0}^{k} \alpha_i\, \mathbf{y}_{j+1-i} + \varepsilon \sum_{i=0}^{k} \alpha_i\, \mathbf{e}_{j+1-i} + \mathcal{O}(\varepsilon^2)$$

$$= h\, \Phi(x_j, \mathbf{y}_{j+1}, \ldots, \mathbf{y}_{j+1-k}, \mathbf{F}, h) + h \sum_{i=0}^{k} \frac{\partial \Phi}{\partial \mathbf{y}_{j+1-i}}\, \varepsilon\, \mathbf{e}_{j+1-i} + \mathcal{O}(h\varepsilon^2).$$

It is seen that the perturbations $\mathbf{e}_j$ satisfy formally a linear difference equation that depends on the solution $\{\mathbf{y}_j\}$ of the original problem:

$$\sum_{i=0}^{k} \alpha_i \, \mathbf{e}_{j+1-i} = h \sum_{i=0}^{k} \frac{\partial \mathbf{\Phi}}{\partial \mathbf{y}_{j+1-i}} \mathbf{e}_{j+1-i} \text{ for } j \geq k - 1,$$

$$\mathbf{e}_r \text{ given for } 0 \leq r < k.$$

Just as with the differential equation itself, this formal analysis tells us how perturbations of the numerical solution propagate, at least for small perturbations. Keep in mind that we are discussing here small perturbations to the numerical solution; the step size h does not have to be small. In principle we solve the equation of first variation so that we understand the stability of the differential equation, solve this analog for the difference equation so that we understand the stability of the numerical method, and then relate the two. In this generality we are unable to carry out this plan even qualitatively, much less in detail. In the next section we restrict the class of $\mathbf{F}$ treated to the point that we can work out the details.

EXAMPLE 1. For specific methods and equations, we can work out enough details to understand the stability of a computation in this way. Again we return to the example of Hubbard and West [1990, 1992] treated as Example 1 in both Chapter 2 and 5. The differential equation $y' = y^2 - x$ is integrated with the midpoint Euler formula and a fixed step size of 0.3. The integration is advanced according to

$$y_{j+1} - y_j = h[(y_j + 0.5h(y_j^2 - x_j))^2 - x_j - 0.5h].$$

A little calculation leads to

$$e_{j+1} - e_j = 2h(y_j + 0.5h(y_j^2 - x_j))(1 + hy_j)e_j.$$

This is more easily understood when e_{j+1} is written as the product of a "growth factor" and e_j:

$$e_{j+1} = (1 + 2h(y_j + 0.5h(y_j^2 - x_j))(1 + hy_j))e_j.$$

Where the growth factor has magnitude greater than 1, the computation is unstable and otherwise, stable. Figure 6.1 shows the integration of the initial value problem with $y(0) = -4$. The stable null isocline $-\sqrt{x}$ is shown. As we saw earlier, the numerical solution approaches this curve and follows it until the stable "isocline" of the discrete problem approximating it comes to an end. Another stable "isocline" emerges later and the numerical solution approaches and then follows it, as may be seen in the figure. Also plotted is the growth factor just derived and the lines $y = \pm 1$. It is seen that the integration is somewhat unstable while the solution is moving from the unstable "isocline" that approximates $-\sqrt{x}$

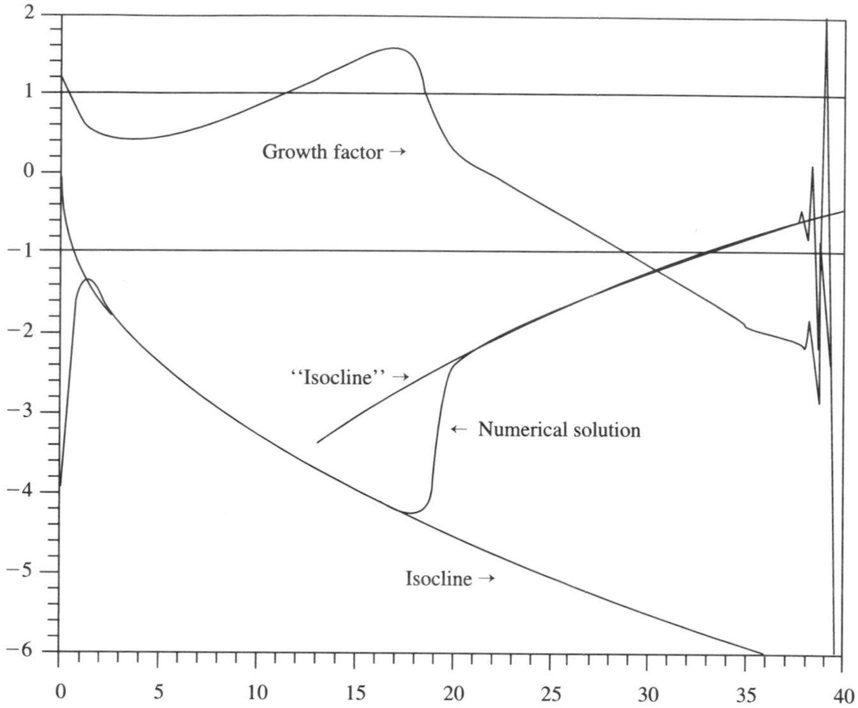

Figure 6.1

to the stable "isocline" and is then stable for some distance. The magnitude of the growth factor is increasing and eventually the integration becomes unstable. Notice that the results are quite acceptable for some time after the growth factor has magnitude greater than 1—it can take some while for the effects of instability to become evident.

§2 A Special Class of Problems

For guidance as to a reasonable class of problems for more detailed study, we recall our study of the stability of the differential equation. Matters were considerably simpler when we treated linear problems and in one respect, the results were stronger—we did not need to limit ourselves to small perturbations because this assumption is needed only for approximating a non-linear problem by a linear one. However, to take full advantage of restricting ourselves to linear differential equations, we must work with numerical methods that are linear when applied to a linear problem. Fortunately this is true of all the standard methods. There is another important simplification due to linearity. Any block of k successive values $\mathbf{y}_j$ can be regarded as initial values for the rest of the integration. If we can analyze what happens to perturbations of initial values, we can use this fact and linearity to understand what happens if there are perturbations at every step—the effects of the successive perturbations just add up.

Even for linear differential equations, it is not easy to understand the stability of the initial value problem. A class of equations that is general enough to be useful and yet simple enough to analyze easily is taken up in the next theorem.

THEOREM 1. *Consider a differential equation of the form*

$$\mathbf{y}' = J\,\mathbf{y} + \mathbf{g} \tag{2.1}$$

when the constant matrix J has a complete set of eigenvectors. Let $\mathbf{u}(x)$ *and* $\mathbf{v}(x)$ *be the solutions of (2.1) with initial values* $\mathbf{u}(a) = \mathbf{A}$ *and* $\mathbf{v}(x) = \mathbf{A} + \boldsymbol{\delta}$. *Let the eigenvalues of J* be $\{\lambda_m\}$. *If $Re(\lambda_m) > 0$ for some m, then there are perturbations* $\boldsymbol{\delta}$ *such that* $\|\mathbf{u}(x) - \mathbf{v}(x)\| \to \infty$ *exponentially fast as $x \to \infty$. If $Re(\lambda_m) \le 0$ for all m, then for any* $\boldsymbol{\delta}$, $\|\mathbf{u}(x) - \mathbf{v}(x)\|$ *is uniformly bounded for all x. Further, if $Re(\lambda_m) < 0$ for all m, then for any* $\boldsymbol{\delta}$, $\|\mathbf{u}(x) - \mathbf{v}(x)\| \to 0$ *exponentially fast as $x \to \infty$.*

Proof The fact that J has a complete set of eigenvectors implies that J can be diagonalized by a similarity transformation,

$$M\,J\,M^{-1} = D = \mathrm{diag}\{\lambda_m\}.$$

The difference $\Delta(x) = \mathbf{v}(x) - \mathbf{u}(x)$ satisfies $\Delta' = J\Delta$. In a new variable $\mathbf{w} = M\,\Delta$, we have

$$\mathbf{w}' = M\,\Delta' = M\,J\,M^{-1}\,M\,\Delta = D\,\mathbf{w}.$$

From this we find that for each m,

$$w_m(x) = e^{\lambda_m(x-a)}w_m(a).$$

The solution component $w_m(x)$ grows exponentially fast as $x \to \infty$ whenever $\mathrm{Re}(\lambda_m) > 0$ and $w_m(a) \neq 0$. When $\mathrm{Re}(\lambda_m) \le 0$, $w_m(x)$ is bounded in magnitude by $|w_m(a)|$ for all x, and when $\mathrm{Re}(\lambda_m) < 0$, $w_m(x) \to 0$ exponentially fast. Returning to the original variables $\Delta = M^{-1}\,\mathbf{w}$, the assertions of the theorem follow easily from these observations.

Let us now return to the discrete problem and suppose that $\mathbf{z}_j$ is the solution of

$$\mathbf{z}_r = \mathbf{y}_r + \boldsymbol{\delta}_r \qquad 0 \le r < k,$$

$$\sum_{i=0}^{k} \alpha_i \, \mathbf{z}_{j+1-i} = h \, \boldsymbol{\Phi}(x_j, \mathbf{z}_{j+1}, \ldots, \mathbf{z}_{j+1-k}, \mathbf{F}, h), \qquad j \ge k - 1.$$

We restrict our attention to problems of the form (2.1). Because these problems are linear, we do not have to suppose that the perturbations $\boldsymbol{\delta}_r$ are small. If we subtract the equation for $\mathbf{y}_j$ from that for $\mathbf{z}_j$ and use the special form of $\mathbf{F}(x, \mathbf{y})$, we find for all the standard methods that the difference $\Delta_j = \mathbf{z}_j - \mathbf{y}_j$ satisfies

$$\Delta_r = \boldsymbol{\delta}_r \ \text{ given } \ 0 \le r < k,$$

$$\sum_{i=0}^{k} \alpha_i \, \Delta_{j+1-i} = hJ \sum_{i=0}^{k} R_i(hJ) \, \Delta_{j+1-i}, \qquad j \ge k - 1.$$

Here each $R_i(hJ)$ is a rational function of the matrix hJ. This special form is crucial to the analysis of stability developed here, so it is important to appreciate that it is true of all the standard methods.

As a simple example of the form, the AB2 formula gives

$$\mathbf{y}_{j+1} = \mathbf{y}_j + h\left[\frac{3}{2}(J\,\mathbf{y}_j + \mathbf{g}) - \frac{1}{2}(J\,\mathbf{y}_{j-1} + \mathbf{g})\right],$$

$$\mathbf{z}_{j+1} = \mathbf{z}_j + h\left[\frac{3}{2}(J\,\mathbf{z}_j + \mathbf{g}) - \frac{1}{2}(J\,\mathbf{z}_{j-1} + \mathbf{g})\right],$$

and subtraction results in

$$(\mathbf{z}_{j+1} - \mathbf{y}_{j+1}) - (\mathbf{z}_j - \mathbf{y}_j) = hJ\left[\frac{3}{2}(\mathbf{z}_j - \mathbf{y}_j) - \frac{1}{2}(\mathbf{z}_{j-1} - \mathbf{y}_{j-1})\right],$$

which is

$$\Delta_{j+1} - \Delta_j = hJ\left[\frac{3}{2}\Delta_j - \frac{1}{2}\Delta_{j-1}\right].$$

In this example we see that $R_0(hJ) = 0$, $R_1(hJ) = \tfrac{3}{2}I$, and $R_2(hJ) = -\tfrac{1}{2}I$ (and $\alpha_0 = 1$, $\alpha_1 = -1$, $\alpha_2 = 0$).

An easy way to identify the functions $R_i(hJ)$ is to apply the formula to the homogenous differential equation ($\mathbf{g} = \mathbf{0}$) because if the initial values $\mathbf{y}_r$ are all $\mathbf{0}$, then all the $\mathbf{y}_j$ are $\mathbf{0}$ and Δ_j is just $\mathbf{z}_j$. For example, in the case of a LMM, we have

$$\sum_{i=0}^{k} \alpha_i \mathbf{z}_{j+1-i} = h \sum_{i=0}^{k} \beta_i \mathbf{F}_{j+1-i} = hJ \sum_{i=0}^{k} \beta_i \mathbf{z}_{j+1-i},$$

from which we can read off that $R_i(hJ) = \beta_i I$ for each $i = 0, 1, \ldots, k$. If such a formula is used as a predictor, we obtain a value

$$\mathbf{z}_{j+1}^* = -\frac{1}{\alpha_0^*} \sum_{i=1}^{k^*} (\alpha_i^* - h\, \beta_i^*\, J)\mathbf{z}_{j+1-i},$$

where, as usual, asterisks have been used to indicate quantities associated with the predictor. We increase k if necessary so that $k \geq k^*$. If k^* is bigger than k, there are indices i for which β_i^* is defined and β_i is not. We define such β_i to be zero and handle similarly the corresponding β_i^* in the case of k^* less than k. Because the predictor is an explicit formula, $\beta_0^* = 0$. With these notational simplifications, it is easy to combine the expression for the predictor with a corresponding expression for a corrector to identify the $R_i(hJ)$ for a PECE implementation:

$$R_i(hJ) = \beta_i I - \frac{\beta_0}{\alpha_0^*} (\alpha_i^*\, I - \beta_i^*\, hJ) \qquad i = 0, 1, \ldots, k.$$

EXERCISE 1. Work out directly the functions $R_i(hJ)$ for AM2 and for the predictor–corrector pairs AB1–AM2 and AB2–AM2 implemented in PECE form.

The functions $R_i(hJ)$ look rather different in the case of Runge–Kutta methods. Recall from Chapter 4 that the form of an explicit Runge–Kutta formula of $s + 1$ stages applied to an autonomous equation is

$$\mathbf{Z}_0 = \mathbf{z}_j, \qquad \mathbf{F}_0 = \mathbf{F}(\mathbf{Z}_0),$$

and for $k = 1, \ldots, s$

$$\mathbf{Z}_k = \mathbf{z}_j + h \sum_{m=0}^{k-1} \beta_{k,m}\, \mathbf{F}_k, \qquad \mathbf{F}_k = \mathbf{F}(\mathbf{Z}_k).$$

Finally,

$$\mathbf{z}_{j+1} = \mathbf{z}_j + h \sum_{m=0}^{s} \gamma_m\, \mathbf{F}_m.$$

For $\mathbf{F}$ of the special form $J\, \mathbf{y}$, we have $\mathbf{F}_0 = J\, \mathbf{z}_j$ and

$$\mathbf{Z}_1 = (I + hJ\, \beta_{1,0})\mathbf{z}_j.$$

Similarly, $\mathbf{F}_1 = J\,\mathbf{Z}_1$ and

$$\mathbf{Z}_2 = \mathbf{z}_j + h\,\beta_{2,0}\,J\,\mathbf{z}_j + h\,\beta_{2,1}\,J(I + hJ\,\beta_{1,0})\mathbf{z}_j$$
$$= (I + (\beta_{2,0} + \beta_{2,1})hJ + \beta_{2,1}\,\beta_{1,0}(hJ)^2)\mathbf{z}_j.$$

Note the form of these expressions. The first intermediate value $\mathbf{Z}_1$ comes from the product of $\mathbf{z}_j$ and a linear polynomial in hJ and the second intermediate value $\mathbf{Z}_2$ comes from the product of $\mathbf{z}_j$ and a quadratic polynomial in hJ. Clearly each intermediate value is given by a similar expression with the degree of the polynomial in hJ successively increased by one. When these expressions are substituted into the recipe for $\mathbf{z}_{j+1}$, the result has the form

$$\mathbf{z}_{j+1} = \mathbf{z}_j + hJ\,P_s(hJ)\mathbf{z}_j,$$

where $P_s(hJ)$ is a polynomial of degree s in the matrix hJ. Because we are considering now only explicit Runge–Kutta formulas, $R_0(hJ) = 0$ and we recognize from this expression that $R_1(hJ) = P_s(hJ)$. Further notice that the result $\mathbf{z}_{j+1}$ of this formula of $s + 1$ stages is the product of $\mathbf{z}_j$ with a polynomial of degree $s + 1$ in the matrix hJ, namely, $I + hJ\,P_s(hJ)$. This polynomial is called the *stability polynomial*. Implicit Runge–Kutta formulas lead to $R_i(hJ)$ that are rational functions of hJ.

EXERCISE 2. Work out the $R_i(hJ)$ and the stability polynomial for the improved Euler formula and for the midpoint Euler formula.

EXERCISE 3. Work out the $R_i(hJ)$ for the trapezoidal rule—an implicit Runge–Kutta formula—by applying the formula to $\mathbf{y}' = J\,\mathbf{y}$.

A change of variables greatly simplified discussion of the stability of the differential equation because it uncoupled the system. Fortunately, the same is true of the difference equation. Multiplying the equation by M, we obtain

$$\sum_{i=0}^{k} \alpha_i\, M\Delta_{j+1-i} = h\,M\,J\,M^{-1} \sum_{i=0}^{k} M\,R_i(hJ)M^{-1}\,M\Delta_{j+1-i}.$$

Now we use the important facts that when R_i is a rational function of a matrix hJ,

$$M\,R_i(hJ)\,M^{-1} = R_i(h\,M\,J\,M^{-1}),$$

and further, because J is similar to the diagonal matrix $D = \text{diag}\{\lambda_m\}$,

$$R_i(h\,M\,J\,M^{-1}) = R_i(hD) = \text{diag}\{R_i(h\lambda_m)\}.$$

In terms of the new variables $\mathbf{w}_j = M\Delta_j$, the difference equation becomes

$$\sum_{i=0}^{k} \alpha_i \, \mathbf{w}_{j+1-i} = \sum_{i=0}^{k} hD \, R_i(hD)\mathbf{w}_{j+1-i},$$

which is uncoupled. A typical component ω_j of the vector $\mathbf{w}_j$ satisfies an equation of the form

$$\sum_{i=0}^{k} (\alpha_i - h\lambda \, R_i(h\lambda))\omega_{j+1-i} = 0, \tag{2.2}$$

where λ is the corresponding eigenvalue of J. This is a linear difference equation with constant coefficients. The main reason for restricting the class of problems we consider is to get to a problem we can analyze. In Chapter 5 we investigated the stability of a constant coefficient difference equation by constructing solutions from the roots of the characteristic polynomial associated with the equation. Recall we found that if the characteristic polynomial $\rho(\theta)$ of the difference equation has a root ζ of magnitude greater than 1, then the equation has a solution that grows exponentially fast. Equation (2.2) is analyzed in the same way. The characteristic polynomial of this equation is

$$\rho(\theta; h\lambda) = \sum_{i=0}^{k} (\alpha_i - h\lambda \, R_i(h\lambda))\theta^{k-i}.$$

The roots ζ of this polynomial depend on the parameter $h\lambda$, so in what follows they are often written as $\zeta(h\lambda)$. Notice that when $h\lambda = 0$, this characteristic polynomial reduces to the usual characteristic polynomial $\rho(\theta)$ of the method that arose when we investigated stability as $h \to 0$. We are assuming that $\rho(\theta) = \rho(\theta; 0)$ satisfies the root condition, so we know something about the roots $\zeta(0)$. The roots of a polynomial are continuous functions of its coefficients. This fact is proved by Ostrowski [1966] in an elementary way and is often taken up in books on complex analysis. In the present context, as $|h\lambda| \to 0$, the roots $\zeta(h\lambda) \to \zeta(0)$. This observation is quite helpful in understanding the relationship of stability for fixed step size to zero-stability and we shall exploit it later to gain insight.

Just as when $h = 0$, for each root $\zeta(h\lambda)$ of the characteristic polynomial, there is a solution of the difference equation of the form ζ^j. Because the problem is linear, the sequence $\omega_j = \varepsilon\zeta^j$ is the solution of (2.2) with initial values $\omega_r = \varepsilon\zeta^r$ for $0 \le r < k$. If the magnitude of $\zeta(h\lambda)$ is greater than one, the ζ^j grow exponentially fast as j increases, with the consequence that no matter how small the initial perturbations are, i.e., no matter how small ε is, the perturbations $\varepsilon\zeta^j$ soon become "large." It is in this sense that we say the scheme is unstable. It should be remarked at this point that many people get the impression that a code cannot use a step size that leads to instability of this kind. This impression is incorrect. The codes can, *and do*, use such step sizes. It is more accurate to say that the codes cannot use such a step size for *many* steps. One of the roles of variation of step size is to recognize and reduce the step size when necessary for stability. Present algorithms work pretty well, but there is not much theory supporting them.

As in Chapter 5, we have found circumstances leading to instability, but we also need to investigate when the integration is stable. It turns out that if all the roots of $\rho(\theta; h\lambda)$ have magnitude less than 1, every solution of the difference equation is bounded in magnitude uniformly in j. This says that perturbations to the initial values do not "explode" as the integration progresses. The result is assumed here so that we may continue with a study of stability in general and a proof is provided in a section at the end of the chapter.

Let us not lose sight of the fact that we have been discussing the stability of the numerical method for just one component of the uncoupled equations. The integration of the system of uncoupled equations is stable if, and only if, the integration of every component is stable. As with the differential equation, the coupling is accomplished by multiplication by a constant matrix. As a consequence, the integration of the original system (2.1) is stable if, and only if, the integration of all of the uncoupled component equations is stable.

The matter of stability in the present situation is not quite as simple as it might appear. This is because we need to relate the stability of the numerical method to the stability of the differential equation—the one should imitate the behavior of the other. We do solve differential equations that are mildly unstable and a difference scheme that is stable for such problems is not completely satisfactory. An important example of this that will be taken up later is provided by the BDFs. Still, we are mostly interested in stable problems and our attention is focussed on schemes that are stable when the problem is. Differential equations in the class that we are at present investigating are stable when $\text{Re}(\lambda) < 0$. For a problem in this class, the growth in perturbations of initial values that occurs when a root $\zeta(h\lambda)$ is greater in magnitude than 1 is not tolerable for very many steps.

We have spoken of perturbations to the initial values, but in fact they could be perturbations to any k successive values because we can regard these values as initial values for the rest of the integration. We have not yet raised the matter of perturbations to the recipe. Suppose the recipe is perturbed only once, say when $\mathbf{z}_{j+1}$ is computed. We then have $\mathbf{z}_m = \mathbf{y}_m$

for $m \leq j$, $\mathbf{z}_{j+1}$ satisfies the perturbed recipe

$$\sum_{i=0}^{k} \alpha_i \, \mathbf{z}_{j+1-i} = h \, \mathbf{\Phi}(x_j, \mathbf{z}_{j+1}, \ldots, \mathbf{z}_{j+1-k}, \mathbf{F}, h) + h \, \boldsymbol{\delta}_j,$$

and $\mathbf{z}_m$ satisfies the unperturbed recipe (1.1) for $m > j + 1$. The perturbation causes $\mathbf{z}_{j+1}$ to differ from the value $\mathbf{y}_{j+1}$ computed from (1.1). The effects of perturbing the recipe at x_j are the same as those of perturbing the solution value $\mathbf{y}_{j+1}$ to $\mathbf{z}_{j+1}$ and using the unperturbed recipe thereafter, a situation we already understand. This equivalence allows us to understand a perturbation to the recipe at a single step and by linearity the effect of a perturbation to the recipe at another step is just additive. If the integration is stable with respect to perturbations of the solution, the effects of disturbances to the recipe might add up unpleasantly, but they will not "explode" like those of an unstable integration. All this, of course, depends on linearity and is only approximately true for nonlinear problems and only for "small" perturbations at that.

A method is said to be *absolutely stable* for a given $h\lambda$ with $\text{Re}(\lambda) < 0$ if all roots of its characteristic polynomial $\rho(\theta; h\lambda)$ lie inside the unit circle in the complex plane. The set of all such $h\lambda$ is said to be the *region of absolute stability* of the method. Sometimes this concept is defined so that roots can have magnitude 1 and sometimes not; it makes little difference provided that the case of roots of magnitude 1 is kept in mind. The qualifier "absolute" is often dropped. To apply the concept of *stability region* to the stability of the integration of systems of the form (2.1), it is clear that we have stability if, and only if, $h\lambda$ is in the (absolute) stability region of the method for all eigenvalues λ of the Jacobian matrix J.

If we had applied the numerical method to the solution of the homogeneous scalar equation $y' = \lambda y$, we would have arrived directly at the constant coefficient, scalar difference equation

$$\sum_{i=0}^{k} (\alpha_i - h\lambda \, R_i(h\lambda)) y_{j+1-i} = 0.$$

Absolute stability for a given $h\lambda$ corresponds to all solutions $\{y_j\}$ of this difference equation being bounded uniformly in j. The equation $y' = \lambda y$ for complex λ is often called the *test equation*. Clearly the stability region for a method can be determined by investigating the test equation alone. Stability can be interpreted as a kind of accuracy requirement for the numerical method when applied to the test equation because it amounts to requiring some qualitative agreement with the true solution: If $h\lambda$ corresponds to being inside the stability region, the numerical solution decreases in magnitude just like the true solution does and if $h\lambda$ corresponds to being outside the region, the numerical solution increases, so does not agree even qualitatively with the true solution.

The class of equations (2.1) is a compromise between generality and our ability to work out enough details to find out what is going on. Certainly it is reasonable to insist that the method perform well on the special class of equations (2.1), but it must be kept in mind that this is only a rough guide to the behavior of the method for more realistic problems. Let us now consider why we might think it provides even a rough guide. For a general problem in autonomous form, $\mathbf{y}' = \mathbf{F}(\mathbf{y})$, we can approximate $\mathbf{F}$ near $(x_j, \mathbf{y}(x_j))$ by a linear function of $\mathbf{y}$,

$$\mathbf{F}(\mathbf{y}) \approx \mathbf{F}(\mathbf{y}(x_j)) + \frac{\partial \mathbf{F}}{\partial \mathbf{y}} (\mathbf{y}(x_j))(\mathbf{y} - \mathbf{y}(x_j)),$$

and hope that the behavior of the numerical solution of the original problem can be modeled near $(x_j, \mathbf{y}(x_j))$ by the solution of an approximating problem, $\mathbf{u}' = J\,\mathbf{u} + \mathbf{g}$, with constant Jacobian

$$J = \frac{\partial \mathbf{F}}{\partial \mathbf{y}} (\mathbf{y}(x_j))$$

and constant forcing function

$$\mathbf{g} = \mathbf{F}(\mathbf{y}(x_j)) - J\,\mathbf{y}(x_j).$$

The approximating problem is of the kind we have analyzed in terms of h and the eigenvalues λ_m of the constant Jacobian. It is assumed that the stability of the method applied to the approximating problem is the same as the stability of the method applied to the general problem. This is not always true, but the approach furnishes guidelines that have been found to be quite useful in practice. Of course, if the approximation is to be a reasonable one, it is obviously necessary to consider an interval (in x) on which $\mathbf{y}(x)$ and the local Jacobian do not change much from their values at x_j. This approach is a classic one in the theory of differential equations—the stability of solutions of a non-linear problem is investigated by determining the stability of solutions of a linear approximating problem. The only new element here is that in addition, we must consider the stability of the numerical scheme for integrating the approximating problem. It is recognized in the theory of ordinary differential equations that linear stability analysis is a valuable tool, but that it has limitations. The same is true in the present application of the technique.

EXERCISE 4. Lambert [1980] presents a series of examples showing that a linearized stability analysis can be misleading, even when the equation has the simple form $\mathbf{y}' = J(x)\mathbf{y}$. It is assumed that the qualitative behavior of solutions of the given problem is the same as that of the problem that arises by "freezing" the coefficients, i.e., replacing the variable Jacobian $J(x)$ in the equation with the constant matrix $J(x_n)$. Suppose that x is positive and

$$J(x) = \begin{pmatrix} 0 & 1 \\ -\dfrac{1}{16x^2} & -\dfrac{1}{2x} \end{pmatrix}.$$

Show that the eigenvalues of $J(x_m)$ are real and negative. This implies that all solutions $\mathbf{v}(x)$ of the approximating problem $\mathbf{v}' = J(x_n)\mathbf{v}$ have $\|\mathbf{v}(x)\| \to 0$ as $x \to \infty$. Show that the solutions of the given problem are

$$\mathbf{y}(x) = \alpha\begin{pmatrix} 4x^{1/4} \\ x^{-3/4} \end{pmatrix} + \beta\begin{pmatrix} 4x^{1/4}\ln(x) \\ x^{-3/4}(4 + \ln(x)) \end{pmatrix}.$$

Argue that all non-trivial solutions $\mathbf{y}(x)$ have $\|\mathbf{y}(x)\| \to \infty$ as $x \to \infty$.

In a linearized stability analysis the eigenvalues λ_m of a local Jacobian are crucial. The behavior of nearby solution curves is approximated after a change of variables by solutions of equations of the form $w_m' = \lambda_m w_m + t_m$. Perturbations to such solutions grow or decay at an exponential rate of $\mathrm{Re}(\lambda_m)$. The *time constant* of such an equation is the time, or here the distance x, required for a perturbation to decay by a factor of $\exp(-1)$. The term is used loosely in the present context to refer to the exponential rates of growth or decay, $\mathrm{Re}(\lambda_m)$. An eigenvalue with positive real part indicates instability of the equation and an eigenvalue with "large" negative real part indicates a "super-stability" of the equation that often imposes a step size restriction on the numerical method. It is important to recognize the presence of such eigenvalues and physical insight is often useful. As an example, think of modeling a fluid flow with a chemical reaction taking place in the fluid. Typically reactions will occur very much faster than the movement of the fluid. The different time scales mean that according to a linear approximation, solutions evolve with different time constants and in this case, very different time constants. Provided that the physical system is stable with respect to small changes of the data, we have reason to expect that local Jacobians will have eigenvalues with non-positive real parts and that some eigenvalues will have real parts that are negative and "large" in magnitude. If, on the other hand, we are modeling an explosion, we would expect that small changes in the data would lead to significant changes in the solution, hence we would expect some eigenvalues with significant positive real parts.

EXAMPLE 2. The solutions of Airy's equation, $y'' = xy$, were discussed in Example 5 of Chapter 2. Let us now consider a linearized stability analysis of the equation. First we write the equation as an autonomous system of equations by introducing the variables $y_1 = y$, $y_2 = y'$, $y_3 = x$, and differentiate with respect to t $(=x)$:

$$\frac{d}{dt} y_1 = y' = y_2,$$

$$\frac{d}{dt} y_2 = y'' = xy = y_1 y_3,$$

$$\frac{d}{dt} y_3 = \frac{d}{dx} x = 1.$$

The Jacobian matrix is then

$$\frac{\partial \mathbf{F}}{\partial \mathbf{y}} = \begin{pmatrix} 0 & 1 & 0 \\ y_3 & 0 & y_1 \\ 0 & 0 & 0 \end{pmatrix},$$

the characteristic equation is

$$|\mathbf{F}_{\mathbf{y}} - \lambda I| = -\lambda(\lambda^2 - y_3) = 0,$$

and the eigenvalues are $\lambda = 0$ and $\lambda = \pm\sqrt{y_3} = \pm\sqrt{x}$. In an exercise this equation was to be integrated over the interval $[0, 1]$. This corresponds to the classical situation, in agreement with the fact that the eigenvalues of the Jacobian at x are no bigger than 1 in magnitude for all x in the interval. Notice that there are both positive and negative eigenvalues. In Chapter 2 we discussed the stability of the equation and found that for large x, it is (very) unstable. This is in agreement with the linearized stability analysis for there is a positive eigenvalue that is large then. The issue of integrating the equation for negative x was not taken up earlier, but it is quite interesting because solutions oscillate increasingly rapidly as x becomes large and negative. Here we observe that for negative x, the two non-trivial eigenvalues are $\lambda = \pm i\sqrt{|x|}$, that is, they are pure imaginary numbers and for large x they have large magnitude.

EXERCISE 5. The two-body problem can be written as the first-order system

$$\begin{aligned} y_1' &= y_2 & y_1(0) &= 1 - \varepsilon \\ y_2' &= -y_1/r^3 & y_2(0) &= 0 \\ y_3' &= y_4 & y_3(0) &= 0 \\ y_4' &= -y_3/r^3 & y_4(0) &= \sqrt{\frac{1+\varepsilon}{1-\varepsilon}} \end{aligned}$$

$$\text{where } r^2 = y_1^2 + y_3^2.$$

Show that the Jacobian evaluated at the initial point is

$$\begin{pmatrix} 0 & 1 & 0 & 0 \\ \dfrac{2}{(1-\varepsilon)^2} & 0 & 0 & 0 \\ 0 & 0 & 0 & 1 \\ 0 & 0 & \dfrac{-1}{(1-\varepsilon)^2} & 0 \end{pmatrix}$$

and then that the eigenvalues are $\pm(2/(1-\varepsilon)^3)^{1/2}$, $\pm(-1/(1-\varepsilon)^3)^{1/2}$. Using the linear stability theory, compare the stability of this two-body problem at its initial point (a time when the two bodies are closest) for the test problems $D1$ and $D5$ which have $\varepsilon = 0.1$ and 0.9, respectively. Some numerical results for the solution of $D5$ are presented in Figure 2.7 in Chapter 2. This exercise will help you to understand why the code had to use a small step size at the initial point.

We have already considered what happens when the forward and backward Euler methods are applied to the test equation in the case of λ that are real and negative. Permitting complex λ, the forward Euler method, alias AB1, applied to the test equation $y' = \lambda y$ is

$$z_{j+1} = z_j + h\lambda\, z_j,$$

or

$$z_{j+1} + (-1 - h\lambda)z_j = 0.$$

From this equation we see that

$$\rho(\theta; h\lambda) = \theta + (-1 - h\lambda),$$

and obviously the only root of the characteristic equation is

$$\zeta = 1 + h\lambda.$$

The stability region is the set of complex $h\lambda$ with $\operatorname{Re}(h\lambda) < 0$ such that

$$|1 + h\lambda| < 1.$$

This is a disk of radius 1 centered about $(-1, 0)$. For an eigenvalue λ with imaginary part large compared to its real part, i.e., λ near the imaginary axis, h has to be rather small for the computation to be stable.

The backward Euler method, alias AM1 and BDF1, has

$$z_{j+1} = z_j + h\lambda\, z_{j+1},$$

hence

$$\rho(\theta; h\lambda) = \theta(1 - h\lambda) - 1.$$

There is a single root

$$\zeta = \frac{1}{1 - h\lambda}.$$

This root has magnitude less than one *outside* a disk of radius 1 centered at $(+1, 0)$, so the stability region includes the whole left half complex plane. Methods that are stable for all λ with $\operatorname{Re}(\lambda) < 0$ are called *A-stable*. This dramatic difference in the stability of the forward and backward Euler methods helped earlier to motivate the consideration of implicit methods. The fact that the implicit Euler method has no restriction on the step size to get stability for the class of problems we consider is not an unqualified blessing. Remember that the differential equation itself is not stable when $\operatorname{Re}(\lambda) > 0$. For such a problem, the step size must be restricted so that $h\lambda$ is in the disk where the method is *un*stable if there is to be qualitative agreement with the solution of the test equation. The backward Euler method is BDF1 and this behavior is also true of the higher order BDF.

EXAMPLE 3. The trapezoidal rule is stable for all $\operatorname{Re}(h\lambda) < 0$, a fact that is left as an exercise. Some people take this too seriously and think that the method never has a restriction on the step size due to stability. One reason for thinking this is that many treatments of stability do not first convert the equations to autonomous form, hence study a problem like $y' = \lambda(x)y$ by means of the approximating equation $y' = \lambda(x_n)y$. It is then concluded that if $\lambda(x) < 0$ for all x, the trapezoidal rule is stable for all step sizes h. This approach is consistent with approximations we have made elsewhere, but a careful look at this simple problem instills a measure of caution in accepting the results of an approximate analysis. As Gourlay [1970] points out, there may be a restriction on the step size if the trapezoidal rule is to be stable. When the trapezoidal rule is applied to the equation $y' = \lambda(x)y$,

$$y_{n+1} = \frac{1 + \frac{h}{2}\lambda(x_n)}{1 - \frac{h}{2}\lambda(x_{n+1})} y_n.$$

For stability the factor must have magnitude no greater than 1. Considering that $h > 0$ and $\lambda(x) < 0$, it is clear that the factor is never greater than $+1$. However, it is smaller than -1 when

$$2 > -\frac{h}{2}[\lambda(x_n + h) - \lambda(x_n)] = -\frac{h^2}{2}\lambda'(\xi),$$

and if $\lambda'(\xi) < 0$, this represents a restriction on the step size due to stability.

The restriction is not surprising when the problem is written in autonomous form,

$$\frac{dy}{dt} = \lambda(x)y,$$

$$\frac{dx}{dt} = 1.$$

A linearization approximates

$$\begin{pmatrix} \lambda(x)y \\ 1 \end{pmatrix} \approx \begin{pmatrix} \lambda(x_n)y_n \\ 1 \end{pmatrix} + \begin{pmatrix} \lambda(x_n) & \lambda'(x_n)y_n \\ 0 & 0 \end{pmatrix} \begin{pmatrix} y - y_n \\ x - x_n \end{pmatrix}$$

or

$$\lambda(x)y \approx \lambda(x_n)y_n + \lambda(x_n)(y - y_n) + \lambda'(x_n)y_n(x - x_n).$$

In this form it is clear that how fast $\lambda(x)$ can change near x_n affects the validity of the linearization and correspondingly the conclusion that there is no restriction on the step size due to stability.

EXERCISE 6. Show that the trapezoidal rule is A-stable. Further show that it is unstable for all $\text{Re}(h\lambda) > 0$. In this sense the trapezoidal rule matches the stability properties of the test equation better than does the backward Euler method. On the other hand, for $\text{Re}(h\lambda)$ that are negative and "large," perturbations are damped out extremely fast in the differential equation. Show that the backward Euler method imitates this behavior (much) better than the trapezoidal rule. (Unpleasant oscillations can appear and persist for long periods when using the trapezoidal rule in certain circumstances.)

EXAMPLE 4. Despite the cautionary note sounded by the last example, a linearized analysis can be very helpful in explaining computational results. To illustrate this let us use Euler's method to integrate

$$y' = -\left(2 + \frac{10}{1 + (x - 10)^4}\right)y + x, \qquad 0 \le x \le 20, \qquad y(0) = 0,$$

with a constant step size h. According to the linear analysis, this method is stable when $|1 + h\lambda| \le 1$ for all eigenvalues λ of the local Jacobian. In the case of this scalar problem, λ is the Jacobian itself. Here the Jacobian has a peak at $x = 10$, and as Figure 6.2a shows, if the step size is 0.25, the integration is unstable near the peak. On the other hand, if the step size is 0.04, the integration is stable for the whole integration. Figure 6.2b shows the results of the two integrations. The integration with the smaller step size provides a good approximation to the solution that is connected by straight line segments in the figure. The integration with the larger step size shows a growing oscillation near $x = 10$ that is typical of instability. On the far side of the peak the problem changes enough for the integration

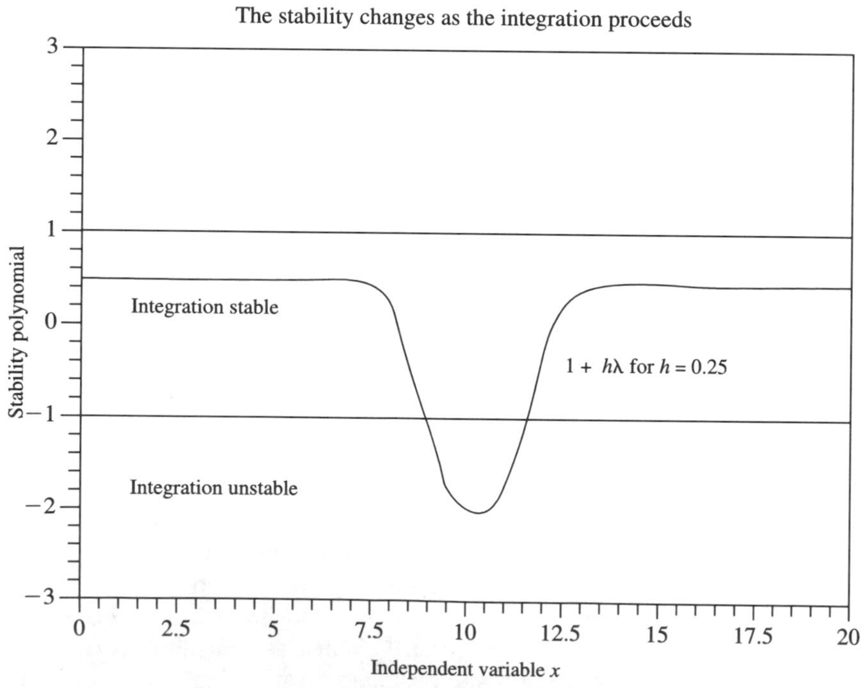

Figure 6.2a

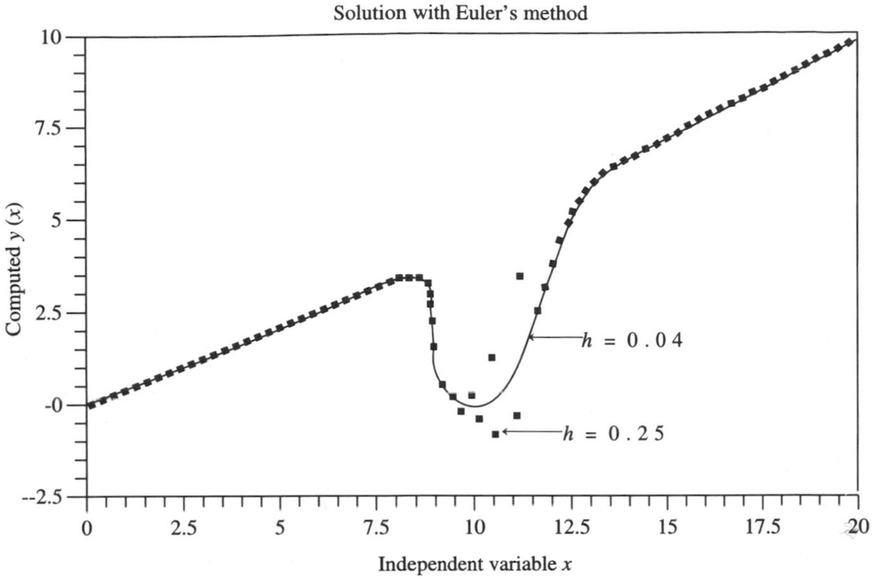

Figure 6.2b

to be stable with this step size and it is observed that the effects of instability are actually damped out. This should serve as a warning: It is quite possible to obtain results that look good, and even be good, in one portion of the integration and still have unacceptable results elsewhere. Notice that where the integration is stable, there is little difference between the two results, confirming that the difficulty with the larger step size is not the accuracy of the formula.

EXERCISE 7. One of the problems in the test set of Hull et al. [1972] is $y' = -0.5y^3$, $y(0) = 1$ which is to be integrated on [0, 20]. It is easily found that the solution is $(x + 1)^{-1/2}$. Although this problem seems innocuous, the comparison of Shampine, Watts and Davenport [1976] reports that when one of the codes was provided the length of the interval of integration as an initial step size—not an unreasonable guess for the particular code being investigated—the integration became unstable. Explain why a code might encounter difficulties with stability when solving this problem if the approximate solution is very poor.

EXERCISE 8. The knee problem presented in Example 3 of Chapter 3 is

$$\varepsilon \frac{dy}{dx} = (1 - x)y - y^2, \qquad 0 \le x \le 2, \qquad y(0) = 1.$$

As pointed out there, the solution is close to the null isocline $u(x) = 1 - x$ on $[0, 1]$ and close to the null isocline (and solution) $u(x) \equiv 0$ on $[1, 2]$. The parameter ε satisfies $0 < \varepsilon \ll 1$. For the computations of the example, $\varepsilon = 10^{-4}, 10^{-6}$. Show that for these values of ε and some $\delta > 0$, the equation is very stable along the isocline $u(x) = 1 - x$ for $x \leq 1 - \delta$ and very unstable for $x \geq 1 + \delta$. Similarly, show that the equation is very unstable along the solution $u(x) \equiv 0$ for $x \leq 1 - \delta$ and very stable for $x \geq 1 + \delta$.

The only examples we have seen so far of methods that are stable for "large" $h\lambda$ are implicit. This is not accidental; there are no standard methods that are explicit and have an infinite stability region. Let us go into this important fact for the classes of methods that we have been studying. First we take up explicit Runge–Kutta methods. The characteristic polynomial is then

$$\rho(\theta; h\lambda) = \theta - 1 - h\lambda\, R_1(h\lambda),$$

which has the single root

$$\zeta_1(h\lambda) = 1 + h\lambda\, R_1(h\lambda).$$

We have seen that $R_1(h\lambda)$ is a polynomial in $h\lambda$, so $h\lambda\, R_1(h\lambda)$ is a nonconstant polynomial in $h\lambda$. As $|h\lambda| \to \infty$, so does $|h\lambda\, R_1(h\lambda)|$. Accordingly, for all sufficiently large $|h\lambda|$, it must be the case that $|\zeta_1| > 1$ and $h\lambda$ is outside the stability region—the stability region is finite.

A LMM has a characteristic polynomial of the form

$$\rho(\theta; h\lambda) = \rho(\theta) - h\lambda\, \sigma(\theta) = \sum_{i=0}^{k} (\alpha_i - h\lambda\beta_i)\theta^{k-i},$$

where $\rho(\theta)$ is the first characteristic polynomial of the method and $\sigma(\theta)$ is the second. Because we consider only explicit methods here, $\beta_0 = 0$ and the coefficient of θ^k is $\alpha_0 \neq 0$. With this observation and denoting the k roots of the characteristic polynomial by $\zeta_1, \ldots, \zeta_k$, we can write

$$\rho(\theta; h\lambda) = \alpha_0 \prod_{m=1}^{k} (\theta - \zeta_m).$$

This implies that

$$|\rho(1; h\lambda)| = |\alpha_0| \prod_{m=1}^{k} |1 - \zeta_m|.$$

For a zero-stable method we have $\rho(1) = 0$ and $\rho'(1) = \sigma(1) \neq 0$ (the principal root is simple). Let us suppose that the method is absolutely stable for a given $h\lambda$, i.e., $|\zeta_m| < 1$ for each m. Then

$$|\rho(1; h\lambda)| < |\alpha_0| 2^k.$$

On the other hand,

$$|\rho(1; h\lambda)| = |\rho(1) - h\lambda\, \sigma(1)| = |0 - h\lambda\, \rho'(1)| = |h\lambda|\, |\rho'(1)|.$$

Because $\rho'(1) \neq 0$, we then have

$$|h\lambda| < \frac{|\alpha_0| 2^k}{|\rho'(1)|}.$$

This bound on the magnitude of the $h\lambda$ for which the formula is stable shows that the stability region is finite.

The argument for PECE predictor–corrector pairs is quite similar to that for explicit LMMs. If all roots ζ_m of $\rho(\theta; h\lambda)$ have magnitude less than one, then as with a LMM,

$$|\rho(1; h\lambda)| < |\alpha_0| 2^k.$$

For these formulas

$$\rho(1; h\lambda) = \rho(1) - h\lambda\, \sigma(1) + h\lambda \frac{\beta_0}{\alpha_0^*} [\rho^*(1) - h\lambda\, \sigma^*(1)].$$

The nature of the method requires $\beta_0 \neq 0$, $\alpha_0^* \neq 0$ and zero-stability requires $\rho(1) = 0$, $\rho^*(1) = 0$, $\sigma(1) = \rho'(1) \neq 0$. There are two cases. If $\sigma^*(1) = 0$, then as before,

$$|\rho(1; h\lambda)| = |h\lambda|\, |\rho'(1)|$$

leads to a bound on stable $|h\lambda|$. If $\sigma^*(1) \neq 0$, then

$$\rho(1; h\lambda) = -h\lambda\, \rho'(1) - (h\lambda)^2 \frac{\beta_0}{\alpha_0^*} \sigma^*(1).$$

We have a bound on $|\rho(1; h\lambda)|$ that does not depend on $|h\lambda|$. On the other hand, this last expression shows that $|\rho(1; h\lambda)|$ grows like $|h\lambda|^2$. The stability region cannot be finite, else we would have stable $h\lambda$ of arbitrarily large magnitude for which the bound is to hold, and this is not possible.

Let us collect these results as a theorem:

Theorem 2. *All zero-stable explicit Runge–Kutta methods, explicit linear multistep methods, and explicit predictor–corrector pairs in PECE implementation have finite regions of absolute stability.*

The continuity of the roots $\zeta_m(h\lambda)$ of $\rho(\theta; h\lambda)$ provides useful insight when $|h\lambda|$ is "small" and when it is "large." As $|h\lambda| \to 0$, $\rho(\theta; h\lambda) \to \rho(\theta)$

in the cases of LMMs and the PECE implementations of predictor–corrector pairs. Because we assume that $\rho(\theta)$ satisfies the root condition, for all sufficiently small $|h\lambda|$, all the non-essential roots of $\rho(\theta)$ are approximated by roots of $\rho(\theta; h\lambda)$ sufficiently well that the $\zeta_m(h\lambda)$ have magnitude less than 1. This illuminates the possibilities for instability, but it does not settle the marginal case of essential roots. Because we always have the principal root of magnitude 1, we cannot ignore this marginal case.

Essential roots *can* cause trouble. The midpoint rule,

$$\mathbf{y}_{j+1} - \mathbf{y}_{j-1} = 2h\,\mathbf{F}(x_j, \mathbf{y}_j),$$

furnishes an example. Obviously, $\rho(\theta) = \theta^2 - 1$ and there are two essential roots ± 1. The characteristic polynomial

$$\rho(\theta; h\lambda) = \theta^2 - 2h\lambda\,\theta - 1$$

has the two roots

$$\pm 1 + h\lambda + \mathcal{O}(h^2).$$

If we write the complex number λ in the form $a + bi$, then

$$\begin{aligned} |-1 + h\lambda + \mathcal{O}(h^2)| &= |(-1 + ha + \mathcal{O}(h^2)) + i(hb + \mathcal{O}(h^2))| \\ &= \sqrt{(-1 + ha + \mathcal{O}(h^2))^2 + (hb + \mathcal{O}(h^2))^2} \\ &= \sqrt{1 - 2ha + \mathcal{O}(h^2)} \\ &= 1 - ha + \mathcal{O}(h^2). \end{aligned}$$

For $a = \mathrm{Re}(\lambda) < 0$, this says that for all sufficiently small h, one of the roots has magnitude bigger than one. Thus all sufficiently small $|h\lambda|$ with $\mathrm{Re}(\lambda) < 0$ correspond to points outside the stability region—the method is not stable. This is disconcerting because the method is zero-stable! This example helps us appreciate that the limits taken are different. In the one case, the step size is fixed and we ask how perturbations propagate as more steps are taken. In the other case, the interval is fixed and we ask about the maximum size of perturbations in a sequence of integrations with the step size tending to zero.

EXERCISE 9. Solve $y' = -y$, $y(0) = 1$ with the midpoint rule and constant step size h. Use the exact solution to get the necessary starting values. Solve the problem on a fixed interval, say [0, 10], with a sequence of step sizes h tending to 0. You will find that the results are poor until the step size is sufficiently small. Explore the reason for the poor results for "large" h by comparing the results obtained with a large h, e.g., $h = 1$, to the results obtained with the same h and an initial value that is a little different from 1. You will find that the initial perturbation grows as the integration proceeds.

All convergent methods have the principal root $+1$. Does this root of magnitude 1 cause the kind of trouble we just saw in the example of the midpoint rule? To get some insight, let us first look informally at the case of LMMs. We are interested in roots of $\rho(\theta; h\lambda) = \rho(\theta) - h\lambda\,\sigma(\theta) = 0$ that are near $+1$, so let us look for a root in the form $\zeta_1(h\lambda) = 1 + \gamma h\lambda + \mathcal{O}(h^2)$. That is, we try to find a constant γ such that this represents a root. Taylor series expansion shows that $\rho(\theta) = \rho(1) + \gamma h\lambda\rho'(1) + \mathcal{O}(h^2)$ and $\sigma(\theta) = \sigma(1) + \mathcal{O}(h)$, hence that

$$\rho(\theta; h_1) = \rho(1) + h\lambda[\gamma\rho'(1) - \sigma(1)] + \mathcal{O}(h^2) = 0.$$

Because the principal root $+1$ is simple, $\rho(1) = 0$ and $\rho'(1) = \sigma(1)$. If we are to have a root of the specified form as $h \to 0$, this expression says that we must have $\gamma = 1$. For small h then, there is one root near $+1$ that has the form $1 + h\lambda + \mathcal{O}(h^2)$. Arguing as in the example of the midpoint rule, it follows that if $\mathrm{Re}(\lambda) > 0$, so that the differential equation is unstable, then for all sufficiently small $|h\lambda|$, the point $h\lambda$ is outside the stability region and the method is unstable, as it should be. Similarly, if $\mathrm{Re}(\lambda) < 0$ so that the differential equation is stable, then for all sufficiently small $|h\lambda| \neq 0$, we have $|\zeta_1(h\lambda)| < 1$ and the method is stable. We see, then, that this root approximating the principal root does not contribute to instability (which, however, is not to say that the method is not unstable due to some other root of the characteristic polynomial).

EXERCISE 10. Investigate predictor–corrector (PECE) methods in a similar way.

When an explicit Runge–Kutta method is applied to the test equation $y' = \lambda y$ with $y(0) = 1$ and we take $y_0 = 1$, we find that $y_1 = (1 + h\lambda\, P_s(h\lambda))y_0 = \zeta_1(h\lambda)$. If the method is accurate of order p, then $y_1 = y(h) + \mathcal{O}(h^{p+1})$. Substituting for y_1 and $y(h)$, we obtain

$$\zeta_1(h\lambda) = \exp(h\lambda) + \mathcal{O}(h^{p+1}) = 1 + h\lambda + \mathcal{O}(h^2),$$

just what we need to conclude that the principal root causes no difficulty with stability for small h. The approach can be generalized to provide a formal proof of the results just derived informally for LMM and predictor–corrector methods. Using the special form that Φ has when applied to $y' = \lambda y$, we find that for a LMM,

$$\sum_{i=0}^{k} \alpha_i\, y(x_{j+1-i}) = h\lambda \sum_{i=0}^{k} R_i(h\lambda) y(x_{j+1-i}) + \mathcal{O}(h^{p+1}).$$

For this differential equation the solution $y(x_{j+1-i}) = \exp((k - i)h\lambda)y(x_{j+1-k})$, so

$$\left[\sum_{i=0}^{k} (\alpha_i - h\lambda\, R_i(h\lambda))(e^{h\lambda})^{k-i}\right] y(x_{j+1-k}) = \mathbb{O}(h^{p+1}).$$

We identify the factor in brackets as $\rho(\exp(h\lambda); h\lambda)$ and conclude that it is $\mathbb{O}(h^{p+1})$. We have already written

$$\rho(\theta; h\lambda) = \alpha_0 \prod_{m=1}^{k} (\theta - \zeta_m(h\lambda))$$

in terms of the roots $\zeta_m(h\lambda)$ of the polynomial $\rho(\theta; h\lambda)$. Combining the results, we have

$$\mathbb{O}(h^{p+1}) = \rho(e^{h\lambda}; h\lambda) = \alpha_0(e^{h\lambda} - \zeta_1(h\lambda)) \prod_{m=2}^{k} (e^{h\lambda} - \zeta_m(h\lambda)).$$

The principal root is simple, meaning that $\zeta_m(0) \neq 1$ for $m > 1$. This implies that each of the factors with $m \geq 2$ has the form $(1 - \zeta_m(0) +$ higher order terms in $h\lambda)$. Accordingly, the only way the right hand side can be $\mathbb{O}(h^{p+1})$ is for the factor $(\exp(h\lambda) - \zeta_1(h\lambda))$ to be $\mathbb{O}(h^{p+1})$. This is an interesting observation in its own right because it shows that the principal root is the one approximating the behavior of the solution of the differential equation. Recasting the result in the form

$$\zeta_1(h\lambda) = \exp(h\lambda) + \mathbb{O}(h^{p+1}) = 1 + h\lambda + \mathbb{O}(h^2)$$

shows that the principal root does not cause any difficulties with absolute stability for small h.

EXERCISE 11. The characteristic polynomials of explicit Runge–Kutta methods have only one root. Further, a method of s stages has a $\rho(\theta; h\lambda)$ that is a polynomial of degree at most s in the variable $h\lambda$. Use the expression just derived for $\zeta_1(h\lambda)$ to conclude that all explicit Runge–Kutta methods of order p of p stages have the *same* region of absolute stability. Hint: Show that they have the same stability polynomial.

The exercise says that all explicit Runge–Kutta methods of p stages and order p have the same absolute stability region. Examples of such methods were given in Chapter 4 for $p = 1, 2, 3, 4$. Their stability regions are displayed in Figure 6.3. It turns out to be impossible to achieve order p with only p stages for $p \geq 5$ (Butcher [1987]).

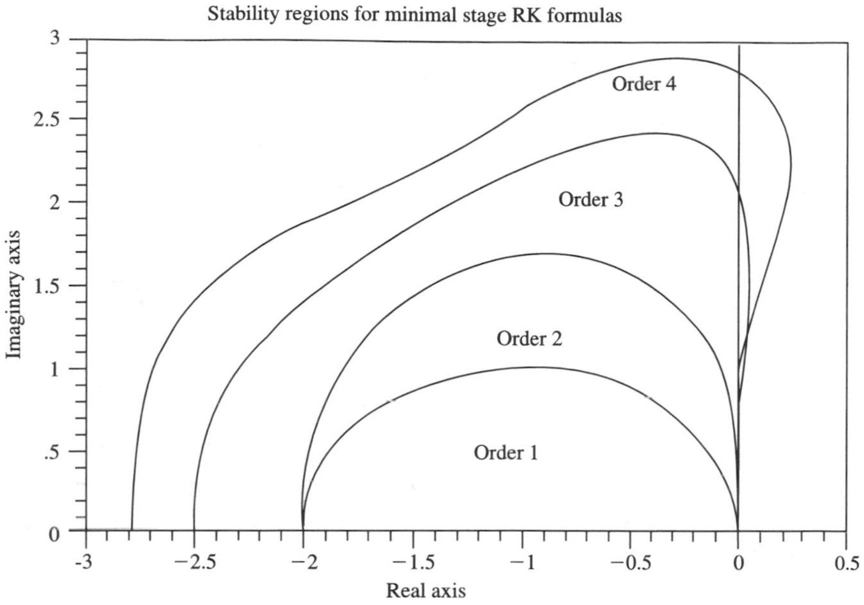

Figure 6.3

Theorem 3. *All strongly stable linear multistep methods are absolutely stable for all sufficiently small $|h\lambda|$. The same is true of all strongly stable predictor–corrector pairs in PECE implementation.*

Proof By definition, a strongly stable method has only the principal root and non-essential roots. We have just seen that for $\mathrm{Re}(\lambda) < 0$ and sufficiently small $|h\lambda|$, the principal root $\zeta_1(h\lambda)$ has magnitude smaller than 1. By continuity, the other roots tend to non-essential roots of $\rho(\theta)$ as $h \to 0$, hence have magnitude smaller than 1 for all sufficiently small $|h\lambda|$.

From the point of view of absolute stability, formulas with $\rho(\theta) = \alpha_0(\theta^k - \theta^{k-1})$ are especially interesting. Because they have only the principal root and 0 as a (multiple) non-essential root, it seems likely that the stability region will be comparatively large for non-zero $h\lambda$ and that perturbations will be strongly damped, at least for small $h\lambda$. We call such methods *Adams-like*. Another situation of great interest is the possibility of an infinite stability region, such as that of the backward Euler formula. In general the BDFs have the form

$$\sum_{i=0}^{k} \alpha_i \, \mathbf{y}_{j+1-i} = h \, \mathbf{F}(x_{j+1}, \mathbf{y}_{j+1}).$$

We call methods with characteristic polynomials of the general form

$$\rho(\theta; h\lambda) = \rho(\theta) - h\lambda\, \beta_0\, \theta^k$$

for $\beta_0 \neq 0$ *BDF-like.*

THEOREM 4. *A BDF-like method is absolutely stable for all sufficiently large $|h\lambda|$.*

Proof The characteristic equation of a BDF-like method can be written as

$$\frac{1}{h\lambda}\, \rho(\theta) - \beta_0\, \theta^k = 0.$$

By continuity, as $|h\lambda| \to \infty$, the roots $\zeta_m(h\lambda)$ of this equation tend to the roots of $-\beta_0\, \theta^k = 0$, which are all zero. This says that for all sufficiently large $|h\lambda|$, all the roots $\zeta_m(h\lambda)$ are less than one in magnitude and the formula is absolutely stable.

As we noted in connection with BDF1, the result of this theorem is not exactly what we would prefer because the equation itself is unstable when $\mathrm{Re}(\lambda) > 0$. For "large" $h\lambda$, the BDFs play the same role that the Adams methods play for "small" $h\lambda$. In a sense, they are as stable as we can get within the family of LMMs.

Generally the region of absolute stability is determined numerically. One way to proceed is to choose a λ and then vary the real parameter h until it is found that the characteristic polynomial has a root of magnitude 1. This locates a point on the boundary of the stability region. By taking $\lambda = e^{i\psi}$ and varying the real angle ψ, the whole boundary can be determined. This is a convenient way to proceed for explicit Runge–Kutta methods because computation of the one root of the characteristic polynomial is trivial. The boundary locus method is a related way to proceed that is more practical for other methods. On the boundary of the stability region, the characteristic polynomial has a root of magnitude 1, i.e., the root has the form $e^{i\psi}$. The idea is to select an angle ψ and then to compute a complex number $h\lambda$ such that $e^{i\psi}$ is a root of the characteristic equation. As ψ is varied from 0 to 2π, these $h\lambda$ provide the boundary of the stability region. Naturally, an efficient computation uses the fact that the boundary point $h\lambda$ for one ψ is a good starting guess for an iterative procedure for computing a boundary point for a value ψ^* close to ψ. The boundary locus method is especially convenient for LMMs because then we seek $h\lambda$ such that

$$0 = \rho(e^{i\psi}; h\lambda) = \rho(e^{i\psi}) - h\lambda\, \sigma(e^{i\psi}),$$

hence

$$h\lambda = \rho(e^{i\psi})/\sigma(e^{i\psi})$$

is the point on the boundary of the stability region. We have no need to go into details about the determination of stability regions, but an example will show that a little care may be required.

Figure 6.4 shows the boundary locus for AB4—the curve labeled "magnitude 1.0." Because the curve intersects itself, it is not clear whether the lobes in the right half complex plane represent regions where the method is stable or not. One way to see that the method is not stable in these lobes is to modify the program for plotting the boundary locus so as to plot the curve corresponding to $1.1e^{i\psi}$ rather than $e^{i\psi}$. This second curve gives us the $h\lambda$ for which the characteristic equation has a root of specific form and magnitude 1.1. The fact that portions of this curve lie within the lobes tells us that the characteristic equation has roots of magnitude 1.1 there, hence

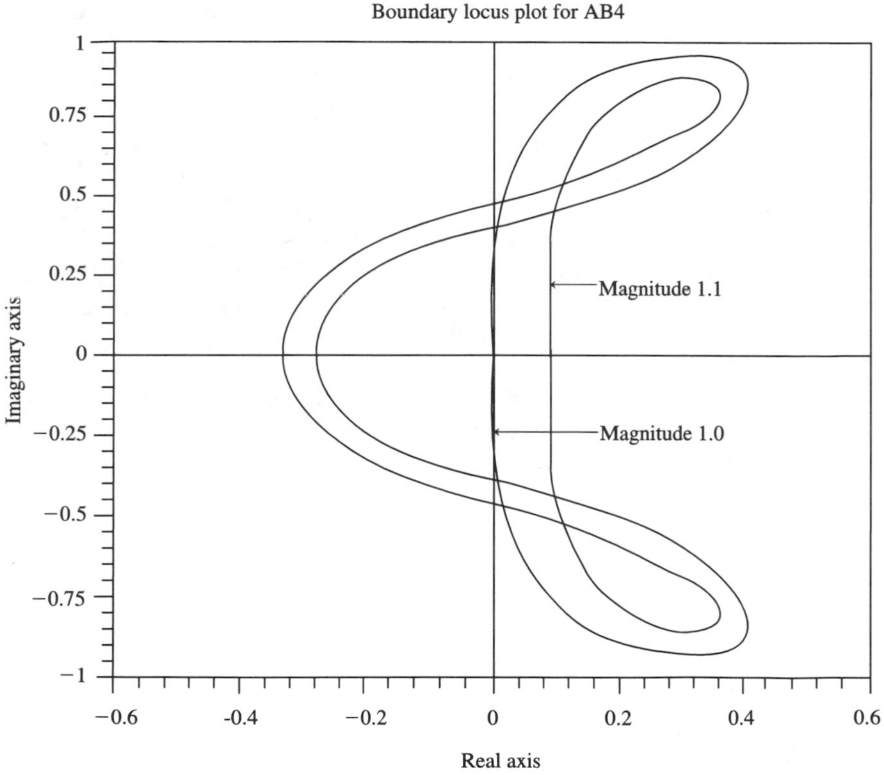

Figure 6.4

that the method is unstable in the lobes. This is what we would prefer, because the differential equation itself is unstable in the right half complex plane. Although it is not clear in the plot, AB4 *is* stable in a (very) small portion of the right half complex plane.

EXERCISE 12. MATLAB is a very convenient tool for computing the stability region of a LMM. The language provides for polynomials, which makes it easy to define the ρ and σ polynomials of the LMM. It is straightforward to write an M-file of just a few lines that will accept these polynomials, compute, say, 100 points on the boundary of the stability region of the LMM, and return these points in a form suitable for plotting. After invoking the M-file, you can see the boundary by making use of the convenient plotting capabilities provided in the language. If you have MATLAB or an equivalent tool at your disposal, compare the stability of AM3 to AB3 by computing their stability regions and plotting them together. Once you are set up to do this exercise, you are in a position to compute easily the stability region of any of the LMMs that we have been studying.

It is perhaps surprising that some methods have stability regions consisting of several disjoint pieces. For practical purposes, only the piece containing the origin is important. The eigenvalue λ is part of the problem. The step size h is part of the solution procedure and in principle we let h tend to zero to get convergence. We want not just that $H\lambda$ be in the region of stability for a given H, but also that $h\lambda$ be in the region for all $0 < h \leq H$. Otherwise, a reduction of the step size might convert a stable computation into an unstable one. If a method has a stability region that is not connected, it is obvious that there are λ and H such that the computation is stable with step size H but unstable for some $h < H$. This disagreeable possibility is not confined to the unusual regions that are not connected. For the description of an infinite stability region the concept of *$A(\alpha)$ stability* is useful. By this it is meant that the sector

$$\{z = re^{i\psi} | r > 0, \pi - \alpha < \psi < \pi + \alpha\}$$

is included in the region of stability. By definition, if $H\lambda$ lies in such a sector, so does $h\lambda$ for all $0 < h < H$. Outside the largest such sector, anomalies are possible. The BDFs are important in practice because of their infinite stability regions, but most are not stable for the entire left half complex plane. As Figures 6.5a and 6.5b show, for BDFs of orders greater than 2, the step size must be restricted to keep the computation stable whenever $h\lambda$ lies near the imaginary axis. It is also quite possible

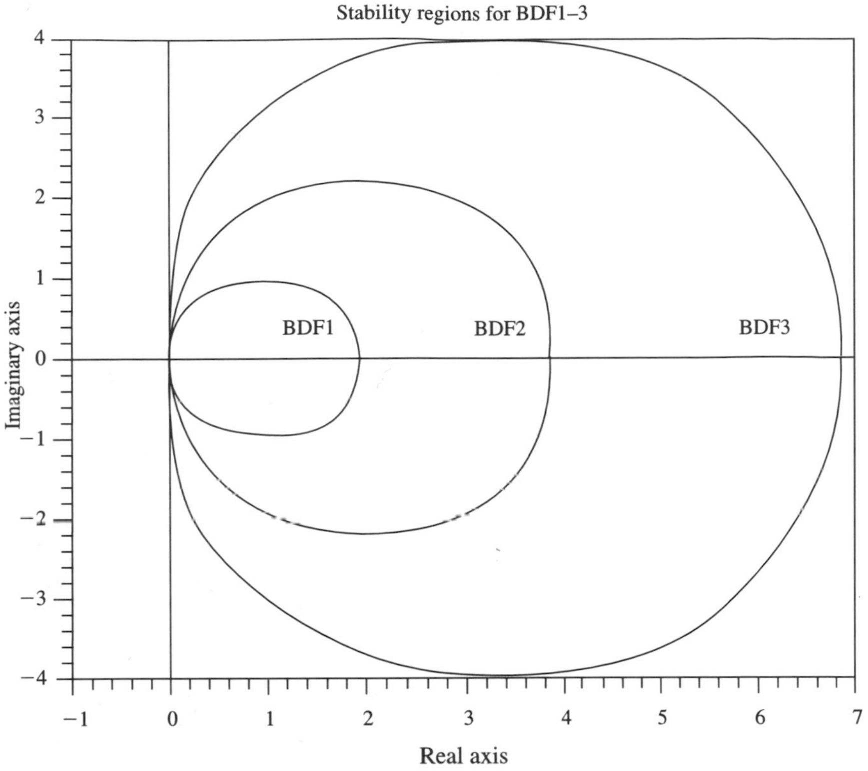

Figure 6.5a

that the computation be stable for a step size H, but not for smaller step sizes h. The angle α in the *$A(\alpha)$ stability of BDFs* is:

order p	1	2	3	4	5	6
α in degrees	90°	90°	88°	73°	52°	18°

There is an important practical point to be gleaned from this table. Many people think that because the stability regions of the BDFs are infinite, codes based on these methods never suffer from the stability restrictions that characterize stiffness. This is not correct. If the problem has a Jacobian with dominant eigenvalues near the imaginary axis, such formulas are no better than popular formulas with a finite stability region. The angle of $A(\alpha)$ stability is so small in the case of BDF6 that general-purpose codes do not use this formula. Some do not even use BDF5. Codes based on the

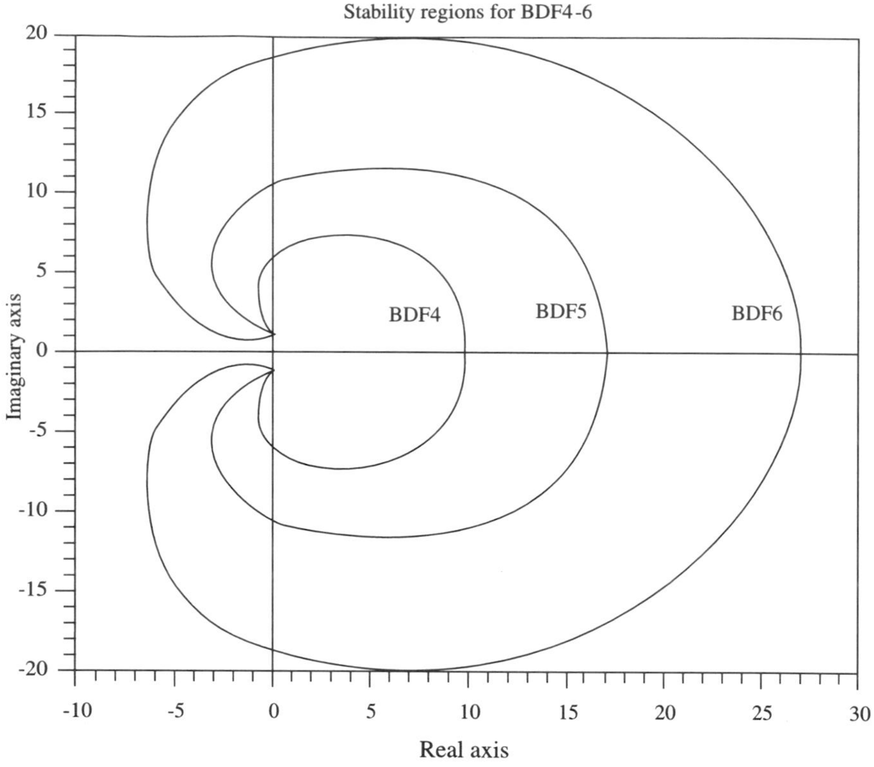

Figure 6.5b

BDFs are intended for problems that lead to stability difficulties, so they trade the efficiency of high orders for the stability of the lower order formulas.

EXAMPLE 5. To emphasize the point about the stability regions of the BDFs, let us integrate the initial value problem (Gaffney [1984])

$$\mathbf{y}' = \begin{bmatrix} -10 & \omega & 0 & 0 & 0 & 0 \\ -\omega & -10 & 0 & 0 & 0 & 0 \\ 0 & 0 & -4 & 0 & 0 & 0 \\ 0 & 0 & 0 & -1 & 0 & 0 \\ 0 & 0 & 0 & 0 & -0.5 & 0 \\ 0 & 0 & 0 & 0 & 0 & -0.1 \end{bmatrix} \mathbf{y},$$

$$\mathbf{y}(0) = \begin{bmatrix} 1 \\ 1 \\ 1 \\ 1 \\ 1 \\ 1 \end{bmatrix},$$

on [0, 64]. The eigenvalues of this block diagonal matrix are $-10 + i\omega$, $-10 - i\omega$, -4, -1, -0.5, -0.1. The real eigenvalues all correspond to $h\lambda$ that lie on the left half of the real axis and when the parameter ω is taken to be 100, the eigenvalues of maximum magnitude correspond to $h\lambda$ that lie on two rays at about $\pm 84°$ from this line. This means that the BDFs of orders 3 and lower will not suffer a stability restriction, but those of higher order will. The code LSODE is a variable order BDF code that permits a user to specify the maximum order to be used. The default value is 5. The problem was solved with a relative error tolerance of 10^{-4} and a scalar absolute error tolerance of 10^{-6}. The code was supplied a subroutine for the (trivial) evaluation of the Jacobian. The following results were obtained for runs made with various maximum orders, MAXORD:

MAXORD	STEPS	NFCN	NJAC
1	3649	3984	218
2	1268	1524	108
3	674	756	44
4	7887	8598	439
5	7537	8746	465

Here STEPS is the number of steps, NFCN is the number of $\mathbf{F}$ evaluations, and NJAC is the number of $\mathbf{F_y}$ evaluations made in the run. As the maximum order permitted rises, the cost goes down because a higher order is more efficient for achieving the accuracy specified. However, when orders 4 and 5 are permitted, the cost goes up a great deal. This is because the code does resort to these orders in an attempt to achieve the specified accuracy more efficiently and the stability regions of the formulas of these orders cause the step size to be restricted to keep the integration stable. This must be interpreted as a weakness in the order selection scheme in LSODE. Only the maximum order is specified, so we might hope that the scheme would recognize that order 3 is the most efficient and use it. Unfortunately, the behavior seen here is typical rather than a defect of this particular code. More research is needed to improve the order selection schemes in this regard. As a practical matter, one should keep in mind that BDF codes can suffer from instability and that restricting the maximum order used in such a code might in some circumstances improve its performance by forcing the code to use more stable formulas.

Stability regions are helpful in selecting which methods to implement. It is found, for example, that an Adams–Bashforth–Moulton PECE predictor–corrector pair of moderate to high order is more than twice as stable as the Adams–Bashforth formula of the same order, and perhaps much more. See, e.g., the plots in Shampine and Gordon [1975]. This means that the predictor–corrector pair is a bargain when stability restricts the step size, even though it is twice as expensive per step. As another example, Figure 6.6 compares the stability of AB4 to that of AM4. The cost of evaluating the implicit formula AM4 will vary during an integration, but two function evaluations per step is representative. Clearly AM4 is substantially more stable than AB4 on an equal cost basis.

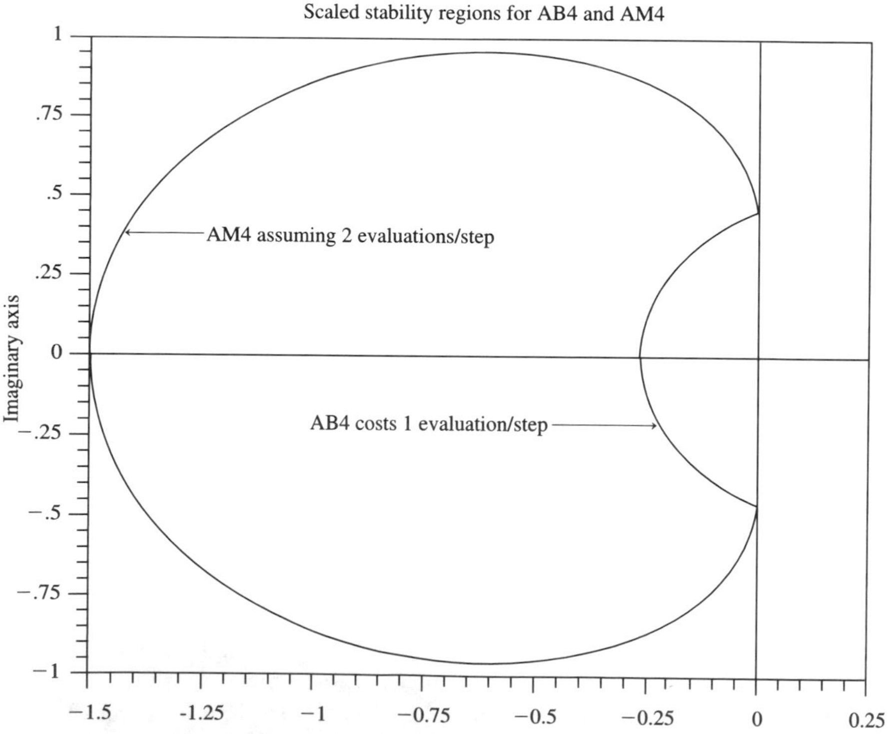

Figure 6.6

In Chapter 8 we take up the stiff problems for which step sizes that would yield the desired accuracy must be reduced drastically to keep the computation stable. For such problems the size and shape of the stability region are vital to efficient solution. When the product of a Lipschitz constant and the length of the interval of integration is not large, the classical situation, the problem is not stiff. When solving non-stiff problems, it is usual that the step size be determined by accuracy. With this in mind, the accuracy considerations of Chapter 4 take precedence in the choice of "optimal" parameters for a family of explicit Runge–Kutta formulas, but attention must be paid to stability as well. Unfortunately, there is usually no simple relation between a set of parameter values and the stability region of the resulting formula. In practice a search is done in which parameters are varied and computed stability regions are compared. It is not even clear how stability regions ought to be compared. It is usual to seek regions that are as big as possible and resemble a half disk. When comparing the sizes of stability regions, it is necessary to scale them by the cost of a step. The older literature contains statements about how much more stable explicit Runge–Kutta formulas are than Adams formulas. This, however, does not take into account the fact that the Runge–Kutta formulas are considerably more expensive per step and must therefore take a much larger step to be competitive. The reason for seeking a boundary that is approximately an arc of a circle is to have a method that will behave much the same when the problem is changed by a small amount. Often the behavior near the imaginary axis is troublesome in this regard. More precisely, for a point $re^{i\psi}$ on the boundary of the stability region, many methods are such that as the angle ψ tends to 90°, the distance r tends to 0, or at least to a value considerably smaller than its average on the boundary.

It is found that in rough terms, the stability of families such as Adams–Moulton, Adams–Bashforth–Moulton PECE, and BDF decreases as the order increases. Effective order selection algorithms recognize when stability is restricting the step size and resort to a more stable, lower order. Some details can be found in Shampine and Gordon [1975]. They exploit this consequence of an effective order selection scheme to recognize stiffness in their Adams code. The higher order BDFs are not used in current codes. This is a statement about the order selection algorithms because if the algorithms could select the "best" order for the task at hand, then the more orders available to the code, the better. Unfortunately, as with Example 5, the popular codes based on BDFs can get "stuck" at an order for which there is a stability restriction even though a lower order formula would not suffer such a restriction.

§3 How Instability Is Manifested

Let us return now to the question about how instability is manifested in the course of a single integration. For the sake of simplicity we restrict ourselves to explicit Runge–Kutta methods and constant step size h. It has been argued that a guide to the behavior of the numerical solution of a general equation $\mathbf{y}' = \mathbf{F}(\mathbf{y})$ is provided by a study of the numerical solution of an approximating linear equation $\mathbf{y}' = J\,\mathbf{y} + \mathbf{g}$. In our earlier study of these special problems we assumed that J is diagonalizable. Now it is technically convenient to suppose in addition that J is non-singular. Assuming this, the differential equation has a solution $\mathbf{v}(x)$ that is constant, namely $\mathbf{v} = -J^{-1}\,\mathbf{g}$. This assumption about J is reasonable in this context of an approximation to $\mathbf{F}_{\mathbf{y}}(\mathbf{y})$ because arbitrarily close to any singular matrix, there is a non-singular matrix. When provided the correct initial value, Runge–Kutta methods integrate exactly a constant solution. That this is true even for non-linear problems is an easy consequence of the form of such methods and the fact that $\mathbf{F}(\mathbf{v}) = \mathbf{0}$.

EXERCISE 13. Suppose that $\mathbf{y}' = \mathbf{F}(\mathbf{y})$ has the constant solution $\mathbf{v}(x) \equiv \mathbf{v}$, i.e., $\mathbf{F}(\mathbf{v}) = \mathbf{0}$. Show that if an explicit Runge–Kutta method is given $\mathbf{y}_j = \mathbf{v}$, then it will produce $\mathbf{y}_{j+1} = \mathbf{v}$. Argue then that if the method is started with the correct value $\mathbf{v}$, it will integrate the equation without error.

Because the special problems are linear, the solution $\mathbf{y}(x)$ can be written as $\mathbf{v} + \mathbf{u}(x)$ where $\mathbf{u}'(x) = J\,\mathbf{u}(x)$. For an explicit Runge–Kutta method we have seen that when solving the homogeneous equation for $\mathbf{u}(x)$, the numerical solution can be expressed as $\mathbf{u}_{j+1} = P(hJ)\mathbf{u}_j$, where $P(hJ)$ is the stability polynomial of the formula. Because these methods are exact for the constant solution $\mathbf{v}$ and because they are linear for linear problems, the numerical solution $\mathbf{y}_j$ of the inhomogeneous problem can be decomposed in a way analogous to the decomposition of $\mathbf{y}(x)$:

$$\mathbf{y}_{j+1} = \mathbf{v} + \mathbf{u}_{j+1} = \mathbf{v} + P(hJ)\mathbf{u}_j.$$

The slope of the local solution at x_j, $\mathbf{F}(\mathbf{y}_j)$, furnishes an approximation to $\mathbf{y}'(x_j)$. This quantity is always computed because it is the first stage of the Runge–Kutta formula. For the special $\mathbf{F}$ that we are studying, it is possible to relate the slope at x_j to that at x_{j+1}:

$$\begin{aligned}\mathbf{F}(\mathbf{y}_{j+1}) = J\,\mathbf{y}_{j+1} + \mathbf{g} = J[\mathbf{y}_{j+1} - \mathbf{v}] &= J\,\mathbf{u}_{j+1} = J\,P(hJ)\mathbf{u}_j \\ &= P(hJ)J\,\mathbf{u}_j = P(hJ)J[\mathbf{y}_j - \mathbf{v}] \\ &= P(hJ)[J\,\mathbf{y}_j + \mathbf{g}] = P(hJ)\mathbf{F}(\mathbf{y}_j).\end{aligned}$$

For this special class of problems, the slope at x_{j+1} is precisely the product of the stability polynomial and the slope at x_j. For general $\mathbf{F}$ this is approximately true because the formula depends continuously on $\mathbf{F}$. Accordingly, if the linearized problem is unstable with the step size h, we expect to see an exponential growth in size of the approximate derivatives $\mathbf{F}(\mathbf{y}_j)$.

Earlier we saw that instability might be revealed when solving a general autonomous problem with constant step size h by a growth in the size of $\mathbf{y}_{j+1} - \mathbf{y}_j$. In the present circumstances we have

$$\begin{aligned}\mathbf{y}_{j+1} - \mathbf{y}_j &= (\mathbf{v} + \mathbf{u}_{j+1}) - (\mathbf{v} + \mathbf{u}_j) = \mathbf{u}_{j+1} - \mathbf{u}_j \\ &= P(hJ)(\mathbf{u}_j - \mathbf{u}_{j-1}) = P(hJ)(\mathbf{y}_j - \mathbf{y}_{j-1}).\end{aligned}$$

We now see that for these special problems the difference is multiplied by the stability polynomial of the formula at each step and for general $\mathbf{F}$ this is approximately so. In this linearization of the general situation treated earlier, we come to a better understanding of the matter. Earlier we had to say that if the perturbations were such as to excite instability, the instability would be observed by a growth in the size of $\mathbf{y}_{j+1} - \mathbf{y}_j$. Now we see that it would be exceptional not to observe an exponential growth of this quantity when the step size is too large for stability. Of course, this is true only to the extent that the linear model describes the non-linear problem.

We have not yet taken up how local error is estimated with Runge–Kutta formulas, but the essence of the matter is easily described. Each step from $(x_j, \mathbf{y}_j)$ is taken with two formulas, one producing $\mathbf{y}_{j+1}$ and the other, $\mathbf{y}^*_{j+1}$. If the result $\mathbf{y}^*_{j+1}$ is of higher order than the result $\mathbf{y}_{j+1}$, it is easy to see that the local error in the result $\mathbf{y}_{j+1}$ can be estimated by

$$\mathbf{est}_j = \mathbf{y}^*_{j+1} - \mathbf{y}_{j+1} \approx \text{local error at } x_j.$$

Our analysis applies to both formulas, hence

$$\begin{aligned}\mathbf{est}_j = \mathbf{y}^*_{j+1} - \mathbf{y}_{j+1} &= [\mathbf{v} + P^*(hJ)\mathbf{u}_j] - [\mathbf{v} + P(hJ)\mathbf{u}_j] = Q(hJ)\mathbf{u}_j \\ &= Q(hJ)P(hJ)\mathbf{u}_{j-1} = P(hJ)Q(hJ)\mathbf{u}_{j-1} \\ &= P(hJ)[\mathbf{y}^*_j - \mathbf{y}_j] = P(hJ)\mathbf{est}_{j-1}.\end{aligned}$$

Here the notation $Q(hJ)$ has been used for $P^*(hJ) - P(hJ)$. This says that if the formula is unstable for the linearized problem, then the estimate of the local error will grow exponentially fast. A growth of this kind will quickly result in a value of $\|\mathbf{est}_j\|$ that is larger than the tolerance on the local error, at which time the code will reduce the step size. Because Runge–Kutta formulas are stable for all sufficiently small h, reducing the step size enough will result in a stable integration. This provides an explanation that is valid in some reasonable circumstances for the experimental observation that control of the local error stabilizes the integration.

Generally we expect that when we ask for more accuracy, a code will need more steps to achieve this accuracy and the cost will go up. This, however, is true only when it is accuracy that determines the step size. It is stability that determines the step size for the stiff problems discussed in Chapter 8. What happens when we apply a method with a finite stability region to a stiff problem? The essence of the matter is simple. As we discuss in Chapter 7, an efficient step size algorithm will increase the step size to the largest value that it predicts will yield the required accuracy. If this step size is too big for stability, the integration will become unstable and the algorithm will reduce the step size to the point that the integration is again stable. This step size is smaller than necessary for achieving an accurate solution, so after the numerical solution smooths out, the code will again increase the step size. The net effect is that an efficient and effective step size selection algorithm will produce step sizes that are on average equal to the largest stable value. The first investigation of this question, Shampine [1975], made strong assumptions about the problem and the numerical methods. The issue is better understood now and some arguments contributing to this were presented above. There is so much experimental evidence in support of this assertion about the behavior of the step size that some authors have characterized it as part of the folklore of the subject. Some interesting consequences pointed out in Shampine [1975] have also passed into the folklore, so let us go into them briefly.

When a good code for non-stiff problems is applied to a stiff problem, that is, to a problem for which stability rather than accuracy governs the choice of step size during a significant portion of the integration, automatic selection of the step size will keep the integration stable. Where stability determines the step size, the average step size will be nearly independent of the accuracy desired and accordingly, the cost will be nearly independent of the accuracy. The code will achieve the desired control of the error, it is just (much) less efficient than one based on a more stable method. To illustrate these experimental observations some computations are quoted here from Shampine [1975] and further computations with the same problem are quoted in Example 2 of Chapter 8 to make other points.

EXAMPLE 6. The initial value problem

$$\mathbf{y}' = \begin{pmatrix} -0.1 & -199.9 \\ 0 & -200 \end{pmatrix}\mathbf{y}, \qquad 0 \le x \le 50, \qquad \mathbf{y}(0) = \begin{pmatrix} 2 \\ 1 \end{pmatrix},$$

has the solution $y_1(x)$ and $\exp(-.1x) + \exp(-200x)$, $y_2(x) = \exp(-200x)$. Suppose this problem is to be integrated with an absolute error tolerance of ε. There is an initial interval on which the solution changes rapidly and the step size is mainly determined by the accuracy requirement. On, say, [1, 50] the solution is easy to approximate and the step size is mainly determined by stability. When this initial value problem was integrated with RKF, a Runge–Kutta code based on the Fehlberg (4, 5) pair, the following results were obtained for the interval [1, 50]:

ε	NFCN	ERROR	HAVG
$1E-1$	21672	$3E-3$	.0156
$1E-2$	21744	$3E-4$	.0156
$1E-3$	21671	$3E-5$	.0157
$1E-4$	21697	$3E-6$	.0157
$1E-5$	21725	$3E-7$	.0157
$1E-6$	21687	$3E-8$	.0156
$1E-7$	21694	$3E-9$	.0156
$1E-8$	21680	$3E-10$	.0156
$1E-9$	21747	$3E-11$	.0157
$1E-10$	21613	$6E-12$	.0155
$1E-11$	21691	$5E-12$	.0156
$1E-12$	21727	$9E-12$	.0156

Here ERROR is the maximum absolute error seen at any step in the interval [1, 50]. Obviously the code is able to integrate this stiff problem accurately. (At the most stringent tolerances, the accuracy is reduced due to roundoff.) It is seen that cost as measured by the number of function evaluations, NFCN, scarcely depends on the tolerance over a wide range of tolerances. Correspondingly, the average step size, HAVG, is remarkably constant. The eigenvalue λ of the Jacobian of maximum magnitude is -200. The product of the average step size and λ is about -3.07. This corresponds very nearly to being on the boundary of the stability region which is at -3.02.

§4 Constant Coefficient Difference Equations

In this section we study the behavior of solutions of the (scalar) problem

$$z_0, z_1, \ldots, z_{k-1} \text{ given}$$

$$\sum_{i=0}^{k} \alpha_i z_{j+1-i} = c_j \qquad j \geq k-1 \tag{4.1}$$

when $\alpha_0 \alpha_k \neq 0$. The forcing terms c_j are given. This problem is closely analogous to an initial value problem for a (scalar) constant coefficient differential equation of order k. We shall repeatedly refer to this more familiar problem for guidance. A solution to the initial value problem for the difference equation is a sequence of numbers z_j for $j = 0, 1, \ldots$. To bring out the analogy with the continuous function that represents a solution of a differential equation, let us introduce notation for the sequences. For example, let us write $z = \{z_j\}$ and $c = \{c_j\}$. It is convenient to introduce an operator $\mathcal{D}$. It acts on a sequence z to produce a new sequence with terms given by

$$(\mathcal{D}z)_j = \sum_{i=0}^{k} \alpha_i \, z_{j+1-i} \quad \text{for} \quad j \geq k - 1.$$

The notation in this section is sometimes a bit clumsy, e.g., c_j and $(\mathcal{D}z)_j$ are not defined for $j < k - 1$. In this notation, (4.1) becomes $\mathcal{D}z = c$.

In some ways the difference equation is easier to analyze than the differential equation. For one thing, existence and uniqueness of a solution is trivial. We have the non-degeneracy assumption that $\alpha_0 \neq 0$. The initial values z_r, $r = 0, 1, \ldots, k - 1$, are given. We are assuming that the forcing function, the sequence c, is given. Because of this, a little manipulation of the difference equation provides an explicit recipe for z_{j+1}:

$$z_{j+1} = \frac{1}{\alpha_0}\left[c_j - \sum_{i=1}^{k} \alpha_i \, z_{j+1-i}\right] \quad \text{for} \quad j \geq k - 1.$$

With the assumptions made, it is clear that there always exists a solution sequence z and it is unique.

Just as with linear differential equations, it is important to study the solution of homogeneous equations because inhomogeneous problems can be solved in terms of solutions of homogeneous problems. The difference equation is said to be homogeneous if $c_j = 0$ for all $j \geq k$, or in more compact notation, $c = 0$. Let z be the solution of the homogeneous problem

$$\mathcal{D}z = 0, \qquad z_0, \ldots, z_{k-1} \text{ given.} \tag{4.2}$$

Recall how we solve the initial value problem for a homogeneous, constant coefficient ordinary differential equation of order k:

$$y(a), y'(a), \ldots, y^{(k-1)}(a) \text{ given,}$$

$$\sum_{i=0}^{k} \alpha_i \, y^{(k-i)} = 0.$$

We look for solutions of the form $y(x) = \exp(\zeta x)$. Substitution into the equation leads to

$$\sum_{i=0}^{k} \alpha_i \, \zeta^{k-i} \exp(\zeta x) = \exp(\zeta x) \, \rho(\zeta) = 0,$$

where $\rho(\zeta)$ is the characteristic polynomial of the differential equation. Each root ζ of the characteristic polynomial provides a solution of the assumed form. If the polynomial has k distinct roots, i.e., all the roots are simple, this results in k distinct solutions. Because the equation is linear, any linear combination of these solutions is also a solution. It turns out that these simple solutions are linearly independent, hence there is a linear combination of these simple solutions that satisfies any specified initial conditions. The matter is more complicated when the characteristic polynomial has a multiple root, so discussion of this case will be deferred for the moment. We followed this general approach earlier when deriving necessary conditions for zero-stability. Specifically, we looked for special solutions of $\mathscr{D}v = 0$ of the form $v_j = \zeta^j$ and found that such a solution would have to satisfy

$$(\mathscr{D}v)_j = \sum_{i=0}^{k} \alpha_i \, y_{j+1-i} = \sum_{i=0}^{k} \alpha_i \, \zeta^{j+1-i} = \zeta^{j+1-k} \rho(\zeta).$$

In general, then, a solution v of this special kind is obtained for any root ζ of $\rho(\zeta)$. Here $\rho(\zeta)$ is the characteristic polynomial of the difference equation just as we defined it in connection with discrete variable methods for constant step size. The characteristic polynomial is of degree k, so it has k roots counting multiplicity. Supposing that all the roots are simple, k distinct solutions $v^{(\mu)}$ for $\mu = 0, \ldots, k-1$ can be defined by

$$v_j^{(\mu)} = \zeta_\mu^j, \qquad j \geq 0,$$

where the $\zeta_0, \zeta_1, \ldots, \zeta_{k-1}$ are the k distinct roots of $\rho(\zeta)$. Our next task is to show that any solution of the initial value problem (4.2) can be found in terms of these special solutions.

First we need to show that just as with linear differential equations, any linear combination of solutions of the homogeneous equation is also a solution of the homogeneous equation. For constants $a_0, a_1, \ldots, a_{k-1}$, let

$$v = a_0 \, v^{(0)} + a_1 v^{(1)} + \cdots + a_{k-1} \, v^{(k-1)}$$

or, in terms of the elements of the sequences,

$$v_j = a_0 \, v_j^{(0)} + a_1 \, v_j^{(1)} + \cdots + a_{k-1} \, v_j^{(k-1)}.$$

On applying the operator $\mathscr{D}$ to v, we have

$$(\mathscr{D}v)_j = \sum_{i=0}^{k} \alpha_i \, v_{j+1-i} = a_0 \sum_{i=0}^{k} \alpha_i \, v_{j+1-i}^{(0)} + \cdots + a_{k-1} \sum_{i=0}^{k} \alpha_i \, v_{j+1-i}^{(k-1)},$$

or

$$\mathscr{D}v = a_0 \, \mathscr{D}v^{(0)} + \cdots + a_{k-1} \, \mathscr{D}v^{(k-1)} = 0.$$

Now we ask if we can find constants $a_0, \ldots, a_{k-1}$ such that the solution of (4.2) can be written as

$$z = a_0 \, v^{(0)} + \cdots + a_{k-1} \, v^{(k-1)}. \tag{4.3}$$

We have just seen that the sequence z defined in this way is a solution of the homogeneous equation, so the only question is whether it has the right initial values. If this representation is to be valid for the initial values, we must have

$$\begin{aligned} z_0 &= a_0 \, v_0^{(0)} + a_1 \, v_0^{(1)} + \cdots + a_{k-1} \, v_0^{(k-1)} \\ &\vdots \\ z_{k-1} &= a_0 \, v_{k-1}^{(0)} + a_1 \, v_{k-1}^{(1)} + \cdots + a_{k-1} \, v_{k-1}^{(k-1)} \end{aligned}$$

This is a set of k linear equations for the k unknown constants a_r. On substituting the definitions of the special solutions, the coefficient matrix is found to be

$$V = \begin{pmatrix} 1 & 1 & \ldots & 1 \\ \zeta_0 & \zeta_1 & & \zeta_{k-1} \\ \vdots & & & \\ \zeta_0^{k-1} & \zeta_1^{k-1} & \ldots & \zeta_{k-1}^{k-1} \end{pmatrix}$$

This is a van der Monde matrix and such matrices are known to be nonsingular when the ζ_μ are distinct, as they are here. (One way to prove this fact is to show that it is equivalent to the uniqueness of polynomial interpolation at k distinct points.) This implies that for any solution z of the homogeneous problem, there is a set of coefficients a_r such that z can be represented in the form (4.3). If the vector $\mathbf{z}$ is defined to be $(z_0, z_1, \ldots, z_{k-1})^T$ and $\mathbf{a}$ is defined to be $(a_0, a_1, \ldots, a_{k-1})^T$, then $\mathbf{a} = V^{-1}\mathbf{z}$. This leads to a bound on the coefficients a_r in terms of the initial values z_r:

$$\max_{0 \le r < k} |a_r| = \|\mathbf{a}\|_\infty \le \|V^{-1}\|_\infty \, \|\mathbf{z}\|_\infty = \|V^{-1}\|_\infty \max_{0 \le r < k} |z_r|.$$

The main thing we want to know about the solution of a homogeneous difference equation is whether it grows or decays. Earlier it was found that the root condition is necessary for stability. The argument was general and it implies that if the roots of the characteristic equation of (4.2) do not satisfy the root condition, there are initial values for which the solution z of (4.2) grows indefinitely as $j \to \infty$. Suppose now that the root condition holds. This implies that for each μ,

$$|v_j^{(\mu)}| = |\zeta_\mu^j| = |\zeta_\mu|^j \le 1 \quad \text{for all } j.$$

These special solutions are all uniformly bounded in magnitude by 1 and from the representation (4.3), any solution z of the homogeneous equation is bounded by

$$|z_j| \leq \sum_{r=0}^{k-1} |a_r| \quad \text{for all } j,$$

where the a_r are determined from the initial conditions. Using the bound on the a_r in terms of the starting values z_r, we have then

$$|z_j| \leq k \, \|V^{-1}\|_\infty \max_{0 \leq r < k} |z_r| \quad \text{for all } j. \tag{4.4}$$

We conclude that when the root condition holds, any solution of the homogeneous problem can be bounded uniformly in j in terms of the size of its initial values—the solution is stable. Let us collect these results in the form of a theorem.

THEOREM 5. *The initial value problem for a difference equation (4.1, 4.2) has a unique solution $\{z_j\}$. For the difference equation to be stable, it is necessary that the roots ζ_j of the characteristic equation satisfy the root condition. If the equation is homogeneous, $c_j = 0$ for all j, and the ζ_j are all distinct and satisfy the root condition, then the z_j are uniformly bounded in j. The inequality (4.4) provides an explicit bound in terms of the initial values and the ζ_j.*

The matter is more complicated when $\rho(z)$ has a multiple root because then there are not enough of these special solutions to satisfy all the initial conditions. In the case of ordinary differential equations, it is found that if ζ_μ is a multiple root, then in addition to the solution $\exp(\zeta_\mu x)$ there is also a solution $x \exp(\zeta_\mu x)$. In our earlier discussion of necessary conditions for zero-stability, the analog of this result was developed for difference equations, viz. if ζ_μ is a multiple root, then in addition to the solution $\{\zeta_\mu^j\}$, there is a solution $\{(j + 1)\zeta_\mu^j\}$. For both differential and difference equations, the additional solution can be derived as a limit of two solutions corresponding to distinct roots ζ_μ and ζ_σ as the root ζ_σ tends to ζ_μ to form a double root. It is not usual in courses on differential equations to sort out the details, and we are not going to do this for the difference equations either. It is not that it is particularly difficult, rather that it is not very illuminating. Besides, all the methods that interest us have characteristic polynomials that have simple roots in nearly all the situations that concern us. There is a matter that does require some comment, though. If ζ_μ is a triple root, there is an additional solution of the form $\{(j + 2)(j + 1)\zeta_\mu^j\}$, and so forth. The general form of one of the additional solutions needed to deal with a multiple root ζ_μ is $\{(j + q) \cdots (j + 1)\zeta_\mu^j\}$. The question is whether these additional solutions are uniformly bounded in j. As with the necessary condition for zero-stability, they are not bounded if $|\zeta_\mu| = 1$. For this reason the root condition was imposed so that all multiple roots ζ_μ would have magnitude less than 1. Now

$$|(j + q) \cdots (j + 1)\zeta_\mu^j| \leq (j + q)^q \, |\zeta_\mu|^j \leq \max_{1 \leq t} f(t)$$

where

$$f(t) = (t + q)^q \, |\zeta_\mu|^t.$$

Logarithmic differentiation leads easily to

$$f'(t) = f(t) \left[\frac{q}{t + q} + \ln |\zeta_\mu| \right].$$

Because $q \geq 1$ and $|\zeta_\mu| < 1$, it is easy to see that $f(t)$ has a unique maximum for all $t \geq 1$. These observations imply that there is a constant K that bounds on the magnitude of such a solution for all j and by increasing it if necessary, K bounds the magnitude of such solutions for all the multiplicities possible. The issue here is the growth of these special solutions. For simple roots, the special solutions do not increase at all as j increases. In the case of multiple roots, the additional special solutions do increase initially, but eventually the power of ζ_μ dominates so that the whole sequence is uniformly bounded. If one works through all the details, the conclusion is that there is a constant Γ such that

$$|z_j| \leq \Gamma \max_{0 \leq r < k} |z_r| \quad \text{for all } j$$

whenever the characteristic polynomial satisfies the root condition. The analysis leads to

Theorem 6. *The root condition is a necessary and sufficient condition for all solutions of a homogeneous, constant coefficient difference equation to be uniformly bounded in terms of their initial values.*

All the non-essential roots of $\rho(\theta)$ have magnitude less than one. There is, then, a constant τ such that

$$|\zeta_\mu| \leq \tau < 1, \qquad \text{all non-essential } \zeta_\mu.$$

In the case of a simple root, it is obvious that the elements of the special solution $\{\zeta_\mu^j\}$ tend to 0 like τ^j as j tends to infinity, which is to say that they decay exponentially fast. An argument along the lines of that given to show boundedness of the additional solutions arising when a root is multiple shows that in general there is a constant γ such that

$$|v_j| \leq \gamma\tau^j$$

for any of the special solutions v associated with non-essential roots. This says that all the special solutions associated with non-essential roots decay very quickly as the integration proceeds. The essential root $\zeta = 1$ must be present if the formula is to imitate the differential equation. The other roots are present only when the method uses previously computed solution values. This result says that the effects of non-essential roots, at least as seen in these special solutions, decay exponentially fast as the integration proceeds. For this reason we are especially interested in methods that satisfy the strong root condition. Indeed, these observations suggest that the smaller τ is, the better because the smaller is the effect of previously computed solution values on the stability of the computation.

EXERCISE 14. Show that if $\rho(\zeta)$ satisfies the strong root condition, any solution of the homogeneous problem $\mathcal{D}z = 0$ tends quickly to a constant value. To do this, recall that such a solution can be written in the form

$$z = a_0\, v^{(0)} + \cdots + a_{k-1}\, v^{(k-1)},$$

where the $v^{(i)}$ are the special solutions and the a_i are chosen to yield the specified initial values. Take in this representation $v^{(0)} = \{1\}$ to be the solution associated with the principal root and show that $z_j \to a_0$ as $j \to \infty$.

EXAMPLE 7. We have seen that the eigenvalues of the Jacobian of a system of equations play an important role in the stability of numerical methods. Some of the examples of semi-discretization of partial differential equations taken up in Example 8 of Chapter 1 are very illuminating. As it happens, their eigenvalues can be determined by solving difference equations. By way of an example, let us consider the Jacobian that arises in the approximation of the one dimensional heat equation by finite differences with equal mesh spacing $\Delta = 1/N$. The matrix is

$$J = \frac{1}{\Delta^2}\begin{bmatrix} -2 & 1 & 0 & \cdots & 0 & 0 \\ 1 & -2 & 1 & \cdots & 0 & 0 \\ 0 & 1 & -2 & & & \vdots \\ \vdots & & & & & \vdots \\ 0 & 0 & 0 & 1 & -2 & 1 \\ 0 & 0 & 0 & 0 & 1 & -2 \end{bmatrix}.$$

Let us work with the scaled matrix $M = \Delta^2 J$. If λ is an eigenvalue of M corresponding to an eigenvector $\mathbf{v}$, then in components

$$v_{j+1} - 2v_j + v_{j-1} = \lambda v_j \qquad j = 2, \ldots, N-1,$$
$$v_2 - 2v_1 = \lambda v_1,$$
$$v_{N-1} - 2v_N = \lambda v_N.$$

We need to determine λ and $\mathbf{v}$, remembering that $\mathbf{v}$ is only determined up to a constant multiple. The components v_j satisfy a difference equation with constant coefficients

$$v_{j+1} - (2 + \lambda)v_j + v_{j-1} = 0,$$

at least for j in the range $2, \ldots, N-1$. If we define $v_0 = 0$ and $v_{N+1} = 0$, the components v_1 and v_N will also satisfy the difference equation. We have seen that solutions of this difference equation have the form

$$v_j = a_0\, \zeta_1^j + a_1\, \zeta_2^j$$

where ζ_1 and ζ_2 are the roots of the characteristic equation

$$0 = \zeta^2 - (2 + \lambda)\zeta + 1 = (\zeta - \zeta_1)(\zeta - \zeta_2).$$

If v_0 and v_{N+1} are to have the correct values, then

$$0 = v_0 = a_0\, \zeta_1^0 + a_1\, \zeta_2^0 = a_0 + a_1,$$
$$0 = v_{N+1} = a_0\, \zeta_1^{N+1} + a_1\, \zeta_2^{N+1}.$$

Because the product of the two roots is 1, the second equation leads to

$$0 = \zeta_1^{N+1}\, v_{N+1} = a_0\, \zeta_1^{2(N+1)} + a_1(\zeta_1\zeta_2)^{N+1} = a_0\, \zeta_1^{2(N+1)} + a_1.$$

The two equations then imply that

$$\zeta_1^{2(N+1)} = 1,$$

which is to say that ζ_1 is a root of unity. Specifically,

$$\zeta_1 = e^{i[2\pi m/2(N+1)]} = e^{i(\pi m/N+1)} \quad \textit{for} \quad m = 1, \ldots, N$$

are the possibilities. The root $\zeta_2 = \zeta_1^{-1}$ is different from ζ_1 with these choices. (Recall that we need distinct roots for the representation postulated.) Now

$$\zeta_1^2 - (2 + \lambda)\zeta_1 + 1 = 0$$

implies that

$$2 + \lambda = \zeta_1 + \zeta_2 = e^{i(\pi m/N+1)} + e^{-i(\pi m/N+1)} = 2\cos\left(\frac{m\pi}{N+1}\right).$$

From this we conclude that the eigenvalues are

$$\lambda_m = -2 + 2\cos\left(\frac{m\pi}{N+1}\right) \qquad m = 1, \ldots, N.$$

Component j of the corresponding eigenvector is given by

$$v_j = a_0(\zeta_1^j - \zeta_2^j) = a_0(e^{i(\pi mj/N+1)} - e^{-i(\pi mj/N+1)}) = 2ia_0 \sin\left(\frac{mj\pi}{N+1}\right).$$

If we take the multiple a_0 to be $1/(2i)$, the eigenvector corresponding to λ_m is

$$v_j = \sin\left(\frac{mj\pi}{N+1}\right) \qquad j = 1, \ldots, N.$$

EXERCISE 15. When solving the advection equation with periodic boundary condition by semi-discretization with finite differences, we encountered a system of differential equations with $N \times N$ Jacobian matrix

$$J = -\frac{c}{2\Delta}\begin{pmatrix} 0 & 1 & 0 & \ldots & 0 & -1 \\ -1 & 0 & 1 & \ldots & 0 & 0 \\ 0 & -1 & 0 & & & \vdots \\ \vdots & & & & & \vdots \\ 0 & 0 & 0 & -1 & 0 & 1 \\ 1 & 0 & 0 & 0 & -1 & 0 \end{pmatrix}.$$

Determine the eigenvalues and eigenvectors of the scaled matrix $M = (-2\Delta/c)J$. Let v_j be component j of an eigenvector for $j = 1, \ldots, N$. Introduce $v_0 = v_N$ and $v_{N+1} = v_1$. This leads to a difference equation for v_j with $j = 1, \ldots, N$. Try a solution of the form ζ^j. Show that the conditions $v_0 = v_N$ and $v_{N+1} = v_1$ are satisfied when ζ is a root of unity, specifically, $\zeta = \exp(i2\pi m/N)$ for $m = 0, \ldots, N-1$. Show that this ζ is a root of the characteristic equation along with $-\zeta^{-1}$. Deduce that the eigenvalues are

$$\lambda_m = 2i\sin\left(\frac{2m\pi}{N}\right) \qquad m = 0, \ldots, N-1.$$

7

Error Estimation and Control

In this chapter we take up a number of issues connected with the estimation and control of the error. For this purpose we must go beyond the simple convergence results of Chapter 5, which merely tell us the order of the error, and actually approximate the error. This hardly seems possible without being much more specific about the methods used, how the integration is started, and how the step sizes are selected. Numerical experiments encourage us to investigate the behavior of the error because they show that it can be quite regular:

EXAMPLE 1. To illustrate the behavior of the error, the problem $y' = \cos(x)y$, $y(0) = 1$ was integrated from 0 to 2π with a constant step size h to approximate the solution $y(x) = \exp(\sin(x))$. Three methods were used, viz. the improved Euler method, AB2, and the midpoint rule. The last two methods require a starting value y_1 in addition to the given initial value $y_0 = 1$. In the case of AB2 this starting value was computed with the improved Euler method and in the case of the midpoint rule, with the forward Euler method. In Chapter 5 it was shown that all these computations are convergent of order 2. Now we are interested in the error itself, so we computed and plotted the *magnified errors* $(y(x_n) - y_n)/h^2$. As seen in Figure 7.1, the magnified errors of the improved Euler method appear to be converging to a smooth function $e(x)$ as $h \to 0$. It is worth comment that often people expect numerical errors will increase steadily as the integration proceeds; this example and others that follow make the point that they can decrease as well as increase. The behavior of the error of AB2 displayed in Figure 7.2 is similar to that of the improved Euler method. However, the behavior of the error of the midpoint rule is more complex. As seen in Figure 7.3, the magnified errors do not themselves appear to converge to a smooth function, but the errors at even numbered steps do

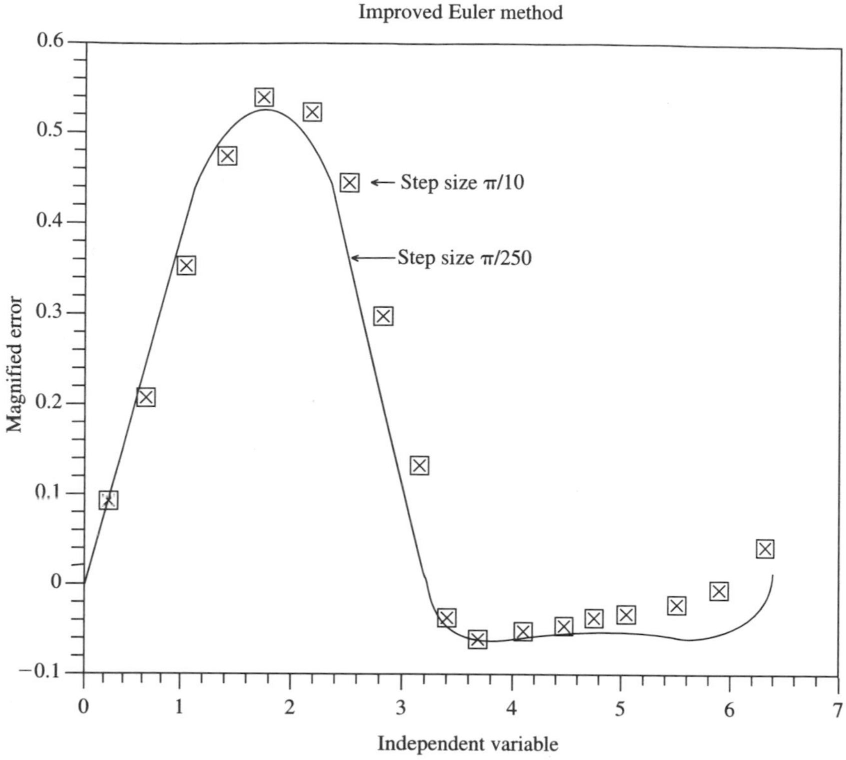

Figure 7.1

appear to converge to a smooth function and those at odd numbered steps appear to converge to a *different* smooth function.

In the first section we develop an asymptotic expansion for the error in useful circumstances. In particular, we shall come to an understanding of the numerical experiments of Figures 7.1 and 7.2, though we do not try to explain fully Figure 7.3. How the error behaves is interesting in its own right and its behavior is fundamental to understanding most of the other issues taken up in this chapter. General purpose codes for the initial value problem estimate the error made at each step and select the step size automatically so as to control this error. In subsequent sections we study how the error is estimated, how an appropriate step size is selected, the popular ways of controlling the error, and the implications of the error controls. These matters are fundamental to understanding modern codes for solving the initial value problem for systems of ordinary differential equations.

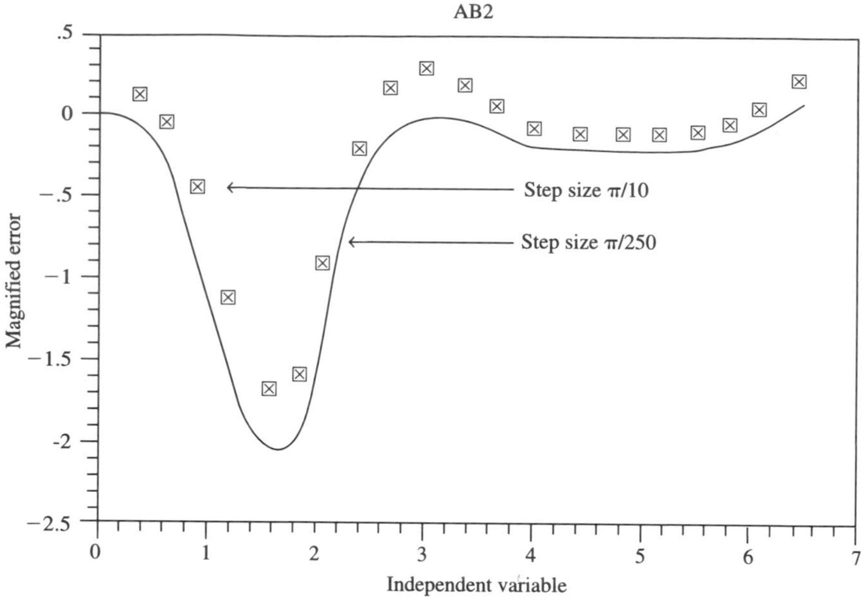

Figure 7.2

§1 Asymptotic Behavior of the Error

A convergence proof rests upon a stability result for the discrete problem defining the numerical solution $\{\mathbf{y}_j\}$ that approximates the solution $\mathbf{y}(x)$ of

$$\mathbf{y}' = \mathbf{F}(x, \mathbf{y}), \qquad a \leq x \leq b, \qquad \mathbf{y}(a) = \mathbf{A}. \tag{1.1}$$

In Chapter 5 stability was shown with certain restrictions on the step sizes used, the specific restrictions depending on the nature of the numerical method investigated. For now let us suppose that the step size is a constant h. Convergence was established by showing that for $0 \leq r < \nu$, $\mathbf{y}(x_r)$ agrees with the starting value $\mathbf{y}_r$ to $\mathcal{O}(h^p)$ and that the sequence $\{\mathbf{y}(x_j)\}$ satisfies the difference equation with a discrepancy that is $\mathcal{O}(h^{p+1})$. Stability of the numerical method then implies that the difference between $\mathbf{y}(x_j)$ and $\mathbf{y}_j$ is uniformly $\mathcal{O}(h^p)$. Now we refine this and show that in reasonable circumstances, we can define a function $\mathbf{e}(x)$ such that the difference between $\mathbf{y}(x_j) + h^p\mathbf{e}(x_j)$ and $\mathbf{y}_j$ is uniformly $\mathcal{O}(h^{p+1})$. We pursue the same line of proof with the sequence $\{\mathbf{y}(x_j) + h^p\mathbf{e}(x_j)\}$ now playing the role of the sequence $\{\mathbf{y}(x_j)\}$. Evidently the result we want follows if we can show that the first ν values in this new sequence differ from the starting values by $\mathcal{O}(h^{p+1})$ and that the sequence satisfies the difference equation with a

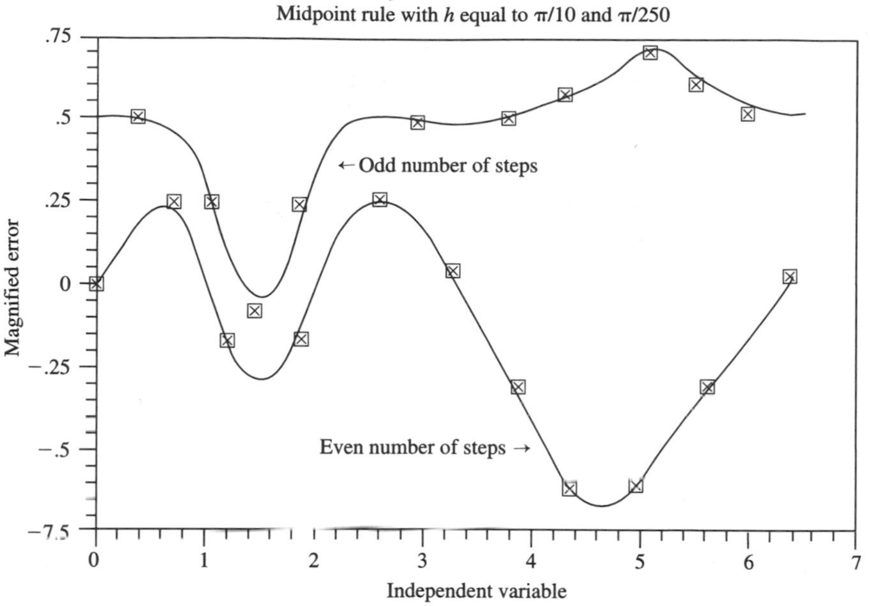

Figure 7.3

discrepancy that is $\mathcal{O}(h^{p+2})$. This is not always possible, as the counter-example of the midpoint rule illustrated in Figure 7.3 makes clear. When it is not possible, this does not mean that the error does not behave is some regular way; like the midpoint rule, it might simply behave in a way more complex than we are postulating here.

Hitherto all we have assumed about the local truncation error of a method is its order p. This error is the amount by which $\{\mathbf{y}(x_j)\}$ fails to satisfy the difference equation for $\{\mathbf{y}_j\}$ and now that we focus our attention on this discrepancy, we must know more about its form. For our present purposes, we suppose that the local truncation error has the form

$$h^{p+1}\,\boldsymbol{\psi}(x_j, \mathbf{y}(x_j)) + \mathcal{O}(h^{p+2}).$$

The function $\boldsymbol{\psi}(x, \mathbf{y})$ here is called the *principal error function*. It depends on $\mathbf{y}(x)$, but often the notation will be simplified by not showing this explicitly. Although we did not draw attention to it at the time, we have already seen expressions for the principal error functions of the methods we study. In the case of a (normalized) LMM of order p, the truncation error was found to be

$$C_{p+1}\,h^{p+1}\,\mathbf{y}^{(p+1)}(x_j) + \mathcal{O}(h^{p+2}),$$

where C_{p+1} is the error constant of the method. Such methods have principal error functions that are quite simple. The Runge–Kutta methods generally have much more complicated $\boldsymbol{\psi}$ involving the elementary differentials of $\mathbf{F}$. Except for the factor of h^{p+1}, the principal error function is just the leading term in the Taylor expansion of the local truncation error. Some examples were worked out in Chapter 4. For instance, we studied two-stage, second-order explicit Runge–Kutta formulas and by Taylor series expansion found the error of these formulas to have the form

$$h^3\left[\left(\frac{1}{6} - \frac{1}{2}\gamma_1\,\beta_{1,0}^2\right)\mathbf{D}_1^3 + \frac{1}{6}\mathbf{D}_2^3\right] + \mathbb{O}(h^4).$$

Here γ_1 and $\beta_{1,0}$ are parameters defining the formula, and $\mathbf{D}_1^3$ and $\mathbf{D}_2^3$ are elementary differentials.

We must also say more about the *starting procedures*. So far we have assumed of the $k - 1$ starting values only that

$$\mathbf{y}_r = \mathbf{s}_r(h) = \mathbf{y}(x_r) + \mathbb{O}(h^p).$$

We hope to find a function $\mathbf{e}(x)$ with a continuous derivative such that

$$\mathbf{y}_j = \mathbf{y}(x_j) + h^p\mathbf{e}(x_j) + \mathbb{O}(h^{p+1}) \tag{1.2}$$

uniformly in j. If this is to be true for all x in $[a, b]$, then because $\mathbf{e}(x_r) = \mathbf{e}(a) + \mathbb{O}(h)$, it must be the case that

$$\mathbf{y}_r = \mathbf{y}(x_r) + h^p\mathbf{e}(a) + \mathbb{O}(h^{p+1}).$$

Unfortunately, this condition, which is necessary for the postulated expansion, is not true of all reasonable starting procedures. If we take $\mathbf{y}_0 = \mathbf{y}(a)$, which is invariably done in practice, the necessary condition requires that $\mathbf{e}(a) = \mathbf{0}$. This, however, then implies that we must have $\mathbf{y}_r = \mathbf{y}(x_r) + \mathbb{O}(h^{p+1})$ for all the initial values—the starting procedures must be more accurate than is required merely to get convergence of order p. The midpoint rule of order two was started with the first-order Euler method in the numerical experiment of Figure 7.3, so this start does not satisfy the necessary condition for the existence of an error expansion of the kind we seek. Although we do not pursue the matter here, a few words about the results seen in Figure 7.3 are appropriate. Roughly speaking, the magnified error has two components because the characteristic polynomial of this formula has two essential roots and the relatively inaccurate starting procedure causes both components to be present and of the same order in the magnified error.

We shall assume that the starting functions produce values for which

$$\mathbf{y}_r = \mathbf{s}_r(h) = \mathbf{y}(x_r) + h^p\boldsymbol{\lambda} + \mathbb{O}(h^{p+1}). \tag{1.3}$$

With such procedures we can satisfy the necessary condition by defining $\mathbf{e}(a) = \boldsymbol{\lambda}$. The question now is whether this describes situations important in practical computation. If we take $\mathbf{y}_0 = \mathbf{y}(a)$, the assumption is valid for one-step methods with $\boldsymbol{\lambda} = \mathbf{0}$, the case of the numerical experiment of Figure 7.1. It is also valid when the starting procedures are more accurate than necessary. This follows because if all the starting functions are of order at least p, then

$$\mathbf{y}_r = \mathbf{s}_r(h) = \mathbf{y}(x_r) + \mathbb{O}(h^{p+1})$$

and the condition holds with $\boldsymbol{\lambda} = \mathbf{0}$. Such a start will be referred to as an *accurate start*. This was the case of the numerical experiment of Figure 7.2 because we used the second-order improved Euler formula to start the second-order formula AB2. In both the situations pointed out, $\boldsymbol{\lambda} = \mathbf{0}$, so why not simply require this? Later we shall see that the assumption about the starting functions is just what we need to account for a change of step size, but in that context $\boldsymbol{\lambda}$ is generally not $\mathbf{0}$.

Proving the asymptotic representation for the error is straightforward, though the details are somewhat tedious. To see how it all goes, let us work through the case of LMMs. Recall that $\{\mathbf{y}_j\}$ satisfies

$$\mathbf{y}_r = \mathbf{s}_r(h), \qquad 0 \le r < k,$$

$$\sum_{i=0}^{k} \alpha_i \, \mathbf{y}_{j+1-i} = h \sum_{i=0}^{k} \beta_i \, \mathbf{F}(x_{j+1-i}, \mathbf{y}_{j+1-i}).$$

We begin our investigation of how well the sequence $\{\mathbf{y}(x_j) + h^p \mathbf{e}(x_j)\}$ satisfies these equations by looking at the starting values. They tell us how to define the initial value of $\mathbf{e}(x)$ because for $0 \le r < k$,

$$\begin{aligned}
\mathbf{y}_r - (\mathbf{y}(x_r) + h^p \mathbf{e}(x_r)) &= \mathbf{s}_r(h) - (\mathbf{y}(x_r) + h^p \mathbf{e}(a) + \mathbb{O}(h^{p+1})) \\
&= (\mathbf{y}(x_r) + h^p \boldsymbol{\lambda} + \mathbb{O}(h^{p+1})) \\
&\quad - (\mathbf{y}(x_r) + h^p \mathbf{e}(a) + \mathbb{O}(h^{p+1})) \\
&= h^p \, (\boldsymbol{\lambda} - \mathbf{e}(a)) + \mathbb{O}(h^{p+1}) = \mathbb{O}(h^{p+1}),
\end{aligned}$$

provided that $\mathbf{e}(a) = \boldsymbol{\lambda}$.

The difference equation will tell us how to define $\mathbf{e}(x)$ at other x. What we have to do is substitute the sequence $\{\mathbf{y}(x_j) + h^p \mathbf{e}(x_j)\}$ into the difference equation satisfied by $\{\mathbf{y}_j\}$ to obtain the discrepancies $\{\mathbf{d}_{j+1}\}$ and then see how we must define the magnified error function to make these discrepancies $\mathbb{O}(h^{p+2})$. The details become tedious when we come to simplifying the expressions enough to see what must be done. Direct substitution gives us

$$\sum_{i=0}^{k} \alpha_i \,(\mathbf{y}(x_{j+1-i}) + h^p \mathbf{e}(x_{j+1-i}))$$
$$= h \sum_{i=0}^{k} \beta_i \,\mathbf{F}(x_{j+1-i}, \mathbf{y}(x_{j+1-i}) + h^p \mathbf{e}(x_{j+1-i})) + \mathbf{d}_{j+1}.$$

At some point we must expect the local truncation error to play a role and it is now time for it to come on stage. The sequence $\{\mathbf{y}(x_j)\}$ satisfies

$$\sum_{i=0}^{k} \alpha_i \,\mathbf{y}(x_{j+1-i}) = h \sum_{i=0}^{k} \beta_i \,\mathbf{F}(x_{j+1-i}, \mathbf{y}(x_{j+1-i}))$$
$$+ C_{p+1} h^{p+1} \mathbf{y}^{(p+1)}(x_{j+1}) + \mathcal{O}(h^{p+2}),$$

where the local truncation error has been written in terms of the principal error function. The idea is to relate the discrepancy $\mathbf{d}_{j+1}$ to the principal error function by subtracting one of these equations from the other, but the fact that $\mathbf{F}$ is generally non-linear complicates the matter. To see more clearly what is going on, let us suppose for a time that the function is linear, i.e., $\mathbf{F}(x, \mathbf{y}) = J(x)\,\mathbf{y} + \mathbf{g}(x)$. In such a case the two expressions become

$$\sum_{i=0}^{k} \alpha_i(\mathbf{y}(x_{j+1-i}) + h^p \mathbf{e}(x_{j+1-i})) = h \sum_{i=0}^{k} \beta_i[J(x_{j+1-i})\,(\mathbf{y}(x_{j+1-i})$$
$$+ h^p \mathbf{e}(x_{j+1-i})) + \mathbf{g}(x_{j+1-i})] + \mathbf{d}_{j+1}$$

and

$$\sum_{i=0}^{k} \alpha_i \,\mathbf{y}(x_{j+1-i}) = h \sum_{i=0}^{k} \beta_i[J(x_{j+1-i})\,\mathbf{y}(x_{j+1-i}) + \mathbf{g}(x_{j+1-i})]$$
$$+ C_{p+1} h^{p+1} \mathbf{y}^{(p+1)}(x_{j+1}) + \mathcal{O}(h^{p+2}).$$

Subtraction gives

$$h^p \sum_{i=0}^{k} \alpha_i \,\mathbf{e}(x_{j+1-i}) = h^{p+1} \sum_{i=0}^{k} \beta_i \,J(x_{j+1-i})\,\mathbf{e}(x_{j+1-i})$$
$$- C_{p+1} h^{p+1} \mathbf{y}^{(p+1)}(x_{j+1}) + \mathbf{d}_{j+1} + \mathcal{O}(h^{p+2}).$$

Expanding $\mathbf{e}(x)$ and $J(x)$ about x_{j+1} leads first to

$$h^p \sum_{i=0}^{k} \alpha_i(\mathbf{e}(x_{j+1}) - ih\,\mathbf{e}'(x_{j+1}) + \mathcal{O}(h^2))$$
$$= h^{p+1} \sum_{i=0}^{k} \beta_i(J(x_{j+1})\,\mathbf{e}(x_{j+1}) + \mathcal{O}(h))$$
$$- C_{p+1} h^{p+1} \mathbf{y}^{(p+1)}(x_{j+1}) + \mathbf{d}_{j+1} + \mathcal{O}(h^{p+2})$$

and then to

$$h^p\left(\sum_{i=0}^{k}\alpha_i\right)\mathbf{e}(x_{j+1}) - h^{p+1}\left(\sum_{i=0}^{k} i\alpha_i\right)\mathbf{e}'(x_{j+1}) + \mathcal{O}(h^{p+2})$$

$$= h^{p+1}\left(\sum_{i=0}^{k}\beta_i\right)J(x_{j+1})\,\mathbf{e}(x_{j+1}) + \mathcal{O}(h^{p+2})$$

$$- C_{p+1}h^{p+1}\mathbf{y}^{(p+1)}(x_{j+1}) + \mathbf{d}_{j+1} + \mathcal{O}(h^{p+2})$$

Obviously we need some properties of the coefficients defining the method to simplify this further. A convergent, normalized LMM has $\rho(1) = 0$ and $\rho'(1) = 1 = \sigma(1)$. In terms of the coefficients this is

$$\rho(1) = \sum_{i=0}^{k}\alpha_i = 0,$$

$$\rho'(1) = 1 = \sum_{i=1}^{k}(k - i)\,\alpha_i = k\rho(1) - \sum_{i=1}^{k} i\alpha_i = -\sum_{i=1}^{k} i\alpha_i,$$

$$\sigma(1) = 1 = \sum_{i=0}^{k}\beta_i.$$

When these facts are used in the expression above, it simplifies to

$$h^{p+1}\mathbf{e}'(x_{j+1}) + \mathcal{O}(h^{p+2}) = h^{p+1}J(x_{j+1})\,\mathbf{e}(x_{j+1})$$
$$- C_{p+1}h^{p+1}\mathbf{y}^{(p+1)}(x_{j+1}) + \mathbf{d}_{j+1} + \mathcal{O}(h^{p+2}),$$

and then a little manipulation gives us

$$\mathbf{d}_{j+1} = h^{p+1}\left[\mathbf{e}'(x_{j+1}) - J(x_{j+1})\,\mathbf{e}(x_{j+1})\right.$$
$$\left. + C_{p+1}h^{p+1}\mathbf{y}^{(p+1)}(x_{j+1})\right] + \mathcal{O}(h^{p+2}).$$

Now we have it. After all the simplifications we see that if we define $\mathbf{e}(x)$ so that

$$\mathbf{e}'(x) = J(x)\,\mathbf{e}(x) - C_{p+1}h^{p+1}\mathbf{y}^{(p+1)}(x),$$

then the discrepancy will be $\mathcal{O}(h^{p+2})$. The definition of $\mathbf{e}(a)$ has already made the starting discrepancy $\mathcal{O}(h^{p+1})$. These facts and stability of the numerical method imply that

$$\mathbf{y}_j = \mathbf{y}(x_j) + h^p\,\mathbf{e}(x_j) + \mathcal{O}(h^{p+1})$$

uniformly in j.

Now let us go back and deal with $\mathbf{F}$ that are not linear. The only place we used linearity was when we simplified the result of subtracting the relation satisfied by $\{\mathbf{y}(x_j)\}$ that involved the principal error function from the relation satisfied by $\{\mathbf{y}(x_j) + h^p\, \mathbf{e}(x_j)\}$ that involved the discrepancy $\mathbf{d}_{j+1}$. In the nonlinear case we must deal with terms of the form

$$\Delta_i = \mathbf{F}(x_{j+1-i}, \mathbf{y}(x_{j+1-i}) + h^p\, \mathbf{e}(x_{j+1-i})) - \mathbf{F}(x_{j+1-i}, \mathbf{y}(x_{j+1-i})).$$

As usual in such situations, we linearize with a mean value theorem to get

$$\Delta_i = \frac{\partial \mathbf{F}}{\partial \mathbf{y}}(x_{j+1-i}, \mathbf{y}(x_{j+1-i}))\, h^p\, \mathbf{e}(x_{j+1-i}) + \mathbb{O}(h^{2p}\, \|\mathbf{e}(x_{j+1-i})\|^2).$$

In the linear case we had

$$\Delta_i = \frac{\partial \mathbf{F}}{\partial \mathbf{y}}\,(x_{j+1-i},\, \mathbf{y}(x_{j+1-i}))\; h^p\; \mathbf{e}(x_{j+1-i})$$

because $J(x_{j+1-i},\, \mathbf{y}(x_{y+1-i}))$ is the Jacobian of $\mathbf{F}$ in this case. The only difference due to $\mathbf{F}$ being nonlinear is the presence of errors due to linearizing the Δ_i. These errors are absorbed in the $\mathbb{O}(h^{p+2})$ term and drop out of sight.

For constant step size the proof of the asymptotic expression for the error of predictor–corrector methods goes the same way, the details just being a little more tedious, and the same is true of Runge–Kutta methods. Let us now state formally the conclusions.

Theorem 1. *A stable LMM of order p is used to approximate the solution* $\mathbf{y}(x)$ *of*

$$\mathbf{y}' = \mathbf{F}(x, \mathbf{y}), \qquad a \leq x \leq b, \qquad \mathbf{y}(a) = \mathbf{A}. \tag{1.1}$$

Suppose that the step size is a constant h in each integration and that the starting values $\mathbf{y}_r$ *satisfy*

$$\mathbf{y}_r = \mathbf{s}_r(h) = \mathbf{y}(x_r) + h^p \boldsymbol{\lambda} + \mathbb{O}(h^{p+1}). \tag{1.3}$$

Then

$$\mathbf{y}_j = \mathbf{y}(x_j) + h^p \mathbf{e}(x_j) + \mathbb{O}(h^{p+1}) \tag{1.2}$$

uniformly in j. Here $\mathbf{e}(x)$ *is the solution of*

$$\mathbf{e}'(x) = \frac{\partial \mathbf{F}}{\partial \mathbf{y}}\,(x,\, \mathbf{y}(x))\; \mathbf{e}(x) - \boldsymbol{\psi}(x,\, \mathbf{y}(x)), \qquad \mathbf{e}(a) = \boldsymbol{\lambda},$$

and $\boldsymbol{\psi}(x, \mathbf{y})$ *is the principal error function of the method. With the same assumptions, the result is also true for a stable predictor–corrector (PECE) pair of order p and for a Runge–Kutta formula of order p.*

The role of the predictor in a predictor–corrector pair was discussed earlier in the context of a bound on the error. With an approximation for the error we can be more precise. Recall that in a PECE implementation, a predictor–corrector pair with a predictor of order at least as high as the corrector has a local truncation error that to leading order is the same as the (implicit) corrector formula used by itself—the principal error functions are the same. The magnified error function $\mathbf{e}(x)$ of Theorem 1 depends on the method only through its principal error function. This implies that to leading order, even one correction of a sufficiently accurate predicted value provides a method that has the same error as the implicit corrector iterated to completion. If the predictor is of order one lower than the corrector, the principal error function of the predictor–corrector formula is different from that of the implicit corrector and because of this, the magnified error functions are different. The error of the predictor–corrector pair is not necessarily worse than that of the implicit formula, merely different.

The numerical results of Figures 7.1 and 7.2 illustrate the theory we have developed. Let us now work through some simple examples.

EXAMPLE 2. The principal error function of the forward Euler method is seen from the Taylor expansion

$$\begin{aligned}\mathbf{y}(x_{j+1}) &= \mathbf{y}(x_j) + h\,\mathbf{y}'(x_j) + \frac{h^2}{2}\,\mathbf{y}''(x_j) + \mathcal{O}(h^3)\\ &= \mathbf{y}(x_j) + h\,\mathbf{F}(x_j, \mathbf{y}(x_j)) + \frac{h^2}{2}\,\mathbf{y}''(x_j) + \mathcal{O}(h^3),\end{aligned}$$

to be $1/2\ \mathbf{y}''(x)$. When the step size is a constant h and we start with the given initial value, the magnified error function is defined by

$$\mathbf{e}'(x) = \frac{\partial \mathbf{F}}{\partial \mathbf{y}}(x, \mathbf{y}(x))\,\mathbf{e} - \frac{1}{2}\,\mathbf{y}''(x), \qquad \mathbf{e}(a) = \mathbf{0}.$$

Suppose now that we wish to solve the scalar problem

$$y' = y, \qquad y(0) = 1.$$

The solution $y(x) = \exp(x)$, so the equation for $e(x)$ is

$$e' = e - \frac{1}{2}\exp(x), \qquad e(0) = 0.$$

It is easily verified that

$$e(x) = -\frac{1}{2}\,x\exp(x).$$

The theory then says that

$$y_j = y(x_j) + h\,e(x_j) + \mathcal{O}(h^2) = \exp(x_j) - \frac{h}{2}x_j\exp(x_j) + \mathcal{O}(h^2).$$

As the integration proceeds, the absolute error

$$|y_j - y(x_j)| = h\,|e(x_j)| + \mathcal{O}(h^2) = \frac{h}{2}x_j\exp(x_j) + \mathcal{O}(h^2)$$

grows exponentially fast. In contrast, the relative error

$$\frac{|y_j - y(x_j)|}{|y(x_j)|} = \frac{h}{2}x_j + \mathcal{O}(h^2)$$

grows at only a modest rate as the integration proceeds. The details for the problem

$$y' = -y, \qquad y(0) = 1$$

are very similar. In this case $y(x) = \exp(-x)$ and

$$e' = -e - \frac{1}{2}\exp(-x), \qquad e(0) = 0.$$

The magnified error function is

$$e(x) = -\frac{1}{2}x\exp(-x),$$

and

$$y_j = y(x_j) - \frac{h}{2}x_j\exp(-x_j) + \mathcal{O}(h^2).$$

In this case the absolute error

$$|y_j - y(x_j)| = \frac{h}{2}x_j\exp(-x_j) + \mathcal{O}(h^2)$$

has a more complex behavior. The error starts with the value zero and (to leading order) increases to a maximum after which it decreases exponentially fast. On the other hand, the relative error

$$\frac{|y_j - y(x_j)|}{|y(x_j)|} = \frac{h}{2}x_j + \mathcal{O}(h^2)$$

behaves just as for the growing exponential. This is, however, a little surprising because the solution is decaying exponentially fast and the step size is not being adapted to the solution.

EXERCISE 1. Solve these two example problems numerically on the interval [0, 1] and compare the results to the theory.

EXAMPLE 3. As a numerical example of the theory we consider the second order BDF2. It is

$$\mathbf{y}_{j+1} = \frac{1}{3}[4\mathbf{y}_j - \mathbf{y}_{j-1}] + \frac{2h}{3}\mathbf{F}(x_{j+1}, \mathbf{y}_{j+1}),$$

and its truncation error is

$$\frac{1}{3}h^3\,\mathbf{y}^{(3)}(x_j) + \mathcal{O}(h^4).$$

If we start with $\mathbf{y}_0 = \mathbf{y}(a)$ and use the improved Euler formula to form $\mathbf{y}_1$, we have an accurate start so that $\boldsymbol{\lambda} = \mathbf{0} = \mathbf{e}(a)$. According to the theory, the leading term in the asymptotic result will be the same whether we use a second-order Runge–Kutta method to get $\mathbf{y}_1$ or take $\mathbf{y}_1 = \mathbf{y}(x_1)$, and this will be verified computationally.

To make matters simple we approximate the solution $\exp(-x)$ of $y' = -y$, $y(0) = 1$ on [0, 1]. With a linear, scalar equation like this one, we can evaluate easily the implicit BDF2. The equation for the magnified error function becomes

$$e'(x) = -e(x) + \frac{1}{3}\exp(-x), \qquad e(0) = 0.$$

It is easy to verify that

$$e(x) = \frac{x}{3}\exp(-x),$$

and the theory then says that

$$y_j = \exp(-x_j) + \frac{h^2}{3}x_j\exp(-x_j) + \mathcal{O}(h^3).$$

In particular, at the end of the interval of integration, $x_N = 1$,

$$\text{ratio} = \frac{y_N - \exp(-x_N)}{h^2} = \frac{1}{3}\exp(-1) + \mathcal{O}(h) = 0.1226\ldots + \mathcal{O}(h).$$

The problem was integrated with a sequence of step sizes h and this ratio calculated. The table presents the values RATIO1 resulting from the second order Runge–Kutta start and the values RATIO2 resulting from starting with the true solution.

h	RATIO1	RATIO2
10^{-1}	0.1013	0.1119
10^{-2}	0.1208	0.1217
10^{-3}	0.1224	0.1225
10^{-4}	0.1227	0.1227

It is seen that even with rather large step sizes, there is good agreement between the theory and the computed results.

EXERCISE 2. In Example 2 of Chapter 5, the problem $y' = -y$, $y(0) = 1$ was solved for $0 \leq x \leq 1$ with AB2 and a constant step size. For each $h = 10^{-M}$, $M = 1, 2, 3, 4$, the maximum error of the integration was computed:

$$E(h) = \max_{0 \leq j \leq 1/h} |y(jh) - y_j|$$

When the method was started accurately, it was found that

$$\lim_{h \to 0} \frac{E(h)}{h^2} \approx 0.15328.$$

Explain this theoretically, and in particular, determine what the theoretical limit is. The expression for the truncation error of AB2 given in Chapter 4 will be useful, as will study of the preceding examples. Pay attention to the fact that it was the *maximum* error on the interval that was measured in the numerical experiment rather than the error at a given point.

The theorems of Chapter 5 say that an inaccurate start might reduce the rate of convergence. Specifically, if the method is of order p and the starting values are accurate only to order $q < p$, convergence is guaranteed only to be as fast as order q. It was asserted that this is not a defect of the analysis, rather the fact of the matter. Now we can be more precise. Suppose that

$$\mathbf{s}_r(h) = \mathbf{y}(x_r) + \boldsymbol{\lambda} h^q + \mathcal{O}(h^{q+1})$$

for some $\boldsymbol{\lambda} \neq \mathbf{0}$. Any method of order p can be regarded as a method of order $q < p$ with a principal error function $\boldsymbol{\psi}(x, \mathbf{y})$ that is identically $\mathbf{0}$:

$$\text{local truncation error} = \mathcal{O}(h^{p+1}) = \mathbf{0} h^{q+1} + \mathcal{O}(h^{q+2}).$$

With a principal error function that is identically zero, we have

$$\mathbf{y}_j = \mathbf{y}(x_j) + h^q \, \mathbf{e}(x_j) + \mathcal{O}(h^{q+1}),$$

where

$$\mathbf{e}' = \frac{\partial \mathbf{F}}{\partial \mathbf{y}} (x, \mathbf{y}(x))\, \mathbf{e}, \qquad \mathbf{e}(a) = \boldsymbol{\lambda}.$$

Wherever $\mathbf{e}(x) \neq \mathbf{0}$, the error is of exact order q. This means that an inaccurate start can spoil the accuracy possible with a high order formula throughout the entire interval of integration.

Starting values

$$\mathbf{s}_r(h) = \mathbf{y}(x_r) + \boldsymbol{\lambda}_r\, h^p + \mathbb{O}(h^{p+1})$$

with different $\boldsymbol{\lambda}_r$ disturb the asymptotic expansion of the error of $\mathbf{y}_j$. Let $\mathbf{z}_j$ be the numerical solution when the starting values are

$$\mathbf{z}_r = \mathbf{y}(x_r) + \boldsymbol{\lambda} h^p + \mathbb{O}(h^{p+1}).$$

We have developed an asymptotic expansion of the error of $\mathbf{z}_j$ that is valid for *any* value of $\boldsymbol{\lambda}$. One might wonder whether it would be possible to choose $\boldsymbol{\lambda}$ so that $\mathbf{z}_j$ and $\mathbf{y}_j$ have the same error to leading order. Of course this cannot be true near $x = a$, but it seems possible that the effect of the start decay out as x increases so that it could be true away from the initial point. For a method that satisfies the strong root condition, a vector $\boldsymbol{\lambda}$ *can* be deduced from the $\boldsymbol{\lambda}_r$ such that the result derived earlier is still true provided that we stay away from the initial point by considering a subinterval $[a', b]$ where $a < a'$. The behavior of the error can also be sorted out when there are essential roots other than the principal one. It is more complicated because as we saw with the midpoint rule, starting effects persist and the behavior does not have the same form as that after an accurate start. Henrici [1962, 1963], Stetter [1973], and Hairer, Nørsett, and Wanner [1987] derive results of this kind. We do not present such results here because the methods that interest us all satisfy the strong root condition and they are normally started accurately. We also do not take up general methods that satisfy the strong root condition, but to expose the ideas, let us look at a few of the details for ABν with constant step size h. Suppose that

$$\mathbf{y}_r = \mathbf{s}_r(h) = \mathbf{y}(x_r) + \boldsymbol{\lambda}_r h^\nu + \mathbb{O}(h^{\nu+1}), \qquad 0 \le r < \nu$$

and

$$\mathbf{y}_{j+1} = \mathbf{y}_j + h \sum_{k=1}^{\nu} \beta_k\, \mathbf{F}(x_{j+1-k}, \mathbf{y}_{j+1-k}) \qquad j \ge \nu - 1.$$

The analysis presented earlier says that for any $\boldsymbol{\lambda}$, if

$$\mathbf{z}_r = \mathbf{y}(x_r) + \boldsymbol{\lambda} h^\nu + \mathbb{O}(h^{\nu+1}), \qquad 0 \le r < \nu$$

and

$$\mathbf{z}_{j+1} = \mathbf{z}_j + h \sum_{k=1}^{\nu} \beta_k \mathbf{F}(x_{j+1-k}, \mathbf{z}_{j+1-k}), \qquad j \geq \nu - 1,$$

then

$$\mathbf{z}_j = \mathbf{y}(x_j) + h^{\nu}\mathbf{e}(x_j) + \mathcal{O}(h^{\nu+1}).$$

Let us now bound the difference $\boldsymbol{\varepsilon}_j = \mathbf{y}_j - \mathbf{z}_j$. In familiar fashion we find that for $j \geq \nu - 1$,

$$\|\boldsymbol{\varepsilon}_{j+1}\| \leq \|\boldsymbol{\varepsilon}_j\| + hL \sum_{k=1}^{\nu} |\beta_k| \, \|\boldsymbol{\varepsilon}_{j+1-k}\|,$$

and in particular,

$$\|\boldsymbol{\varepsilon}_{\nu}\| \leq \|\boldsymbol{\varepsilon}_{\nu-1}\| + hL \sum_{k=1}^{\nu} |\beta_k| \, \|\boldsymbol{\varepsilon}_{\nu-k}\|.$$

For any choice of $\boldsymbol{\lambda}$ we have $\|\boldsymbol{\varepsilon}_r\| = \|(\boldsymbol{\lambda}_r - \boldsymbol{\lambda})\, h^{\nu} + \mathcal{O}(h^{\nu+1})\| = \mathcal{O}(h^{\nu})$ for $0 \leq r < \nu$. Let us choose $\boldsymbol{\lambda} = \boldsymbol{\lambda}_{\nu-1}$ so that $\|\boldsymbol{\varepsilon}_{\nu-1}\| = \mathcal{O}(h^{\nu+1})$. Because of the factor of h in the last inequality, we then have $\|\boldsymbol{\varepsilon}_{\nu}\| = \mathcal{O}(h^{\nu+1})$. Repetition of the argument for $j = \nu, \nu + 1, \ldots$ shows that there is a block of ν successive $\boldsymbol{\varepsilon}_j$ starting at $j = \nu - 1$ for which $\|\boldsymbol{\varepsilon}_j\| = \mathcal{O}(h^{\nu+1})$. Lemma 1 of Chapter 5 can now be applied to show that all subsequent $\|\boldsymbol{\varepsilon}_j\|$ are $\mathcal{O}(h^{\nu+1})$—this is like an integration with starting errors that are $\mathcal{O}(h^{\nu+1})$ and no local truncation errors. Except for the first few steps we have

$$\mathbf{y}_j = \mathbf{z}_j + \mathcal{O}(h^{\nu+1}) = \mathbf{y}(x_j) + h^{\nu}\mathbf{e}(x_j) + \mathcal{O}(h^{\nu+1}),$$

the result we wanted.

EXERCISE 3. Exercise 2 is devoted to a theoretical explanation of the numerical experiment of Example 2 of Chapter 5 that involved an integration with AB2 and an accurate start. This exercise takes up another experiment of the example that involved an integration with AB2 started inaccurately with the forward Euler method.

In general, when $\mathbf{y}_0 = \mathbf{y}(a)$ and $\mathbf{y}_1$ comes from the forward Euler method, what are $\boldsymbol{\lambda}_0$ and $\boldsymbol{\lambda}_1$? The result just established says that AB2 started in this way has an asymptotic expansion for its error that to leading order is the same as what we would get if we had started with

$$\mathbf{y}_r = \mathbf{s}_r(h) = \mathbf{y}(x_r) + h^2\boldsymbol{\lambda} + \mathcal{O}(h^3), \qquad r = 0, 1$$

where $\boldsymbol{\lambda} = \boldsymbol{\lambda}_1$. It was found numerically that

$$\lim_{h \to 0} \frac{E(h)}{h^2} \approx 0.49998.$$

Prove that the theoretical limit is 0.5.

It was not difficult to extend convergence results for constant step size to a quasi-constant step size integration, so let us try to do the same for the asymptotic expansion. In this model, when integrating (1.1) from a to b, the step size is permitted to change at a finite number of fixed points $a < c_1 < c_2 < \ldots < c_s < b$, and nowhere else. In the convergence analysis we supposed that a step size h_1 is used in the integration from a to c_1, a step size $h_2 \leq \Gamma h_1$ is used in the integration from c_1 to c_2, and so forth. Naturally, if we wish to discuss the actual behavior of the error, we must specify precisely how h_2 is related to h_1 and not just in the current integration, but in all the integrations of our convergence model. This will be done by means of a step size selection function.

A *step size selection function* is a function $\theta(x)$ that is piecewise continuous and differentiable on $[a, b]$. It has at most a finite number of discontinuities in (a, b) and at each discontinuity it has both right and left hand limits. It satisfies $0 < \Delta \leq \theta(x) \leq 1$ for all x in $[a, b]$ and $\theta(\zeta) = 1$ for some $\zeta \in [a, b]$. In a particular integration with maximum step size H, the step size at a mesh point x_j is $\theta(x_j+)H$. A sequence of integrations is considered for which $H \to 0$.

A *quasi-constant step size model* is described by a piecewise constant step size selection function that has discontinuities at the c_m. When integrating from a to c_1, the step size is the constant $h_1 = \theta(a)H = \theta(c_1-)H$ and when integrating from c_1 to c_2, it is $h_2 = \theta(c_1+)H = \mu h_1$ where $\mu = \theta(c_1+)/\theta(c_1-)$, and so forth. The asymptotic expansion of the error for constant step size can be used much like the convergence results for constant step size to obtain an asymptotic expansion in a quasi-constant step size model. Indeed, we need only refine the argument to show that the starting values after a change of step size satisfy the hypotheses of Theorem 1. The integration from a to $x_N = c_1$ is done with the step size h_1. It yields the numerical solution values $\mathbf{y}_{N-r} \approx \mathbf{y}(c_1 - rh_1)$ for $r = 0, 1, \ldots, m$. The integration from c_1 to c_2 is done with the constant step size $h_2 = \mu h_1$. We change to this step size at c_1 by evaluating the interpolating polynomial

$$\mathbf{P}(x) = \sum_{r=0}^{m} \mathbf{y}_{N-r} L_r(x)$$

to obtain the starting values

$$\mathbf{Y}_{N-j} = \mathbf{P}(c_1 - jh_2), \qquad 0 \leq j < k.$$

At this point in the proof of convergence in Chapter 5, we applied a convergence theorem for constant step size to the interval $[a, c_1]$ to conclude that $\mathbf{y}_{N-r} = \mathbf{y}(c_1 - rh_1) + \mathcal{O}(h_1^{p+1})$. Here we apply Theorem 1 to refine this to

$$\mathbf{y}_{N-r} = \mathbf{y}(c_1 - rh_1) + h_1^p\, \mathbf{e}(c_1 - rh_1) + \mathcal{O}(h_1^{p+1}).$$

As in the convergence proof, we substitute this expansion into the expression for the interpolant to find

$$\begin{aligned}\mathbf{Y}_{N-j} &= \sum_{r=0}^{m} \mathbf{y}(c_1 - rh_1)\, L_r(c_1 - jh_2) \\ &\quad + \sum_{r=0}^{m} h_1^p\, \mathbf{e}(c_1 - rh_1)\, L_r(c_1 - jh_2) + \mathcal{O}(h_1^{p+1}).\end{aligned}$$

The first sum arose earlier and we showed using interpolation theory that

$$\sum_{r=0}^{m} \mathbf{y}(c_1 - rh_1)\, L_r(c_1 - jh_2) = \mathbf{y}(c_1 - jh_2) + \mathcal{O}(h_1^{m+1}).$$

A similar argument applied to the function $\mathbf{e}(x)$ leads to

$$\sum_{r=0}^{m} h_1^p\, \mathbf{e}(c_1 - rh_1)\, L_r(c_1 - jh_2) = h_1^p\, \mathbf{e}(c_1 - jh_2) + \mathcal{O}(h_1^{p+1}).$$

(The result is different because the function $\mathbf{e}(x)$ may not be as smooth as $\mathbf{y}(x)$). Using $h_2 = \mu h_1$ to relate these results to the new step size and then expanding $\mathbf{e}(c_1 - jh_2)$, we find that

$$\begin{aligned}\mathbf{Y}_{N-j} &= \mathbf{y}(c_1 - jh_2) + h_2^p\, \mu^{-p}\, \mathbf{e}(c_1 - jh_2) + \mathcal{O}(h_2^{p+1}) \\ &= \mathbf{y}(c_1 - jh_2) + h_2^p\, \mu^{-p}\, \mathbf{e}(c_1 -) + \mathcal{O}(h_2^{p+1}).\end{aligned}$$

We see now that these starting values for the integration from c_1 to c_2 satisfy (1.3) with the constant $\boldsymbol{\lambda} = \mu^{-p}\, \mathbf{e}(c_1 -)$. Theorem 1 then gives us the asymptotic expansion

$$\mathbf{y}_j = \mathbf{y}(x_j) + h_2^p\, \mathbf{E}(x_j) + \mathcal{O}(h_2^{p+1})$$

on the interval $[c_1, c_2]$. Here

$$\begin{aligned}\mathbf{E}(c_1 +) &= \mu^{-p}\, \mathbf{e}(c_1 -), \\ \mathbf{E}' &= \frac{\partial \mathbf{F}}{\partial \mathbf{y}}(x, \mathbf{y}(x))\, \mathbf{E} - \boldsymbol{\psi}(x), \qquad c_1 + \leq x < c_2.\end{aligned}$$

A tidier way to write this is to extend the definition of $\mathbf{e}(x)$ to $[c_1, c_2]$ by

$$\mathbf{e}(c_1+) = \mathbf{e}(c_1-),$$
$$\mathbf{e}' = \frac{\partial \mathbf{F}}{\partial \mathbf{y}}(x, \mathbf{y}(x))\,\mathbf{e} - \mu^p \boldsymbol{\psi}(x), \qquad c_1+ \le x < c_2,$$

since then $\mathbf{E}(x) = \mu^{-p}\,\mathbf{e}(x)$ and

$$\begin{aligned}\mathbf{Y}_j &= \mathbf{y}(x_j) + h_2^p\,\mu^{-p}\,\mathbf{E}(x_j) + \mathbb{O}(h_2^{p+1})\\ &= \mathbf{y}(x_j) + h_1^p\,\mathbf{e}(x_j) + \mathbb{O}(h_1^{p+1}) \qquad c_1+ \le x < c_2.\end{aligned}$$

With the extended definition of $\mathbf{e}(x)$, this last form of the expansion is valid for the whole integration from a to c_2. Repetition of the argument deals with a finite number of changes of step size at fixed points $a < c_1 < c_2 < \ldots < c_s < b$. On writing the expressions in terms of the step size selection function, we have

Theorem 2. *A stable LMM of order p is used to approximate the solution* $\mathbf{y}(x)$ *of*

$$\mathbf{y}' = \mathbf{F}(x, \mathbf{y}), \qquad a \le x \le b, \qquad \mathbf{y}(a) = \mathbf{A}. \tag{1.1}$$

Suppose that the starting values $\mathbf{y}_r$ *satisfy*

$$\mathbf{y}_r = \mathbf{s}_r(h) = \mathbf{y}(x_r) + H^p\boldsymbol{\lambda} + \mathbb{O}(H^{p+1}). \tag{1.3}$$

The step size is specified by a piecewise constant step size selection function $\theta(x)$. *The step size is changed by means of interpolating at least p consecutive solution values. Let the function* $\mathbf{e}(x)$ *be defined on each subinterval* $[a', b')$ *where* $\theta(x)$ *is constant as the solution of*

$$\mathbf{e}' = \frac{\partial \mathbf{F}}{\partial \mathbf{y}}(x, \mathbf{y}(x))\,\mathbf{e} - \theta^p(x)\,\boldsymbol{\psi}(x, y(x)) \qquad a' \le x < b'$$

with initial value

$$\begin{cases}\mathbf{e}(a) = \boldsymbol{\lambda} & \text{if } a' = a,\\ \mathbf{e}(a'+) = \mathbf{e}(a'-) & \text{otherwise.}\end{cases}$$

Here $\boldsymbol{\psi}(x, \mathbf{y})$ *is the principal error function of the method. Then*

$$\mathbf{y}_j = \mathbf{y}(x_j) + H^p\,\mathbf{e}(x_j) + \mathbb{O}(H^{p+1})$$

uniformly on $[a, b]$. *With the same assumptions, the result is also true for a stable predictor–corrector (PECE) pair of order p and for a Runge–Kutta formula of order p.*

In Chapter 5 we proved that Runge–Kutta and Adams methods converge even when the step size might be changed at any step. If we are to hope to determine the asymptotic behavior of the error, we must be more specific about the step sizes used. Once again we suppose that the step sizes are given in terms of a maximum step size H and a step size selection function $\theta(x)$. The existence of a positive lower bound Δ for $\theta(x)$ is trivial in the quasi-constant step size model, but not in the present circumstances. It guarantees that a finite number of steps will suffice for each integration. Our convergence theorems for the Adams methods require that the ratio of successive steps be uniformly bounded. This is true for the present model. Indeed, where θ is differentiable, the step sizes change smoothly because

$$\frac{h_{j+1}}{h_j} = \frac{\theta(x_{j+1})H}{\theta(x_j)H} = \frac{\theta(x_j) + h_j\theta'(\xi)}{\theta(x_j)} = 1 + \mathcal{O}(H).$$

With the step sizes specified in the manner described, Theorem 1 is true of fully variable step size implementations of the Adams methods. It is easy to prove this directly by following the proof given for constant step size and using the stability and convergence results for variable step size instead of those for constant step size. For these methods the expansion of Theorem 2 is just a special case resulting from a piecewise constant step size selection function, but, of course, the way that the step size is changed is different in the two implementations.

THEOREM 3. *The Adams–Bashforth method of order ν, ABν, is used to approximate the solution $\mathbf{y}(x)$ of*

$$\mathbf{y}' = \mathbf{F}(x, \mathbf{y}), \qquad a \le x \le b, \qquad \mathbf{y}(a) = \mathbf{A}. \tag{1.1}$$

Suppose that the starting values $\mathbf{y}_r$ satisfy

$$\mathbf{y}_r = \mathbf{s}_r(H) = \mathbf{y}(x_r) + H^\nu \boldsymbol{\lambda} + \mathcal{O}(H^{\nu+1}). \tag{1.3}$$

The step size is specified by a step size selection function $\theta(x)$ and a fully variable step size implementation is used. Let the function $\mathbf{e}(x)$ be defined as the solution of

$$\mathbf{e}' = \frac{\partial \mathbf{F}}{\partial \mathbf{y}}(x, \mathbf{y}(x))\,\mathbf{e} - \theta^p(x)\,\boldsymbol{\psi}(x, y(x)) \qquad a \le x \le b,$$
$$\mathbf{e}(a) = \boldsymbol{\lambda}.$$

Here $\boldsymbol{\psi}(x, \mathbf{y})$ is the principal error function of the method. Then

$$\mathbf{y}_j = \mathbf{y}(x_j) + H^\nu\,\mathbf{e}(x_j) + \mathcal{O}(H^{\nu+1})$$

uniformly on [a, b]. With the same assumptions, the same result is true of the Adams–Moulton method of order ν, AMν. With the same assumptions, the same result is true of an ABν^*–AMν predictor–corrector (PECE) pair with $\nu^* \geq \nu - 1$.

These results suppose that the step sizes used are specified in advance. We shall see that the results can be used to describe integrations done when codes select their step sizes automatically, but then the step size h_j chosen at a point x_j may not be precisely $\theta(x_j)H$. Because of this we need to observe that the results developed remain true if the step size is "almost" that given by a step size selection function. Specifically, the expansions of Theorems 2 and 3 remain true if only

$$h_j = \theta(x_j)H + \mathcal{O}(H^2).$$

§2 Estimation of the Global, or True, Error

Just knowing that there is an asymptotic expansion of the form

$$\mathbf{y}_j = \mathbf{y}(x_j) + H^p\, \mathbf{e}(x_j) + \mathcal{O}(H^{p+1})$$

allows us to do something that is quite interesting. Suppose we carry out two integrations using the same step size selection function, one integration with maximum step size H and the other with maximum step size $H/2$. At any point $b = x_N$,

$$\mathbf{y}_N = \mathbf{y}(b) + H^p\, \mathbf{e}(b) + \mathcal{O}(H^{p+1}),$$

$$\mathbf{y}_{2N}^* = \mathbf{y}(b) + \left(\frac{H}{2}\right)^p \mathbf{e}(b) + \mathcal{O}(H^{p+1}),$$

where $\mathbf{y}_{2N}^*$ is the result of $2N$ steps, each of half the length of the step taken in the primary integration. Subtracting these expressions results in

$$\mathbf{y}_N - \mathbf{y}_{2N}^* = (2^p - 1)\left(\frac{H}{2}\right)^p \mathbf{e}(b) + \mathcal{O}(H^{p+1}),$$

whence it is obvious that we can approximate the true, or global, error of the more accurate result by

$$\frac{(\mathbf{y}_{2N}^* - \mathbf{y}_N)}{2^p - 1} = -\left(\frac{H}{2}\right)^p \mathbf{e}(b) + \mathcal{O}(H^{p+1}).$$

(The error of the less accurate result can be estimated, too.) This principle of *global extrapolation* is the basis for some production-grade codes that not only solve the initial value problem, but also estimate the global error as they do it. The earliest, GERK, (pronounced "jerk") was written by Watts and Shampine and is now available from a number of sources. Another early code written by Gladwell [1979] is available in the NAG library. These codes control the local error in a conventional way and just report the approximate global error. Although the basic idea is simple, there are many important practical details that are taken up in the paper of Shampine and Watts [1976b]. For one thing, it is not easy to decide automatically the validity of the asymptotic expansion of the error that underlies the estimate. The results of computations with $H = 0.3$ and $H/2 = 0.15$ displayed in Figure 5.1b show how the expansion might fail to be valid both because the step size is too large and because one of the integrations becomes unstable.

EXERCISE 4. Augment the experiments of Exercise 1 to include estimation of the true error by global extrapolation.

EXERCISE 5. As described, global extrapolation uses two independent integrations, a primary integration and a secondary integration with half the step size of the primary, to estimate the global error of the more accurate integration. In GERK these integrations are carried out simultaneously. This code is based on the Fehlberg (4, 5) pair and it adapts the step size to the solution in the usual way so as to keep the local error smaller than a specified tolerance. In principle the step size could be chosen in either of the two integrations. For a given tolerance it would be cheaper to select the step size in the secondary integration and use a bigger step size in the primary integration that is used only for the estimation of the error. GERK does not do this because the computations can be unstable. How? What is going on?

§3 Error Control, Step Size Adjustment, and Efficiency

So far in this chapter it has been assumed that somehow the step sizes have been specified in advance. The main reason for monitoring the local error in the course of an integration is to have some assurance that the problem has been solved in a reasonable sense. However, it also permits an adaptation of the step size to the problem. In the first instance, this provides an efficient solution of the problem, but more important, it greatly increases the reliability of solution and makes the solution of difficult problems possible. Monitoring the error and adjusting the step size both raise questions and afford new possibilities that we take up now.

There is a distinction between control of local error and control of local truncation error. Let us clarify the distinction by discussing explicit one-step methods. These methods have the form

$$\mathbf{y}_{j+1} = \mathbf{y}_j + \mathbf{h}_j \,\mathbf{\Phi}(x_j, \mathbf{y}_j, \mathbf{F}, h).$$

The local solution $\mathbf{u}(x)$ is the solution of the problem $\mathbf{u}' = \mathbf{F}(x, \mathbf{u})$, $\mathbf{u}(x_j) = \mathbf{y}_j$ and the local error is

$$\mathbf{le}_j = \mathbf{u}(x_j + h_j) - \mathbf{y}_{j+1}.$$

In the particular case of $\mathbf{y}_j = \mathbf{y}(x_j)$, the local solution coincides with the true solution $\mathbf{y}(x)$, and the local error is called the local truncation error,

$$\mathbf{lte}_j = \mathbf{y}(x_j + h_j) - \mathbf{y}_{j+1}.$$

In terms of the principal error function, these quantities for a method of order p have the forms

$$\mathbf{le}_j = h_j^{p+1}\, \boldsymbol{\psi}(x_j, \mathbf{y}_j) + \mathcal{O}(h_j^{p+2}),$$

and

$$\mathbf{lte}_j = h_j^{p+1}\, \boldsymbol{\psi}(x_j, \mathbf{y}(x_j)) + \mathcal{O}(h_j^{p+2}).$$

The two expressions show that

$$\mathbf{le}_j = \mathbf{lte}_j + \mathcal{O}(h_j^{p+1}\, \|\mathbf{y}_j - \mathbf{y}(x_j)\|) + \mathcal{O}(h_j^{p+2}),$$

hence that ordinarily there is little difference between the two quantities. Previous convergence results have been based on the local truncation error. It is possible to give a simple and illuminating proof of convergence for one-step methods based on control of the local error. Before presenting it we must begin our discussion of the control of error.

The user of a code specifies a norm $\|\cdot\|$ in which the error is to be measured and a tolerance $\tau > 0$. In §3 of Chapter 3 we took up the matter of defining a reasonable norm and we need discuss it no further here. Two kinds of control on the local error (local truncation error) are seen. The criterion of *error per step*, EPS, accepts the step to x_{j+1} only when

$$\|\mathbf{le}_j\| \leq \tau. \tag{3.1}$$

The criterion of *error per unit step*, EPUS, accepts the step only when

$$\|\mathbf{le}_j\| \leq h_j\tau. \tag{3.2}$$

The role of the step size is quite different in the present view of a numerical integration. In the classical theory the step sizes are specified and the accuracy of the integration is investigated as the largest step size tends to zero. In modern practice, a tolerance is specified and a code is to select step sizes so that a norm of the local error is bounded by the tolerance. The accuracy is then studied as the tolerance tends to zero.

EXERCISE 6. We have assumed throughout that the order p of the formula satisfies $p \geq 1$. Show that if the problem is smooth, it is always possible to find a step size h small enough to pass the test (3.1). Further show that when $p \geq 2$, the same is true for the test (3.2), but not when $p = 1$. In §8 of Chapter 5, it is pointed out that the order of the local error is reduced on steps where the problem is not smooth. A severe lack of smoothness can reduce the order to 1. The problem

$$y'' = -y - \operatorname{sgn}(y) - 3\sin(2x), \qquad 0 \leq x \leq 8\pi, \qquad y(0) = 0, y'(0) = 3,$$

involves a discontinuous function

$$\operatorname{sgn}(y) = \begin{cases} +1 & \text{if } y \geq 0, \\ -1 & \text{if } y < 0, \end{cases}$$

that causes the second derivative of $y(x)$ to have a jump discontinuity when $y(x)$ changes sign. This simplified model of a system with a relay was used by Shampine et al. [1976] to test how well codes cope with a lack of smoothness. Among the codes tested were two implementations of the Fehlberg (4, 5) Runge–Kutta pair. RK4 has an error per unit step control and RFK45, an error per step control. Do you understand now why RK4 could not integrate this problem past the first change of sign in $y(x)$ and RKF45 could? You might try your favorite code on this problem to see if it can cope with the repeated lack of smoothness. (An analytical solution can be found in Exercise 18 of Chapter 5.)

It is comparatively easy to prove that control of the local error by the criterion of EPUS implies a control of the true error. Let $\mathbf{y}(x)$ be the solution of $\mathbf{y}' = \mathbf{F}(x, \mathbf{y})$ that we seek to approximate and let $\mathbf{u}(x)$ be the local solution at x_j. To relate the global error at x_{j+1} to the global error at x_j and the local error of the step, we write

$$\mathbf{y}(x_{j+1}) - \mathbf{y}_{j+1} = (\mathbf{u}(x_{j+1}) - \mathbf{y}_{j+1}) + (\mathbf{y}(x_{j+1}) - \mathbf{u}(x_{j+1})).$$

The first term here is the local error of the step. The second term is the difference at x_{j+1} of two solutions of the differential equation, so we can bound it in familiar fashion in terms of the difference at x_j, the step size, and the Lipschitz constant L. Along with the EPUS control of the local error this leads to

$$\begin{aligned} \|\mathbf{y}(x_{j+1}) - \mathbf{y}_{j+1}\| &\leq \|\mathbf{le}_j\| + \|\mathbf{y}(x_j) - \mathbf{y}_j\|\, e^{Lh_j}, \\ &\leq h_j\tau + \|\mathbf{y}(x_j) - \mathbf{y}_j\|\, e^{Lh_j}. \end{aligned}$$

In Chapter 5 we learned that such a bound implies that for any x_n,

$$\|\mathbf{y}(x_n) - \mathbf{y}_n\| \leq e^{L(x_n - a)} [\|\mathbf{y}(a) - \mathbf{y}_0\| + \tau(x_n - a)].$$

When $\mathbf{y}_0 = \mathbf{y}(a)$ we have a bound on the global error in terms of the tolerance τ, the length of the interval of integration, and the stability of the differential equation itself. Notice that this *bound* on the error is proportional to the tolerance τ.

The situation with an error per step control is rather different; it is easy to write down examples for which there is no convergence as the tolerance tends to zero. When this was first appreciated, it disturbed the people writing codes because they had found that the better codes based on EPS perform about as well as the ones based on EPUS in every respect, and considerably better in some. Eventually (Shampine [1977]) it was realized that quality codes select an *efficient* step size and this leads to a relation between the tolerance τ and the step size selected that does guarantee convergence. Regardless of whether we use EPS or EPUS, we are certainly interested in selecting a step size that is efficient, or at any rate, not too inefficient. Let us then discuss briefly some of the issues in the selection of an efficient step size.

When a code takes a step from x_j to $x_j + h_j$, an estimate $\mathbf{est}_j$ is made of the local error incurred in the step:

$$\mathbf{est}_j \approx \mathbf{le}_j = \mathbf{u}(x_j + h_j) - \mathbf{y}_{j+1} = h_j^{p+1}\boldsymbol{\psi}(x_j, \mathbf{y}_j) + \mathcal{O}(h_j^{p+2}).$$

Just how this is done need not concern us right now. According to the expression for the local error, if we had taken the step from x_j with a step size h^* to get a result $\mathbf{y}_{j+1}^*$, the local error would have been

$$\mathbf{u}(x_j + h) - \mathbf{y}_{j+1}^* = (h^*)^{p+1}\, \boldsymbol{\psi}(x_j, \mathbf{y}_j) + \mathcal{O}((h^*)^{p+2}).$$

From this it is seen that the error would have been about

$$(h^*/h_j)^{p+1}\, \mathbf{est}_j,$$

a computable quantity. If the estimated local error is bigger than the tolerance, this idea is used to estimate the *optimal step size*, the largest step size that is predicted to pass the error test. When the control is EPS, we are interested in h^* such that

$$\tau = \|\mathbf{le}_j^*\| \approx (h^*/h_j)^{p+1}\, \|\mathbf{est}_j\|,$$

hence

$$h^* \approx h_j(\tau/\|\mathbf{est}_j\|)^{1/(p+1)}.$$

In terms of the principal error function, this optimal step size is

$$h^* \approx (\tau/\|\boldsymbol{\psi}(x_j, \mathbf{y}_j)\|)^{1/(p+1)}.$$

A step size that does not pass the error test, $\|\mathbf{est}_j\| > \tau$, is rejected. In the manner described, a step size that is predicted to pass the test is estimated and the step is tried again. This new step size is necessarily smaller, but the optimal step size is predicted to yield a local error with norm exactly τ. There are many uncertainties in this prediction process and a local error that is even a little too big will result in another rejected step. Rejected steps are expensive, so we should be conservative when selecting the step size. One scheme is to use a fraction of the optimal step size. A value of 0.9 is representative of those seen in the codes. A scheme that is equivalent in codes of fixed order is to aim at a fraction β of the tolerance τ. As an example, some codes select a step size that is predicted to yield a local error of size $\tau/2$. In what follows, let us suppose this second scheme is employed so that with EPS control,

$$h^* \approx h_j(\beta\tau/\|\mathbf{est}_j\|)^{1/(p+1)} \approx (\beta\tau/\|\boldsymbol{\psi}(x_j, \mathbf{y}_j)\|)^{1/(p+1)}. \tag{3.3}$$

Later we shall discuss the selection of β.

We have been considering what to do when a step is rejected. If the step from x_j is a success, we want to predict what step size might be used for the next step. The local error of this step will be

$$\begin{aligned}\mathbf{le}_{j+1} &= \mathbf{u}(x_{j+1} + h_{j+1}) - \mathbf{y}_{j+2} \\ &= h_{j+1}^{p+1}\, \boldsymbol{\psi}(x_{j+1}, \mathbf{y}_{j+1}) + \mathcal{O}(h_{j+1}^{p+2}).\end{aligned}$$

This can be approximated using the estimated local error for the current step in much the same way as for a rejected step. We have $x_{j+1} = x_j + h_j$ and $\mathbf{y}_{j+1} = \mathbf{y}_j + \mathcal{O}(h_j)$, which along with differentiability of the principal error function allows us to write

$$\boldsymbol{\psi}(x_{j+1}, \mathbf{y}_{j+1}) = \boldsymbol{\psi}(x_j, \mathbf{y}_j) + \mathcal{O}(h_j).$$

With this observation, we find that we can predict a suitable h_{j+1} as

$$h_{j+1} = h_j\, (\beta\tau/\|\mathbf{est}_j\|)^{1/(p+1)}.$$

Notice that this is the same recipe that we use after an unsuccessful step.

Returning to the issue of the criterion of error per step, with the assumption that the step size is given by (3.3), we can already see why we should expect convergence; later an asymptotic analysis will firm up the analysis. With a criterion of EPS, the optimal step size h_j is chosen so that

$$h_j \approx (\beta\tau/\|\boldsymbol{\psi}(x_j, \mathbf{y}_j)\|)^{1/(p+1)}.$$

This implies that

$$\tau^{1/(p+1)} \approx h_j(\|\boldsymbol{\psi}(x_j, \mathbf{y}_j)\|/\beta)^{1/(p+1)}.$$

The approximation suggests that for small τ (small step sizes), there will be a constant C such that

$$\tau^{1/(p+1)} \leq C\, h_j.$$

A little manipulation of the EPS criterion then leads to

$$\|\mathbf{le}_j\| \leq \tau = \tau^{1/(p+1)}\, \tau^{p/(p+1)} \leq h_j(C\, \tau^{p/(p+1)}).$$

We recognize this as an error per unit step control with a tolerance $C\, \tau^{p/(p+1)}$. The bound developed for EPUS applies and shows that when the optimal step size is used at each step, the error tends to zero as the tolerance does. Indeed, the bound on the error is proportional to $\tau^{p/(p+1)}$. This is not as nice as being proportional to τ, but for the moderately high orders p seen in popular codes, it is not very different. This is only a bound, but an asymptotic analysis says that the error itself behaves this way, as was observed experimentally in the early codes.

The selection of a suitable step size involves considerable art. Details can be found in Shampine and Watts [1979] and Shampine [1985a] and here just some of the issues will be raised. Even the evaluation of the recipe just presented must be done carefully, e.g., it is perfectly possible that $\|\mathbf{est}_j\|$ vanish. In writing software, one must always consider the possibility that the hypotheses are not valid. If all the functions are smooth, the arguments made imply that the step size will change slowly from step to step. Accordingly, an estimated error that leads to a prediction of a "large" change of step size—increase or decrease—must be a consequence of some hypothesis being invalid. For example, if it should happen that $\|\boldsymbol{\psi}(x_j, \mathbf{y}_j)\|$ vanishes, or is atypically small, the leading term in the expansion of the local error will not dominate and several of the arguments made would be affected. Recall, too, that for some of the numerical methods investigated in this book, stability requires a bound on the rate of change of the step size. For these reasons, the rate of increase of the step size is bounded in all the codes, bounds of 2 or 5 being representative. The rate of decrease is also bounded, but the situation is somewhat different and the step size is often allowed to decrease faster than it is allowed to increase. When a step fails, it is usual to insist on a substantial reduction of the step size so as to avoid repeated failures to take a step.

It is often not appreciated that asymptotic arguments do not justify a "large" increase of step size. To understand better this matter, suppose that we estimate an error of the form $\mathbf{c}\, h^{p+1}$ by neglecting terms like $\mathbf{d}\, h^{p+2}$. If the terms neglected are no larger than a tenth of the error in norm, we would be quite happy with the quality of the estimator. Thus we might have

$$\frac{\|\mathbf{d}\ h^{p+2}\|}{\|\mathbf{c}\ h^{p+1}\|} \approx 0.1.$$

In estimating what new step size might be used, we predict that on changing h to αh, the error will change to $\mathbf{c}\ (\alpha h)^{p+1}$. However, on an increase in step size, the terms neglected become relatively more important:

$$\frac{\|\mathbf{d}\ (\alpha h)^{p+2}\|}{\|\mathbf{c}\ (\alpha h)^{p+1}\|} \approx 0.1\alpha.$$

If we were to contemplate increasing h by a factor of 10 or more, we see that the terms neglected would be larger than the predicted error. This general argument shows that we must be very cautious about making "large" increases in the step size on the basis of the leading term in an asymptotic approximation.

EXAMPLE 4. To illustrate how art plays a role, let us consider an example the author constructed to illuminate a common defect in the early step size selection algorithms. The problem is

$$\begin{aligned} y' &= 0 \qquad x < 0, \\ &= x^6 \qquad x \geq 0 \end{aligned}$$

with initial value $y(-1) = 0$. The solution of this problem is

$$\begin{aligned} y(x) &= 0 \qquad x < 0, \\ &= \frac{x^7}{7} \qquad x \geq 0, \end{aligned}$$

which is quite a smooth function. When asked to integrate this problem with a pure absolute error test of 10^{-7} and started with a step size of 10^{-1}, the code DIFSUB took 2362 function evaluations to pass the origin! Any reasonable method will integrate $y' = 0$ exactly, and this is true of the methods in DIFSUB, so why does the code work so hard? The difficulty lies in the bounds placed on the step size. The code tries the given initial step size of 10^{-1} and estimates (correctly) that the error is 0. Accordingly, it increases the step size as much as possible—a factor of 10^4 in this code. When it tries this step size, it attempts to step far past the origin. On the basis of what happened with the first step, it predicts that the solution is zero there, but in fact the solution grows pretty rapidly for $x > 0$. This situation is revealed by a very large estimated local error. This causes the code to reduce the step size drastically, so drastically that it predicts that quite a small step size is appropriate. This new step size falls (far) short of the origin and when it is attempted, the code again estimates the result to be exact. The sequence of events is repeated and the code creeps up to the origin. Although contrived, this example makes clear that bounds on the rate of change of the step size can greatly improve the efficiency of solution of some problems. If all is going well, such bounds have no effect on the step size used.

EXERCISE 7. Apply your favorite code to this problem and see whether its step size selection algorithm is up to the task.

The initial step size is crucial because it indicates the scale of the problem. In early codes the user guessed an initial step size. That is why we stated the initial step size used by DIFSUB in the last example. All the algorithms in a code depend on the step size being small enough that the leading term dominate in asymptotic expansions. Because of this, a guess that is too big can lead to nonsense. A guess that is much too small is inefficient unless the first step is treated differently from the rest because the usual bound on the rate of increase of step size will cause a number of unnecessarily small steps to be taken before the code is able to use the "optimal" step size. It is not easy to provide a good guess because it depends on the problem and the numerical method(s) used by the code. For these reasons there has been considerable effort devoted to the automatic selection of the initial step size, a matter taken up in another section.

There is a general idea in numerical analysis called *extrapolation*. The idea is that if one has a good estimate of the error in a quantity, the estimate can be added to the quantity to get a more accurate result. In §2 we used this idea with an asymptotic expansion of the error of an integration to estimate the true, or global, error in a process called *global extrapolation*. It can also be used in the course of a single step, *local extrapolation* (Shampine [1973]). This device is of considerable practical value, so we go into it now. Suppose that the error estimator is asymptotically correct, meaning that

$$\mathbf{est}_j = \mathbf{le}_j + \mathcal{O}(h_j^{p+2}) = \mathbf{u}(x_j + h_j) - \mathbf{y}_{j+1} + \mathcal{O}(h_j^{p+2}).$$

It is surprising, but not all the estimators that have been used for the control of error are correct in this sense—an example will be given shortly. The idea of extrapolation is to define a new solution approximation by

$$\mathbf{y}_{j+1}^* = \mathbf{y}_{j+1} + \mathbf{est}_j.$$

Clearly the local error of this approximation is

$$\mathbf{u}(x_j + h_j) - \mathbf{y}_{j+1}^* = \mathcal{O}(h_j^{p+2})$$

so that the result is of order $p + 1$. Viewing this differently, we see how to go about deriving an estimate of the local error. Take each step with two formulas, one of order p and the other of order $p + 1$, and then estimate the error in the lower order result by comparison:

$$\mathbf{est}_j = \mathbf{y}^*_{j+1} - \mathbf{y}_{j+1}.$$

To emphasize the fact that there are two formulas involved, we speak of a *(p, p + 1) pair of formulas*.

Each step is taken with two formulas. If the error estimate is to be of any use, the higher order formula must be the more accurate. But if it is more accurate, why discard it and advance with the lower order result? Advancing with the higher order result is called *local extrapolation*. There are reasons why one might not want to do local extrapolation. A different formula is used and it is the stability of this formula that governs the stability of the integration. Indeed, it was observed in experiment that when solving problems with DIFSUB and its descendants, adding in the error estimate resulted in a formula with stability unsuitable for stiff problems. Because of this, none of these codes does local extrapolation. This, however, does not mean that local extrapolation cannot be done with the BDFs that are the basis of these codes, just that it cannot be done with the local error estimator of this line of codes.

To the author's knowledge, the first to derive Runge–Kutta formulas with local extrapolation in mind was Zonneveld [1964]. He did not describe what he was doing in these terms, writing instead of controlling the step size used with the basic formula according to the error made by a lower order formula. One of his schemes was taken up in Exercise 11 of Chapter 4. The scheme adds an extra stage to the classic four-stage, fourth-order formula to get a measure of the error. The exercise was to show that this measure could be viewed as the difference between the result of the fourth order formula and a third order formula. Nowadays we would describe his formula and error estimate as a (3, 4) pair implemented with local extrapolation.

The issue of local extrapolation is particularly important for Runge–Kutta formulas because it is relatively expensive to get an estimate of the local error. For example, it is possible to find fourth-order Runge–Kutta formulas of four stages, but at least six stages are required for order five. An asymptotically correct error estimator for a fourth-order formula is equivalent to a fifth-order formula. This implies that no matter how the computations are combined, at least six stages are necessary to evaluate both a formula of four stages and an estimate of its error. Clearly estimating the error is expensive if one is to advance with the fourth-order formula. On the other hand, by advancing the integration with the fifth-order formula, the equivalent fourth-order formula might be obtained without forming any extra stages and the error estimate might then be fairly regarded as "free." When local extrapolation is contemplated, the stability of the higher order formula is an important design consideration. Although the error of an integration done with local extrapolation is not known, it is smaller than if local extrapolation had not been done. Indeed, this provides a useful margin of safety in meeting the user's accuracy request. It is the local error of the lower order formula that is controlled, so the step size is optimal for the lower order formula rather than the one used to advance the integration. One should not be too fastidious about the "optimal" step size. It must be kept in mind that the connection between control of the local error and the global error of interest to the user is indirect. The asymptotic analysis shows that any reasonable selection of step sizes will be successful, and the schemes studied do provide a reasonable way to select them. A good argument against local extrapolation is that if one wants to provide a capability for estimation of the global error, the matter is made much more difficult by local extrapolation. There is room for a difference of opinion, but nowadays all the popular codes based on explicit Runge–Kutta formulas do local extrapolation, and many of the Adams codes do.

EXAMPLE 5. Exercise 12 of Chapter 4 sets the task of verifying that the Merson formula,

$$\begin{array}{c|ccccc}
0 & & & & & \\
1/3 & 1/3 & & & & \\
1/3 & 1/6 & 1/6 & & & \\
1/2 & 0 & 1/8 & 3/8 & & \\
\hline
1 & 1/2 & 0 & -3/2 & 2 & \\
\hline
 & 1/6 & 0 & 0 & 2/3 & 1/6
\end{array}$$

provides an approximate solution $\mathbf{y}_{j+1}$ of order 4 and another, $\mathbf{Y}_4$, of order 3. Merson argued that if $\mathbf{F}(x, \mathbf{y})$ has the special form $\mathbf{F}(x, \mathbf{y}) = \boldsymbol{\alpha}\, x + J\, \mathbf{y}$, then

$$\mathbf{y}(x_{j+1}) - \mathbf{y}_{j+1} = \frac{1}{720} h^5\, \mathbf{y}^{(5)}(x_j) + \mathbb{O}(h^6),$$

and

$$\mathbf{y}(x_{j+1}) - \mathbf{Y}_4 = \frac{1}{120} h^5\, \mathbf{y}^{(5)}(x_j) + \mathbb{O}(h^6).$$

(For details, see §5.4.2 of Gear [1971c].) This observation suggests the estimate

$$\mathbf{est}_j = \frac{1}{5}(\mathbf{y}_{j+1} - \mathbf{Y}_4) = \mathbf{y}(x_{j+1}) - \mathbf{y}_{j+1} + \mathcal{O}(h^6).$$

For problems with $\mathbf{F}$ of this special form, this is an easy way to estimate the local error and a number of codes are based on it.

This estimate cannot be asymptotically correct for general $\mathbf{F}$ because it does not have the right order. In general $\mathbf{Y}_4$ is only of order 3, so the estimate is generally $\mathcal{O}(h^4)$. Put differently, if the estimate were asymptotically correct, $\mathbf{y}^*_{j+1} = \mathbf{y}_{j+1} + \mathbf{est}_j$ would have to be a formula of order 5, and it is easily verified that it is not. And yet, there is a great deal of computational experience that shows the Merson pair to be an effective way to solve the initial value problem. How can this be? After the concept of local extrapolation was formalized, Shampine and Watts [1977, 1979] were able to answer this question. All one need do is to forget about $\mathbf{Y}_4$ and regard the Merson scheme as a pair consisting of $\mathbf{y}^*_{j+1}$ and $\mathbf{y}_{j+1}$. This is a (3, 4) pair with an asymptotically correct estimate of the error of the third order formula that is implemented with local extrapolation. In the measures of quality usual nowadays, the Merson (3, 4) pair is an excellent pair at this order.

The notation EPS and EPUS will be extended to account for local extrapolation. We now take *XEPS* to mean error per step control with local extrapolation, and reserve *EPS* for error per step control without local extrapolation. Similarly *XEPUS* is to mean error per unit step control with local extrapolation, and *EPUS* to mean error per unit step control without local extrapolation. Although some of these four possibilities are (much) more popular than others, all four have been used in important codes.

When an efficient step size is selected, the criterion of XEPS is closely related to EPUS. To see this, we need the principal error function $\boldsymbol{\psi}^*$ for the higher order result $\mathbf{y}^*_{j+1}$ used to advance the integration when local extrapolation is done:

$$\mathbf{le}^*_j = \mathbf{u}(x_j + h_j) - \mathbf{y}^*_{j+1} = h_j^{p+2}\,\boldsymbol{\psi}^*(x_j, \mathbf{y}_j) + \mathcal{O}(h_j^{p+3}).$$

When XEPS is used, the local error of $\mathbf{y}_{j+1}$ is controlled by the criterion of EPS, hence the step size h_j is selected according to (3.3) as

$$h_j \approx (\beta\tau/\|\boldsymbol{\psi}(x_j, \mathbf{y}_j)\|)^{1/(p+1)}.$$

This implies that the actual local error is about

$$\|\mathbf{le}^*_j\| \approx h_j^{p+2}\|\,\boldsymbol{\psi}^*(x_j, \mathbf{y}_j)\| \approx h_j\,\tau\left(\frac{\beta\,\|\boldsymbol{\psi}^*(x_j, \mathbf{y}(x_j))\|}{\|\boldsymbol{\psi}(x_j, \mathbf{y}(x_j))\|}\right).$$

If C is a bound on the quantity in parentheses here, this says that the local error of the actual numerical solution is controlled according to the criterion of EPUS with a tolerance of $C\tau$. For this reason XEPS is sometimes described as a *generalized error per unit step*. This observation helps explain why some of the early codes that use error per step have a behavior that is just as satisfactory as those that use error per unit step; they do local extrapolation, hence implement a generalized error per unit step.

A matter of great importance is measuring the efficiency of software. This is not the simple matter that it might seem. A natural way to measure the efficiency of a code is to solve a problem for a number of tolerances and then to plot the cost against the tolerance. Comparing such plots for codes with different error criteria or different choices of parameters allows one to decide which variant is more efficient for the problem solved. A standard measure of cost is the number of times **F** is evaluated. It has the virtue of being independent of the computer used and for **F** that are expensive to evaluate, it gives a fair measure of the cost. To be sure, when solving stiff problems there is the additional issue of accounting for the cost of the Jacobian and the other costs, the overhead, can be significant for stiff problems. This *first measure of efficiency* has been widely used. A *second measure of efficiency* is obtained in a way similar to the first with the difference that it is the global error rather than the tolerance that is plotted against the cost in function evaluations. This is not so easy. It is necessary to have reference values for the true solution that are obtained from an analytical solution or a very accurate preliminary computation. One must consider carefully the norm in which the error is to be measured if the quality of the numerical solution is to be seen. Where should the error to be measured? Because the error can decrease as well as increase in the course of an integration, a better understanding of the success of a code is obtained by measuring the error at several points. There are a number of subtleties. It is natural to look at the error at mesh points selected by the code, but output at specific points influences the integration in a number of ways and it may be desirable to assess this effect. Furthermore, output between mesh points is obtained by some kind of interpolation process and one might, or might not, want to assess the quality of this process. In any case, variation of step size and variation of order cause answers at a single point to depend in a somewhat irregular way on the tolerance with the consequence that measuring error at a number of points provides smoother statistics.

The distinction between the two measures of efficiency is clear in principle: One considers the tolerance given the code and the other the actual errors made by the code. The main argument favoring the first is that it corresponds exactly to the task given the code. On the other hand, in its simplest form, the first measure does not even verify that the code solved the problem in any reasonable way. Further, the tolerance is only indirectly related to the true error and in practice users develop some feel for the relation in their applications. The second measure corresponds better to what users want. Of course, the developers of codes are likely to take both measures into account. The point here is that they are different and as we shall see, they lead to different answers to important practical questions.

With these two ways of measuring efficiency in mind, let us consider the various error controls. When the step size is smaller than 1, the criterion of EPUS requires the local error to be smaller than the criterion of EPS does. This means that the step size is smaller with EPUS and more steps are required. For all sufficiently small tolerances τ, the step sizes will be smaller than 1 and by the first measure of efficiency, EPUS will be less efficient than EPS. The first measure takes no notice of the fact that smaller step sizes result in a smaller error. The second does, which suggests that the criteria are more comparable in this measure. Later we shall see that in a sense the criteria are equally efficient in the second measure of efficiency. The efficiency of local extrapolation also depends on the measure. Later we can be more precise, but the essence of the matter is simple. Suppose one integration is done with local extrapolation and the other without. When EPS is used, the step size is given by (3.3). The argument $\mathbf{y}_j$ is different in the two integrations, but both arguments are close to $\mathbf{y}(x_j)$, so in both cases the step size is approximately

$$h^* \approx (\beta\tau/\|\boldsymbol{\psi}(x_j, \mathbf{y}(x_j))\|)^{1/(p+1)}.$$

In this approximation, the step size chosen is not affected by local extrapolation, hence the cost of the two integrations is essentially the same. The same argument leads to the same conclusion for EPUS. This argument says that in the first measure of efficiency, there is no advantage to local extrapolation. However, local extrapolation raises the order, hence results in a more accurate integration at the same cost. This makes it significantly more efficient in the second measure of efficiency.

The choice of the parameter β in (3.3) is an important practical matter. Suppose we consider codes A and B that differ only in that they use β_A and β_B, respectively, with $0 < \beta_A < \beta_B < 1$. According to (3.3), the code B selects the larger step size. Granted that the step size is actually chosen in this way, that is, that other rules do not come into play, this implies that code B will be the more efficient in the first measure of efficiency. We conclude that we should take β as close to 1 as possible. For sufficiently small τ, the asymptotic arguments show that (almost always) the step size *is* given by (3.3). The difficulty here is that the closer β is to 1, the smaller τ has to be for this to be true. This is because the closer β is to 1, the more likely that a predicted step will be rejected. The traditional procedure has been to increase β until experiments with a substantial set of test problems shows a further increase to be counterproductive. A surprisingly different conclusion is drawn when we consider the second measure of efficiency. When assessing efficiency in this way, a tolerance τ is selected, a problem is integrated, and, say, the maximum error seen is plotted against the cost in function evaluations. Notice that only the product $\beta\tau$ appears in (3.3). This means that for each tolerance τ_A with code A, there is a corresponding tolerance $\tau_B = \tau_A\beta_A/\beta_B$ that will lead to precisely the same step sizes, accuracy, and cost in code B. Again assuming that the step size is determined by (3.3), we conclude that the efficiency in the second measure does not depend on β at all. In view of the fact that reducing β will reduce the number of step failures and the likelihood of other rules coming into play, we are led to choosing a comparatively small value of β, nearly the opposite of the traditional action. As long as β is not particularly small, the two measures do not differ greatly. It would seem reasonable to compromise by choosing β close to 1, but not so close that failures are at all likely.

§4 Asymptotic Analysis of Step Size Selection

In §1 we developed an asymptotic expansion for the error when the step sizes are specified in advance by a step size selection function $\theta(x)$. In this section we argue that by appropriate definition of $\theta(x)$, it is possible to obtain a reasonable model of practical computation with the better codes implementing some of the most popular methods. This model then provides valuable insight for the choices necessary when writing mathematical software. For a model to be useful, the analyst must be conscious of its assumptions and how well they reflect practice, so as we proceed, these issues will be discussed. The analysis is restricted to a fixed order and it is not evident how to model practical computation with variable order codes. Any asymptotic analysis like this one is applicable only for "small" tolerances τ. A key assumption is that the code uses the optimal step size. This requires the adjustment of the step size at every step, hence is directly applicable only to fully variable step size implementations. The assumption is realistic for one-step methods. Practical implementations of methods with memory do not adjust the step size at every step. Still, this is not a bad assumption for fully variable step size implementations of methods with memory and it is certainly good enough to provide guidelines.

As usual it is assumed that $\mathbf{F}$ is as smooth as necessary. This is a good assumption in practice, at least on subintervals of the original problem. The assumption implies that the principal error function of the method is continuous, hence that $\|\boldsymbol{\psi}(x, \mathbf{y}(x))\|$ is bounded on $[a, b]$. Let

$$\min_{a \le x \le b} \|\boldsymbol{\psi}(x, \mathbf{y}(x))\| = \|\boldsymbol{\psi}(\zeta, \mathbf{y}(\zeta))\|.$$

An important assumption is that this minimum value is positive. This is a natural assumption because if this minimum were to vanish at $x = \zeta$, the formula would be of higher order at ζ.

The codes are given a tolerance τ and seek to achieve an accuracy of $\beta\tau$ at each step. For definiteness we treat EPS; results for the other cases are immediate. Define H by

$$\beta\tau = H^{p+1} \|\boldsymbol{\psi}(\zeta, \mathbf{y}(\zeta))\|.$$

For later use, note that H is a multiple of $\tau^{1/(p+1)}$. Further define

$$\theta(x) = \left(\frac{\|\boldsymbol{\psi}(\zeta, \mathbf{y}(\zeta))\|}{\|\boldsymbol{\psi}(x, \mathbf{y}(x))\|}\right)^{1/(p+1)}.$$

This function is a step size selection function because this continuous function satisfies $0 < \Delta \le \theta(x) \le 1$ on $[a, b]$ and $\theta(\zeta) = 1$. The step size specified at x_j by the maximum step size H and the step size selection function $\theta(x)$ is

$$\begin{aligned}\theta(x_j)H &= \left(\frac{\|\boldsymbol{\psi}(\zeta, \mathbf{y}(\zeta))\|}{\|\boldsymbol{\psi}(x_j, \mathbf{y}(x_j))\|}\right)^{1/(p+1)} \left(\frac{\beta\tau}{\|\boldsymbol{\psi}(\zeta, \mathbf{y}(\zeta)\|}\right)^{1/(p+1)} \\ &= \left(\frac{\beta\tau}{\|\boldsymbol{\psi}(x_j, y(x_j))\|}\right)^{1/(p+1)}.\end{aligned}$$

If integrations are done with this way of specifying the step size, we have seen that

$$\mathbf{y}_j = \mathbf{y}(x_j) + \mathbf{e}(x_j) H^p + \mathcal{O}(H^{p+1}).$$

Because $\mathbf{y}_j$ agrees with $\mathbf{y}(x_j)$ to $\mathcal{O}(H^p)$, the step size $\theta(x_j)H$ is very well approximated by

$$\left(\frac{\beta\tau}{\|\boldsymbol{\psi}(x_j, \mathbf{y}_j)\|}\right)^{1/(p+1)},$$

the value given in (3.3) for the step size selected automatically at $(x_j, \mathbf{y}_j)$. Recall that the asymptotic analysis does not require the step size to be exactly $\theta(x_j)H$—a discrepancy of $\mathcal{O}(H^2)$ does not alter the conclusions. It appears, then, that a reasonable model of integration with automatic selection of the step size is provided by integration with step sizes specified by this particular H and $\theta(x)$. For this to be a reasonable model, we are supposing that the step size satisfies (3.3), hence that other rules do not come into play. As we explained in connection with (3.3), this is a good assumption for stringent tolerances.

There is a matter that has been glossed over, namely the role of the first step. The older codes require the user to guess the initial step size. If it is much smaller than necessary, the bound on the rate of increase of step size will determine the first few step sizes rather than (3.3). Some codes pay attention to getting on scale and they do not begin the integration proper until the first step size is determined by (3.3). This is characteristic of recent codes that automatically select the initial step size. Clearly, how the initial step size is handled may influence how well the model describes practical computation at the beginning of the integration. Still, codes recover from bad guesses pretty quickly, so any computation done with a reasonable guess should be described adequately by the model. Codes that attend properly to getting on scale are modeled well.

EXAMPLE 6. Figure 7.4 shows an experimental step size selection function obtained by integrating the two-body problem with eccentricity 0.9 using the Runge–Kutta code DO2PAF of the NAG library. This code is based on the Merson pair. For each tolerance all the step sizes h_j were stored, the maximum H found, and the ratios h_j/H plotted. The step size shows

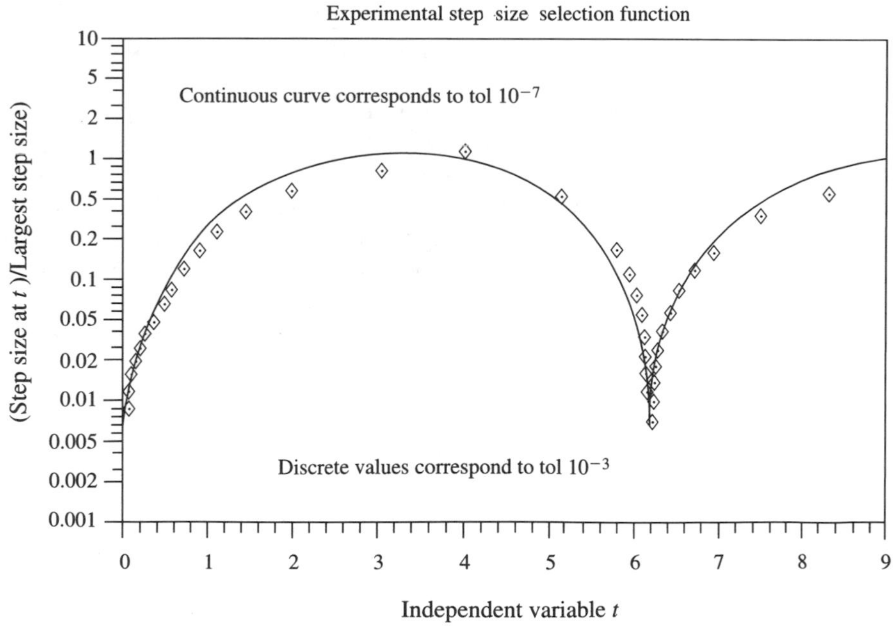

Figure 7.4

a considerable variation, being much smaller as one body approaches the other, so that a logarithmic scale is appropriate for the ratios. It is seen that as $H \to 0$, the automatic step size selection algorithm is producing step sizes that appear to come from a step size selection function.

Accepting the validity of the model, let us now derive some guidelines from it. The asymptotic analysis is couched in terms of the maximum step size H, but codes are given a tolerance τ, so we need to restate the asymptotic results in terms of τ. Because

$$\mathbf{y}_j = \mathbf{y}(x_j) + \mathbf{e}(x_j)\, H^p + \mathcal{O}(H^{p+1}),$$

the true error at x_j is

$$\mathbf{e}_j = \mathbf{y}(x_j) - \mathbf{y}_j \approx - \mathbf{e}(x_j)\, \tau^{p/(p+1)}.$$

The first observation is that control of the local error by the criterion of EPS does lead to a control of the global error, at least when an efficient step size is chosen as we assume. It is easy to write down the other cases and we do so to contrast them. The functions $\mathbf{e}(x)$ in the various expansions are all different, so only the proportionality with respect to τ is stated.

$\mathbf{e}_j \sim \tau^{p/(p+1)}$	EPS
$\sim \tau$	EPUS
$\sim \tau^{(p+1)/p}$	XEPUS
$\sim \tau$	XEPS

Earlier we developed bounds on the error in the cases of EPUS and XEPS that were proportional to the tolerance τ. With suitable assumptions we now see that the errors themselves are proportional to τ. The behavior with respect to a change of tolerance is not so nice when the other two criteria are used, but for moderately high orders, it is not greatly different. To be sure, if the order is low, the model says that the behavior will be noticeably different and this is observed with popular codes.

The cost of the integration is easily approximated. The number of steps taken in the integration from a to b is

$$N = \sum \frac{1}{h_j} h_j = \int_a^b \frac{dx}{\theta(x)H} + \mathcal{O}(1).$$

Here the sum has been manipulated to regard it as a Riemann sum for an integral. If a single step costs c_F evaluations of $\mathbf{F}$, the total cost of the integration is then

$$\text{cost} = c_F N = H^{-1} c_F \int_a^b \theta^{-1}(x)\, dx + \mathcal{O}(1).$$

In the approximations made for the analysis, the cost is not affected by local extrapolation because the step size sequence chosen is the same. On general grounds we have argued that in the first measure of efficiency, local extrapolation does not alter the efficiency. In the second, it certainly does because it provides more accuracy at the same cost. Now we can quantify this matter. In the case of EPS the global error μ is proportional to $\tau^{p/(p+1)}$. Put the other way around, the local error tolerance τ that yields a global error of μ is proportional to $\mu^{(p+1)/p}$. The cost of achieving this accuracy is proportional to the number of steps N which is proportional to $\tau^{-1/(p+1)}$, hence the cost is proportional to $(\mu^{(p+1)/p})^{(-1/(p+1))} = \mu^{-1/p}$. For EPUS, τ is proportional to μ and N is proportional to $\tau^{-1/p}$, hence the cost is proportional to $\mu^{-1/p}$. This says that the two schemes are equally efficient when we take into account the accuracy achieved. In the same way it is found that XEPS and XEPUS are equally efficient in the second measure. They have an efficiency proportional to $\mu^{-1/(p+1)}$. Local extrapolation is seen to be more efficient in the second measure because it gets more accuracy at the same cost.

The crucial issue in our analysis is whether the step sizes selected automatically can be represented by a step size selection function. Figure 7.4 is a simple test of this. Computational experiments require the same care as any other experiments and some details of more elaborate experiments are illustrative:

EXAMPLE 7. The computations of Shampine [1985a] were performed with the code DERKF which uses the Fehlberg (4, 5) pair of Runge–Kutta formulas in an XEPS implementation. The initial step size is selected automatically, but the first step is not repeated, so if the automatic scheme were to propose a step size that is too small, the code would need a few steps to get up to the optimal step size. The computations were performed on the set of 25 test problems of Hull et al. [1972] that are posed on the interval [0, 20]. This interval-oriented code has an intermediate output mode. To suppress the effects of approaching the end of the interval, the code was told to integrate the problem on a longer interval and when intermediate output showed that the code had stepped past $x = 20$, the integration was terminated and only mesh points in [0, 20] were considered. This complicated way of proceeding is needed because many problems in this set are relatively easy for this code and if the code is simply told to integrate to $x = 20$, the asymptotic behavior of the error is affected near the end of the integration as the code adjusts the step size to produce a result at the end. The problems in the set were solved with a mixed relative-absolute error test. The machine-dependent absolute error tolerance of 2.0284×10^{-20} was set as a threshold to prevent difficulties due to solution components that vanish. The relative error tolerances were 10^{-1}, 10^{-2}, $\ldots$, 10^{-11}, and 2.0284×10^{-12}. This last value is the most stringent tolerance recommended for this code on the computer used. Only the results for tolerances such that the maximum step size was no more than a tenth of the length of the interval of integration were used. This is a natural requirement for testing the validity of an analysis based on the maximum step size tending to zero, but some of the problems are so easy that this caused only relative error tolerances of 10^{-7} and smaller to be used for them. The function $\theta(x)$ was approximated by storing the mesh $\{x_j\}$ that was selected at the most stringent tolerance, finding the largest step size H, and then defining $\theta(x_j) = (x_{j+1} - x_j)/H$. At other x, the function $\theta(x)$ was defined by linear interpolation. The hypothesis to be tested is that for a given tolerance τ, the step sizes h_j selected automatically by the code satisfy $h_j = \theta(x_j)H + \mathbb{O}(H^2)$. For each tolerance τ the step sizes selected by the code were saved, the maximum step size H determined, and the smallest constant c_τ found for which

$$|h_j - \theta(x_j)H| \leq c_\tau H^2$$

for all the x_j in the interval $[0, 20]$. The quantity c was defined as the largest of the c_τ. The issue here is whether the experimental step sizes can be described in the manner assumed in the model, so we are more interested in the existence of the quantity c than its size. It was found, for example, that when integrating the problem called A3, $y' = y \cos(x)$, $y(0) = 1$, for its solution $\exp(\sin(x))$, tolerances 10^{-3} and smaller resulted in a maximum step size no more than a tenth of the interval of integration. The constant c turned out to be about 0.5. The problem called B4 is

$$\begin{aligned} y_1' &= -y_2 - y_1 y_3/r & y_1(0) &= 3, \\ y_2' &= y_1 - y_2 y_3/r & y_2(0) &= 0, \\ y_3' &= y_1/r & y_3(0) &= 0, \end{aligned}$$
$$\text{where } r = \sqrt{y_1^2 + y_2^2}.$$

The solution components are $y_1(x) = (2 + \cos(x))\cos(x)$, $y_2(x) = (2 + \cos(x))\sin(x)$, and $y_3(x) = \sin(x)$. In this case the condition on the maximum step size was satisfied for tolerances 10^{-1} and smaller. The constant c was found to be about 0.3.

EXERCISE 8. Test the applicability of the model of this section to a code available to you. This might take the form of the computations just described or those of Figure 7.4. If you have graphical output available, you might want to estimate a step size selection function using the results at a stringent tolerance and then plot values of h_j/H obtained at less stringent tolerances against a plot of your $\theta(x)$. An easier way to study the applicability of the model is to study how the global error behaves as the tolerance is changed. For example, the theory developed says that a code based on XEPS should result in global errors that are reasonably proportional to the tolerance.

§5 Error Estimators

In order to control the local error, we must be able to estimate it. Some estimators use previously computed solution values and others use only information from the current step. This classification does not quite correspond to using methods with memory and one-step methods. Certainly when using a method with memory, it is natural to exploit this memory to estimate the error. Although not so natural, error estimators with memory have been proposed for one-step methods. Because they have not proved very satisfactory in practice, they are not used in the popular codes and we do not discuss them here. At first general principles were used to estimate the error of Runge–Kutta formulas. The estimators were relatively expensive, giving codes based on LMMs an advantage over those based on Runge–Kutta methods. Later it was realized that effective and efficient error estimators could be obtained by deriving them along with the basic formulas. Although this complicates greatly the development of effective Runge–Kutta methods, the best are quite competitive with LMMs.

§5.1 Error Estimators with Memory

Whenever we investigate the use of previously computed solution values, we must specify the mesh on which they were computed. To justify the estimators we need the asymptotic expansion of §1, hence need to assume that the mesh is obtained from a step size selection function. With this assumption the analysis is the same whether or not the step size varies, so to simplify the notation, the error estimators will be developed for a constant step size h. Some estimators of the error made in a step from x_j to x_{j+1} arise naturally from a consideration of the local truncation error when it is assumed that the previously computed solution values $\mathbf{y}_j, \ldots, \mathbf{y}_{j+1-k}$ are *exact*, i.e., $\mathbf{y}_j = \mathbf{y}(x_j), \ldots, \mathbf{y}_{j+1-k} = \mathbf{y}(x_{j+1-k})$. This assumption is completely unrealistic, so after deriving the local truncation error estimators we resort to the asymptotic expansion of §1 to show that the estimators are valid in circumstances that have some relevance to practical computation.

An implicit LMM is implemented by predicting the result at x_{j+1} with an explicit formula and then correcting with the implicit formula. One of the first schemes for estimating the local truncation error, Milne's scheme, exploits this fact. The idea is that if the previously computed values are exact, the local truncation error of a predictor of order p has the form

$$\mathbf{y}(x_{j+1}) - \mathbf{y}^*_{j+1} = C^* \, h^{p+1} \, \mathbf{y}^{(p+1)}(x_j) + \mathbb{O}(h^{p+2}).$$

Similarly, the local truncation error of an implicit LMM, a corrector, of order p has the form

$$\mathbf{y}(x_{j+1}) - \mathbf{y}_{j+1} = C \, h^{p+1} \, \mathbf{y}^{(p+1)}(x_j) + \mathbb{O}(h^{p+2}).$$

Subtraction of these two expressions provides the computable quantity

$$\mathbf{y}_{j+1} - \mathbf{y}^*_{j+1} = (C^* - C) \, h^{p+1} \, \mathbf{y}^{(p+1)}(x_j) + \mathbb{O}(h^{p+2}),$$

and the error estimator

$$\frac{C}{(C^* - C)} (\mathbf{y}_{j+1} - \mathbf{y}^*_{j+1}) = \mathbf{y}(x_{j+1}) - \mathbf{y}_{j+1} + \mathbb{O}(h^{p+2}). \qquad (5.1)$$

Milne's estimator is simple and it can be very convenient. To illustrate the latter attribute, let us look at the form the estimator assumes when Adams methods are implemented in terms of backward differences. Recall that the Adams–Moulton method of order p, AMp, is

$$\mathbf{y}_{j+1} = \mathbf{y}_j + h \sum_{k=1}^{p} \gamma^*_{k-1} \, \nabla^{k-1} \, \mathbf{F}_{j+1},$$

and its local truncation error is

$$\mathbf{y}(x_{j+1}) - \mathbf{y}_{j+1} = \gamma^*_p \, h^{p+1} \, \mathbf{y}^{(p+1)}(x_j) + \mathbb{O}(h^{p+2}).$$

The Adams–Bashforth formula of order p, ABp, is

$$\mathbf{y}^*_{j+1} = \mathbf{y}_j + h \sum_{k=1}^{p} \gamma_{k-1} \, \nabla^{k-1} \, \mathbf{F}_j,$$

and its local truncation error is

$$\mathbf{y}(x_{j+1}) - \mathbf{y}^*_{j+1} = \gamma_p \, h^{p+1} \, \mathbf{y}^{(p+1)}(x_j) + \mathbb{O}(h^{p+2}).$$

The Adams–Moulton formula of order p involves

$$\mathbf{F}_{j+1} = \mathbf{F}(x_{j+1}, \mathbf{y}(x_{j+1})),$$

which is replaced in a predictor–corrector pair in PECE implementation by

$$\mathbf{F}^*_{j+1} = \mathbf{F}(x_{j+1}, \mathbf{y}^*_{j+1}),$$

and the result of a step is then

$$\mathbf{y}_{j+1} = \mathbf{y}_j + h \sum_{k=1}^{p} \gamma_{k-1}^* \nabla^{k-1} \mathbf{F}_{j+1}^*.$$

There is a convenient identity,

$$\mathbf{y}_{j+1} = \mathbf{y}_{j+1}^* + h\, \gamma_{p-1} \nabla^p \mathbf{F}_{j+1}^*,$$

relating the predicted and corrected values in the case of the ABp–AMp pair that was derived in §2.3 of Chapter 4. Using it, the Milne estimator is seen to be

$$\frac{\gamma_p^*}{(\gamma_p - \gamma_p^*)} (\mathbf{y}_{j+1} - \mathbf{y}_{j+1}^*) = \frac{\gamma_p^*}{\gamma_{p-1}} (\mathbf{y}_{j+1} - \mathbf{y}_{j+1}^*)$$

Here the identity $\gamma_p^* = \gamma_p - \gamma_{p-1}$ (see Exercise 21 of Chapter 4) is used to simplify a coefficient. The same estimator can also be used for the implicit Adams–Moulton formula—to leading order, the error of the value $\mathbf{y}_{j+1}$ obtained from iteration to completion can be estimated after a single iteration.

Useful savings are possible by exploiting the fact that the local truncation error is estimated by comparing the predicted value to the first corrected value. In the case of a PECE implementation, if the step is to be rejected, there is no point to making the final evaluation. By avoiding it, we can reduce the cost of a failed step by up to a half. When evaluating an implicit formula, if the step is to be rejected, there is no point to computing $\mathbf{y}_{j+1}$ accurately, so the iteration can be terminated and some function evaluations saved. In the same manner it is possible to reduce the cost of a failed step with other schemes for estimating the local truncation error that we take up.

A different way to estimate the local truncation error of a formula is to compare the result of the formula to the result of a formula of higher order. Continuing to assume that previously computed solution values are exact, let $\tilde{\mathbf{y}}_{j+1}$ be the result of a formula of order $p + 1$ and $\mathbf{y}_{j+1}$ the result of a formula of order p. The local truncation error of $\mathbf{y}_{j+1}$ can then be estimated by

$$\begin{aligned} \tilde{\mathbf{y}}_{j+1} - \mathbf{y}_{j+1} &= (\mathbf{y}(x_{j+1}) - \mathbf{y}_{j+1}) - (\mathbf{y}(x_{j+1}) - \tilde{\mathbf{y}}_{j+1}). \\ &= \mathbf{y}(x_{j+1}) - \mathbf{y}_{j+1} + \mathbb{O}(h^{p+2}), \end{aligned}$$

because the local truncation error of $\tilde{\mathbf{y}}_{j+1}$ is $\mathbb{O}(h^{p+2})$. To apply this to Adams formulas, recall the convenient identity connecting the Adams–Bashforth predictor of order p to the Adams–Moulton corrector of order $p + 1$:

$$\tilde{\mathbf{y}}_{j+1} = \mathbf{y}_{j+1}^* + h\, \gamma_p \nabla^p \mathbf{F}_{j+1}^*.$$

Along with the identity used above connecting $\mathbf{y}^*_{j+1}$ and the Adams–Moulton corrector of order p, we have

$$\begin{aligned} \tilde{\mathbf{y}}_{j+1} - \mathbf{y}_{j+1} &= h\, \gamma_p\, \nabla^p\, \mathbf{F}^*_{j+1} - h\, \gamma_{p-1}\, \nabla^p\, \mathbf{F}^*_{j+1} \\ &= h\, \gamma^*_p\, \nabla^p\, \mathbf{F}^*_{j+1}. \end{aligned} \tag{5.2}$$

This is the estimator derived in Chapter 4 in analogy to Taylor series expansion of the solution.

Another way to derive an error estimator is to approximate directly the derivative in

$$C\, h^{p+1}\, \mathbf{y}^{(p+1)}(x_j) + \mathbb{O}(h^{p+2})$$

by backward differences. As stated in the appendix, for any function $\mathbf{f}(x)$ with $q + 1$ continuous derivatives on $[x - qh, x]$,

$$\nabla^q\, \mathbf{f}(x) = h^q\, \mathbf{f}^{(q)}(*),$$

where the asterisk indicates that the components are evaluated at different points in the interval. This suggests the estimator

$$C\, \nabla^{p+1}\, \mathbf{y}_{j+1} = C\, h^{p+1}\, \mathbf{y}^{(p+1)}(x_j) + \mathbb{O}(h^{p+2}). \tag{5.3}$$

This estimator is particularly well suited to the BDFs. Adams methods are based on interpolation to the values $\mathbf{F}_{j+1-i} = \mathbf{F}(x_{j+1-i}, \mathbf{y}_{j+1-i})$, so a more natural estimator for such methods is

$$\begin{aligned} C\, h\, \nabla^p\, \mathbf{F}_{j+1} &= C\, h\, \nabla^p\, \mathbf{F}(x_{j+1}, \mathbf{y}(x_{j+1})) = C\, h\, \nabla^p\, \mathbf{y}'(x_{j+1}), \\ &= C\, h^{p+1}\, \mathbf{y}^{(p+1)}(x_j) + \mathbb{O}(h^{p+2}). \end{aligned} \tag{5.4}$$

In the case of the Adams methods, this last estimator is the same as (5.2), but it could be used for other LMMs as well.

These are the usual estimators and the usual ways that they are derived. In introductory texts on numerical analysis, it is often the case that no more is said. Such a superficial treatment leaves out issues of the most fundamental nature. The derivations presented are fine as far as they go because estimators that do not work properly when presented with exact data cannot be of any interest at all. The trouble with them is the assumption that the data is exact. This assumption is certainly not true in practical computation. Indeed, if it were true for the step to x_{j+1}, it would obviously not be true for the step to x_{j+2}. Furthermore, we cannot hope to estimate correctly the local truncation error if the data available to the estimator is too inaccurate. Let us suppose then that a formula of order p is being used and that all has gone well in the integration up to x_j in the sense that the numerical solution is related to the true solution by an asymptotic expansion of the form

$$\mathbf{y}_j = \mathbf{y}(x_j) + \mathbf{e}(x_j)\, h^p + \mathbb{O}(h^{p+1}).$$

Considering that the errors in the values computed prior to x_j are $\mathcal{O}(h^p)$, it is far from clear that we can estimate a local error that is $\mathcal{O}(h^{p+1})$. Indeed, unless the previously computed solutions have errors that are related to one another, the estimators may *not* be correct. Anything that disturbs the smooth behavior of the error might ruin the error estimate. In particular, the estimator might not be valid just after the starting values are formed and just after a point where **F** is not smooth. It does not seem to be generally appreciated that the error estimates for LMMs might not be reliable. The derivation of error estimators for Runge–Kutta methods is (much) more complicated, but the estimates are better justified theoretically.

Let us now justify Milne's scheme for the estimation of local truncation error in reasonable circumstances. It is supposed that the (normalized) corrector formula

$$\sum_{i=0}^{k} \alpha_i \mathbf{y}_{j+1-i} = h \sum_{i=0}^{k} \beta_i \mathbf{F}(x_{j+1-i}, \mathbf{y}_{j+1-i})$$

is of order p. The solution $\mathbf{y}(x)$ satisfies

$$\sum_{i=0}^{k} \alpha_i \mathbf{y}(x_{j+1-i}) = h \sum_{i=0}^{k} \beta_i \mathbf{F}(x_{j+1-i}, \mathbf{y}(x_{j+1-i})) + C h^{p+1} \mathbf{y}^{(p+1)}(x_j) + \mathcal{O}(h^{p+2}).$$

For now let us suppose that the (normalized) predictor formula is also of order p. It is likely to require more previously computed solution values than does the corrector. To keep the notation simple, the sums in the corrector formula are extended, if necessary, by introducing zero coefficients so that the number of terms is the same for both formulas. The predicted solution y_{j+1}^* is obtained from

$$\alpha_0^* \mathbf{y}_{j+1}^* + \sum_{i=1}^{k} \alpha_i^* \mathbf{y}_{j+1-i} = h \sum_{i=1}^{k} \beta_i^* \mathbf{F}(x_{j+1-i}, \mathbf{y}_{j+1-i}).$$

The solution $\mathbf{y}(x)$ satisfies

$$\sum_{i=0}^{k} \alpha_i^* \mathbf{y}(x_{j+1-i}) = h \sum_{i=1}^{k} \beta_i^* \mathbf{F}(x_{j+1-i}, \mathbf{y}(x_{j+1-i})) + C^* h^{p+1} \mathbf{y}^{(p+1)}(x_j) + \mathcal{O}(h^{p+2}).$$

In the typical case of Adams methods, an Adams–Bashforth formula is used to predict and an Adams–Moulton formula is used to correct. For such pairs the characteristic polynomials of the two formulas are the same, i.e., $\rho(\theta) = \rho^*(\theta)$. For simplicity we assume this and refer the reader to

Stetter [1973] for the case of more general predictors. Suppose that

$$\mathbf{y}_j = \mathbf{y}(x_j) + h^p\mathbf{e}(x_j) + \mathcal{O}(h^{p+1})$$

for a continuously differentiable function $\mathbf{e}(x)$. For the present purpose it is not necessary to assume that this $\mathbf{e}(x)$ is the same function defined in §1; all that is needed is a regular behavior of the error. Subtracting the equation satisfied by the $\mathbf{y}(x_j)$ from that satisfied by the $\mathbf{y}_j$, we obtain

$$\begin{aligned}\sum_{i=0}^{k} \alpha_i(\mathbf{y}_{j+1-i} - \mathbf{y}(x_{j+1-i}))\\ &= h \sum_{i=0}^{k} \beta_i[\mathbf{F}(x_{j+1-i}, \mathbf{y}_{j+1-i}) - \mathbf{F}(x_{j+1-i}, \mathbf{y}(x_{j+1-i}))]\\ &\quad - h^{p+1}\, C\, \mathbf{y}^{(p+1)}(x_j) + \mathcal{O}(h^{p+2}).\end{aligned}$$

Here

$$\mathbf{F}(x_{j+1-i}, \mathbf{y}_{j+1-i}) - \mathbf{F}(x_{j+1-i}, \mathbf{y}(x_{j+1-i})) = \frac{\partial \mathbf{F}}{\partial \mathbf{y}} h^p\, \mathbf{e}(x_{j+1-i}) + \mathcal{O}(h^{2p}),$$

where the Jacobian is evaluated at $\mathbf{y}(x_{j+1-i})$. Using the differentiability of $\mathbf{e}(x)$ and expanding further, the right hand side of the equation above becomes

$$h^{p+1}\left[\left(\sum_{i=0}^{k} \beta_i\right)\frac{\partial \mathbf{F}}{\partial \mathbf{y}}(x_j, \mathbf{y}(x_j))\,\mathbf{e}(x_j) - C\mathbf{y}^{(p+1)}(x_j)\right] + \mathcal{O}(h^{p+2}).$$

Making use of the assumption that $\rho^*(\theta) = \rho(\theta)$, similar manipulations for the predictor lead to

$$\begin{aligned}\alpha_0(\mathbf{y}^*_{j+1} - \mathbf{y}(x_{j+1})) + \sum_{i=1}^{k} \alpha_i(\mathbf{y}_{j+1-i} - \mathbf{y}(x_{j+1-i}))\\ &= h^{p+1}\left[\left(\sum_{i=1}^{k} \beta_i^*\right)\frac{\partial \mathbf{F}}{\partial \mathbf{y}}(x_j, \mathbf{y}(x_j))\,\mathbf{e}(x_j)\right.\\ &\quad \left. - C^*\mathbf{y}^{(p+1)}(x_j)\right] + \mathcal{O}(h^{p+2}).\end{aligned}$$

By consistency and normalization,

$$\sigma^*(1) = \sum_{i=1}^{k} \beta_i^* = \rho^{*\prime}(1) = 1, \qquad \sigma(1) = \sum_{i=0}^{k} \beta_i = \rho'(1) = 1.$$

Subtraction of the expression involving the predictor formula from the one involving the corrector then leads to

$$\alpha_0(\mathbf{y}_{j+1} - \mathbf{y}^*_{j+1}) = h^{p+1}(C^* - C)\,\mathbf{y}^{(p+1)}(x_j) + \mathbb{O}(h^{p+2}).$$

From this we see that the computable expression

$$\frac{C}{(C^* - C)}\alpha_0(\mathbf{y}_{j+1} - \mathbf{y}^*_{j+1}) = C h^{p+1}\mathbf{y}^{(p+1)}(x_j) + \mathbb{O}(h^{p+2})$$

is an asymptotically correct estimate of the local truncation error of the corrector formula. This is just Milne's estimate (5.1) as derived earlier with the assumption that α_0 was equal to 1.

It is actually possible to estimate the error of any predictor of the kind postulated that has an order lower than p. To see this, retrace the proof with the assumption that the predictor is of order $q < p$. The first change occurs where we now have

$$\begin{aligned}\alpha_0(\mathbf{y}^*_{j+1} - \mathbf{y}(x_{j+1})) &+ \sum_{i=1}^{k} \alpha_i(\mathbf{y}_{j+1-i} - \mathbf{y}(x_{j+1-i})) \\ &= h^{p+1}\left(\sum_{i=1}^{k} \beta_i^*\right)\frac{\partial \mathbf{F}}{\partial \mathbf{y}}(x_j, \mathbf{y}(x_j))\,\mathbf{e}(x_j) \\ &\quad - C^*\,h^{q+1}\,\mathbf{y}^{(q+1)}(x_j) + \mathbb{O}(h^{q+2}).\end{aligned}$$

Subtraction of the equivalent result for the corrector of order p then results in

$$\alpha_0(\mathbf{y}_{j+1} - \mathbf{y}^*_{j+1}) = C^*\,h^{q+1}\,\mathbf{y}^{(q+1)}(x_j) + \mathbb{O}(h^{q+2}),$$

which is a computable estimate of the error of the predictor. This amounts to estimating the error in $\mathbf{y}^*_{j+1}$ by comparing it to the more accurate result $\mathbf{y}_{j+1}$. This result tells us that if we are integrating with an Adams–Moulton formula of order p, we can estimate the local truncation error of any Adams–Bashforth formula of order no greater than p. In backward difference form this is very easy. In view of the fact that the local truncation error of the Adams–Moulton formula of order p is a simple multiple of the local truncation error of the Adams–Bashforth formula of the same order, it is also possible to estimate the local truncation error of all Adams–Moulton formulas of orders no greater than p. This can be seen more directly by examining the argument for Milne's scheme once again. No use was made of the fact that the predictor is an explicit formula—this was assumed only because it is always the case when Milne's scheme is used. Automatic selection of the most efficient order depends on being able to estimate the errors that would have been made at other orders.

A variation of the argument presented can be used to demonstrate the validity of the estimator (5.4). The idea is first to observe that the explicit formula used for estimation of the error can be different from the predictor and then to define an explicit formula for which (5.4) represents an estimator of Milne type. We shall have no need for the details.

The estimator (5.3) based on differencing the $\mathbf{y}_k$ is just as plausible as the estimator (5.4) based on differencing the $\mathbf{F}_k$, but it is more difficult to justify. To do so we must be more specific about the form of the global errors. Let us define vectors $\boldsymbol{\varepsilon}_k$ by

$$\mathbf{y}_k = \mathbf{y}(x_k) + \mathbf{e}(x_k)\, h^p + \boldsymbol{\varepsilon}_k\, h^{p+1} + \mathbb{O}(h^{p+2}).$$

To see better what is happening, let us be specific and estimate the local truncation error of one of the Euler formulas for a scalar equation. In this simple case

$$\frac{1}{2}\nabla^2 y_{j+1} \approx \frac{1}{2}\, h^2 y''(x_j).$$

The difference operator is linear, so with the assumption about the form of the errors it is the case in general that the estimator (5.3) yields

$$\begin{aligned} C\,\nabla^{p+1}\mathbf{y}_{j+1} &= \nabla^{p+1}\mathbf{y}(x_{j+1}) + \nabla^{p+1}\mathbf{e}(x_{j+1})\,h^p \\ &\quad + \nabla^{p+1}\boldsymbol{\varepsilon}_{j+1}\,h^{p+1} + \mathbb{O}(h^{p+2}), \end{aligned}$$

and in the case of the specific example,

$$\begin{aligned} \frac{1}{2}\nabla^2 y_{j+1} &= \frac{1}{2}\nabla^2 y(x_{j+1}) + \nabla^2 e(x_{j+1})\,h + \nabla^2 \varepsilon_{j+1} h^2 + \mathbb{O}(h^4), \\ &= \frac{1}{2}h^2 y''(x_j) + \nabla^2 e(x_{j+1})\,h + \nabla^2 \varepsilon_{j+1} h^2 + \mathbb{O}(h^3). \end{aligned}$$

Now

$$\nabla^2 \varepsilon_{j+1} \equiv \varepsilon_{j+1} - 2\varepsilon_j + \varepsilon_{j-1},$$

from which it is obvious that the magnitude of the second difference is bounded by four times the maximum magnitude of the ε_k. Likewise, the norm of $\nabla^{p+1}\,\boldsymbol{\varepsilon}_{j+1}$ is bounded by a constant multiple of the largest of the norms of the $\boldsymbol{\varepsilon}_k$. More generally, differences of quantities that are uniformly $\mathbb{O}(h^{p+2})$ are themselves generally $\mathbb{O}(h^{p+2})$.

If we make no further assumption about the errors $\boldsymbol{\varepsilon}_k$, the most we can claim for $\nabla^{p+1}\,\boldsymbol{\varepsilon}_{j+1}$ is that it is $\mathbb{O}(1)$. The global errors in the $\mathbf{y}_k$ spoil the estimate of the local truncation error because their effect in the estimator is just as large as the local truncation error itself. Fortunately, the estimate has the correct order so that the step size that might be selected using it is not unreasonable. To justify the estimator, we must assume a stronger relationship between the errors at successive steps than hitherto. The asymptotic analysis of §1 can be extended in suitable circumstances to further terms in an expansion of the error. Suppose now that

$$\mathbf{y}_j = \mathbf{y}(x_j) + h^p\mathbf{e}(x_j) + h^{p+1}\,\boldsymbol{\varepsilon}(x_j) + \mathbb{O}(h^{p+2})$$

for a continuously differentiable function $\boldsymbol{\varepsilon}(x)$ and a twice continuously differentiable function $\mathbf{e}(x)$. It is the size of the quantity $\nabla^{p+1}\,\boldsymbol{\varepsilon}(x_{j+1})$ that is of immediate concern. In the case of the example we have

$$\nabla^2\,\varepsilon(x_{j+1}) = \nabla\,\varepsilon(x_{j+1}) - \nabla\,\varepsilon(x_j) = h\,\varepsilon'(\xi_{j+1}) - h\,\varepsilon'(\xi_j),$$

on using a mean value theorem and the fact that $\varepsilon(x)$ is continuously differentiable. With the assumptions made, this difference is $\mathbb{O}(h)$. Likewise, $\nabla^{p+1}\,\boldsymbol{\varepsilon}(x_{j+1})$ is $\mathbb{O}(h)$. A similar argument shows that for the smoother function $\mathbf{e}(x)$, the difference $\nabla^{p+1}\,\mathbf{e}(x_{j+1})$ is $\mathbb{O}(h^2)$. In the case of the example, this is simply

$$\nabla^2\,e(x_{j+1}) = h^2\,e''(\xi).$$

With these observations about the sizes of terms, it is found that the effects of the global errors in the $\mathbf{y}_k$ are of higher order than the local truncation error and the estimator is asymptotically correct.

In the variable order codes, it is necessary to estimate what the error would have been had the current step been taken at a higher order. If it is assumed that a method of order p has been used throughout and if the error is smooth in the sense that there is an asymptotic expansion of the error with at least a couple of terms, then an argument much like the one just given can be used to justify an estimate of the error of a formula of order $p + 1$. Details for Adams methods can be found in Shampine and Gordon [1975].

§5.2 Error Estimators without Memory

In this subsection the basics of estimating the local error of one-step methods and some general principles for accomplishing this are taken up. Because we are mainly interested in explicit Runge–Kutta methods, we keep them in mind while discussing the estimators. In the next subsection the general considerations will be made specific to explicit Runge–Kutta methods and the derivation of "optimal" formulas will be discussed.

A variety of ideas have been proposed for the estimation of local error, but only one is in general use. It is, however, seen in different guises. The basic idea is to take each step with two formulas independently. Thus, a pair of formulas of orders p and (at least) $p + 1$ are used. Suppose the formula of order $p + 1$ yields the result $\mathbf{y}_{j+1}^*$ when stepping from x_j to x_{j+1} and the formula of order p yields $\mathbf{y}_{j+1}$. The local error of the less accurate formula satisfies

$$\begin{aligned}\mathbf{le}_j &= \mathbf{u}(x_{j+1}) - \mathbf{y}_{j+1} = (\mathbf{y}_{j+1}^* - \mathbf{y}_{j+1}) + (\mathbf{u}(x_{j+1}) - \mathbf{y}_{j+1}^*)\\ &= \mathbf{y}_{j+1}^* - \mathbf{y}_{j+1} + \mathbb{O}(h^{p+2}).\end{aligned}$$

Because $\mathbf{le}_j$ is $\mathbb{O}(h^{p+1})$, the quantity

$$\mathbf{est}_j = \mathbf{y}_{j+1}^* - \mathbf{y}_{j+1}$$

provides an asymptotically correct, computable estimate of the local error.

The question now is, given a formula of order p, how do we construct a higher order formula? This might be done without reference to the original formula, but the first sound estimators were derived using one of two general principles for constructing the error-estimating companion. Because the principles are usually described as ways of estimating the local error, it is often not appreciated that they correspond to ways of forming a higher order formula. Nevertheless, any asymptotically correct estimate of the local error of a formula of order p yields a companion formula of order $p + 1$ by simply defining

$$\mathbf{y}_{j+1}^* = \mathbf{y}_{j+1} + \mathbf{est}_j.$$

The more important general principle is called *doubling*, *halving*, or *(Richardson) extrapolation*, depending on how it is viewed. It is a local version of the way described in §2 for estimating the global error. A step of length h is taken from $(x_j, \mathbf{y}_j)$ to get $\mathbf{y}_{j+1}$. Then two steps of length $h/2$ are taken from $(x_j, \mathbf{y}_j)$ to construct successively

$$\tilde{\mathbf{y}}_{j+1/2} = \mathbf{y}_j + \frac{h}{2}\,\mathbf{\Phi}(x_j, \mathbf{y}_j),$$

$$\tilde{\mathbf{y}}_{j+1} = \tilde{\mathbf{y}}_{j+1/2} + \frac{h}{2}\,\mathbf{\Phi}(\mathrm{x}_{j+1/2}, \tilde{\mathbf{y}}_{j+1/2}).$$

Exercise 4 of Chapter 4 shows that in the case of Runge–Kutta methods, this amounts to another Runge–Kutta formula of twice as many stages that takes a step of length h from $(x_j, \mathbf{y}_j)$ to form $\tilde{\mathbf{y}}_{j+1}$. The local error of the basic formula is given in terms of its principal error function as

$$\mathbf{u}(x_{j+1}) - \mathbf{y}_{j+1} = h^{p+1}\,\boldsymbol{\psi}(x_j, \mathbf{y}_j) + \mathbb{O}(h^{p+2}) = \mathbf{le}_j.$$

It should not be surprising that the local error of $\tilde{\mathbf{y}}_{n+1}$ can be expressed in terms of the same principal error function. Naturally the first half step from $(x_j, \mathbf{y}_j)$ has an error that can be expressed in this way,

$$\mathbf{u}(x_{j+1/2}) - \tilde{\mathbf{y}}_{j+1/2} = \left(\frac{h}{2}\right)^{p+1} \boldsymbol{\psi}(x_j, \mathbf{y}_j) + \mathbb{O}(h^{p+2})$$
$$= \frac{1}{2^{p+1}} \mathbf{le}_j + \mathbb{O}(h^{p+2}).$$

The local error of the half step from $x_{j+1/2}$ involves a different local solution $\mathbf{v}(x)$ that has the initial value

$$\mathbf{v}(x_{j+1/2}) = \tilde{\mathbf{y}}_{j+1/2}$$

In terms of this local solution, the local error of the second half step is

$$\mathbf{v}(x_{j+1}) - \tilde{\mathbf{y}}_{j+1} = \left(\frac{h}{2}\right)^{p+1} \boldsymbol{\psi}(x_{j+1/2}, \tilde{\mathbf{y}}_{j+1/2}) + \mathbb{O}(h^{p+2}).$$

Because

$$x_{j+1/2} = x_j + \frac{h}{2} = x_j + \mathbb{O}(h) \quad \text{and} \quad \tilde{\mathbf{y}}_{j+1/2} = \mathbf{y}_j + \mathbb{O}(h),$$

the principal error function can be expanded in this expression to get

$$\mathbf{v}(x_{j+1}) - \tilde{\mathbf{y}}_{j+1} = \left(\frac{h}{2}\right)^{p+1} \boldsymbol{\psi}(x_j, \mathbf{y}_j) + \mathbb{O}(h^{p+2}).$$

Subtracting the two expressions involving answers at x_{j+1}, we find

$$(\tilde{\mathbf{y}}_{j+1} - \mathbf{y}_{j+1}) + (\mathbf{u}(x_{j+1}) - \mathbf{v}(x_{j+1})) = \left(1 - \frac{1}{2^{p+1}}\right) \mathbf{le}_j + \mathbb{O}(h^{p+2}).$$

To proceed further we must relate the two local solutions, or more specifically, their values $\mathbf{v}(x_{j+1})$ and $\mathbf{u}(x_{j+1})$. We have had to do this kind of thing before where we found that, roughly speaking, nearby solutions of the differential equation are parallel for a single step. Specifically, for Lipschitzian $\mathbf{F}$, the bound on how fast solutions can spread apart implies that

$$\mathbf{v}(x_{j+1}) - \mathbf{u}(x_{j+1}) = \mathbf{v}(x_{j+1/2}) - \mathbf{u}(x_{j+1/2})$$
$$+ \mathbb{O}\left(\frac{h}{2} \|\mathbf{v}(x_{j+1/2}) - \mathbf{u}(x_{j+1/2})\|\right).$$

Recalling that

$$\mathbf{v}(x_{j+1/2}) - \mathbf{u}(x_{j+1/2}) = \tilde{\mathbf{y}}_{j+1/2} - \mathbf{u}(x_{j+1/2}) = \frac{1}{2^{p+1}} \mathbf{le}_j + \mathbb{O}(h^{p+2}),$$

we then get

$$\tilde{\mathbf{y}}_{j+1} - \mathbf{y}_{j+1} = \left(1 - \frac{1}{2^{p+1}} - \frac{1}{2^{p+1}}\right) \mathbf{le}_j + \mathbb{O}(h^{p+2}),$$
$$= \left(\frac{2^p - 1}{2^p}\right) \mathbf{le}_j + \mathbb{O}(h^{p+2}).$$

This provides the computable estimate of the local error

$$\mathbf{u}(x_{j+1}) - \mathbf{y}_{j+1} = \mathbf{le}_j = \left(\frac{2^p}{2^p - 1}\right)(\tilde{\mathbf{y}}_{j+1} - \mathbf{y}_{j+1}) + \mathbb{O}(h^{p+2}).$$

Estimating the error in $\mathbf{y}_{j+1}$ in this way is called *halving*. It is to be expected that the approximation $\tilde{\mathbf{y}}_{j+1}$ resulting from two half steps will be a better approximation to $\mathbf{y}(x_{j+1})$ than $\mathbf{y}_{j+1}$. A modification of the analysis of halving leads to the estimate

$$\mathbf{u}(x_{j+1}) - \tilde{\mathbf{y}}_{j+1} = \left(\frac{1}{2^p - 1}\right)(\tilde{\mathbf{y}}_{j+1} - \mathbf{y}_{j+1}) + \mathbb{O}(h^{p+2}).$$

It is natural to advance the integration with the more accurate result $\tilde{\mathbf{y}}_{j+1}$. When this is done, the error estimation procedure is called *doubling*. The practical distinction is that with doubling, two steps are taken before the local error is estimated. This means that the estimate is cheaper per step, but the step size is adjusted less frequently. It should be appreciated that advancing with $\tilde{\mathbf{y}}_{j+1}$ is *not* local extrapolation; $\tilde{\mathbf{y}}_{j+1}$ is more accurate than $\mathbf{y}_{j+1}$, but it is not of higher order.

Let us be more specific about the cost of doubling. Suppose that we are using an explicit Runge–Kutta formula of s stages. Each of the two half steps costs s stages. The full step costs only $s - 1$ stages because the stage $\mathbf{F}(x_j, \mathbf{y}_j)$ is already available from taking the first half step. The whole process then costs a total of $3s - 1$ stages per step. Runge–Kutta formulas of high order involve many stages and the absolute cost of a step implemented in this way can become significant. One reason for being concerned about the absolute cost of a step is the cost of a failed step. With error estimation by doubling, if the step is rejected, generally only $\mathbf{F}(x_j, \mathbf{y}_j)$ can be reused for the next try, so there is a loss of $3s - 2$ stages. Many early codes reduced the cost of a failed step by taking advantage of the special structure of doubling. If the step size used for a second try is taken to be half of the step size that failed, the first half step of the failed try can be reused as the whole step with the smaller step size. This is appealing, but experience showed it to be counterproductive. Halving the step size may be too big a reduction, which is inefficient and provides more accuracy than is wanted, or it may not be big enough, leading to another failure.

Present practice is to ignore this possibility and to select always the "optimal" step size for another try. When estimating the local error by doubling, the estimate costs $s - 1$ stages out of a total of $3s - 1$. Although this seems high compared to other schemes that will be described below, the estimate is of high quality and with local extrapolation it is possible for such schemes to be competitive.

Local extrapolation amounts to advancing with the higher order result

$$\mathbf{y}_{j+1}^{*} = \tilde{\mathbf{y}}_{j+1} + \left(\frac{1}{2^p - 1}\right)(\tilde{\mathbf{y}}_{j+1} - \mathbf{y}_{j+1}) = \tilde{\mathbf{y}}_{j+1} + \mathbf{est}_j$$

With this principle for estimating the local error, if the basic formula is an explicit Runge–Kutta formula, then this formula is an explicit Runge–Kutta formula of order $p + 1$. It is often thought that Richardson extrapolation is somehow different from taking a step with a pair of formulas. It is only the viewpoint that is different. We see now that the construction can be viewed as a general way of producing a higher order formula with the local error estimated by comparison. Only recently (Shampine [1985c]) was it noticed that because the error

$$\begin{aligned}\mathbf{u}(x_{j+1/2}) - \tilde{\mathbf{y}}_{j+1/2} &= \frac{1}{2^{p+1}}\,\mathbf{le}_j + \mathbb{O}(h^{p+2}) \\ &= \frac{1}{2}\left(\frac{1}{2^p - 1}\right)(\tilde{\mathbf{y}}_{j+1} - \mathbf{y}_{j+1}) + \mathbb{O}(h^{p+2}),\end{aligned}$$

we can get a result of order $p + 1$ at $x_{j+1/2}$, too, by

$$\mathbf{y}_{j+1/2}^{*} = \tilde{\mathbf{y}}_{j+1/2} + \frac{1}{2}\left(\frac{1}{2^p - 1}\right)(\tilde{\mathbf{y}}_{j+1} - \mathbf{y}_{j+1}).$$

EXERCISE 9. The code of Gear [1972, p. 83] is based on the classic four-stage, fourth-order formula with error estimation by doubling implemented with local extrapolation. How might this last observation be used to provide this code with an interpolant?

The other general principle for estimating the local error is to step to x_{j+1} and then step from $(x_{j+1}, \mathbf{y}_{j+1})$ back to x_j with a step size of $-h$ to form $\tilde{\mathbf{y}}_j$. To leading order the magnitude of the local error of the step from x_{j+1} to x_j is the same as that of the step from x_j to x_{j+1} and for a formula of odd order, the two errors have the same sign. The local error can then be estimated by comparing $\tilde{\mathbf{y}}_j$ to the known value $\mathbf{y}_j$:

$$\mathbf{le}_j = \frac{1}{2}(\mathbf{y}_j - \tilde{\mathbf{y}}_j) + \mathbb{O}(h^{p+2}).$$

Because the local error estimate obtained using this general principle is expensive compared to doubling, the details are left to an exercise.

EXERCISE 10. Justify this estimate of the local error for one-step methods of odd order. Count the cost of the estimate for a formula of s stages. When the step succeeds, one stage can be used for the next step. Because most steps succeed, it can be neglected fairly in your count.

§5.3 Choosing a Runge–Kutta Formula—Error Estimate

It was seen in the last subsection that in the case of Runge–Kutta formulas, a formula plus an asymptotically correct estimate of the local error is equivalent to a pair of formulas. As soon as it was considered desirable to estimate and control local errors, the classical aim of producing formulas of a given order in as few stages as possible lost its relevance. The aim then became to produce a pair of formulas of orders p and $p + 1$ in as few stages as possible. However, it has been pointed out that formulas should be considered on a basis of equal cost and nowadays there is a growing appreciation that minimal stage pairs may not be the most effective. In this section we discuss the derivation of effective pairs of formulas. Because all the popular explicit Runge–Kutta codes do local extrapolation, the result of the higher order formula will be denoted in this section by $\mathbf{y}_{j+1}$ and the result of the lower order member of a pair will be denoted by $\mathbf{y}_{j+1}^*$.

If we are to take each step with two formulas, it seems obvious that we should have as many stages as possible common to the formulas. Stated in this way, this is a natural goal. However, it was a long time before it was stated in this way because when attention is focussed on the formula and an estimate of its error, the goal is less obvious. At first the general principles described in the last section were used to estimate the local error. Eventually Sarafyan and Fehlberg independently derived local error estimators to go along with the basic formula that were considerably cheaper than those provided by the general principles. In some cases researchers approached this task as one of devising a pair of formulas, but presented the fruits of their investigations as a formula plus error estimate because they did not have local extrapolation in mind.

The way England [1969] went about deriving one of his formulas shows clearly why one might hope to do better than using the general principle of doubling. He studied four-stage, fourth-order formulas. If any one of the possibilities is implemented with doubling error estimate, two half steps are taken with the formula to get the result $\mathbf{y}_{j+1}^*$ used to advance the integration and a full step is taken with the formula to get a result used in estimating the local error of $\mathbf{y}_{j+1}^*$. The step itself involves 8 stages and the extra work of estimating the error amounts to 3 stages. The question is whether it is possible to estimate the error using fewer additional stages. Exercise 13 of Chapter 4 takes up an unsuccessful attempt to use Simpson's rule to provide an estimate for the classic four-stage, fourth-order formula. England succeeded by considering other four-stage, fourth-order formulas. He found that by proper choice of the basic formula, he could get an error estimate involving only one extra stage. Because he did not have local extrapolation in mind, he computes exactly the same result $\mathbf{y}_{j+1}^*$ as with doubling, but the cost of a successful step is reduced from 11 to 9 stages. One might reasonably ask if this new estimate is as good as that provided by doubling. Shampine and Watts [1971] investigated this matter and at the same time studied experimentally the effect of changing a production-grade code, RUNKUT, to use the new scheme. This code is based on a four-stage, fourth-order Runge–Kutta formula different from that chosen by England. Its truncation error coefficients are close in size to those of the scheme used by England and replacing the basic formula in RUNKUT with the one used by England had, on average, no effect on the accuracy nor on the cost, though for individual problems a difference was seen. This supports statements made in §1.2 of Chapter 4 where the selection of accurate formulas was discussed in terms of the truncation error coefficients. Changing to the England estimator reduces the cost of a successful step from 11 stages to 9. The cost of a failed step appears to be reduced from 10 to 8, but there is a special savings with England's formula like that seen earlier with PECE implementations of predictor–corrector methods: The extra stage and the estimate of the local error are formed before the last stage of the step is formed, so that if the step is rejected, the last stage is never formed and the cost is only 7 evaluations of $\mathbf{F}$. The tests of Shampine and Watts took advantage of this, but the savings are not very important because most steps succeed. In experiments it was found that England's way of estimating the local error did, on average, reduce the total number of evaluations of $\mathbf{F}$ to solve the problem in the proportion 9/11. The accuracy of the estimator was shown to be less than that of doubling, but perfectly adequate. England's scheme provides a significant increase in efficiency when local extrapolation is not done, as it was not in those days, and in particular, was not in RUNKUT. The matter is less clear when local extrapolation is done because the more accurate estimate of the error with doubling corresponds to a more accurate higher order formula.

Although based on a minimal stage formula, England's pair involves 9 stages. What is the minimum number of stages of a (4, 5) *pair*? Clearly it cannot be less than the number of stages required for the higher order formula alone. As it happens, we have already seen an example of a minimal stage pair of lower order, namely the predictor–corrector pair AB1–AM2, the first-order Euler formula and the second-order trapezoidal rule:

$$\mathbf{y}^*_{j+1} = \mathbf{y}_j + h\,\mathbf{F}(x_j, \mathbf{y}_j),$$

$$\mathbf{y}_{j+1} = \mathbf{y}_j + \frac{h}{2}\,[\mathbf{F}(x_j + h, \mathbf{y}^*_{j+1}) + \mathbf{F}(x_j, \mathbf{y}_j)].$$

This was first derived in Chapter 4 as a two-stage, second-order Runge–Kutta formula called the improved Euler formula. In the general notation for an autonomous system this is

$$\mathbf{F}_0 = \mathbf{F}(\mathbf{y}_j),$$

$$\mathbf{Y}_1 = \mathbf{y}_j + h\,\mathbf{F}_0, \qquad \mathbf{F}_1 = \mathbf{F}(\mathbf{Y}_1),$$

$$\mathbf{y}_{j+1} = \mathbf{y}_j + h\left[\frac{1}{2}\,\mathbf{F}_0 + \frac{1}{2}\,\mathbf{F}_1\right].$$

A pair of formulas of orders 1 and 2 is referred to as a (1, 2) pair and this one is seen to cost two evaluations of **F** per step, which is as cheap as possible.

This little example of a (1, 2) pair does not exploit fully the possibilities for the second order formula. As was pointed out in Exercise 5 of Chapter 4, for a constant $\zeta \neq 0, 1$, the result $\tilde{\mathbf{y}}_{j+1} = \zeta\mathbf{y}^*_{j+1} + (1 - \zeta)\mathbf{y}_{j+1}$ represents another explicit Runge–Kutta formula. It amounts to forming a different linear combination of the stages. In the case of the example, the first-order formula of the pair uses only the first stage. By developing a first order formula that uses both, additional flexibility is obtained. For one thing, a more accurate formula is possible. In general the local error of the higher order formula is

$$\mathbf{u}(x_j + h) - \mathbf{y}_{j+1} = h^{p+2}\,\boldsymbol{\psi}(x_j, \mathbf{y}_j) + \mathbb{O}(h^{p+3}),$$

and that of the lower is

$$\mathbf{u}(x_j + h) - \mathbf{y}^*_{j+1} = h^{p+1}\,\boldsymbol{\psi}^*(x_j, \mathbf{y}_j) + \mathbb{O}(h^{p+2}).$$

Because $\zeta \neq 0$,

$$\mathbf{u}(x_j + h) - \tilde{\mathbf{y}}_{j+1} = h^{p+1}\,\zeta\,\boldsymbol{\psi}^*(x_j, \mathbf{y}_j) + \mathbb{O}(h^{p+2}).$$

Using ζ, we can adjust the size of the truncation error coefficients (not the individual coefficients, rather the norm of the vector with these coefficients as components) in the lower order formula to be whatever we want. Of course, as $\zeta \to 0$, these coefficients tend to 0 because the formula is of order $p + 1$ when $\zeta = 0$. This fact makes it plain that some thought must be devoted to the issue of the quality of the error estimate.

The absolute stability of the formula used to advance the integration is of considerable practical importance. It is not obvious that any attention ought to be paid to the stability of the companion formula. A stability region can be interpreted as a measure of the accuracy of the formula for a quite restricted class of differential equations. Only qualitative agreement of the numerical and true solutions is required, but the agreement is to hold for "large" step sizes, so it is of particular interest. When comparing the results of two formulas, should a step size correspond to being in the stability region of one member of a pair but not the other, the difference indicates a large error and an unacceptable step size. The applicability of these arguments to general equations is unclear because the class of equations for which the analysis is possible is so restricted, but for at least some kinds of equations, it is the intersection of the two stability regions that governs the acceptable step sizes. Experiments show that it is of general value to match the stability regions of the two formulas, insofar as this is possible. The (1, 2) pair used as an example has both stability regions fully determined by the fact that each formula is a minimal stage formula. As Figure 6.3 shows, the stability region of one-stage, first-order Runge–Kutta formulas does not match well that of two-stage, second-order formulas and the extra flexibility of two stages for the first-order formula can be used to advantage in matching the stability regions.

Sarafyan, Fehlberg, and England all derived (4, 5) pairs using the minimal number of stages, namely six. There are several families of fifth-order formulas involving six stages. The idea is to search among the parameters defining a family to find a set that permits a result of order four to be formed. This is complicated because the equations of condition that must be satisfied are non-linear and the matter is made even more complicated by a search for a good pair. There is another way to get more flexibility in deriving formulas that was mentioned earlier in connection with one of the general principles for error estimation. On stepping to x_{j+1}, the first stage of the next step is always $\mathbf{F}(x_j + h, \mathbf{y}_{j+1})$. Since most steps succeed, this stage can be evaluated early as an additional stage in the current step. In the case of the example (1, 2) pair, we write

$$\mathbf{Y}_2 = \mathbf{y}_{j+1} = \mathbf{y}_j + h\left[\frac{1}{2}\mathbf{F}_0 + \frac{1}{2}\mathbf{F}_1\right],$$

and form $\mathbf{F}_2 = \mathbf{F}(\mathbf{Y}_2)$. If the step is accepted, the $\mathbf{F}_0$ of the next step is taken to be this last stage of the current step. Because most steps are accepted, it is fair to regard this extra stage as "free." Of course, if a step is rejected, this is truly an extra stage, hence an extra cost. Although the idea was used earlier, Dormand and Prince popularized it with the name *First Same As Last*—FSAL. It is seen in some of the most widely used pairs at this time. For example, RKSUITE implements (2, 3), (4, 5), and (7, 8) pairs. The two lower order pairs use FSAL and the highest does not (even though it is due to Prince and Dormand). So far we have not had to specify whether we plan to do local extrapolation. The FSAL technique requires a decision about this because the extra stage must be formed using the value that will be used to advance the integration. Either choice might be made, but there are few examples of FSAL being done without local extrapolation, a pair due to Fehlberg [1970] being one. Shampine and Watts [1977, 1979] examined many (4, 5) pairs and chose one of six stages due to Fehlberg as the best for implementation in their widely used code RKF45. Tests by other workers confirmed that this was a very good choice and for a long time it was generally agreed that this was the most effective pair of this order. Fehlberg did not have in mind local extrapolation when he derived the pair, but the pair is significantly more efficient when local extrapolation is done and all effective implementations, including in particular RKF45, do it. The Fehlberg pair does not use the FSAL technique, so by exploiting the extra flexibility gained in this way, Dormand and Prince [1980] subsequently derived a more efficient (4, 5) pair. (Shampine [1986] argues that they were unduly conservative and recommends a modification that is still more efficient in one measure.) The flexibility gained by assuming FSAL is largely confined to the fourth-order formula. Still more flexibility is gained by allowing more stages. With seven stages, it is possible to obtain families of sixth-order formulas. This suggests that with one extra stage, it is possible to obtain very accurate fifth-order formulas and still retain a great deal of flexibility for optimizing both the fourth and fifth order formulas. Bogacki and Shampine [1989b] exploited this fact to derive a (4, 5) pair that is about as much more efficient than the Dormand–Prince pair as the Dormand–Prince pair is more efficient than the Fehlberg pair.

EXERCISE 11. The approach England took to improving doubling was exemplified in Shampine [1973] by deriving estimators for second-order formulas. In present day terms, the most attractive defines $\mathbf{y}_{j+1}^*$ as the result of two half steps with the improved Euler formula (AB1–AM2). It is possible to obtain a third-order result using only the four stages formed in the computation of $\mathbf{y}_{j+1}^*$. Comparison of this result to $\mathbf{y}_{j+1}^*$ then provides a "free" error estimate. Write out the Butcher array for $\mathbf{y}_{j+1}^*$ and use the table of truncation error coefficients to verify that this formula is of order two. There is exactly one formula of order 3 that can be obtained as a linear combination of the four stages—derive this formula.

More flexibility in the construction of a third-order companion for $\mathbf{y}^*_{j+1}$ is obtained by using FSAL to get another stage $\mathbf{F}_4 = \mathbf{F}(x_{j+1}, \mathbf{y}^*_{j+1})$ (which is $\mathbf{F}_0$ for the next step). Derive a family of third order companions for the second-order formula. If this were not merely an exercise, a search among the members of the family would be made to find a formula that provides a pair of high quality.

For the remaining discussion in this section we need to recall some notation from Chapter 4. When the differential equation is autonomous, the general form of an explicit Runge–Kutta formula is

$$\mathbf{F}_0 = \mathbf{F}(\mathbf{y}_j)$$

and for $k = 1, \ldots, s$,

$$\mathbf{Y}_k = \mathbf{y}_j + h \sum_{m=0}^{k-1} \beta_{k,m} \mathbf{F}_m, \qquad \mathbf{F}_k = \mathbf{F}(\mathbf{Y}_k).$$

The higher order solution is

$$\mathbf{y}_{j+1} = \mathbf{y}_j + h \sum_{k=0}^{s} \gamma_k \mathbf{F}_k.$$

In addition, we now suppose there is a lower order solution

$$\mathbf{y}^*_{j+1} = \mathbf{y}_j + h \sum_{k=0}^{s} \gamma^*_k \mathbf{F}_k.$$

There may be stages used by one formula and not the other; this is taken into account by introducing zero coefficients as necessary. The evaluation of one formula can be regarded as "embedded" in the evaluation of the other, and error estimates derived in this way are called *embedded error estimates*. It is supposed that the lower order formula is of order p so that its local error has an expansion

$$\mathbf{u}(x_j + h) - \mathbf{y}^*_{j+1} = \sum_{i=p+1}^{\infty} h^i \left(\sum_{n=1}^{\lambda_i} T^{*i}_n \mathbf{D}^i_n \right).$$

The higher order formula is of order $p + 1$ so that its local error is

$$\mathbf{u}(x_j + h) - \mathbf{y}_{j+1} = \sum_{i=p+2}^{\infty} h^i \left(\sum_{n=1}^{\lambda_i} T^i_n \mathbf{D}^i_n \right).$$

In discussing the size of the truncation error coefficients T^i_n for a given i, it is convenient to write them as components in a vector $\mathbf{T}^i$ of γ_i elements so that a norm of the vector provides a natural and convenient measure of size.

There is an implementation matter that might be discussed at this point. In a code there is no need actually to form $\mathbf{y}^*_{j+1}$. The integration will be advanced with $\mathbf{y}_{j+1}$, so what we need is the error estimate

$$\mathbf{est}_j = \mathbf{y}_{j+1} - \mathbf{y}^*_{j+1} = h \sum_{k=0}^{s} (\gamma_k - \gamma_k^*) \mathbf{F}_k.$$

In many of the popular codes, the coefficients $\delta_k = \gamma_k - \gamma_k^*$ are stored and the norm of the estimated error is computed as

$$\|\mathbf{est}_j\| = |h| \left\| \sum_{k=0}^{s} \delta_k \mathbf{F}_k \right\|.$$

On the other hand, some codes actually form the two approximate solutions, subtract one from the other, and compute the norm of the result. The two approaches are not equivalent in finite precision arithmetic. It is perfectly possible that the two approximate solutions agree in all digits so that there is a significance failure on subtracting and the error is estimated to be exactly zero. The other scheme avoids this by subtracting out the dominant term analytically. This is not entirely a good thing, because when the first approach has a significance failure, the second estimates the error in $\mathbf{y}_{j+1}$ to be less, perhaps much less, than that in the correctly rounded representation of $\mathbf{y}(x_{j+1})$, and this simply cannot be. Either form might be used in practice; it is just a matter of appreciating that they do not behave quite the same. The principal advantage of the first form is that the effects of finite precision arithmetic are revealed more clearly. In the first instance this is seen on a significance failure, a clear statement that the solution cannot be made more accurate. It is also seen when both approximations are very accurate because then arithmetic errors made in forming the approximate solutions may be revealed by the subtraction and they would not be revealed with the second form. The principal advantage of the second form is that the behavior of the estimate is a smoother function of the step size and this leads to a smoother behavior of the predicted, "optimal" step size.

Let us now briefly consider some of the issues of quality in the derivation of an effective pair of formulas in the context of a simple (2, 3) pair. The derivation is relatively simple in this case, but the pair itself is of some importance. It is used by the TI-85 graphics calculator, it is one of the pairs available in the package Differential Systems of Gollwitzer [1991], and it is one of the pairs available in RKSUITE. More details of the derivation of the pair are found in Bogacki and Shampine [1989b]. The derivation started with the premise that the third order formula would be minimal stage. In §1.2 of Chapter 4 it was stated that the "optimal" formula of Ralston is about as accurate as any and in Exercise 14 it is found that two of the four truncation error coefficients are zero. This formula samples at distinct points in the span of the step. The coefficients are simple. For these reasons this formula was chosen as the higher order member of the pair. Its Butcher array is

$$\begin{array}{c|ccc} 0 & & & \\ \frac{1}{2} & \frac{1}{2} & & \\ \frac{3}{4} & 0 & \frac{3}{4} & \\ \hline & \frac{2}{9} & \frac{1}{3} & \frac{4}{9} \end{array}$$

More comments about some of these points can be made now. In line with present practice, it is planned to advance with the third-order formula, local extrapolation, and to use the FSAL technique. This technique adds to the stages presented the stage $\mathbf{F}_3 = \mathbf{F}(x_j + h, \mathbf{y}_{j+1})$. Written for non-autonomous equations, as done here, it is clear that the last stage is evaluated at the end of the step and this is another sample in the span of the step for the pair of formulas. There are as many distinct samples as possible and they are spread throughout the interval. Because the integration is to be done with the third-order formula, zero truncation error coefficients are welcome—they just mean that the integration might be more accurate than usual for some classes of problems. Shortly we shall see that such coefficients in the companion formula are undesirable. This example is much simpler than that of even the (4, 5) pairs: When Fehlberg derived his (4, 5) pair, he started with a family of six-stage formulas of order five. For a given choice of parameters there might not be an embedded formula of order four. However, he found a choice that did lead to a family of pairs and the parameters defining the family could then be chosen to optimize accuracy (and other things). The stability region was not fully determined by the order, so it could be optimized as well. Dormand and Prince had additional flexibility due to the FSAL technique and Bogacki and Shampine had still more due to another stage. The derivations of these formulas were much more complicated but not different in spirit from that of this simple example. The main difference is that derivation of a pair of formulas is coupled in a way that is not exemplified here because the third-order formula is chosen independently of its companion.

The error of the second-order result is estimated by comparing it to the third-order result. It might seem desirable, then, to choose an inaccurate second-order formula because its error will be estimated very well and a relatively inaccurate second-order result does no harm because it is not used to advance the integration. All this is true, but if we were to do this, the two measures of efficiency discussed in §2 would lead to substantially different conclusions. Recall that one measure considers the cost of the integration as a function of the accuracy requested (the tolerance) and the other, the cost as a function of the accuracy achieved. A notable example of a pair inefficient by one measure and efficient by the other is Zonneveld's (4, 5) pair [1964]. It is desirable that codes be efficient by both measures of efficiency and for this to be the case, it is necessary that the lower order formula be accurate.

It is important that neither of the two truncation error coefficients T_2^{*3} and T_1^{*3} vanish. If one were to vanish, there would be a class of problems for which the formula is of order three rather than the order two expected. This causes obvious difficulties with error estimation and control. This point was not appreciated by workers in the field until Babuŝka pointed out that some high-order pairs derived by Fehlberg have quite a special behavior when solving quadrature problems—the formula of order p is then of order $p + 1$ and the estimated error is zero. Nowadays people try to derive formulas for which the truncation error coefficients not only do not vanish, but are all of the same size, insofar as this is possible. The aim is to obtain a uniform behavior in the sense that equations that are similar will be solved similarly. In their derivation of the (2, 3) pair being discussed, Bogacki and Shampine went so far as to make the two truncation error coefficients the same. (They would not have gone to this extreme if they had had to compromise on any other measure of quality.) It is seen that quite a different attitude is taken to the presence of zero truncation errors in a formula depending on whether it is the lower or the higher order member of a pair.

As was observed in §1.2 of Chapter 4, we want the formula of order p to "look" like a formula of order p for even comparatively large step sizes h. A conventional measure of this is in the case of a second-order formula

$$B = \frac{\|\mathbf{T}^{*4}\|}{\|\mathbf{T}^{*3}\|}.$$

A small value of B means that the leading term in the expansion of the local error is likely to dominate for comparatively large h. This is a gross measure of size and the constraints of the last paragraph may be thought of as part of avoiding a small denominator here. Very recent investigations go rather further and compare the size of the leading term in the expansion to that of several successive terms. For the example this amounts to $\|\mathbf{T}^{*3}\|$ not being too small compared not just to $\|\mathbf{T}^{*4}\|$, but also to $\|\mathbf{T}^{*5}\|$, $\|\mathbf{T}^{*6}\|$, ... To a considerable degree this requirement limits how small $\|\mathbf{T}^{*3}\|$ can be made, which is to say that it prevents making the formula so accurate that it is essentially of third order. A (2, 3) pair of Fehlberg [1970] illustrates the point. For this pair $\|\mathbf{T}^{*3}\|_2$ is 6.70×10^{-4} and $\mathbf{B}$ is 72.5. Experiments confirm that the step size has to be quite small before the formula behaves like a second-order formula. By way of comparison, the Bogacki–Shampine pair has the much smaller value of $\mathbf{B} = 1.35$ (and, correspondingly, a less accurate second-order formula with $\|\mathbf{T}^{*3}\|_2 = 2.94 \times 10^{-2}$).

The two formulas are related by the estimate of the local error. Subtracting the expansions of the local errors, we find that the estimate can be expanded as

$$\mathbf{y}_{j+1} - \mathbf{y}^*_{j+1} = (\mathbf{u}(x_j + h) - \mathbf{y}^*_{j+1}) - (\mathbf{u}(x_j + h) - \mathbf{y}_{j+1})$$

$$= h^{p+1} \sum_{n=1}^{\lambda_{p+1}} T^{*(p+1)}_n \mathbf{D}^{p+1}_n + \sum_{i=p+2}^{\infty} h^i \left(\sum_{n=1}^{\lambda_i} (T^{*i}_n - T^i_n)\, \mathbf{D}^i_n \right).$$

The leading term in this expansion is the local error that we wish to estimate, so we want it to dominate. As with the truncation error, we would like to have

$$C = \frac{\|\mathbf{T}^{*4} - \mathbf{T}^4\|}{\|\mathbf{T}^{*3}\|}$$

small so as to have a good estimate of the local truncation error for h as large as possible. For the Fehlberg (2, 3) pair C is 4.66 and for the Bogacki–Shampine pair it is 1.38.

The stability region of the second-order formula needs to be matched to that of the third-order formula. Bogacki and Shampine selected parameters yielding a second-order formula with a stability region that includes that of the third-order formula, but is not a lot bigger.

Certain of the constraints described can be translated into mathematical constraints on the parameters, but most are sufficiently vague that one must explore the space of parameters in a heuristic way. After finding a "best" formula, it is usual to modify the coefficients slightly to obtain a pair that involves only "nice" coefficients, but this modification is not allowed to compromise the measures of quality. This is not entirely a matter of esthetics. A difficult derivation like the Prince and Dormand [1981] derivation of an effective (7, 8) pair will result in "optimal" coefficients that cannot be represented in a few decimal digits. Computers do not "care" about this, but it does cause some software difficulties. Obviously we want the coefficients to be specified as accurately as possible in a code, but the fact that computers have such different word lengths (which causes codes to be written in single precision for some computers and double for others) makes the provision of accurate coefficients an annoying matter. It is complicated by the fact that codes are often moved from one machine to another. Indeed, moving a code to a machine with a longer word length cannot be done properly unless the documentation provides explicitly for this task by giving the requisite additional digits. Present practice approximates coefficients by ratios of integers that can be entered exactly in all the popular computers. An investigation taking all these issues into account led Bogacki and Shampine to the choice

$$\mathbf{y}_{j+1}^{*} = \mathbf{y}_j + h\left[\frac{7}{24}\mathbf{F}_0 + \frac{1}{4}\mathbf{F}_1 + \frac{1}{3}\mathbf{F}_2 + \frac{1}{8}\mathbf{F}_3\right]$$

for the companion formula of order two to go with Ralston's formula in a FSAL, local extrapolation implementation.

We have still not taken up an important consideration in the derivation of effective Runge–Kutta formulas, namely interpolation. Early work on this task had the aim of providing an interpolation capability for existing pairs. For example, the capability was provided for Fehlberg's (4, 5) pair by Horn [1983], then for the Dormand–Prince (4, 5) pair by Shampine [1985b], and still later improved interpolants were derived as a better understanding of the issues emerged. Now when a new pair is derived, this capability is taken into account from the beginning. This was the case, for example, with the (4, 5) pair of Bogacki and Shampine. It makes the derivation still more complicated because it is coupled to the choice of the pair. At the lowest orders it is easy to obtain interpolants. Indeed, in the case of the (2, 3) pair whose derivation was sketched above, Bogacki and Shampine had in mind from the beginning the use of Hermite cubic interpolation to the value, $\mathbf{y}_j$, and slope, $\mathbf{F}(x_j, \mathbf{y}_j)$, at the beginning of the step and to the value, $\mathbf{y}_{j+1}$, and slope, $\mathbf{F}(x_{j+1}, \mathbf{y}_{j+1})$, at the end of the step. In Chapter 4 it was shown that this provides approximate solutions throughout $[x_j, x_{j+1}]$ that are as accurate as the solution at the end of the step, $\mathbf{y}_{j+1}$. Interpolation at higher orders such as (4, 5) is much more troublesome and at still higher orders such as (7, 8), no completely satisfactory scheme is yet known.

§6 Starting a Code

The methods discussed in this book for the solution of the initial value problem

$$\mathbf{y}' = \mathbf{F}(x, \mathbf{y}) \qquad a \le x \le b, \qquad \mathbf{y}(a) \text{ given},$$

proceed by steps from the known value of $\mathbf{y}(x)$ at $x = a$ to the final point $x = b$. The first step is critical. If it is much too large, the estimated error may be completely wrong and the whole integration ruined, perhaps without any indication from the code that the results are inaccurate. Because the algorithms for the estimation of error and for the adjustment of step size depend on making "small" changes to the step size, it is necessary to impose a limit on the rate of change of step size. As a consequence, if the first step is much too small, an unnecessarily small (inefficient) step size is used for some while as the code increases the step size to an *on-scale* value.

The experience of software developers is that users do not wish to become involved with the specification of an initial step size and even if they are willing, they are not usually in a position to provide a good value. The difficulty here is that the initial step size depends on the method and just how it is implemented. These are matters of little interest to the user, especially to the user of an interval-oriented code who just wants to obtain accurate, reliable answers at certain points with as little trouble as possible. (Practically everybody!) A situation easily appreciated is that of variable order, variable step size implementations of methods with memory like Adams methods and BDFs. The first step is taken with a method with no memory and low order. This is ordinarily a much lower order than would be used for most steps with the consequence that the first step is very much shorter than subsequent steps. This fact may come as a surprise to the user who expects the code to use, say, order 5 for most of the integration. Guessing an initial step size appropriate to a formula of order 5, or observing a typical step size in a previous run, may result in a starting step size that is much too large for a formula of order 1. Many problems are not set up directly by a user, e.g., the integrator might be a module in a larger code, and in such a situation it is essential that the initial step size be determined in an automatic and reliable way.

Algorithms have been devised for the automatic selection of the intial step size that are quite successful at providing an on-scale value both inexpensively and reliably. Example 4 of Chapter 3 demonstrates that they are not always successful. Because the algorithms are very convenient and generally provide (much) better starting values than users do, they have become a part of all quality software for the initial value problem. When continuing an integration after a small change to the problem and when integrating a family of very similar problems, an appropriate initial step size may be known to the user. Sometimes it is important to make use of this value because the task involves so many restarts that the cost of an automatic start becomes significant in aggregate. Occasionally it is necessary to assist the code by providing in a more direct way some scale information in the form of an initial step size. For these reasons, careful designs permit the user to specify an initial step size, with, however, due attention paid to the possibility that the value provided is not as good as the user thinks. Gladwell, Shampine, and Brankin [1987] describe some of the approaches taken to the task and present developments of their own. In this section some of the key ideas are described briefly.

There are two basic issues. One is to take a step small enough that the error estimate is credible. The other is to determine efficiently an "optimal" step size. Perhaps the most natural and reliable way to accomplish the first task is the way proposed by Sedgwick [1973]: Try a step size that is about as small as makes sense in the precision available. If it fails, the problem cannot be solved in the precision available and if it succeeds, it will provide a credible estimate of the error that can be used to increase the step size to an on-scale value. The trouble with this approach is that with modern computers, this starting step size is very small indeed. With the usual restrictions of the rate of increase of step size, it takes many steps to work up to an on-scale value. It is a temptation to permit very large changes in the step size to make this phase of the computation less inefficient, but there is then the danger that after a large increase the error estimate will be unreliable. The difficulty with reliability is compounded when the order is being increased rapidly along with the step size. Despite the appeal of Sedgwick's approach, the popular codes try to start with a step size that is closer to being on-scale.

The first non-trivial scheme for the automatic selection of the initial step size seems to have been that of Shampine as presented in Shampine and Gordon [1972]. Variations of the idea have been used in many codes over many years and experience has shown that it provides a step size H that is almost certainly not disastrously bad and is often on-scale. Recent schemes use something of the kind to get started and then test the step size while attempting to take a step. The principal difficulty in getting started is that very little information is available. In the specification of the problem, the user provides implicitly some information about the scale of the problem. Just how this is done depends on the design of the integrator. It may be done with a maximum step size, a guessed initial step size, a first output point, or a final point in the integration. Even the person who uses the integrator as a "black box" provides valuable scale information in the form of a first output point because often this point is chosen to reflect the (expected) behavior of the solution, hence how fast the solution changes. This information is used to place limits on the attempted initial step size.

The tolerance τ and the norm $\|\cdot\|$ used for measuring the local error affect the selection of the step size. It is not easy to handle the error control in a reliable and robust manner. One difficulty is that it involves a measure of the SIZE of the solution over the step, this before even trying to approximate the solution. It is not unusual that a zero value be specified at the initial point. A zero initial value presents a technical problem for a relative error control, but the more fundamental issue is that there is then no natural indication of the scale of the solution component.

The approach of Gladwell, Shampine, and Brankin involves several phases. The first involves two ideas. One is to estimate an initial step size for a specific formula and then use it as a crude guess for an appropriate step size for the formula of the code. A Taylor series formula is used for this purpose. The formula of order m has

$$\mathbf{y}(a + h) = \mathbf{y}(a) + h\,\mathbf{y}'(a) + \frac{h^2}{2!}\,\mathbf{y}''(a) + \ldots + \frac{h^m}{m!}\,\mathbf{y}^{(m)}(a).$$

The local truncation error of this formula is asymptotically

$$E_m = \left\| \frac{h^{m+1}}{(m + 1)!}\,\mathbf{y}^{(m+1)}(a) \right\|.$$

The other idea is to postulate that the norm of the local error of the formula of order $p \geq m$ can be approximated by

$$(E_m)^{(p+1)/(m+1)}.$$

In this approximation, the largest step size H that leads to a local error no larger than the given tolerance τ is

$$H = \left(\frac{(m + 1)!}{\|\mathbf{y}^{(m+1)}(a)\|} \right)^{1/(m+1)} \tau^{1/(p+1)}.$$

Although other m have been used, c.f. Curtis [1978] and Watts [1983], Shampine originally chose $m = 0$ and this choice is by far the most common. When $m = 0$, the derivative needed here is immediately available:

$$\|\mathbf{y}'(a)\| = \|\mathbf{F}(a, \mathbf{y}(a))\|.$$

The rule then is

$$H = \min(|b - a|, \tau^{1/(p+1)}\|\mathbf{F}(a, \mathbf{y}(a))\|^{-1}).$$

The defects of this procedure leap to the eye. If $\|\mathbf{y}'(a)\| = 0$, or is very small, the local error of the Taylor series formula either does not have the form given for E_0 or is not well represented by it except for very small h. This situation would be revealed by a very large, or even infinite, step size H from the formula. All the information supplied about the solution holds at $x = a$ and so may not be helpful even as far away as $x = a + H$, hence not helpful when we actually try a step. In particular, the weights of the norm depend on the result of the step and can at best be approximated from information available at a. The only information supplied about the method is its order p, and even this assumes that $\mathbf{F}$ is smooth enough that the usual order holds. Obviously this lack of information precludes finding a step size tailored to the formula. The rule of thumb connecting the error of a formula of order m to one of order p can at most be useful in a semi-quantitative way. Still, the procedure does have its virtues. It produces a step size H that changes in the expected manner for a formula of order p when the tolerance τ is changed and when the independent and dependent variables are scaled. All the information at our disposal is used—the distance to the first output point, the initial value $\mathbf{y}(a)$, the equation (through the use of $\mathbf{F}(a, \mathbf{y}(a))$), the nature of the error control (through the norm), the tolerance, and a little information about the method (its order). It is not obvious how one could do better without more information.

After predicting an initial step size H, a step is taken cautiously. One of the key ideas for testing H is to insist that the product of H and a Lipschitz constant L be "not large." This is necessary for a number of reasons that include stability and a rapid rate of convergence of simple iteration for evaluating an implicit formula. Most starting procedures can be described as taking a step with an explicit Runge–Kutta formula. When the equation is in autonomous form, a lower bound $\mathcal{L}$ for L can be computed from the stages:

$$\|\mathbf{F}_k - \mathbf{F}_m\| = \|\mathbf{F}(\mathbf{Y}_k) - \mathbf{F}(\mathbf{Y}_m)\| \le L\,\|\mathbf{Y}_k - \mathbf{Y}_m\|,$$

hence

$$\mathcal{L} = \frac{\|\mathbf{F}_k - \mathbf{F}_m\|}{\|\mathbf{Y}_k - \mathbf{Y}_m\|} \le L.$$

It is required that $H\mathcal{L} \le c$ for a constant c that is "not large." The Lipschitz constant L is a bound on the magnitude of any eigenvalue of the Jacobian of $\mathbf{F}$ and this suggests a way of selecting an appropriate c. If the stability region of the formula is approximated by a half-disk of radius c, then a step size H such that $HL \le c$ implies that the computation is stable. The practical test $H\mathcal{L} \le c$ is a kind of necessary condition because if the test is not passed, we are certain that $HL > c$. As explained in Gladwell et al.

[1987], there is good reason to expect that if the step size is out of scale, a large lower bound will be computed from the later stages and the test will be effective at recognizing H that are much too big. There are a good few practical details that we leave to the article cited.

After leaving the second phase of the computation, a successful step has been taken with a step size H that has passed a considerable number of tests on $H\mathcal{L}$. With some confidence in the estimate of the local error, it can be used to adjust the step size to an "optimal" value. It is conventional on finding a successful initial step simply to begin the integration and let the code find an on-scale value as the integration proceeds. Ordinarily the rate of increase of the step size is limited severely because prediction of a large increase cannot be trusted. The situation now is rather different because we expect occasionally to have to increase a trial step size quite a lot. One way to handle this that is seen for example in DIFSUB is to alter the usual limit on the rate of increase so as to permit faster increase in the start. A representative approach would be to permit an increase by as much as a factor of 1000 after the first step, 100 after the second step, and 10 thereafter. Gladwell et al. prefer to repeat the first step until an on-scale value is determined. In this they permit rather larger increases than they would after the start. Proceeding in this way, all steps of the integration proper are on-scale and subject to the same restrictions on the rate of increase. It might seem wasteful to discard a successful step, but it is not because the only trial steps that are discarded are those that are much too small and they would not advance the integration further than the much larger trial steps that follow them.

8

Stiff Problems

The better codes based on Runge–Kutta and Adams methods are generally very effective, but they are inefficient when solving a class of problems called stiff, so inefficient that they are impractical for these problems. Constant step size implementations of the methods produce numerical results that oscillate and grow explosively. Implementations that vary the step size so as to control local errors can solve stiff problems, but they use step sizes that seem ridiculously small. This behavior is both puzzling and frustrating.

In the classical situation that $(b - a)L$ is not large, the step size is ordinarily determined by the desired local accuracy. When we leave this situation, other factors become important and they may require the step size to be very much smaller than accuracy would permit. Early work on the numerical solution of ordinary differential equations relied upon a result to the effect that in the classical situation the initial value problem is well posed and up to now this has been our basic assumption. Of course, for us to attempt the numerical solution of a problem, it must be well posed, so it is not immediately obvious that there are any interesting problems that do not correspond to the classical situation. The fact of the matter is that *many* important problems are of this kind. That these problems are not included in the classical theory is partly a question of the analytical tools that we have brought to bear, but it is mainly due to the problems being more difficult. We have already encountered some problems of this kind. In our study of

$$\mathbf{y}' = J\,\mathbf{y} + \mathbf{g}(x), \tag{1}$$

we found that the initial value problem is stable if $\mathrm{Re}(\lambda) \le 0$ for all eigenvalues λ of the constant matrix J. It is easy to see that a Lipschitz constant for such a problem must be at least as big as $|\lambda|$ for any eigenvalue λ. Specifically, if $\mathbf{w} = \mathbf{u} - \mathbf{v}$ is an eigenvector corresponding to the eigenvalue λ, then

$$\|(J\,\mathbf{u} + \mathbf{g}) - (J\,\mathbf{v} + \mathbf{g})\| = \|J\,\mathbf{w}\| = \|\lambda \mathbf{w}\| = |\lambda|\,\|\mathbf{u} - \mathbf{v}\| \le L\|\mathbf{u} - \mathbf{v}\|.$$

If there is an eigenvalue λ such that $\mathrm{Re}(\lambda)$ is negative and large compared to $(b - a)$, then the equation is stable, but the situation is not classical because $(b - a)L$ is large. If the step size is such that hL is of modest size, the situation is essentially the classical one, but it is perfectly possible that the initial value problem have a solution that is easy to approximate and a step size appropriate to the solution is such that hL is "large." In such a situation the step size that would yield the desired local accuracy may have to be restricted drastically to keep the computation stable and to evaluate implicit methods efficiently. Such problems are called *stiff*.

In this chapter we try to understand what stiffness is and how to use methods taken up earlier, mainly the BDFs, in such a way that we can solve stiff problems effectively. This is an active area of research and there is much about both the theory and practice of solving stiff problems that is not well understood. For this reason the chapter places special emphasis on illustrative examples.

§1 What Is Stiffness?

There are a variety of constraints on the step size. An obvious one is to achieve the desired local truncation error. Another is to keep the computation stable. When we employ an implicit formula, a third is to ensure that the iteration scheme used to evaluate the formula converges at an acceptable rate. For simple (functional) iteration we found that if the product of the step size h and the Lipschitz constant L is of moderate size, then the iteration would converge at an acceptable rate. Roughly speaking, the same condition yields stability for all the popular methods. In the classical situation that $(b - a)L$ is "not large," these constraints on the step size cannot be severe, so normally it is accuracy that determines h. Although these are the major constraints, in some circumstances there are others that determine the step size. For example, if the method does not have an interpolation capability, the step size must be shortened to obtain answers at specified points. When output is "frequent," this can determine the step sizes used. Quality software will either provide an interpolation capability, or will diagnose this situation, so no more will be said about the possibility. Often the functions defining the problem are only piecewise smooth and it is not rare that the problem is so smooth on each piece that it is the length of the piece that constrains the step size rather than accuracy. Some examples have been presented in other chapters and several others will be presented here. Some possible remedies will be mentioned and then no more will be said about this kind of constraint.

To better understand the possibility of stiffness, let us consider the scalar problem

$$y' = J(y - p(x)) + p'(x), \qquad y(0) = A, \tag{2}$$

for constant J. The analytical solution is

$$y(x) = (A - p(0)) \exp(Jx) + p(x).$$

When J is negative the solution curves come together and the problem is stable, no matter what the size of the Lipschitz constant $L = |J|$. If $|J|$ is large compared to the length of the interval, the curves come together quickly. In such a case $y(x)$ is virtually identical to the particular solution $p(x)$ after a short distance called the *initial transient* or *boundary layer*. If this initial value problem is integrated with the forward Euler method, the local truncation error is

$$\frac{h^2}{2} y''(\xi) = \frac{h^2}{2} [(A - p(0)) J^2 \exp(J\xi) + p''(\xi)].$$

Near the initial point, that is, for small x, the first term dominates because of the large value of J^2 and because the exponential is nearly one:

$$\frac{h^2}{2} y''(\xi) \approx \frac{h^2}{2} (A - p(0)) J^2.$$

From this it is clear that to obtain an accurate solution in the transient, it is necessary that $hL = h|J|$ be small. On an interval for which these approximations are good, the initial value problem corresponds to the classical situation. The interesting situation is outside the transient where

$$\frac{h^2}{2} y''(\xi) \approx \frac{h^2}{2} p''(\xi).$$

Now the step size that will provide an accurate solution is nearly independent of J. Provided that $p(x)$ is "slowly varying," meaning here that the second derivative is not large compared to $(b - a)$, an accurate solution can be obtained with a step size h such that $|hJ| = hL \gg 1$. On the other hand, the theory of absolute stability applies to this problem directly and we have already seen that the explicit Euler method is not stable unless $|hJ| < 2$ for such problems. Outside an initial transient, it is the stability requirement that determines the step size and it is a severe constraint when J is large and negative. This is surprising and frustrating to a person who has not looked into the matter because all our experience with non-stiff problems says that the easier the solution is to approximate, the more efficient the integration.

EXERCISE 1. The initial value problem (2) is somewhat oversimplified. Discuss in the same way the solution of the (uncoupled) pair of equations

$$y_1' = J_1\,(y_1 - p_1(x)) + p_1'(x), \qquad y_1(0) = A_1,$$
$$y_2' = J_2\,(y_2 - p_2(x)) + p_2'(x), \qquad y_2(0) = A_2.$$

With a pair of equations there are two transients to consider; the situation you need to think about is when one transient has decayed out, but the other has not.

Continuing with this illuminating example, we might resort to the backward Euler method (BDF1) to overcome the stability restriction. Recalling that its local truncation error has the same form as the forward Euler method, it would appear that we can solve the example problem easily because with BDF1 there is no restriction of the step size due to stability. Unfortunately, this is not so. It is necessary to consider how the implicit formula is evaluated. For this trivial equation, it is possible to write down immediately the solution of the algebraic equation for y_{j+1}. In general, though, the formula would be evaluated by simple iteration. Recall that we found simple iteration does not generally converge when solving this equation with BDF1 unless $|hJ| < 1$. This constraint on the step size is just as severe as the stability constraint on the forward Euler method. For this simple equation, it is easy to evaluate the formula, but obviously we must come up with a more effective scheme for evaluating the implicit formula for general problems.

Stiffness is a complex matter. The essence is that a step size that would yield the desired local accuracy must be reduced greatly to carry out the integration. The principal reasons for such a reduction are to stabilize the integration and to secure an adequate rate of convergence when evaluating an implicit formula by simple iteration. In the classical situation these reasons cannot lead to a severe reduction and people simply forget about them with the consequence that they are surprised when a problem with a slowly varying solution cannot be integrated with a large step size.

In many treatments of stiffness attention is focussed solely on the eigenvalues of the local Jacobian and the possibility of having to restrict the step size to stabilize the integration. This is superficial. In the first place, the length of the interval of integration must be taken into account. If the interval is short enough that we are in the classical situation, the problem cannot reasonably be described as stiff. In particular, within the initial transient(s) the problem may not be stiff. An ancedote might make the point. While working at the Sandia National Laboratories, the author was asked by a colleague about dealing with the storage required for the solution of some very large systems of equations. The equations resulted from semi-discretization of partial differential equations that arose in a study of the possibility of suppressing hydrogen gas explosions in a reactor containment vessel by means of foams. Workers in the field assumed that such systems are stiff because of the greatly different time scales of chemical reactions and fluid flows (leading to eigenvalues of local Jacobians of greatly different size). Generally speaking, this is correct, but in this instance an *explosion* is modeled and the physical situation suggests strongly that for the period of time being studied, the differential equations are *un*stable, rather than *very* stable, as is characteristic of stiff problems. If this suggestion is correct, turning to a code intended for stiff problems would be counterproductive, as well as straining the rapid memory capacity of even the Cray computer being used. Advised to try a code intended for non-stiff problems that checks for the possibility of stiffness, specifically a Runge–Kutta code of moderate order because of its modest storage requirements, it was confirmed that the problem was not, in fact, stiff.

Shampine and Gear [1979] relate an anecdote about a laser model described by a speaker at a conference in which they participated. The model involved some 250 energy levels and its solution was consuming large amounts of computer time. Simpler models with these states aggregated into a smaller number of states were clearly stiff (meaning that they would be solved far more efficently with a good code intended for stiff problems than good codes intended for non-stiff problems), but the refined model was solved more efficiently by codes intended for non-stiff problems. Each solution component had the same qualitative behavior. After a while the level corresponding to a solution component would become populated and the population would grow rapidly to a level about which it varied slowly. The computational difficulty originates in the fact that the various levels were populated successively and there were a great many of them. As far as the codes were concerned, they were always resolving a transient for some energy level. Eventually, of course, the whole system would reach a more-or-less steady state and methods appropriate to stiff systems would show their worth, but the cost of reaching the steady state was prohibitive and the scientist was not especially interested in the long-time behavior. Methods appropriate to non-stiff systems perform better in a transient because that is what they are intended for.

An initial transient is only one situation when a small step size must be used to achieve the desired accuracy. One of the author's colleagues at Sandia, H.A. Watts, was asked about the integration of some equations describing an accelerometer that was proving to be unexpectedly expensive.

There was some reason to think that the mechanical system might lead to stiffness and many people think that any problem difficult for a good code intended for non-stiff problems must be stiff. When advised to try a code that diagnoses stiffness, the researcher found that the code did not believe the problem to be stiff. On closer examination, it was clear that the difficulty was due to one of the coefficients in the equation being defined by interpolation to telemetry data. Because the interpolant was only smooth in pieces, the solution of the system of differential equations was smooth only between data points and its approximation by a high-order formula required the code to locate the data points automatically due to the interpolants not connecting smoothly across these points. In this case the step size was restrained by the smoothness of the problem to a degree not anticipated by the person posing the problem. An appropriate remedy is to use a smoother interpolant. It should be appreciated that it is not the *accuracy* of the interpolant that is the issue, rather how *smooth* it is.

One of the complexities of stiffness is that it depends not just on the problem, but also on the method used to integrate the problem. Problems with local Jacobians having eigenvalues rather close to the imaginary axis are sometimes called *stiff oscillatory* problems. The higher order methods in popular codes suffer from stiffness when applied to such problems. In particular, the stability of the BDFs near the imaginary axis deteriorates swiftly as the order is increased. Example 5 of Chapter 6 shows what can happen when solving a stiff oscillatory problem with the variable order BDF code LSODE. This code allows the user to specify the maximum order. The low-order BDFs are quite stable and when the maximum order was restricted to the point that only stable formulas were considered, the problem was integrated efficiently. However, when an order was permitted for which the step size had to be restricted on grounds of stability, the integration was inefficient. Although LSODE is intended for stiff problems, it *can* suffer from stiffness and whether this happens depends on both the problem and just which formulas are used. These considerations are by no means limited to BDFs and they make it difficult to speak simply of a problem being stiff. Gaffney [1984] compares a number of codes when solving stiff oscillatory ODEs, including several codes based on methods that have received little attention in this text. In this context another kind of method should be mentioned. Some important codes are based on extrapolation. The approach is a natural development of ideas presented in this text and we refer to the paper of Deuflhard [1985] for this development. We have chosen not to pursue these methods because those intended for non-stiff problems amount to a highly structured family of explicit Runge–Kutta formulas and they do not appear to warrant separate treatment, c.f. Shampine and Baca [1986]. Those intended for stiff problems really do merit study, especially in the present context, because they provide high order methods with excellent stability. Because their justification would be rather lengthy, we refer the reader to the literature and to more advanced texts. It is useful to have an effective extrapolation code like METAN1 (Deuflhard [1983]) available as an alternative to a BDF code.

EXAMPLE 1. Example 3 of Chapter 2 discusses the numerical solution of an unstable problem that has been mistakenly labeled as stiff. The practical definition of a stiff problem is that a code intended for stiff problems will solve the problem *much* more efficiently than one intended for non-stiff problems. So, let us see whether the problem is stiff by this definition. The problem is

$$y'' = 100y, \qquad 0 \le x \le 3, \qquad y(0) = 1,\, y'(0) = -10.$$

The matter is easily explored with LSODE because it implements both the Adams–Moulton formulas, suitable for non-stiff problems, and the BDFs, suitable for stiff problems. A mixed error control was used with a relative error tolerance of 10^{-4} and scalar absolute error tolerance of 10^{-14}. This is basically a relative error control, so we might hope to track the decreasing solution with some modest accuracy. Output was required at 100 equally spaced points for the plots of Figure 8.1. The run with the Adams–Moulton formulas, $mf = 10$, took 135 steps at a cost of 184 function evaluations.

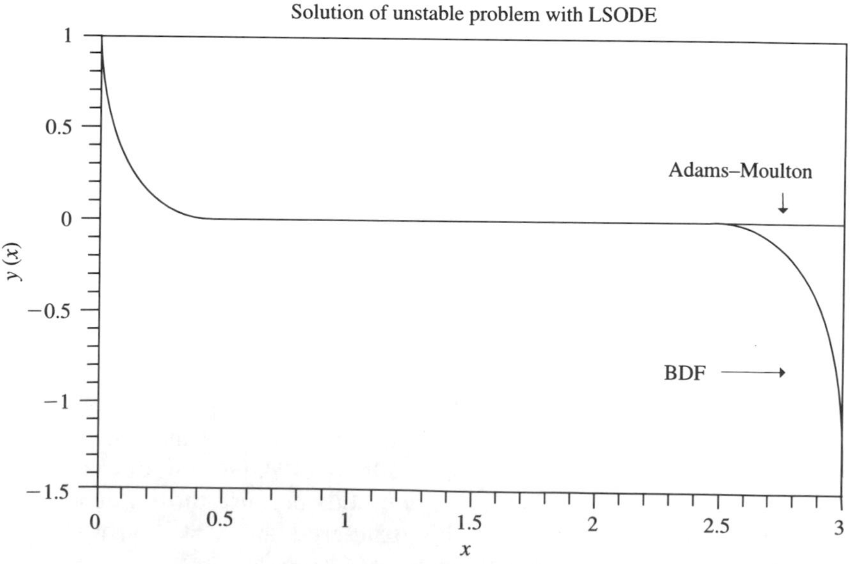

Figure 8.1

The run with the BDFs used a differenced Jacobian (the analytical Jacobian is easy to write down, but differencing makes the measure of work more comparable), $mf = 22$. With these choices the code took 140 steps at a cost of 177 function evaluations. The integration with the BDFs is less expensive, but the unimportant difference is not what is meant by stiffness. (In this particular case, the computation is less expensive with the BDFs because the results are less accurate!)

EXAMPLE 2. Example 6 of Chapter 6 quotes some results from an example that we discuss further here. The initial value problem is

$$\mathbf{y}' = \begin{pmatrix} -.1 & -199.9 \\ 0 & -200 \end{pmatrix}\mathbf{y}, \qquad 0 \le x \le 50, \qquad \mathbf{y}(0) = \begin{pmatrix} 2 \\ 1 \end{pmatrix}.$$

The solution components are $y_1(x) = \exp(-.1x) + \exp(-200x)$ and $y_2(x) = \exp(-200x)$. Suppose this problem is to be integrated with an absolute error tolerance of ε. There is an initial interval on which the solution changes rapidly and we expect the step size to be determined mainly by the accuracy requirement. Let us define T_ε as the first point where $|y_2(T_\varepsilon)| = \varepsilon$. With the error control specified, the solution is easy to approximate on $[T_\varepsilon, 50]$ and we expect the step size to be determined mainly by stability. In Shampine [1975] two variable order codes were used to integrate this problem and the costs of the integrations of the two subintervals $[0, T_\varepsilon]$, $[T_\varepsilon, 50]$ were measured. STEP is an Adams–Bashforth–Moulton code in PECE implementation with local extrapolation. It selects an initial step size automatically. DIFSUB has both Adams–Moulton formulas and BDFs. The BDFs were used and the code was supplied with an analytical expression for the Jacobian. The cost of obtaining the Jacobian was neglected. This code must be supplied an initial step size. At each tolerance, a number of runs were first made to find the largest initial step size that would succeed and then this "optimal" value was used in the comparison. Obviously DIFSUB has been given a considerable advantage with respect to the cost of integrating over $[0, T_\varepsilon]$. The following results were obtained:

	STEP			DIFSUB		
ε	NFCN1	NFNC2	ERROR	NFCN1	NFCN2	ERROR
$1E-1$	13	13022	$9E-1$	20	33	$6E-2$
$1E-2$	27	13071	$9E-2$	48	43	$1E-2$
$1E-3$	45	12999	$1E-2$	66	106	$1E-3$
$1E-4$	67	13068	$1E-3$	96	125	$1E-4$
$1E-5$	87	12914	$1E-4$	133	162	$2E-5$

ε	STEP NFCN1	STEP NFNC2	STEP ERROR	DIFSUB NFCN1	DIFSUB NFCN2	DIFSUB ERROR
$1E-6$	112	13307	$1E-5$	190	147	$2E-6$
$1E-7$	141	13283	$5E-7$	232	245	$2E-7$
$1E-8$	167	13190	$6E-8$	344	213	$4E-8$
$1E-9$	205	13326	$7E-9$	447	247	$4E-9$
$1E-10$	261	13441	$4E-10$	575	304	$6E-10$
$1E-11$	294	14191	$9E-11$	733	385	$4E-11$
$1E-12$	352	14734	$6E-12$	992	464	$7E-12$
$1E-13$	426	15022	$1E-12$	1298	607	$4E-12$

Here ERROR is the maximum absolute error seen at any step in the interval [0, 50]. NFCN1 is the number of function evaluations required for the integration over $[0, T_\varepsilon]$ and NFCN2 the cost for the integration over $[T_\varepsilon, 50]$. On the first subinterval where the step size is determined by accuracy, STEP is able to overcome its disadvantages and perform the integration more efficiently than DIFSUB. The difference is unimportant except perhaps at the more stringent tolerances. The main reason for the difference seen then is that the Adams code implements formulas of order as high as 12 and the BDF code implements formulas of order only as high as 5—as pointed out in Chapter 4, if you ask for enough accuracy, higher order formulas will be more efficient. The dramatic difference in cost is seen on the second subinterval where the step size is determined by stability. Here the infinite stability regions of the BDFs implemented in DIFSUB allow the code to work with a very much larger step size than that allowed by the finite stability regions of the predictor–corrector methods. This interval of the integration shows the kind of difference that we call stiff. This problem is only *mildly* stiff—later we shall see examples for which solution using a code intended for non-stiff problems is impractical.

The results presented here for STEP complement those given for a Runge–Kutta code in Chapter 6. There it was pointed out that when the step size is determined by stability, the cost is nearly independent of the accuracy desired (and achieved). The picture is blurred with a variable order code: On average the step size corresponds to being on the boundary of the stability region of the formula being used, but from time to time, the code changes the order, that is, the formula, and the various formulas have different stability regions. In the case of the methods implemented in STEP, these regions are not so terribly different and STEP does not vary the order much when the problem is stiff, so in a very rough way, the cost is independent of the tolerance on the subinterval where the problem is stiff.

EXAMPLE 3. Before effective codes for the solution of stiff problems became generally available, some stiff problems were solved by using singular perturbation theory and codes for non-stiff problems. As a technique of applied mathematics, perturbation theory is valuable for the insight it provides into the structure of solutions as a function of a small parameter ε. This kind of insight is difficult to obtain from a set of numerical solutions for various ε. On the other hand, a numerical method can provide an accurate solution of a problem for given ε, whereas perturbation theory may not provide a solution at all or may not provide a solution of the desired accuracy. The approaches complement one another.

Some interesting problems can be put into the form

$$\begin{aligned} \mathbf{y}' &= \mathbf{f}(\mathbf{y}, \mathbf{z}), \qquad && \mathbf{y}(a) \text{ given}, \\ \varepsilon\mathbf{z}' &= \mathbf{g}(\mathbf{y}, \mathbf{z}), \qquad && \mathbf{z}(a) \text{ given}, \end{aligned}$$

where the scalar parameter ε is small, $0 < \varepsilon << 1$. Of course this could be written as a single set of equations, but the form makes clearer the special role of the components that have been grouped as the vector $\mathbf{z}(x)$. As an example, Chapter 10 of Lin and Segel [1988] is devoted to the study of a problem in biochemical kinetics that leads after considerable preparation to the initial value problem

$$\begin{aligned} s' &= s + (s + \kappa - \lambda)c, \qquad && s(0) = 1, \\ \varepsilon c' &= s - (s + \kappa)c, \qquad && c(0) = 0. \end{aligned}$$

When written in standard form, it is seen that the Jacobian of the system has the form

$$\begin{pmatrix} \dfrac{\partial \mathbf{f}}{\partial \mathbf{y}} & \dfrac{\partial \mathbf{f}}{\partial \mathbf{z}} \\[2ex] \varepsilon^{-1}\dfrac{\partial \mathbf{g}}{\partial \mathbf{y}} & \varepsilon^{-1}\dfrac{\partial \mathbf{g}}{\partial \mathbf{z}} \end{pmatrix}.$$

The sum of the eigenvalues of the Jacobian is equal to its trace and the trace here is normally $\mathcal{O}(\varepsilon^{-1})$ because of the entries in the lower right hand block of the matrix. Accordingly, these singularly perturbed systems normally have at least one eigenvalue that is large. Generally the components of a system of this form change very rapidly for x near the initial point. This is because $z'(a) = \varepsilon^{-1}\mathbf{g}(\mathbf{y}(a), \mathbf{z}(a))$ is large for all components of $\mathbf{g}(\mathbf{y}(a), \mathbf{z}(a))$ that do not vanish. In suitable circumstances the rapid change of the solution is confined to a small interval about the initial point that is called a boundary layer in this context. As may be found in, e.g., O'Malley [1988], with suitable hypotheses that include the stability of the solution,

$$\mathbf{y}(x) = \mathbf{y}_0(x) + \varepsilon\, \mathbf{y}_1(x) + \cdots + \varepsilon^N \mathbf{y}_N(x) + \mathbb{O}(\varepsilon^{N+1}),$$
$$\mathbf{z}(x) = \mathbf{z}_0(x) + \varepsilon\, \mathbf{z}_1(x) + \cdots + \varepsilon^N \mathbf{z}_N(x) + \mathbb{O}(\varepsilon^{N+1}),$$

Here the coefficient functions $\mathbf{y}_j(x)$ and $\mathbf{z}_j(x)$ are smooth and do not depend on ε. This is an *outer expansion* of the solution that is valid only for x bounded away from the initial point a, that is, outside an initial boundary layer. Where it is valid, we expect the original problem to be stiff because the smooth solution permits a large step size, the solution is stable, and local Jacobians have an eigenvalue that is $\mathbb{O}(\varepsilon^{-1})$.

Inserting the expansions into the differential equations and equating the coefficients of like powers of ε leads to the equations defining the $\mathbf{y}_j(x)$ and $\mathbf{z}_j(x)$. In particular,

$$\mathbf{y}_0'(x) = \mathbf{f}(\mathbf{y}_0(x), \mathbf{z}_0(x)), \qquad \mathbf{y}_0(a) = \mathbf{y}(a),$$
$$\mathbf{0} = \mathbf{g}(\mathbf{y}_0(x), \mathbf{z}_0(x)).$$

In the case of the example, we have

$$s_0' = -s_0 + (s_0 + \kappa - \lambda)c_0, \qquad s(0) = 1,$$
$$0 = s_0 - (s_0 + \kappa)c_0.$$

The problems that arise in this way are examples of differential-algebraic systems, DAEs, and more specifically, semi-explicit DAEs. In the Problems chapter, a class of DAEs that includes the ones arising here are solved by reducing them to ODEs. The algebraic equations are solved to obtain $\mathbf{z}_0(x) = \mathbf{R}(\mathbf{y}_0(x))$. This function is substituted into the equation for $\mathbf{y}_0(x)$ to end up with an initial value problem for an ODE:

$$\mathbf{y}_0' = \mathbf{f}(\mathbf{y}_0, \mathbf{R}(\mathbf{y}_0)), \qquad \mathbf{y}_0(a) = \mathbf{y}(a).$$

In the case of the example, we have $c_0 = s_0/(s_0 + \kappa)$ and

$$s_0' = -s_0 + \frac{s_0 + \kappa - \lambda}{s_0 + \kappa} s_0, \qquad s(0) = 1.$$

In contrast to the Jacobian of the original system, all the entries of the Jacobian of this reduced system are $\mathbb{O}(1)$, so the initial value problem for $\mathbf{y}_0(x)$ is a classical one that can be integrated with comparative ease.

It is generally not possible to satisfy all the initial conditions with the outer solution. The solution changes rapidly in the boundary layer from the given initial values to values that lead to a slowly varying solution. The problem is not stiff in this initial transient. More specifically, on an interval of length ε, a problem with a Jacobian that is $\mathbb{O}(\varepsilon^{-1})$ is in the classical situation. One possibility is to integrate the original system until it can be approximated by the reduced system. Formally the first term in the outer expansion arises by setting ε to 0 in the original system. One way to describe this is to say that the system is in a quasi steady-state. In a steady state, $\mathbf{z}'(x)$ would be $\mathbf{0}$; here it is assumed that $\varepsilon\mathbf{z}'(x)$ is small enough to neglect. The approximations $\mathbf{y}_0(x)$, $\mathbf{z}_0(x)$ are sometimes called the *pseudo steady-state approximation*. One of the goals of singular perturbation theory is to sort out the behavior in the boundary layer both to approximate the solution there by an *inner expansion* and to connect properly the two expansions. The theory provides insight about stiff problems, but it is no longer necessary to approximate the problem to obtain systems that are not stiff. As we discuss how methods appropriate for the original stiff system work, it is illuminating to ask what happens as $\varepsilon \to 0$.

EXERCISE 2. Lapidus, Aiken, and Liu [1973] discuss a number of interesting examples solved by the pseudo steady-state approximation. One is the transition of a proton to various energy levels in a hydrogen-hydrogen bond. The initial value problem is

$$\frac{dx_1}{dt} = -k_1 x_1 + k_2 y, \qquad x_1(0) = 0,$$

$$\frac{dx_2}{dt} = -k_4 x_1 + k_3 y, \qquad x_2(0) = 1,$$

$$\frac{dy}{dt} = k_1 x_1 + k_4 x_2 - (k_2 + k_3)y, \qquad y(0) = 0,$$

with

$$k_1 = 8.4303270 \times 10^{-10}, \qquad k_2 = 2.9002673 \times 10^{11},$$

$$k_3 = 2.4603642 \times 10^{10}, \qquad k_4 = 8.7600580 \times 10^{-6}.$$

The rate constants are such that the intermediate quantity y reacts very quickly. Notice the linear conservation law satisfied by the solution. The problem is to be solved for times as large as $t = 8 \times 10^5$. The Jacobian of this linear problem is obvious, so it is easy to calculate a Lipschitz constant. Is this the classical situation? As posed the problem is not written in the form of a singular perturbation problem. To put it into this form, rescale it by introducing $x_3 = y \times 10^{10}$. Write out the differential-algebraic equations of the pseudo steady-state approximation. Because of linearity it is easy to eliminate the algebraic variable x_3 from the differential equations. Does the solution of the reduced system satisfy the conservation law satisfied by the solution of the full system? Integrate the reduced problem with a code intended for non-stiff problems and compare your solution to that obtained for the full problem using a code intended for stiff problems.

EXERCISE 3. The knee problem presented in Example 3 of Chapter 3 is

$$\varepsilon \frac{dy}{dx} = (1 - x)y - y^2, \qquad 0 \le x \le 2, \qquad y(0) = 1.$$

As pointed out there, the solution is close to the null isocline $u(x) = 1 - x$ on $[0, 1]$ and close to the null isocline (and solution) $u(x) \equiv 0$ on $[1, 2]$. The parameter ε satisfies $0 < \varepsilon << 1$. For the computations of the example, $\varepsilon = 10^{-4}, 10^{-6}$. This is an example of a singularly perturbed problem and the null isoclines represent outer solutions. Exercise 8 of Chapter 6 studies the stability of this problem. If you have not done this exercise, do it now and use the results to argue that the problem is stiff. Reproduce the computations of the example. Except for the software issue pointed out in the example, you will find that a code appropriate for stiff problems will integrate this stiff problem quite easily.

EXAMPLE 4. The systems that arise when approximating partial differential equations by semi-discretization are often, perhaps even usually, stiff. A number of examples were given in Example 8 of Chapter 1. Recall that when approximating the advection equation

$$u_t + cu_x = 0$$

with backward differences replacing the derivative with respect to the space variable x, we obtained a system of ordinary differential equations with components $u_i(t)$ approximating $u(i\Delta, t)$ for $\Delta = N^{-1}$ and $i = 1, \ldots, N$. The system had the Jacobian matrix

$$J = -\frac{c}{\Delta}\begin{pmatrix} 1 & 0 & \cdots & 0 \\ -1 & 1 & \cdots & 0 \\ \vdots & \ddots & \ddots & \vdots \\ 0 & \cdots & -1 & 1 \end{pmatrix}.$$

Because this is a triangular matrix, the eigenvalues are obvious, namely $-c/\Delta = -c\,N$. The constant c is positive so these eigenvalues are negative. In this context the number of equations N might be large, so the eigenvalue of greatest magnitude might be large. It cannot be concluded from this statement about the eigenvalues alone that the differential equations are stiff. For that to be true, the solution components $u_i(t)$ would have to be smooth. However, the solution $u(x, t)$ of the partial differential equation does not itself have to be smooth. As is easily verified, the solution is constant along lines for which $x - ct$ is constant, the characteristics. Accordingly, if the initial data, $u(x, 0)$, or the boundary data, $u(0, t)$, has a discontinuity, it will propagate to the right in $u(x, t)$ with speed c. This will be reflected in a lack of smoothness of the approximating functions $u_i(t)$, so it is by no means obvious that the step size will be constrained by stability rather than accuracy. If there is a single discontinuity in otherwise smooth data $u(x, 0)$ and $u(0, t)$, each $u_i(t)$ approximates a smooth function $u(i\Delta, t)$ until the discontinuity reaches $x = i\Delta$. After a short period of rapid change, $u_i(t)$ should once again approximate a smooth function $u(i\Delta, t)$. The difficulty is that all the $u_i(t)$ are integrated simultaneously and there is no long period of time in which all of the components are approximating smooth functions $u(i\Delta, t)$—there is a restriction on the time step. When the restriction is severe, there is no point to high-order methods because they normally involve many function evaluations per step or a long memory: If the step size is not restricted enough, the formula will not exhibit the expected order because the problem is not smooth over the interval spanned by the formula. If the step size is restricted to the point that the high order is evident, it will be so small that the formula will deliver (much) more accuracy than is desired. This is a situation in which a low order method involving just a few function evaluations per step and either no, or a very short, memory is cost-effective.

Some plots help make this matter clear. With suitable data specified, the solution of the advection equation is $u(x, t) = 5/(1 + 100((x - t) - 0.5)^2)$. Figure 8.2a shows the solution plotted as a function of x at selected times. The initial data ($t = 0$) has a peak at $x = 0.5$ and it is seen that the peak is moving to the right at a uniform speed. When the solution is plotted in Figure 8.2b as a function of t at selected distances x, we see the functions that are approximated by the method of lines. Each function has a single peak, but the peaks occur later for x further from the origin. These functions are to be approximated simultaneously and the plot shows how it is possible that the step size be restricted because of sharp changes even though each function has only one peak.

When solving the advection equation with periodic boundary conditions by central differences, we encountered a system of ordinary differential equations with Jacobian

$$J = -\frac{c}{2\Delta}\begin{pmatrix} 0 & 1 & 0 & \cdots & 0 & -1 \\ -1 & 0 & 1 & \cdots & 0 & 0 \\ 0 & -1 & 0 & & & \vdots \\ \vdots & & & & & \vdots \\ 0 & 0 & \cdots & -1 & 0 & 1 \\ 1 & 0 & \cdots & 0 & -1 & 0 \end{pmatrix}.$$

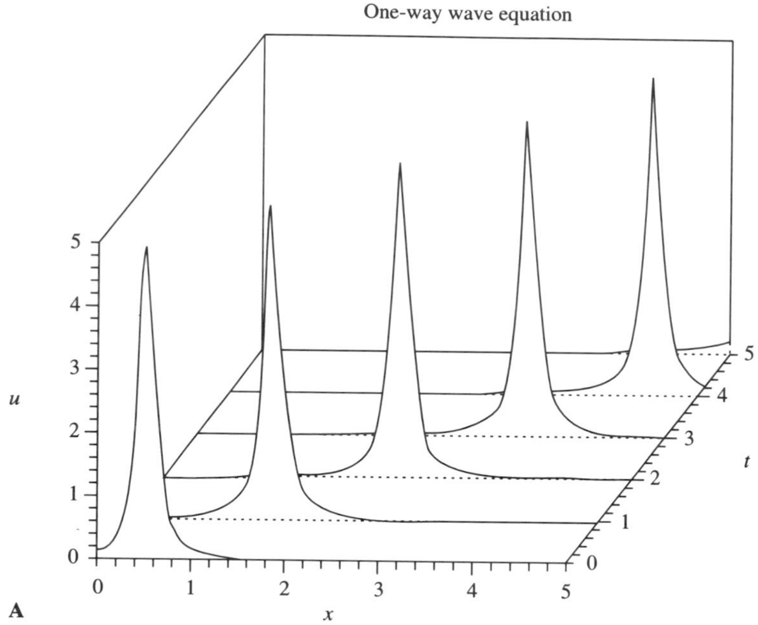

Figure 8.2a

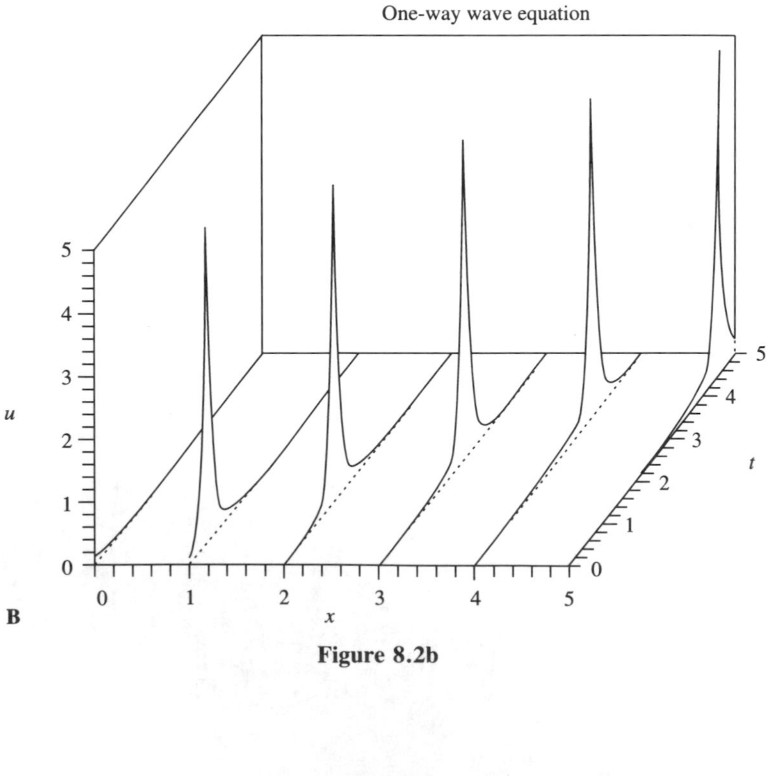

Figure 8.2b

The matrix is skew symmetric, so the eigenvalues must be zero or pure imaginary. Exercise 15 of Chapter 6 sketches a derivation of the eigenvalues. They are found to be

$$\lambda_m = -\frac{c}{\Delta} i \sin\left(\frac{2m\pi}{N}\right) \qquad m = 0, \ldots, N-1.$$

Here $i = \sqrt{-1}$. From this it is easy to see that for large N, the eigenvalues of largest magnitude are about $\pm i\, cN$. This is quite different from the situation with backward differences. In a way it is (much) more troublesome because popular formulas like the BDFs have poor stability for purely imaginary eigenvalues so that a restriction of the step size due to stability might be severe even for modest values of N. Of course, it must be kept in mind that as with semi-discretization by backward differences, it is not necessarily true that the solution components are smooth for such problems.

Approximation of the heat equation in one space variable,

$$u_t = u_{xx},$$

by finite differences for $0 \le x \le 1$ led to a system of equations with Jacobian

$$J = \frac{1}{\Delta^2}\begin{pmatrix} -2 & 1 & 0 & \cdots & 0 & 0 \\ 1 & -2 & 1 & \cdots & 0 & 0 \\ 0 & 1 & -2 & & & \vdots \\ \vdots & & & & & \vdots \\ 0 & 0 & \cdots & 1 & -2 & 1 \\ 0 & 0 & \cdots & 0 & 1 & -2 \end{pmatrix}.$$

The eigenvalues were worked out in Example 7 of Chapter 6 where they were found to be

$$\lambda_m = \frac{1}{\Delta^2}\left(-2 + 2\cos\left(\frac{m\pi}{N+1}\right)\right) \qquad m = 1, \ldots, N.$$

This is understood more easily for large N by expanding in a Taylor series,

$$\cos\left(\frac{\pi}{N+1}\right) = 1 - \frac{1}{2}\left(\frac{\pi}{N+1}\right)^2 + \mathcal{O}(N^{-4}),$$

so that

$$\lambda_1 = N^2\left(-2 + 2 - \left(\frac{\pi}{N+1}\right)^2 + \mathcal{O}(N^{-4})\right)$$

$$= -\pi^2\left(\frac{N}{N+1}\right)^2 + \mathcal{O}(N^{-2}) \approx -\pi^2.$$

Similarly it is found that

$$\lambda_N = -4N^2 + \pi^2\left(\frac{N}{N+1}\right)^2 + \mathbb{O}(N^{-2}) \approx -4N^2.$$

Notice that the eigenvalues are spread throughout the interval $[-4N^2, -\pi^2]$. Often stiff problems have eigenvalues that form clusters, but the example shows that this is not always the case. Because all the eigenvalues are negative real numbers, the problem is stable. Indeed, because N appears as a quadratic in λ_N, there are eigenvalues of considerable magnitude for even modest values of N, hence some error components are heavily damped in the differential equations. The solutions of parabolic equations become smoother in time, so it is reasonable to anticipate that all the components $u_i(t)$ will be relatively smooth functions. In these circumstances, stiffness is to be expected for even moderate values of N such as 50 and it is found in practice that a code based on a method appropriate for stiff problems is dramatically more efficient than one based on a method with a finite stability region.

EXAMPLE 5. The examples presented so far are linear and even have constant Jacobians. Let us now examine the nonlinear van der Pol equation,

$$\ddot{x} + \varepsilon(x^2 - 1)\dot{x} + x = 0.$$

For $\varepsilon > 0$, all non-trivial solutions converge to a limit cycle, a periodic solution. This equation was taken up in Example 6 of Chapter 1 and the limit cycle for $\varepsilon = 10$ was plotted in Figure 1.3. Let us convert the equation to a first order system by introducing $y = \dot{x}$:

$$\dot{x} = y,$$
$$\dot{y} = \varepsilon(1 - x^2)y - x.$$

As discussed in Jordan and Smith [1987], for $\varepsilon \ll$ the limit cycle is approximately

$$x(t) = 2\cos(t), \qquad y(t) = -2\sin(t).$$

It is clear from the differential equations that for x and y near the limit cycle, the derivatives $\dot{x}$ and $\dot{y}$ are of modest size. This suggests that near the limit cycle, the solution components are "slowly varying." The Jacobian matrix for the system evaluated at (t, x, y) is

$$\mathcal{J} = \begin{pmatrix} 0 & 1 \\ -2\varepsilon xy - 1 & \varepsilon(1 - x^2) \end{pmatrix}.$$

Any of the computable norms of $\mathcal{J}$ furnishes a bound for the eigenvalues that is $\mathbb{O}(1)$, but to relate this to subsequent arguments, let us examine the eigenvalues more directly. The eigenvalues are the roots of the characteristic equation $\det(\mathcal{J} - \lambda I) = 0$, which in this case is

$$\lambda^2 + (x^2 - 1)\varepsilon\lambda + 1 + 2\varepsilon xy = 0.$$

It is easy enough to write down the roots of this quadratic and sort out their behavior, but a short cut is to note that for x and y near the limit cycle and small ε, the equation is approximately $\lambda^2 + 1 = 0$, which has the two roots $\lambda_{1,2} = \pm\sqrt{-1}$. The eigenvalues, then, are for small ε close to the imaginary axis and of magnitude about 1. At least for solutions of the equation that are near to the limit cycle and for intervals of interest that are not long, this problem is not stiff.

Van der Pol's equation is more interesting when $\varepsilon >> 1$ because the non-linearity becomes important then. It is known that a full cycle of the limit cycle is composed of a segment on which $x(t)$ changes slowly, a short segment on which it changes very quickly, another segment of slow change, and a final short segment of rapid change. Because of its simplicity and importance, it is frequently treated in textbooks as an example in the theory of singular perturbations. It is convenient to rewrite the equation in terms of the parameter $\delta = \varepsilon^{-1} << 1$:

$$\ddot{x} + \delta^{-1}(x^2 - 1)\dot{x} + x = 0.$$

The system is then

$$\dot{x} = y,$$
$$\dot{y} = \delta^{-1}(1 - x^2)y - x,$$

the Jacobian is

$$\mathcal{J} = \begin{pmatrix} 0 & 1 \\ -2\delta^{-1}xy - 1 & \delta^{-1}(1 - x^2) \end{pmatrix},$$

and the characteristic equation is

$$\lambda^2 + (x^2 - 1)\delta^{-1}\lambda + 1 + 2\delta^{-1}xy = 0.$$

Figure 8.3 shows the solution of the initial value problem when the parameter $\delta = 10^{-3}$ and the initial values are $x(0) = 2$, $y(0) = 0$. The initial values correspond roughly to starting on the limit cycle. With a parameter δ this small, it is seen that the solution can change very rapidly.

In §11.4 of Jordan and Smith [1987] the behavior of the limit cycle is analyzed. It is shown that for $y > 0$, the limit cycle can be described in

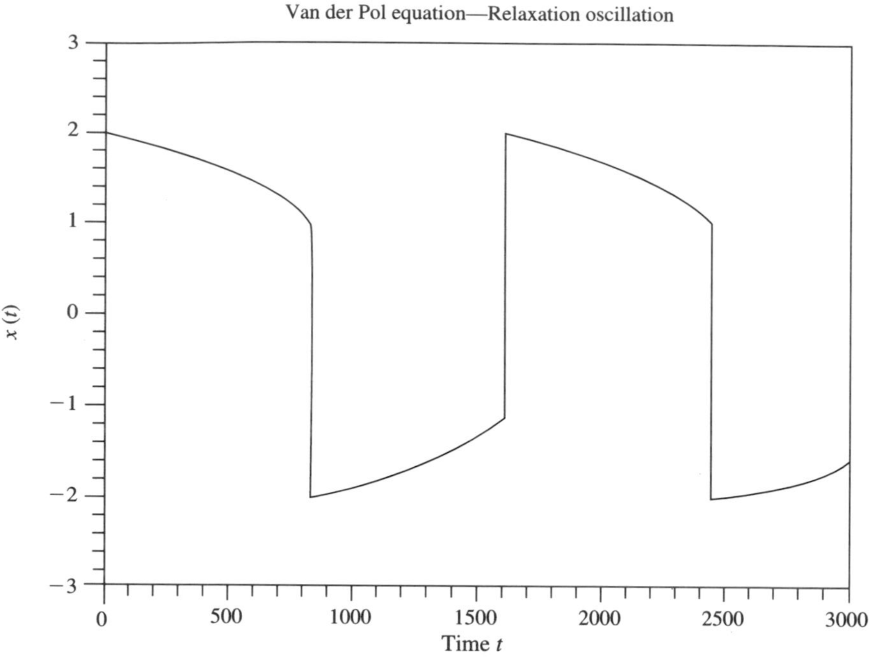

Figure 8.3

terms of three regions:

(i) $-2 \le x \le -2 + \mathcal{O}(\delta^2)$: $\quad \frac{1}{3}\delta y + \frac{2}{9}\delta^2 \log\left(1 - \frac{3}{2}\delta^{-1} y\right) = -(x + 2)$

(ii) $-2 + \mathcal{O}(\delta^2) < x < -1 - \mathcal{O}(\delta^2)$: $\quad y = \delta x/(1 - x^2)$

(iii) $-1 + \mathcal{O}(\delta^{2/3}) < x \le 2$: $\quad \delta y = x - \frac{1}{3}x^3 + \frac{2}{3}$

It is not necessary for us to go into the derivation of such results, but a few comments will be helpful. Let us introduce the scaled time variable $\tau = \delta t$. In this variable the equation becomes

$$\delta^2 \frac{d^2x}{d\tau^2} + (x^2 - 1)\frac{dx}{d\tau} + x = 0.$$

This is approximated for small δ by

$$(x^2 - 1)\frac{dx}{d\tau} + x = 0.$$

Solving for the derivative and returning to the original time variable t leads to

$$y = \frac{dx}{dt} = \delta \frac{dx}{d\tau} = \frac{\delta x}{1 - x^2},$$

the form of the solution in region (ii). For this to be a reasonable approximation, the solution x must be slowly varying in the sense that the product of δ^2 and the second derivative of x with respect to τ is small enough to be neglected in the differential equation. Obviously it is also necessary that we consider only values of x^2 that are not too close to 1. Considerations of this nature lead to the different regions of validity given above.

In region (ii) the approximate solution components are such that x is $\mathcal{O}(1)$ and y is $\mathcal{O}(\delta)$. The equation shows that $\dot{x} = y$ is $\mathcal{O}(\delta)$ and

$$\dot{y} \approx \frac{(1 - x^2)}{\delta} \frac{\delta x}{(1 - x^2)} \qquad x \approx 0.$$

This suggests that near the limit cycle, solutions are slowly varying. Recall from the derivation of this approximation that the second derivative of x has to be small in this region. Substituting the approximations into the characteristic equation leads to

$$\lambda\left(\lambda + \frac{x^2 - 1}{\delta}\right) + \frac{1 + x^2}{1 - x^2} = 0.$$

The roots of this equation can be written down and their behavior studied without difficulty, but the form given suggests an easier way to proceed. For small δ, there is a root

$$\lambda_1 \approx -\frac{x^2 - 1}{\delta}.$$

Using the fact that the constant term in the quadratic equation is the product of the two roots, we then find that there is a root

$$\lambda_2 \approx \delta \frac{1 + x^2}{(1 - x^2)^2}.$$

In this region x ranges from about -2 to about -1. This implies that λ_1 is negative and because it is $\mathcal{O}(\delta^{-1})$, it is large. The other root is positive and because it is $\mathcal{O}(\delta)$, it is small. Near this segment of the limit cycle, solutions are slowly varying. It is known on general grounds that the limit cycle is stable. However, we can see this here from linear stability theory by examination of the eigenvalues. Notice that one of the eigenvalues is positive, but because it is small, the equation is stable. The presence of the eigenvalue with large negative real part leads to stiffness. Often theoretical treatments of stiffness ignore the possibility of eigenvalues with positive real part. Of course, if an eigenvalue with positive real part is present, the real part cannot be large compared to the length of the interval of integration if the problem is to be stable, but we have here a realistic example of a stiff problem with an eigenvalue of positive real part.

In region (iii), x is of moderate size, $\mathbb{O}(1)$, but y is $\mathbb{O}(\delta^{-1})$, which is large. Correspondingly, $\dot{x}$ $(= y)$ is large and

$$\dot{y} = \frac{(1 - x^2)}{\delta} \frac{\left(x - \frac{1}{3}x^3 + \frac{2}{3}\right)}{\delta} - x$$

is generally $\mathbb{O}(\delta^{-2})$, which is even larger. The solution changes very rapidly in this region, an internal boundary layer. To gain an appreciation for the stability of the equation in this region, suppose that $x = 0$. For this particular value of x, the characteristic equation reduces to

$$0 = \lambda^2 - \delta^{-1}\lambda + 1,$$

and the two (real) roots are about

$$\lambda_1 \approx \delta^{-1}, \qquad \lambda_2 \approx \delta.$$

Because one of the roots is large and positive, the equation is unstable. It is interesting that the situation is rather different for $x = 1$. For this particular value of x,

$$y \approx \frac{4}{3}\delta^{-1},$$

and the characteristic equation reduces to

$$0 = \lambda^2 + 1 + 2\,\delta^{-1}y \approx \lambda^2 + \frac{8}{3}\delta^{-2}.$$

The two roots are now complex,

$$\lambda_{1,2} \approx \pm i\sqrt{\frac{8}{3}\delta^{-2}},$$

though still $\mathbb{O}(\delta^{-1})$ in magnitude. In this region the problem is not stiff. The step size must be $\mathbb{O}(1/|\lambda_1|)$ for stability, so this region is hard to integrate, but the difficulty is due to a rapidly changing solution, not stiffness. The limit cycle is periodic. This region corresponds to the initial transient discussed earlier; because of periodicity, it recurs on a regular basis. One lesson to be learned from this is that a problem can change from stiff to non-stiff and back again repeatedly. Often people equate the terms "hard" and "stiff." A problem that is stiff will certainly be hard for a code intended for non-stiff problems, but its solution with an appropriate code may be routine. A non-stiff problem that is hard for a code intended for non-stiff problems may be very hard for a code intended for stiff problems because the formulas and their implementation may be less effective than those in a code intended for such problems.

This example makes another important point. In region (ii) the derivative y is small, but it becomes large in the boundary layer of region (iii) and subsequently becomes small again. This is an example of the situation described in §3 of Chapter 3 that presents difficulties for the specification of accuracy. Clearly control relative to the largest approximation to y seen so far in the integration is reasonable only as far as the first boundary layer. After y decreases from a large value to a small value, such a control is so lax as to be meaningless.

EXERCISE 4. Integrate the initial value problem

$$\dot{x} = y, \qquad x(0) = 2,$$
$$\dot{y} = 1000\,(1 - x^2)y - x, \qquad y(0) = 0$$

to reproduce Figure 8.3. As pointed out in connection with the figure, the initial conditions correspond roughly to starting on the limit cycle. This strains step size selection algorithms because there is little in the way of an initial transient and the codes are likely to be too optimistic about the initial step size. The period of a cycle is about 1615, so an interval [0, 3000] is long enough to show several internal boundary layers. If is found, e.g., that $x(t)$ is about -2 when t is about 807 and $x(t)$ increases to about -1 when t is about 1614. The solution $x(t)$ is again about -2 when t is about 2422 and increases to about -1.5106 at $t = 3000$. Compute some approximate solutions in these intervals where you can use the asymptotic expression (ii) of Example 4 to check your computed values for $y(t)$.

§2 Methods Suitable for Stiff Problems

To solve stiff problems efficiently, we need a method with an unbounded stability region. We saw in Chapter 6 that this requirement forces us to use implicit methods. The BDFs are an example and codes based on them provide the most popular general-purpose codes that are suitable for stiff problems. It is unfortunately the case that the stability regions of the BDFs are increasingly less satisfactory as the order increases, so much so that they become unacceptable at relatively low orders. It is also unfortunate that good stability is bought at the price of accuracy, so that, for example, the Adams–Moulton formula of order p has a considerably smaller error constant than that of the BDF of the same order. It is a temptation to use the same formulas for all problems and this is possible. The reduced maximum order available and the bigger error constants mean that the integration of a non-stiff problem with a BDF code is relatively inefficient. However, the greatest source of inefficiency in this situation is due to the way the implicit formulas are evaluated rather than to the formulas themselves. Simple iteration cannot be used to evaluate implicit formulas when solving stiff problems because the constraint on the step size to get convergence is every bit as severe as the stability constraint on formulas with finite stability regions. It is so complicated and expensive to evaluate implicit formulas in the presence of stiffness that it is natural to ask if it is worth the trouble. This is a good question and the answer has two parts. One is that for many problems that arise in practice, the stability constraint is so severe that techniques appropriate for non-stiff problems are impractical. The other part is that these problems may be solved inexpensively in absolute terms—although each step is expensive compared to a step of an integration of a non-stiff problem, the solutions are easy to approximate and relatively few steps are needed.

§2.1 Evaluating Implicit Formulas

With a stable formula available, the crucial issue in practice is to evaluate the implicit method efficiently. As an example of the task, the backward differentiation formula of order 2 is

$$\mathbf{y}_{j+1} = \frac{1}{3}[4\mathbf{y}_j - \mathbf{y}_{j-1}] + \frac{2}{3}h\,\mathbf{F}(x_{j+1}, \mathbf{y}_{j+1}).$$

When solving this algebraic system for $\mathbf{y}_{j+1}$, the previously computed approximations, the step size, and the mesh point x_{j+1} are all fixed. Accordingly, the algebraic equations for $\mathbf{y}_{j+1}$ have the form

$$\mathbf{y} = h\gamma\,\mathbf{F}(\mathbf{y}) + \boldsymbol{\psi}.$$

Here we have suppressed in the notation the dependence on x in $\mathbf{F}(x, \mathbf{y})$ because it is held fixed at x_{j+1}. The constant γ is characteristic of the method and the vector $\boldsymbol{\psi}$ represents previously computed data that are constant while computing $\mathbf{y}$. Although we shall always have the BDFs in mind, this form is representative of LMMs in general and a little generalization accommodates implicit Runge–Kutta methods as well.

First let us recall why simple iteration cannot be used when the problem is stiff. Suppose that $\mathbf{y}^*$ is the solution of the algebraic equations and $\mathbf{y}^m$ the current iterate. The next iterate is then

$$\mathbf{y}^{m+1} = h\gamma\, \mathbf{F}(\mathbf{y}^m) + \boldsymbol{\psi}.$$

Subtracting the equation satisfied by $\mathbf{y}^*$ leads to an equation for the error,

$$\mathbf{y}^{m+1} - \mathbf{y}^* = h\gamma\, [\mathbf{F}(\mathbf{y}^m) - \mathbf{F}(\mathbf{y}^*)].$$

Suppose now that the differential equation is (1) so that this expression is

$$(\mathbf{y}^{m+1} - \mathbf{y}^*) = h\gamma J(\mathbf{y}^m - \mathbf{y}^*).$$

If the initial vector $\mathbf{y}^0$ is such that the initial error is an eigenvector $\mathbf{v}$ of the matrix J corresponding to an eigenvalue λ, then it is clear that for any m, the error of $\mathbf{y}^m$ is $(h\gamma\lambda)^m\, \mathbf{v}$. If $|h\gamma\lambda| \geq 1$, the iteration obviously does not converge. Accordingly, for these problems it is necessary that $|h\gamma\lambda| < 1$ for all eigenvalues λ of J if there is to be convergence from any initial guess. Clearly simple iteration leads to an unacceptable restriction of the step size when solving stiff problems.

The easier part of evaluating the implicit formula is finding an initial guess $\mathbf{y}^0$ that is accurate. As usual this is done with an explicit predictor. It should approximate $\mathbf{y}(x_{j+1})$ about as well as the value $\mathbf{y}^*$ from the implicit formula; we integrate with an implicit formula because it is more stable, not because it is more accurate. The function $\mathbf{F}$ is linearized about the current iterate $\mathbf{y}^m$,

$$\mathbf{F}(\mathbf{y}) \approx \mathbf{F}(\mathbf{y}^m) + J(\mathbf{y} - \mathbf{y}^m),$$

and the next iterate is obtained by solving the approximating equation

$$\mathbf{y}^{m+1} = \boldsymbol{\psi} + h\gamma\, \mathbf{F}(\mathbf{y}^m) + h\gamma J(\mathbf{y}^{m+1} - \mathbf{y}^m).$$

Although something like this is necessary for the solution of stiff problems, it has serious consequences. If there are N differential equations, the matrix J is $N \times N$. For systems of even moderate size, say $N = 50$, the storage required to evaluate an implicit method in this way is very much larger than that required to evaluate with simple iteration. For problems involving hundreds of equations, and thousands are not unusual, it is necessary to take full advantage of structure even to hold the matrix in rapid-access memory. It is necessary to solve a linear system with matrix $I - h\gamma J$, the *iteration matrix*, for each iterate. This can be quite expensive. So far, J has not been specified. Newton's method would take here the Jacobian evaluated at the current iterate,

$$J = \frac{\partial \mathbf{F}}{\partial \mathbf{y}}\, (\mathbf{y}^m).$$

Although this would (usually) provide rapid convergence, it has very serious disadvantages. One is that it is often inconvenient and/or expensive to form the Jacobian and with Newton's method this is done for each iterate. It is well-known that when solving linear systems by direct methods, the bulk of the work is due to a matrix factorization and solution of several linear systems with the same matrix involves comparatively little more work than solving just one system. For these reasons it is generally better to hold J constant while computing $\mathbf{y}$ and all the popular codes do so. This J is to be an approximation to the Jacobian evaluated at the solution $\mathbf{y}^*$ of the algebraic equations:

$$J \approx \frac{\partial \mathbf{F}}{\partial \mathbf{y}} (\mathbf{y}^*).$$

Such an iteration is called a *simplified Newton iteration* or a *chord method.*

The iteration matrix depends on the step size h as well as J. This means that a change of step size has a serious impact on the overhead of the integration because it implies that a new iteration matrix will have to be formed and factored. The matrix J does not have to be equal to the current local Jacobian; it is better regarded as a matrix chosen to enhance the rate of convergence of the iteration. Because of this and because the solution does not change rapidly when the problem is stiff, acceptable convergence can normally be obtained for a number of steps using the same matrix J. This furnishes a strong reason for working with a constant step size because then the iteration matrix is constant and one factorization suffices for a number of steps. A change of step size should be contemplated only if it represents a substantial increase, or if it is necessary because the step is rejected or the rate of convergence is not acceptable. The quantity γ is characteristic of the formula. In the case of a family of formulas such as the BDFs, typical codes vary the formula used as well as the step size. A change of order (formula) corresponds to changing γ, hence has the same effect on the iteration matrix as a change of step size.

The observations made about the effect of a change of step size are in the context of a quasi-constant step size formulation. If a fully variable step size implementation is used, as it is for example in EPISODE, a change of step size affects the leading coefficient as long as the place where the change occurred is within the memory of the formula. The fixed leading coefficient implementation was described briefly in Chapter 5. This approach keeps the leading coefficient γ constant throughout a change of step size. It provides the advantage enjoyed by the quasi-constant step size formulation of only one change to the iteration matrix due to a change of step size and it also provides in considerable measure the more stable handling of severe and/or repeated change of step size enjoyed by the fully variable step size formulation. An important recent code based on this formulation is VODE (Brown, Byrne, and Hindmarsh [1989]).

When J is held constant, the iteration is only linearly convergent. A little analysis will give us a better idea what is involved. The iterate satisfies

$$\mathbf{y}^{m+1} = \boldsymbol{\psi} + h\gamma\, \mathbf{F}(\mathbf{y}^m) + h\gamma J\, \mathbf{y}^{m+1} - h\gamma J\, \mathbf{y}^m,$$

and the desired solution satisfies

$$\mathbf{y}^* = \boldsymbol{\psi} + h\gamma\, \mathbf{F}(\mathbf{y}^*) + h\gamma J\, \mathbf{y}^* - h\gamma J\, y^*.$$

The iteration error satisfies

$$\mathbf{y}^{m+1} - \mathbf{y}^* = h\gamma \mathcal{J}(\mathbf{y}^m - \mathbf{y}^*) + h\gamma J(\mathbf{y}^{m+1} - \mathbf{y}^*) - h\gamma J(\mathbf{y}^m - \mathbf{y}^*).$$

Here a mean value theorem has been used for $\mathbf{F}$. The matrix $\mathcal{J}$ consists of the Jacobian of $\mathbf{F}$ with its entries evaluated at different points between $\mathbf{y}^m$ and $\mathbf{y}^*$. A little manipulation then gives

$$\mathbf{y}^{m+1} - \mathbf{y}^* = (I - h\gamma\, J)^{-1}\, h\gamma(\mathcal{J} - J)(\mathbf{y}^m - \mathbf{y}^*).$$

If there is a number $\rho < 1$ such that

$$\|(I - h\gamma J)^{-1}\, h\gamma(\mathcal{J} - J)\| \le \rho,$$

then the error is decreased by at least a factor of ρ at each iteration. Consideration of simple cases shows that in general the rate is no faster than ρ. This expression for the effect of an iteration shows clearly what must be done to improve the rate of convergence. There are basically two possibilities: reduce h or make J a better approximation to the local Jacobian evaluated at the solution. There are some unpleasant consequences. For the computation of the iterates the iteration matrix must be retained in factored form. Storage was such a critical problem for the early codes that they did not retain a copy of J. Storage is not a problem for moderate sized systems today, but it remains an issue for very large systems. In view of the fact that when capabilities improve, scientists solve bigger problems, there will presumably always be systems for which doubling the storage required is unpalatable, if not unacceptable. In any event, if convergence is too slow, a new iteration matrix must be formed and factored. If storage was not allocated for a copy of J, a new J must first be formed. If a copy of J is retained, it might be used for a number of steps, but at some point it may be necessary to form a current approximate Jacobian. The tactics are complicated because one must decide whether to enhance convergence by reducing the step size, forming a new Jacobian, or both.

The next iterate $\mathbf{y}^{m+1}$ is not usually computed by the recipe given earlier. Instead, the current iterate is subtracted to get

$$\mathbf{y}^{m+1} - \mathbf{y}^m = \boldsymbol{\psi} + h\gamma\, \mathbf{F}(\mathbf{y}^m) - \mathbf{y}^m + h\gamma\, J(\mathbf{y}^{m+1} - \mathbf{y}^m).$$

Let us define the residual

$$\mathbf{r}^m = \boldsymbol{\psi} + h\gamma\,\mathbf{F}(\mathbf{y}^m) - \mathbf{y}^m.$$

It is better numerically to calculate the small change $\mathbf{y}^{m+1} - \mathbf{y}^m$ and add it to $\mathbf{y}^m$ as needed rather than to calculate the iterate $\mathbf{y}^{m+1}$ directly. The change satisfies

$$(I - h\gamma J)(\mathbf{y}^{m+1} - \mathbf{y}^m) = \mathbf{r}^m.$$

Scaling this system to

$$\left(J - \frac{1}{h\gamma} I\right)(\mathbf{y}^{m+1} - \mathbf{y}^m) = -\frac{1}{h\gamma}\mathbf{r}^m$$

has some advantages. It is cheaper to form the iteration matrix and emphasis is placed in the linear system on J, the important part when the system is stiff. There is another important reason for computing iterates in terms of small corrections rather than directly—the matrices that arise are normally very ill-conditioned. Of course, when the step size h is small, the iteration matrix $I - h\gamma J$ is well conditioned because it is not too different from I. This situation corresponds to a non-stiff portion of the integration. With reasonable assumptions about what constitutes a stiff problem, it can be shown (Shampine [1993]) that the condition of $I - h\gamma J$ is bounded below by a quantity that is approximately equal to $2/3|h\gamma\lambda_N|$ where λ_N is the eigenvalue of J of greatest magnitude. In Example 7 below some results are presented for the integration of Robertson's problem. As equilibrium is approached, $|\lambda_N| \approx 10^4$ and $h \approx 10^9$. According to the bound, the iteration matrix is extremely ill-conditioned. The concept of the conditioning of a matrix does not take into account the role of the right hand side in the system to be solved. It turns out that in this context, the linear systems are not as badly conditioned as the matrices. Organizing the computation in terms of small corrections is essential to the evaluation of the implicit formulas. If the linear systems for the corrections can be solved with a modest accuracy, even one digit correct, then the iteration will eventually converge to an accurate result. Difficulty with the accurate solution of linear systems slows down the rate of convergence, but it does not prevent convergence. Reducing the step size to improve the rate of convergence of the iteration also serves to reduce the ill-conditioning of the iteration matrix and to improve the accuracy of the solution of linear systems.

EXERCISE 5. If you have the source code for a modern integrator intended for stiff problems, you can probably investigate this matter of ill-conditioning quite easily. This is because the code is likely to make use of the linear equation solvers from LINPACK (Dongarra et al. [1979]). For example, the solver DGEFA would be used to decompose an iteration matrix that is treated as a full matrix. LINPACK also has a solver that estimates a condition number of a matrix as it decomposes it. In the case of a full matrix, this solver is DGECO. The code DGECO does an internal call to DGEFA to do the factorization and then does further computations to estimate a condition number. Because of this, it is easy to alter the integrator to call DGECO rather than DGEFA and print out the estimated condition number each time that a new iteration matrix is formed and factored. You should find that the matrix is estimated to be well conditioned in the transient and ill-conditioned where the problem is stiff.

EXAMPLE 6. In a note in the SIGNUM Newsletter, A.R. Curtis suggested a family of problems for testing codes intended for stiff problems and presented some results for a code in the Harwell Subroutine Library that he wrote. The family and the results help us understand one of the factors that affect the cost of solution of stiff problems. The family is specified in terms of a 2×2 matrix A depending on two parameters u, v:

$$A = M\begin{pmatrix} u & 0 \\ 0 & 0 \end{pmatrix}M^{-1} \quad \text{where} \quad M = \begin{pmatrix} \cos(vt) & \sin(vt) \\ -\sin(vt) & \cos(vt) \end{pmatrix}.$$

Note that the eigenvalues of A are u and 0 and that $M^{-1} = M^T$. To create a test problem, a solution $\mathbf{s}(t)$ of two components is selected and a vector $\mathbf{f}(t)$ is defined by

$$\mathbf{f}(t) = \dot{\mathbf{s}}(t) + (I + A)\mathbf{s}(t).$$

Obviously the initial value problem

$$\dot{\mathbf{y}} + (I + A)\mathbf{y} = \mathbf{f}(t), \qquad \mathbf{y}(0) = \mathbf{s}(0)$$

has $\mathbf{s}(t)$ as its solution. The Jacobian of this system is $-(I + A)$; its eigenvalues are -1 and $-(1 + u)$. For a smooth solution $\mathbf{s}(t)$, the stiffness of the problem is governed by the parameter u. The parameter v governs the rate of rotation of the eigenvectors, or equivalently, how fast the Jacobian changes in time.

Curtis presents some results for integrating $\mathbf{s}(t) = (\cos(t), \sin(t))^T$ on $[0, 10\pi]$ for a range of parameter values. The BDF code he used is probably the first production-grade code to save the Jacobian so as not to recompute it merely because of a change of step size. A portion of his results for $u =$

1000 follows:

	$v = 0$	1	10	100	1000
	269	1885	8206	18037	47433
$u = 1000$	306	6369	24636	77081	243182
	1	477	1370	5826	20720

The numbers in each entry are arranged vertically to correspond to:

number of steps
number of function evaluations
number of Jacobian evaluations

The solution is the same for all values of the parameters. For fixed u, the eigenvalues of the Jacobian are the same and in particular, for $u = 1000$, they are -1 and -1001. With its smooth solution, a Jacobian with these eigenvalues, and an interval of $[0, 10\pi]$, this problem is stiff for all v. As v increases, the solution is the same and the stiffness is the same, yet the cost of the integration increases dramatically. What is going on? The answer is one of the factors determining the step size that is often forgotten. The step size must be small enough that the Newton iteration for evaluating the formula converge at an acceptable rate. In the codes there is a tradeoff between reducing the step size and forming a new iteration matrix (hence a new Jacobian). As the parameter v increases, the Jacobian changes more rapidly as a function of t, implying that a new iteration matrix must be formed more often and the step size must be smaller—exactly what is seen in the numerical results.

§2.2 Computing Jacobians

When solving stiff problems we need the Jacobian of $\mathbf{F}$—a great inconvenience compared to solving non-stiff problems. For a large system, writing out all the partial derivatives of $\mathbf{F}$ is often so tedious as to cause mistakes. To take advantage of the structure of large systems, the Jacobian matrix must be stored in an appropriate way and this may also lead to mistakes. For these reasons, numerical approximation of the Jacobian by differences is very popular. However, it is difficult to approximate the Jacobian accurately, in part because it would cost too much. Inaccurate Jacobians increase the cost of the integration, e.g., more iterations are necessary, and the codes are made somewhat less reliable. Because of this, if it is not inconvenient to form a Jacobian analytically, it is often advantageous to do so. The prologue to LSODE puts the matter well: "If the problem is stiff, you are encouraged to supply the Jacobian directly . . . , but if this is not feasible, LSODE will compute it internally by difference quotients. . . ." Despite this good advice, it is probably fair to say that most problems are solved with numerical approximations to the Jacobian.

Until recently, having the codes take care of the Jacobian for you meant that they would approximate it numerically. Now, however, there are programs that start with a subroutine for the evaluation of **F** and generate a subroutine for the efficient evaluation of the exact Jacobian. Juedes [1991] surveys tools for this purpose. An even more recent example is the ADIFOR program of Bischof et al. [1992]. This exciting work is very promising, but it is premature for us to pursue it here, so in what follows we explain what is being done at present in the codes, namely numerical approximation of the Jacobian by differences.

Suppose J is to approximate the Jacobian at $(x, \mathbf{y})$. Typically the code will have available the value of $\mathbf{F}$ at this argument. The function $\mathbf{F}$ is evaluated at nearby arguments and the Jacobian is approximated by differences. Let $\mathbf{e}_k$ be column k of the identity matrix. A scalar δ is chosen, $\mathbf{F}(\mathbf{y} + \delta\mathbf{e}_k)$ is formed, and column k of J is defined by

$$\delta^{-1}[\mathbf{F}(\mathbf{y} + \delta\mathbf{e}_k) - \mathbf{F}(\mathbf{y})].$$

In more detail, for each $i = 1, 2, \ldots, N$,

$$\begin{aligned} F^i(\mathbf{y} + \delta\mathbf{e}_k) &= F^i(y^1, \ldots, y^k + \delta, \ldots, y^N) \\ &= F^i(y^1, \ldots, y^k, \ldots, y^N) + \delta \frac{\partial F^i}{\partial y^k}(\mathbf{y}) + \mathcal{O}(\delta^2), \end{aligned}$$

hence

$$\frac{F^i(\mathbf{y} + \delta\mathbf{e}_k) - F^i(\mathbf{y})}{\delta} \approx \frac{\partial F^i}{\partial y^k}(\mathbf{y}).$$

Proceeding in this way, N evaluations of $\mathbf{F}$ are required to form J. (If central differences were used to get a more accurate approximation or if some kind of test of the accuracy of the approximation were made, more evaluations would be required.) In terms of function evaluations, this is very expensive compared to simple iteration, so it should be done only when necessary. To be sure, if Jacobians are saved and reused properly, a non-stiff problem could be solved with very few Jacobians. However, the early codes for stiff problems form a new Jacobian nearly every time the step size or order is changed, so they are relatively inefficient for the solution of non-stiff problems. More recent codes handle this more efficiently by reusing Jacobians.

As with any difference scheme for approximating derivatives, there are difficulties with scaling. One difficulty is choosing an appropriate increment

δ. A better job could be done if individual components of J were computed, but this is not practical because it would require that the code have access to individual components of $\mathbf{F}$. It is so tedious to program $\mathbf{F}$ in a manner that makes it possible to evaluate specified individual components that it is universal practice to evaluate the whole vector at once. This consideration of a convenient design for the user forces us to compute a whole column of J at a time. The usual difficulty with choosing an increment is that it must be small enough that the difference quotient approximate the derivative accurately, yet not so small that the result consist only of roundoff error. The trouble in this context is that if δ is small enough that the biggest partial derivative in the column is approximated accurately, it may be so small that some other partial derivative is approximated very inaccurately. To illustrate the point, suppose that some component F^i changes so rapidly that a small δ is needed to approximate its partial derivative with respect to y^k. Let us then ask what happens when we approximate the partial derivative of another component F^j. In this let us be very specific and suppose that F^j has the simple form of $\alpha y^k + \beta$ for constants α and β. With this simple form, differencing is exact in principle:

$$\frac{\partial}{\partial y^k} F^j(\mathbf{y}) = \frac{(\alpha y^k + \beta + \delta) - (\alpha y^k + \beta)}{\delta} = \alpha.$$

Nevertheless, if δ is smaller than a unit roundoff in $\alpha y^k + \beta$, then in floating point arithmetic $fl(\alpha y^k + \beta + \delta) = fl(\alpha y^k + \beta)$ and the partial derivative is computed to be 0 rather than α. This is merely an extreme case to make concrete the frequent observation that some of the partial derivatives are computed inaccurately. However, let us not lose sight of the fact that we are not interested in the Jacobian *per se*, we are interested in solving a system of nonlinear algebraic equations. Our expression for the error of an iterate shows that for convergence, it is not necessary that all entries in J approximate well the corresponding entries in $\mathcal{J}$. It is difficult to be precise about the requirement that $J \approx \mathcal{J}$, but it appears that it is the larger entries that need to be approximated with some degree of accuracy. Indeed, advantage is sometimes taken of this observation by setting small entries in J to zero so as to get a sparse matrix.

Computer arithmetic can be troublesome when approximating Jacobians by differences. Generally it is best to use double precision for the integration of initial value problems on computers with word lengths no greater than the IEEE standard. It is obviously advantageous to have a relatively long word length for coping with the loss of significance when differences are formed. However, the matter is not as delicate as it might seem because it is not usually necessary to have a very accurate approximate Jacobian.

More troublesome in this context is the exponent range, which in the IEEE standard is smaller in single precision that in double. Some computers have exponent ranges smaller than the IEEE single precision standard even in their double precision arithmetic. The difficulty is that stiff problems typically involve numbers that span a very large range. This is true, for example, of entries in Jacobian matrices. Also, quite often there are solution components that tend to zero rapidly. Overflows, and especially underflows, are very real possibilities when solving stiff problems using some common computer arithmetics. The case for using double precision when solving stiff problems on a machine conforming to the IEEE standard is a strong one. Besides, the commonly held notion that double precision is more than twice as expensive as single is simply not true for hardware double precision.

For the solution of medium to large stiff problems it is important to pay attention to structure. One place where this is advantageous is in the formation of Jacobians by differencing. By *structure* here is meant information about which entries in $\mathbf{F_y}$ are zero. If, say, the function F^i does not depend on the component y^k, then the partial derivative of F^i with respect to y^k must be zero, and we know to place a 0 in the (i, k) entry of J. This is useful, but more important is the fact that the difference scheme can exploit this information to reduce the number of evaluations of $\mathbf{F}$ required to form J. The original idea is due to Curtis, Powell, and Reid [1974]. Coleman, Garbow and Moré [1984a, 1984b], Salane and Shampine [1985], and Salane [1986] describe further developments and provide software. The expansion

$$\begin{aligned} F^i(\mathbf{y} + \delta\mathbf{e}_k + \delta\mathbf{e}_m) &= F^i(y^1, \ldots, y^k + \delta, \ldots, y^m + \delta, \ldots, y^N) \\ &= F^i(y^1, \ldots, y^k, \ldots, y^N) + \delta\frac{\partial F^i}{\partial y^k} + \delta\frac{\partial F^i}{\partial y^m} + \mathcal{O}(\delta^2) \end{aligned}$$

shows that

$$\frac{F^i(\mathbf{y} + \delta\mathbf{e}_k + \delta\mathbf{e}_m) - F^i(\mathbf{y})}{\delta} \approx \frac{\partial F^i}{\partial y^k} + \frac{\partial F^i}{\partial y^m}.$$

If we knew that $\partial F^i/\partial y^k = 0$, this difference quotient would provide an approximation to $\partial F^i/\partial y^m$. Similarly, if we knew that $\partial F^i/\partial y^m = 0$, it would provide an approximation to $\partial F^i/\partial y^k$. Suppose columns k and m of $\mathbf{F_y}$ are such that if an entry in one column might not be zero, we are sure that the corresponding entry in the other column *is* zero. In such a case, we can extract all non-zero entries in both columns from $\delta^{-1}[\mathbf{F}(\mathbf{y} + \delta\mathbf{e}_k + \delta\mathbf{e}_m) - \mathbf{F}(\mathbf{y})]$. An example will help make the idea clearer. Suppose that the Jacobian $\mathbf{F_y}$ has the structure

$$\begin{pmatrix} X & 0 & 0 & 0 \\ X & X & 0 & 0 \\ 0 & X & X & 0 \\ 0 & 0 & X & X \end{pmatrix},$$

where an X means that the entry might not be 0. The non-zero entries in both the columns 1 and 3 can be approximated using the difference quotient $\delta^{-1}[\mathbf{F}(\mathbf{y} + \delta\mathbf{e}_1 + \delta\mathbf{e}_3) - \mathbf{F}(\mathbf{y})]$ and those in columns 2 and 4 using $\delta^{-1}[\mathbf{F}(\mathbf{y} + \delta\mathbf{e}_2 + \delta\mathbf{e}_4) - \mathbf{F}(\mathbf{y})]$. Ordinarily the value $\mathbf{F}(\mathbf{y})$ is available from other computations, so only two extra evaluations are needed to approximate this Jacobian when its structure is taken into account; the usual scheme would need four. In practice the idea is extended in a straightforward way to obtain all the non-zero entries in a group of columns at a time and the increment is chosen to have different values for the different groups.

The cost of forming J depends on the structure, so generally it cannot be determined without knowledge of the specific Jacobian to be approximated. A particularly simple and quite common case is that of a *banded system*. A system is banded with half bandwidth m_b, if for each k, the kth equation does not involve variables y^j with $|j - k| > m_b$. For such a system, the Jacobian has non-zero elements only within a band about the main diagonal, whence the name. It is not hard to see that the idea just exposed permits a Jacobian of half bandwidth m_b to be approximated using only $2m_b + 1$ evaluations of $\mathbf{F}$, *independent of the size N of the system*. This kind of reduction of cost is typical of large sparse Jacobians so that the cost of forming a Jacobian is usually much smaller than the N evaluations of $\mathbf{F}$ required when no attention is paid to structure.

EXERCISE 6. Work out the details of approximating a tridiagonal Jacobian ($m_b = 1$) by differences using only three function evaluations.

The structure of J is also important to the solution of the linear system. To store a general $N \times N$ matrix requires N^2 memory locations. If a large proportion of the elements of J are zero, the matrix is *sparse*, it is advantageous to store only the non-zero elements. Of course, to do this involves recording the indices of the non-zero elements, so there is a net savings of storage only when the matrix is rather sparse. Again the simplest important case is that of a band matrix. In this case the data structure is relatively simple, e.g., the diagonals that might contain non-zero entries can be kept as vectors. It is worth remark that even in this relatively simple case, forming a Jacobian analytically and storing the non-zero entries properly is a nuisance for the user of a code. There are schemes for doing elimination that pay attention to the presence of zero entries in the matrix and attempt to reduce as much as practical the new non-zero entries that arise during the elimination. Again this is particularly simple for banded matrices. With row partial pivoting it is easy to specify in advance the diagonals that might *fill in*, that is, have non-zero entries created in the elimination process. The gain is easy to understand in the case of a band matrix. When elimination with row pivoting is done, it is necessary to eliminate only the non-zero elements below the diagonal. If there are $m_b \ll N$ such elements, the cost of the elimination is much less than in the general case. In view of the fact that the matrix factorization is the most expensive part of the solution of the linear system, taking account of structure is what makes feasible the solution of large systems by direct means. Some systems that arise in practice are too large for direct methods to be practical and some kind of iterative scheme is used. There are then two iterations going on—an *outer iteration* corresponding to the simplified Newton iteration and an *inner iteration* used to compute each of these iterates. It has been found important to make some use of direct methods, *preconditioning*, to accelerate the inner iterations. This is all rather complicated and is the subject of intensive research at present, so no more will be said about it here. In summary, codes that treat banded systems in a special way are common and when the problem is banded, it is important to take advantage of this. Codes that treat general sparse systems are much less readily available, but anyone contemplating the solution of large systems is likely to need one. Two that are widely available are LSODES and SPRINT, and references to others are found in Curtis [1978].

EXERCISE 7. Example 6 of Chapter 1 takes up the use of arc length as independent variable. Although a valuable technique, it was remarked that this variable is inconvenient for some tasks. Now we are able to appreciate that it presents difficulties when solving stiff problems. Consider the effect of changing from a given independent variable x to arc length on the structure of the Jacobian of the system of equations. Explain why arc length may be impractical for the solution of even moderately large stiff initial value problems.

§2.3 Choice of Unknowns

The most successful early codes for stiff problems take

$$\mathbf{z}_{j+1} = h\, \mathbf{F}(x_{j+1}, \mathbf{y}_{j+1})$$

as the unknown in the algebraic equations rather than $\mathbf{y}_{j+1}$. There is no obvious reason to prefer one unknown over the other. However, this choice makes natural an important action when solving stiff problems that we take up in this section. Once we understand the issue, we can work with either choice of unknown.

Suppressing the subscripts and the independent variable as usual, the algebraic equations

$$\mathbf{y} = h\gamma\, \mathbf{F}(\mathbf{y}) + \boldsymbol{\psi}$$

can be reformulated in terms of the other unknown as

$$\mathbf{z} = h\, \mathbf{F}(\mathbf{y}) = h\, \mathbf{F}(\gamma \mathbf{z} + \boldsymbol{\psi}).$$

A linearization about $\gamma\mathbf{z} + \boldsymbol{\psi}$ leads to

$$\begin{aligned} \mathbf{z}^{m+1} &= h\, \mathbf{F}(\gamma \mathbf{z}^m + \boldsymbol{\psi}) + hJ[(\gamma \mathbf{z}^{m+1} + \boldsymbol{\psi}) - (\gamma \mathbf{z}^m + \boldsymbol{\psi})], \\ &= h\, \mathbf{F}(\gamma \mathbf{z}^m + \boldsymbol{\psi}) + h\gamma J(\mathbf{z}^{m+1} - \mathbf{z}^m), \end{aligned}$$

and then to

$$(I - h\gamma J)(\mathbf{z}^{m+1} - \mathbf{z}^m) = -[\mathbf{z}^m - h\, \mathbf{F}(\gamma \mathbf{z}^m + \boldsymbol{\psi})].$$

This is just like the system to be solved when $\mathbf{y}$ is taken as a variable. The iteration matrix is the exactly the same and the way of proceeding is the same: the change in the current iterate is computed and the right hand side is the residual of the current iterate.

We are computing $\mathbf{y}^m$ as an approximation to $\mathbf{y}^*$ that is itself to approximate $\mathbf{y}(x_{j+1})$, but there is a question about how to compute a corresponding approximation $\mathbf{z}^m$ to the scaled derivative $h\mathbf{y}'(x_{j+1})$. When taking $\mathbf{z}$ as the basic variable, it is clear that we

$$\text{compute } \mathbf{z}^m \text{ and define } \mathbf{y}^m = \gamma \mathbf{z}^m + \boldsymbol{\psi}.$$

Notice that this amounts to defining the pair $\mathbf{y}^m$, $\mathbf{z}^m$ so that the *difference* equation is satisfied exactly. When taking $\mathbf{y}$ as the basic variable, what we did for non-stiff problems was

$$\text{compute } \mathbf{y}^m \text{ and define } \mathbf{z}^m = h\, \mathbf{F}(\mathbf{y}^m).$$

This defines the pair $\mathbf{y}^m$, $\mathbf{z}^m$ so that the *differential* equation is satisfied exactly. Alternatively, the pair could be defined so that the difference equation is satisfied exactly with this choice of basic variable, too, by solving

$$\mathbf{y}^m = \gamma\, \mathbf{z}^m + \boldsymbol{\psi}$$

to get the recipe

$$\text{compute } \mathbf{y}^m \text{ and define } \mathbf{z}^m = \frac{1}{\gamma}(\mathbf{y}^m - \boldsymbol{\psi}).$$

If we were to iterate to completion, there would be no distinction, but in practice, the distinction is very important when the problem is stiff. To see this let

$$\mathbf{z}^* = h\,\mathbf{F}(\mathbf{y}^*).$$

Then with the one definition of the pair,

$$\mathbf{z}^m - \mathbf{z}^* = h\,\mathbf{F}(\mathbf{y}^m) - h\,\mathbf{F}(\mathbf{y}^*) = h\mathcal{J}(\mathbf{y}^m - \mathbf{y}^*),$$

where $\mathcal{J}$ is the matrix arising from application of a mean value theorem to $\mathbf{F}$. With the other definition of the pair,

$$\mathbf{z}^m - \mathbf{z}^* = \frac{1}{\gamma}(\mathbf{y}^m - \boldsymbol{\psi}) - \frac{1}{\gamma}(\mathbf{y}^* - \boldsymbol{\psi}) = \frac{1}{\gamma}(\mathbf{y}^m - \mathbf{y}^*).$$

For a stiff problem, we expect to have $\| h\mathcal{J} \| \gg 1$, so this second definition can provide a very much better approximate derivative. When the problem is not stiff, either definition is acceptable and the first may well provide the better approximation. One of the reasons for the popularity of the $\mathbf{z}$ variable when solving stiff problems was that it led naturally to the better definition. We now see that it is just as easy to work with $\mathbf{y}$ as the basic variable provided that this issue about approximating the derivative is understood. It might be remarked that this is another example of the difference between mathematical theorems and the art of numerical analysis. Although valid as the step size tends to zero, approximating the derivative of the solution by $\mathbf{F}(\mathbf{x}_{j+1}, \mathbf{y}_{j+1})$ is unsatisfactory in the practical solution of stiff problems.

EXERCISE 8. Experiment with some codes available to you to see whether the approximation $\mathbf{y}_j'$ to $\mathbf{y}'(x_j)$ satisfies $\mathbf{y}_j' = \mathbf{F}(x_j, \mathbf{y}_j)$. You will find that this will not be true of codes for stiff problems; it might be true of a code for non-stiff problems, or it might not. Solve a stiff problem with an appropriate code and compare the derivative returned by the code to the value $\mathbf{F}(x_j, \mathbf{y}_j)$ that you compute. Which seems to be the better approximation?

This issue of approximating the derivative is particularly important to the estimation of the local truncation error. In §5.1 of Chapter 7 we investigated the approximation of the local truncation error by two difference schemes that in the case of Euler's method are

$$\frac{1}{2}h^2\,y''(\xi) \approx \frac{1}{2}\nabla^2\,y_{j+1} \approx \frac{1}{2}h\,\nabla\,F(x_{j+1}, y_{j+1}).$$

For sufficiently small step sizes h, one estimator is as accurate as the other. However, when solving a stiff problem, the $hF(x_k, y_k)$ can be such bad approximations to the $hy'(x_k)$ that an error estimator based on them is unusable. It is necessary to difference the y_k or else the smoother approximations to the scaled derivative, the z_k.

§3 Examples

The next two examples are based on the paper, "Applications of EPISODE: An Experimental Package for the Integration of Systems of Ordinary Differential Equations," by A.C. Hindmarsh and G.D. Byrne [1976]. These authors have made many contributions to the practice of solving stiff ODEs and their expertise is evident in their selection of examples and the numerical results they present. Besides the obvious value of studying informative examples, it is useful here to see some numerical results obtained with a quality code. EPISODE is certainly a quality code, but any of a good many other codes would have served our purposes as well; it just happens to be the subject of the particular paper we cite. In a later paper with Jackson and Brown (Byrne et al. [1977]), Hindmarsh and Byrne compare EPISODE to the very popular code GEAR. Anyone interested in how details of implementation and use can affect computations would benefit from a study of this paper. The paper presents other illuminating examples, one of which is given here as an exercise.

EXAMPLE 7. The Robertson problem describes the concentrations of three components in a chemical reaction. It is

$$\begin{aligned} y_1' &= -0.04\, y_1 + 10^4\, y_2 y_3, & y_1(0) &= 1,\\ y_2' &= 0.04\, y_1 - 10^4\, y_2 y_3 - 3 \times 10^7\, y_2^2, & y_2(0) &= 0,\\ y_3' &= 3 \times 10^7\, y_2^2, & y_3(0) &= 0. \end{aligned}$$

The Jacobian matrix of this system is easy to calculate:

$$J = \begin{pmatrix} -.04 & 10^4\, y_3 & 10^4\, y_2 \\ .04 & -10^4\, y_3 - 6 \times 10^7\, y_2 & -10^4\, y_2 \\ 0 & 6 \times 10^7\, y_2 & 0 \end{pmatrix}.$$

Because the initial data has a simple form, it is easy to verify that the eigenvalues of the Jacobian evaluated there are $-0.04, 0, 0$. This suggests that on an interval of length $\mathcal{O}(1)$, the problem is not stiff. On physical grounds it is expected that the concentrations will tend to constant equilibrium values. As seen in Figure 8.4, these values are $y_1(\infty) = y_2(\infty) = 0$ and $y_3(\infty) = 1$ and they obviously provide a constant solution to the differential equation. All standard methods integrate constants exactly, so it is expected that the step size h will become unbounded as $t \to \infty$. A little

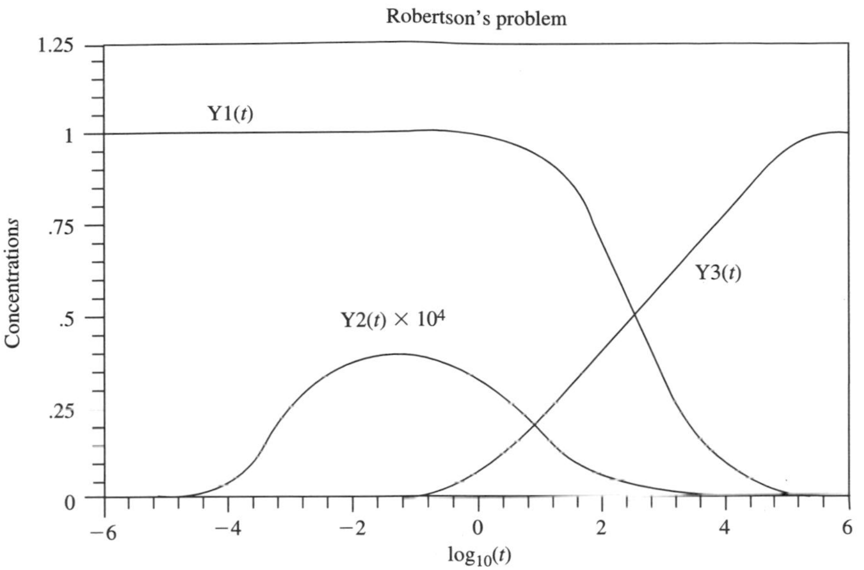

Figure 8.4

calculation shows that the eigenvalues of the Jacobian evaluated at the equilibrium values are 0, 0, $-10^4 - 0.04$. Clearly the problem is extremely stiff near equilibrium. It should be appreciated that the problem is not stiff initially and that the approach to equilibrium is not fast.

Because of the scaling of the variables, it is reasonable to control the estimated local error with a pure absolute error criterion. Hindmarsh and Byrne used the moderately stringent tolerance of 10^{-6}. EPISODE solves stiff problems with a fully variable step size implementation of the BDFs. It permits the user to supply an analytical Jacobian or have the code compute Jacobians by differences. When solved with analytical Jacobian, some representative solution values and the costs as measured by NSTEP, the number of steps, NFCN, the number of function evaluations, and NJAC, the number of evaluations of the Jacobian, are

T	NSTEP	NFCN	NJAC	Y_1	Y_2	Y_3
$4.0E-01$	27	38	19	$9.8517E-01$	$3.3864E-05$	$1.4794E-02$
$4.0E+01$	89	152	30	$7.1582E-01$	$9.1851E-06$	$2.8417E-01$
$4.0E+03$	180	327	50	$1.8320E-01$	$8.9423E-07$	$8.1680E-01$
$4.0E+05$	256	474	71	$4.9409E-03$	$1.9839E-08$	$9.9506E-01$
$4.0E+07$	319	595	91	$5.2329E-05$	$2.0934E-10$	$9.9995E-01$
$4.0E+09$	341	621	105	$5.4561E-07$	$2.1824E-12$	$1.0000E+00$
$4.0E+10$	352	652	115	$-1.1761E-06$	$-4.7043E-12$	$1.0000E+00$

Physically all the solution components must be non-negative, but it is observed here that as equilibrium is approached, negative solution values are formed. This particular integration was terminated at the last point reported and the accuracy of the last result is acceptable. We shall return to this matter shortly, but first notice that this integration is not expensive in any sense. EPISODE implements Adams–Moulton formulas as well as BDFs. If this were not a stiff problem, it would be more efficient to use the Adams–Moulton formulas evaluated with simple iteration. An attempt to do this resulted in

T	NSTEP	NFCN	NJAC	Y_1	Y_2	Y_3
$4.0E-01$	1130	1928	0	$9.8517E-01$	$3.3837E-05$	$1.4793E-02$
$4.0E+00$	11285	19333	0	$9.0552E-01$	$2.2432E-06$	$9.4457E-02$

Again we see the importance of using methods that are appropriate to stiff problems when the problem is stiff—if the problem is at all stiff, its solution is impractical with methods intended for non-stiff problems.

It is not easy to tell whether a problem is stiff and codes intended for non-stiff problems are very inefficient when applied to a stiff problem. Because of this, modern software intended for non-stiff problems attempts to diagnose whether stiffness is the reason that a problem appears to be "difficult." To illustrate this capability, we integrated the problem with UT from RKSUITE. Of the three explicit Runge–Kutta pairs available in this code, the (4, 5) pair was chosen. This code controls the local error according to a scalar relative error tolerance and a vector of threshold values. Because the solution components range from 0 to 1, setting the threshold values to 1 and taking the tolerance to be 10^{-6} results in a pure absolute error control of 10^{-6} as in the computations of Hindmarsh and Byrne. The code was told to integrate from 0 to 4×10^{10}. It returned at $T = 9.4E-2$ with a diagnosis of stiffness. This code attempts to estimate how much it would cost to finish the run if the situation does not improve. It predicts that the cost to finish would be of the order of 3×10^{11} times the cost to reach the current point! A call to the STAT routine revealed that to reach the current point, the code took 50 successful steps and NFCN = 462. The code permits the integration to continue even though it has warned that this will be very inefficient. When UT was told to continue, it returned again at $T = 1.7E-1$ with another diagnosis of stiffness and a prediction that it would cost about 2×10^{11} times as much to complete the integration as it cost to get to this point, namely a total of 90 successful steps at a total cost of NFCN = 845. Continuing the integration leads to more messages of the same kind.

As mentioned, EPISODE can work with internally generated numerical Jacobians. This is so much more convenient that many users do not even consider supplying analytical Jacobians. (They should!) When the problem is solved with this option, the corresponding results are

T	NSTEP	NFCN	NJAC	Y_1	Y_2	Y_3
4.0E − 01	27	95	19	9.8517E − 01	3.3864E − 05	1.4794E − 02
4.0E + 01	77	225	29	7.1582E − 01	9.1849E − 06	2.8417E − 01
4.0E + 03	148	408	45	1.8321E − 01	8.9426E − 07	8.1679E − 01
4.0E + 05	231	633	65	4.9394E − 03	1.9854E − 08	9.9506E − 01
4.0E + 07	327	982	117	3.2146E − 05	1.2859E − 10	9.9997E − 01
4.0E + 09	861	2747	312	−1.8616E + 06	−4.0000E − 06	1.8616E + 06
4.0E + 10	923	2954	330	−1.9142E + 07	−4.0000E − 06	1.9142E + 07

Most of the integration is done just as well as when an analytical Jacobian is provided. However, a negative concentration is introduced earlier in the integration and it is observed that the integration continues long enough for instability to be evident. The two zero eigenvalues of the Jacobian at the equilibrium solution suggest the possibility that the Jacobian might have positive eigenvalues for some arguments that are near to the equilibrium values. MATLAB is a convenient tool for exploring this. When the Jacobian is evaluated at the values obtained for $t = 4.0E + 10$ in the first integration, values that are close to the equilibrium values, it is found that the eigenvalues are about -10^4, $+1.13 \times 10^{-9}$, and -1.59×10^{-19}. There is a positive eigenvalue that seems small, but it must be appreciated that the code took only 11 steps in getting from $t = 4.0E + 9$ to $t = 4.0E + 10$ so the average step size is about 3.3×10^9! Over distances of this size, the effect of the positive eigenvalue is not small and linear stability analysis says that the problem will be unstable with respect to some perturbations. This is an example of the possibility pointed out in Chapter 1 of non-physical approximations leading to disaster.

EXAMPLE 8. A simple problem arising in atmospheric simulation describes ozone kinetics. The concentrations of the oxygen singlet, ozone, and oxygen, are denoted by y_1, y_2, y_3, respectively. Two of the reaction processes are photochemical, hence involve diurnal rate coefficients. In the model the oxygen concentration y_3 is held at the constant value $y_3 = 3.7 \times 10^{16}$. The initial value of the other two are

$$y_1(0) = 10^6, \qquad y_2(0) = 10^{12}.$$

The rate equations are

$$\dot{y}_1 = -q_1 y_1 y_3 - q_2 y_1 y_2 + 2q_3(t) y_3 + q_4(t) y_2,$$
$$\dot{y}_2 = q_1 y_1 y_3 - q_2 y_1 y_2 - q_4(t) y_2.$$

The q_i are rate coefficients. Two are constant:

$$q_1 = 1.63 \times 10^{-16}, \qquad q_2 = 4.66 \times 10^{-16}.$$

Two vary diurnally: for $i = 3, 4$,

$$q_i(t) = \begin{cases} \exp(-c_i/\sin(\omega t)), & \sin(\omega t) > 0, \\ 0 & \sin(\omega t) \leq 0, \end{cases}$$

where

$$c_3 = 22.62, \qquad c_4 = 7.601, \qquad \omega = \pi/43200,$$

and the time t is measured in seconds. Examination of these diurnal rate coefficients shows that they are very nearly square waves, rising sharply at dawn ($t = 0$), remaining near their peak (noontime) values during the day ($\sin(\omega t) > 0$), falling sharply at sunset ($t = 43200$) to a nighttime value of zero, and continuing periodically.

Some care is necessary if sensible results are to be obtained for this problem. Physical reasoning and/or experimentation provide insight. The initial values show that the solution components can be of very different size. The difference can be even more dramatic because during the night, the concentration y_1 is almost 0. It is clear that the two components must be treated differently in the error control, so either a relative control or a scaled vector absolute error control will be required. Hindmarsh and Byrne point out that if a pure relative error control is used throughout the integration, the integrator must work very hard to compute the concentration y_1 even when it is negligible. In our terminology, they use a scalar relative error control with a threshold on each component of 10^{-20}. They also point out that controlling the error relative to the largest value seen so far is not sensible for this problem. This is because the peak value of y_1 is about 10^7 so that their moderately stringent relative error tolerance of 10^{-6} would amount during the night merely to keeping the error in y_1 smaller than $10^{-6} \times 10^7$ in magnitude—y_1 is itself very much smaller than this so there is then no effective control of the error in this component at all. It should be expected that the very sharp changes in $q_3(t)$ and $q_4(t)$ will cause difficulties; remarks about this will be made after some numerical results are presented. Figure 8.5 shows the logarithm of the concentration of the oxygen singlet, $y_1(t)$.

The Jacobian of the system is

$$J = \begin{pmatrix} -q_1y_3 - q_2y_2, & -q_2y_1 + q_4(t) \\ q_1y_3 - q_2y_2, & -q_2y_1 - q_4(t) \end{pmatrix}.$$

When evaluated at the initial data, it is found that the eigenvalues of the Jacobian are about -6 and -10^{-9}. The eigenvalue of -6 does not look large, but it is because time is measured in seconds and the model is to be integrated for a period of days. Put differently, it represents a "time constant"

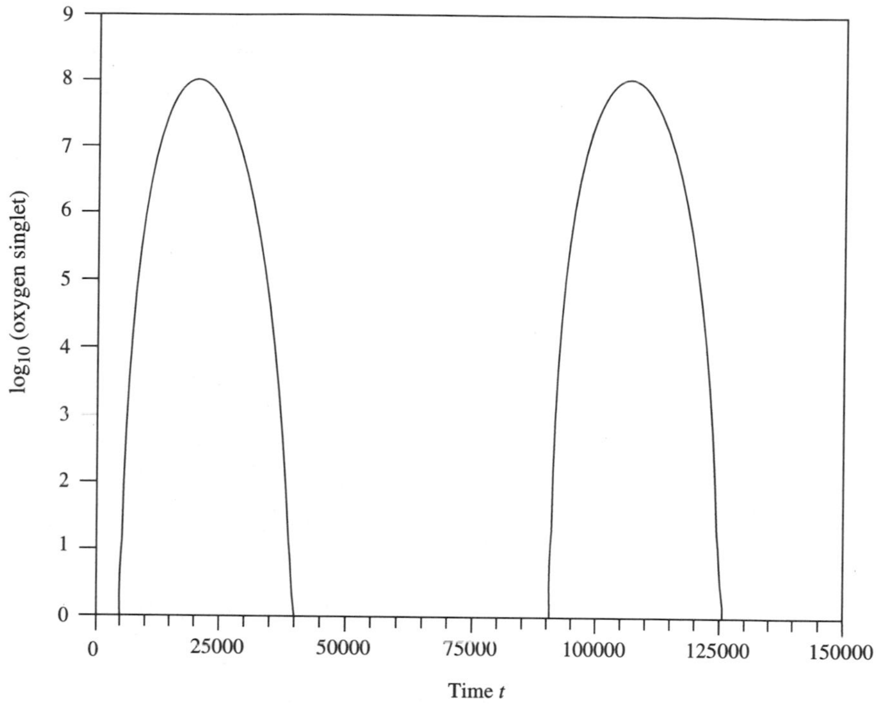

Figure 8.5

of 1/6 sec, which is short compared to the time period of interest. Hindmarsh and Byrne integrated the model for 10.25 days, corresponding to $0 \le t \le 8.856 \times 10^5$. They report results every 2 hours during the first 12 hours, then only at midnight and noon thereafter. The first part of the integration resulted in

T	NSTEP	NFCN	NJAC	Y_1	Y_2
$7.200E + 03$	934	1169	89	$4.1428E + 04$	$1.0000E + 12$
$1.440E + 04$	966	1227	98	$2.5635E + 07$	$1.0002E + 12$
$2.160E + 04$	988	1272	105	$8.7935E + 07$	$1.0386E + 12$
$2.880E + 04$	1008	1315	110	$2.7607E + 07$	$1.0772E + 12$
$3.600E + 04$	1044	1389	118	$4.4663E + 04$	$1.0774E + 12$
$4.320E + 04$	1455	2077	155	$3.0927E - 33$	$1.0774E + 12$
$6.480E + 04$	1462	2084	162	$-1.2178E - 57$	$1.0774E + 12$
$1.080E + 05$	1963	2837	230	$9.4344E + 07$	$1.1160E + 12$
$1.512E + 05$	2491	3698	315	$3.7062E - 66$	$1.1547E + 12$
$1.944E + 05$	2981	4432	377	$1.0074E + 08$	$1.1932E + 12$
$2.376E + 05$	3501	5340	437	$7.2072E - 29$	$1.2318E + 12$

Notice that one of the approximate concentrations is negative. Although a negative value is not physical, the magnitude of the concentration is far smaller than the threshold, so the code has produced all the accuracy that was asked for in this solution component. The components of the solution are about $y_1 = 0$, $y_2 = 10^{12}$, and $y_3 = 3.7 \times 10^{16}$. This is during the night so that the diurnal rate coefficients are 0. With these values of the parameters, the eigenvalues of the Jacobian are 0 and $-q_1y_3 - q_2y_2 \approx -6.031$. Just as in the preceding example, there is a possibility that for nearby values, there is a positive eigenvalue and an unstable problem. Let us look at this matter more closely. The characteristic polynomial is

$$\lambda^2 + \lambda(q_2y_1 + q_1y_3 + q_2y_2) + 2q_1q_2y_1y_3 = 0.$$

The quantity $q_1y_3 \approx 6.031$. From the computations we are led to consider values of y_2 of the general size of 10^{12} and values of y_1 with $|y_1| << 1$. In view of the sizes of q_1 and q_2, we consider values of the solution such that q_2y_2 is of the general size of 10^{-4} and q_2y_1 is extremely small. In these circumstances it is easy to see that one eigenvalue is about $-q_1y_3 \approx -6.031$. The product of the two eigenvalues is the constant term in the quadratic equation, so the other root is about $-2q_2y_1$. This second root is extremely small in magnitude when y_1 is small in magnitude. If y_1 is positive, as we might expect, the problem is stable according to a linear stability analysis. If it is negative but small, the Jacobian has a positive eigenvalue, but it is so small that on the interval of interest the integration is stable. This helps us understand the observed stability with respect to computing values of y_1 that are negative but small in magnitude.

Monitoring of the step size shows that it varies enormously. This might have been expected because of the very sharp changes in $q_3(t)$ and $q_4(t)$, but it is accentuated by the behavior of the solution. During the night the solution components are effectively constant. The code recognizes that they are easy to approximate and increases the step size very rapidly. At sunrise the solution components are suddenly very much more difficult to approximate and the code must, in effect, locate the point where the coefficients change so sharply. Computing without paying due attention to the physical problem can lead to nonsense in the following way: It can happen that a code will increase the step size so rapidly during the night that it steps over the daylight hours completely. It predicts, and then verifies, that the solution components are nearly constant, so it believes the step to be a success. A code like this one obtains results at intermediate points by interpolation, so if an answer were required at, say, noon, the code would compute it to be nearly the same constant value as during the night. This is not a hypothetical possibility; it actually happens with certain specifications of the problem. The difficulty here is one of informing the code about the scale of the problem. The scientist knows that there will be important changes in the concentrations between day and night and in the presence of what amounts to discontinuities in the coefficients, it may be necessary to inform the code about this. Codes that obtain answers by interpolation have limits on how far past an output point they can step internally. By requiring answers frequently enough to reveal the behavior of the solution, they can be forced to recognize the scaling with respect to the independent variable. This is vague because the internal limits are not readily available to the user. A more certain way to proceed is to take advantage of the fact that quality codes with an interpolation capability provide an option to prevent stepping past output points. Some codes allow a user to convey a measure of scale in the independent variable by specifying a maximum step size.

EXERCISES For some of the exercises it is useful to be able to create simple problems that are stiff. One way to do this follows. We understand the solutions of systems of the form

$$w_1' = \lambda_1(w_1 - p_1(x)) + p_1'(x),$$
$$w_2' = \lambda_2(w_2 - p_2(x)) + p_2'(x),$$

in which we choose the eigenvalues λ_i and the functions $p_i(x)$. To couple the equations, we need only introduce a change of variable $\mathbf{y} = M\mathbf{w}$ for a constant matrix M that we can invert analytically. As an example, one might use

$$M = \frac{1}{\sqrt{2}}\begin{pmatrix} 1 & 1 \\ 1 & -1 \end{pmatrix}$$

for which $M^{-1} = M$. This leads to a coupled system of the form

$$\mathbf{y}' = J\,\mathbf{y} + \mathbf{G}(x).$$

EXERCISE 9. A concept often seen in discussions of stiffness is the *stiffness ratio*,

$$\frac{\max_i \operatorname{Re}(-\lambda_i)}{\min_j \operatorname{Re}(-\lambda_j)},$$

where only eigenvalues with negative real parts are considered. This is a plausible measure of the stiffness of a problem, but it must be used rather carefully as the following examples show:

(1) Suppose that the interval of integration is [0, 1] and that $\lambda_1 = -10^4 - 1$, $\lambda_2 = -10^4$. The stiffness ratio is

$$\frac{10^4 + 1}{10^4} = 1 + 10^{-4} \approx 1,$$

which says that the problem is not stiff. However, our analysis looks at the product of the length of the interval and the largest magnitude of eigenvalues with negative real part. Here this is $1 \times (10^4 + 1)$ which says the problem is stiff. Using the scheme suggested at the beginning of these exercises, select slowly varying $p_i(x)$ and try the resulting problem with a code appropriate for stiff problems and one appropriate for non-stiff problems. The practical definition of stiffness is that the code appropriate for stiff problems is much more efficient than the code appropriate for non-stiff problems. Is this problem stiff by the practical definition?
(2) Suppose that the interval of integration is [0, 1] and that $\lambda_1 = -1$, $\lambda_2 = -10^{-4}$. The stiffness ratio is

$$\frac{1}{10^{-4}} = 10^4,$$

which says that the problem is stiff. Comparing the magnitude of the eigenvalue with largest negative real part to the length of the interval of integration leads to 1×1, which says that the problem is not stiff. As in part (1), explore this matter experimentally to decide for yourself.

Rather than provide a third part to this exercise, we quote a statement by Dahlquist et al. [1982] about another issue: "The "*stiffness ratios*" used by some authors . . . may be of some use when one estimates the amount of work needed, if a stiff problem is to be solved by an *explicit* method, but they are fairly *irrelevant in connection with implicit methods*. . . ."

EXERCISE 10. In Byrne and Hindmarsh [1975] a mockup of the photochemical reaction of Example 8 is presented and numerical results computed with GEAR and EPISODE are given. The point of the mockup is to have an extremely simple problem that has a behavior characteristic of more realistic models. The mockup is

$$\dot{y}(t) = \dot{H}(t) - B[y(t) - H(t)], \qquad 0 \le t \le 432{,}000,$$
$$y(0) = H(0),$$

where

$$H(t) = [D + A\,E(t)]/B, \qquad A = 10^{-18}, \qquad B = 10^8, \qquad D = 10^{-19},$$

and

$$E(t) = \begin{cases} \exp(-c\omega/\sin(\omega t)), & \sin(\omega t) > 0, \\ 0 & \sin(\omega t) \leq 0, \end{cases}$$

where

$$c = 4 \quad \text{and} \quad \omega = \pi/43{,}200.$$

By construction, the solution of this initial value problem is $H(t)$. It is nearly a square wave of period 86,400 seconds (24 hours) that starts at 0 at $t = 0$, abruptly attains a large value that it holds for 12 hours, and then abruptly drops to its minimum value that it holds for 12 hours. The cycle is then repeated. The first 12 hours correspond to the 12 daylight hours, and the second 12 to nighttime. It is appropriate to report answers every 43,200 seconds. For a scalar problem like this one, it is easy to obtain the eigenvalues of the Jacobian. According to a linear stability analysis, would you expect this problem to be stiff where the solution is slowly varying? Try solving it with a code intended for non-stiff problems. (Put a short time limit on this run or use your code in such a way that you do not run up a large computing bill.) Try solving it with a code intended for stiff problems. Consult Example 8 for advice on how to proceed. This is a hard problem because of the very sharp changes in the solution at sunrise and sunset and your code might well fail there unless you assist it by providing information about where the difficulties occur.

Problems

The problems provided here amplify the text, so even reading them is worthwhile. They should be solved as the reader progresses through the book. As more is learned about the design of differential equation solvers and about the numerical methods implemented in them, more can be learned from the solution of these problems. There is a considerable amount of art to the successful solution of the initial value problem and some experience with non-trivial problems is essential.

PROBLEM 1. One way that ordinary differential equations arise from partial differential equations is through symmetry. H. T. Davis [1962, p. 371 ff.] describes how Emden's equation

$$y'' + \frac{2}{x} y' + y^n = 0$$

arises in modeling the thermal behavior of a spherical cloud of gas acting under the gravitational attraction of its molecules and the classical laws of thermodynamics. The basic partial differential equation is a Poisson equation for the gravitational potential ϕ of the gas, $\nabla^2\phi = -a^2\phi^n$. Here a and n are physical constants. (The constant n is not necessarily an integer.) It is assumed that $\phi = 0$ on the surface of the sphere of radius R. Because of physical symmetry the partial differential equation reduces to an ordinary differential equation with the distance from the center of the sphere, r, as the independent variable. This is a boundary value problem for ϕ in which one boundary condition is $\phi(R) = 0$. The physical requirement that ϕ be well behaved at the center of the sphere and symmetry lead to the boundary condition $\phi'(0) = 0$. Emden's equation arises on introducing a new dependent variable y by $\phi = \phi_0 y$, where $\phi_0 = \phi(0)$, and a new independent variable x by

$$r = x/(a\phi_0^{(n-1)/2}).$$

In the new variables we have $y'(0) = 0$ and $y(0) = 1$. The potential ϕ is to decrease from its value ϕ_0 at $r = 0$ to the value 0 at $r = R$. This is equivalent to y decreasing from the value 1 at $x = 0$ to the value 0 at $x_0 = Ra\phi_0^{(n-1)/2}$. By integrating the initial value problem for $y(x)$ to the first point x_0 where $y(x_0) = 0$, we find a radius R for which we have a solution of the original partial differential equation.

Davis describes how solutions of Emden's equation are used by astrophysicists to estimate the density and internal temperature of stars. All we do here is consider the numerical solution of the equation itself. Interesting values of n range from 0 to 5. You are to find x_0, the first point where the solution $y(x)$ of Emden's equation vanishes. The analytical solution for $n = 1$,

$$y(x) - \frac{\sin x}{x},$$

and that for $n = 5$,

$$y(x) = (1 + x^2/3)^{-1/2},$$

will help you check out your programs. Figure P.1 displays these two solutions. For $n = 1$, it is obvious that the first root x_0 of $y(x)$ is π. For $n = 1.5$, Davis reports that x_0 is about 3.65375, and for $n = 3$, that x_0 is about 6.89685. For $n = 5$, $y(x)$ does not have a finite root, but $y(x) \to 0$ as $x \to \infty$, so we might say that $x_0 = \infty$ in this case.

When written as a system by introducing the usual variables $y_1 = y$ and $y_2 = y'$, the equation becomes

$$\begin{aligned} y_1' &= y_2 && = F_1(x, y_1, y_2), && y_1(0) = 1, \\ y_2' &= -\frac{2}{\mathrm{X}} y_2 - y_1^n && = F_2(x, y_1,, y_2), && y_2(0) = 0. \end{aligned}$$

If this is programmed in a mechanical fashion and supplied to an integrator, the code will try to evaluate F_2 at the initial point $x = 0$. This will result in an indeterminate value of 0/0. The proper value is specified by a limiting argument using l'Hospital's rule,

$$\lim_{x \to 0} \frac{y'(x)}{x} = \lim_{x \to 0} \frac{y''(x)}{1} = y''(0).$$

The value of $y''(0)$ is obtained by letting x tend to 0 in the differential equation to get

$$0 = y''(0) + 2\, y''(0) + 1^n,$$

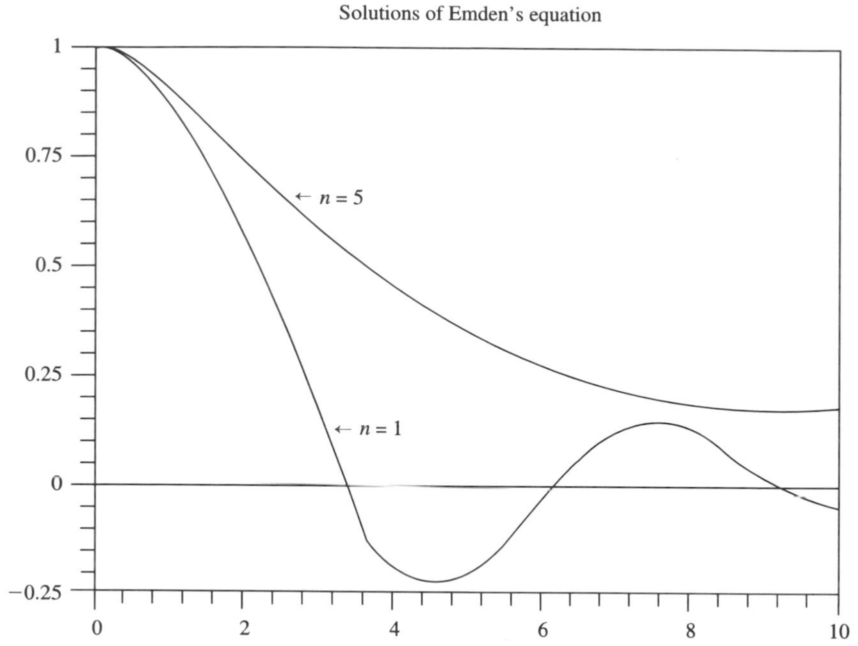

Figure P.1

hence $y''(0) = -1/3$. It is true for this problem that if one programs the subroutine for **F** so that the proper value of F_2 is returned at the initial point, most quality codes will integrate the equation accurately and without difficulty. The general theory does not justify this because it assumes that **F** satisfies a Lipschitz condition and this **F** does not satisfy such a condition on any interval including $x = 0$. It is hardly unusual to have a gap between what we can justify and what apparently "works." Although the situation is murky, de Hoog and Weiss [1977, 1985] have shown that some standard methods do work for certain kinds of singular problems, including this one.

For this particular problem the techniques of Chapter 1 allow us to start the integration in a manner that can be justified. The physical symmetry that gives rise to $y(x)$ assures us that it is an even function of x, hence that its Taylor series expansion about the origin has the form

$$y(x) = y(0) + \frac{y^{(2)}(0)}{2!} x^2 + \frac{y^{(4)}(0)}{4!} x^4 + \ldots .$$

The initial value $y(0)$ is given and we have already found $y^{(2)}(0)$. This is enough to get started, but more terms in the series will make the task easier. Davis reports that

$$y(x) = 1 - \frac{x^2}{3!} + n\frac{x^4}{5!} + (5n - 8n^2)\frac{x^6}{3 \cdot 7!} + (70n - 183n^2 + 122n^3)\frac{x^8}{9 \cdot 9!} + \cdots .$$

You are to derive the first three terms of this Taylor series. To see how it goes, suppose that we want to find α in

$$y(x) = 1 + \alpha x^2 + \mathcal{O}(x^4).$$

Using the binomial series as in Chapter 1,

$$y^n(x) = (1 + \alpha x^2 + \mathcal{O}(x^4))^n = 1 + n\alpha x^2 + \mathcal{O}(x^4).$$

Substituting this expression and corresponding ones for the derivatives of y into the differential equation leads to

$$0 = y'' + \frac{2}{x}y' + y^n = [2\alpha + \mathcal{O}(x^2)] + [4\alpha + \mathcal{O}(x^2)] + [1 + \mathcal{O}(x^2)]$$
$$= [6\alpha + 1] + \mathcal{O}(x^2).$$

For this to be true, it is necessary that $\alpha = -1/6$.

For exploratory computations you may use the simple approach of evaluating $2y'/x$ properly at $x = 0$, but for the computation of x_0 you are to use the Taylor series to start the integration. The series and its derivative are used to approximate $y(\delta)$ and $y'(\delta)$ for some "small" $\delta > 0$. Estimate the errors in the two approximations by the sizes of the first terms omitted in the series. You will have to use some judgment in selecting δ and the number of terms in the series. The accuracy you seek at $x = \delta$ should be consistent with the accuracy you ask of the code for the integration of the non-singular problem for $x \geq \delta$. However you start the integration, integrate Emden's equation for $n = 1.5$ and 3 to locate x_0 approximately. Use the analytical solution for $n = 1$ to check out your program. You are to locate roots only very roughly. Depending on the integrator you select, you might want to print out $y(x)$ at each step or print out values on a set of equally spaced x. The analytical solution for $n = 5$ makes clear the value of doing this kind of exploratory computation. As exemplified by the $n = 1$ case, there can be more than one root. Locate approximately the second root of $y(x)$ when $n = 3$.

There is a difficulty in the computation of $y(x)^n$. The parameter n need not be an integer, so it seems natural to take it to be a REAL or DOUBLE PRECISION variable. However, if you do this, the exponentiation will fail when $y(x) < 0$. It is better practice to take n to be an INTEGER variable when it is an integer and if you do this for $n = 1$ and 3, you will have no difficulty with the integration. The case $n = 1.5$ brings to light the fundamental difficulty that **F** may not be defined for negative values of y. It might appear that in calculating the first root we would not have to concern ourselves with this matter. However, we must integrate to x_0 where $y(x_0) = 0$, a point on the boundary of the region where **F** is defined when $n = 1.5$. A ploy that can be used in such situations is to define an artificial value for **F** for arguments outside its region of definition. For the task at hand we need a value for F_2 when y_1 is negative. We might, for example, take

$$F_2 = -\frac{2}{x} y_2 - (\max(y_1, 0))^n.$$

This is easy to do and it usually works well enough, but it does place a strain on the integrator. The difficulty is that this new **F** is not smooth where y_1 changes sign. Because **F** is not smooth, stepping past the argument x_0 where y_1 changes sign is difficult, so difficult that the typical code will need a small step size to accomplish it and there may be a number of step failures as the code repeatedly attempts to step past with step sizes that are too large. In effect, the typical code will locate the point x_0 and step to it. Note that when this ploy is used, integration past x_0 is meaningless.

You are to find x_0 when $n = 1.5$ and 3 by interchanging independent and dependent variables as described in Example 5 of Chapter 1. Check out your program with $n = 1$. If you move far enough away from the origin using the Taylor series, say to $\delta = 0.1$, you can interchange variables at this point and obtain x_0 by a single call to an interval-oriented code. A more general approach that is necessary if you want to locate subsequent roots of $y(x)$ is to monitor the solution at each step by using an interval-oriented code with intermediate output or a step-oriented code. When a step from (x_j, y_j, y_j') results in $y_{j+1} \leq 0$, a change of variables is made at x_j and an integration in the new independent variable y is done from y_j to 0 to find the corresponding value of x that constitutes a root of $y(x)$. After finding the root, the integration of Emden's equation resumes until another sign change of $y(x)$ is noted. Use whichever procedure you prefer to find the first root. If you are keen, calculate the second root in the case $n = 3$. Davis reports this root to be about 35.96194.

PROBLEM 2. Lighthill's technique of straining coordinates to obtain perturbation solutions is presented in §3.2 of Chapter 3 in Nayfeh [1973]. A specific example treated by Lighthill is

$$(x + \varepsilon y)\frac{dy}{dx} + (2 + x)y = 0, \qquad y(1) = e^{-1}.$$

Here $0 < \varepsilon << 1$ and the integration is to go from $x = 1$ to $x = 0$. The problem is regular on this interval, but it is singular just to the left of $x = 0$ and this causes $y(0)$ to be "large." Nayfeh develops the expansion

$$y(0; \varepsilon) = \left(\frac{3}{\varepsilon}\right)^{2/3} - \frac{3}{10}\left(\frac{3}{\varepsilon}\right)^{1/3} + \mathcal{O}(1).$$

The technique can lead to *erroneous* expansions for problems that resemble this one. Confirm the validity of the expansion for this problem by computing $y(0; \varepsilon)$ numerically for several "small" ε and comparing the values to those given by the expansion.

PROBLEM 3. The one-dimensional earth–moon–spaceship problem can be reduced to the initial value problem

$$\frac{1}{2}\left(\frac{dx}{dt}\right)^2 = \frac{1-\mu}{x} + \frac{\mu}{1-x}, \qquad x(0) = 0.$$

Here $x(t)$ is the distance of the spaceship from the earth at time t. The constant μ is the ratio of the mass of the moon to the sum of the masses of the moon and earth. It is about 1/82.45. Clearly the derivative of x with respcct to t is infinite at $t = 0$, so the equation is singular there. The equation is also singular when the spaceship reaches the moon ($x = 1$ in the units used). For the solution of this problem it is convenient to compute t as a function of x. Find a differential equation for $t(x)$ and put it into standard form. The derivative of t with respect to x is zero at $x = 0$, but you will need to do some analytical work to start the integration. Find values α and p in $t(x) \approx \alpha x^p$ that will give you an approximation to the solution valid for small x. Use this approximation to start the numerical integration at some small x. Nayfeh [1973] illustrates a number of perturbation methods by applying them to the approximation of $t(x)$ for small μ. The lowest order approximation is

$$t(x) = \frac{\sqrt{2}}{3}x^{3/2} + \mathcal{O}(\mu).$$

How does this compare with the approximation you use to start the integration? Compare this perturbation result to your numerical results.

PROBLEM 4. Erdélyi [1956] uses the solution of a singular initial value problem studied by Euler to illustrate asymptotic expansions. The initial value problem is

$$x^2\phi'(x) + \phi(x) = x, \qquad \phi(0) = 0.$$

Attempting a series solution about $x = 0$ leads to

$$\phi(x) \sim x - x^2 + 2x^3 - 6x^4 + \ldots = \sum_{n=0}^{\infty} (-1)^n n! x^{n+1}.$$

This series does not converge for *any* $x \neq 0$; it is an asymptotic expansion. For a given $x_0 > 0$, successive terms in the series decrease at first and then increase. It is usual with asymptotic series to assume that the partial sums represent the function to an accuracy of about the size of the first term omitted. In this particular case, it is not hard to prove that this assumption is true by using the integral representation for the solution

$$\phi(x) = \int_0^{\infty} \frac{xe^{-t}}{1 + xt} \, dt.$$

Specifically, for real x the partial sum

$$\phi_m(x) = \sum_{n=0}^{m} (-1)^n n! x^{n+1}$$

satisfies

$$|\phi(x) - \phi_m(x)| \leq (m + 1)! x^{m+2}.$$

According to this bound, what is the maximum accuracy that you can obtain using the asymptotic expansion at $x = 10^{-1}$?

The asymptotic expansion provides an accurate numerical approximation to $\phi(x)$ for all sufficiently small x, i.e., near the singular point of the differential equation. To approximate, say, $\phi(1)$ we use the expansion to approximate $\phi(\gamma)$ for some $\gamma > 0$ and then integrate from γ to 1. This is not entirely straightforward because of the stability properties of the equation. Writing the equation in standard form, we have

$$\phi' = \frac{x - \phi}{x^2} = f(x, \phi).$$

To investigate the stability of the problem, let ψ be another solution of the differential equation and let $\Delta(x) = \psi(x) - \phi(x)$. Show that for $x \geq \gamma$,

$$\Delta(x) = \Delta(\gamma) \exp(x^{-1} - \gamma^{-1}).$$

When γ is "small," it is seen that $|\Delta(x)|$ decreases very rapidly so that all solution curves approach one another rapidly—the initial value problem is very stable. The asymptotic solution suggests that $\phi(x)$ is a slowly varying function of x and it is clear from the differential equation that it is smooth. All these things put together suggest that the initial value problem is stiff for small x. In Chapter 8 this concept is studied. For a scalar problem like this one the product of the magnitude of the local Jacobian, $\partial f/\partial\phi$, and the length of the interval of integration is an important measure of how stiff the problem is. Here if $\gamma = 10^{-1}$, this product is 100 and if $\gamma = 10^{-3}$, it is 10^6. Of course, as the integration progresses from $x = \gamma$ to $x = 1$, the magnitude of the local Jacobian decreases rapidly and the problem becomes much less stiff. As we see in Chapter 8, a code intended for non-stiff problems is satisfactory for $\gamma = 10^{-1}$, but a code intended for stiff problems is considerably more efficient for $\gamma = 10^{-3}$. You have worked out how much accuracy you can achieve in $\phi(\gamma)$ using the asymptotic series. Use a standard integrator to obtain $\phi(1)$ by starting with the approximation obtained from the series and requiring the integrator to control the local error to the same accuracy during the integration.

PROBLEM 5. The description of some physical problems by differential equations involves algebraic equations, too. A great variety of forms is seen and here only the semi-explicit form

$$\begin{aligned} \mathbf{y}' &= \mathbf{f}(\mathbf{y}, \mathbf{z}), \qquad \mathbf{y}(a) \text{ given}, \\ \mathbf{0} &= \mathbf{g}(\mathbf{y}, \mathbf{z}), \end{aligned}$$

is considered. Such problems have a strong resemblance to a differential equation with conservation laws, but these *differential-algebraic equations*, *DAEs*, are, in fact, quite different in nature. A conservation law is a consequence of the differential equation and the fact that the solution satisfies the algebraic equation is automatic. In the case of a DAE, the algebraic equation helps determine the solution.

Semi-explicit DAEs of index one can be treated as ordinary differential equations. These are problems for which the "algebraic" variables $\mathbf{z}(x)$ can be determined for given $\mathbf{y}(x)$ by solving $\mathbf{0} = \mathbf{g}(\mathbf{y}, \mathbf{z})$. With suitable smoothness assumptions and an initial $\mathbf{z}(a)$ such that $\mathbf{0} = \mathbf{g}(\mathbf{y}(a), \mathbf{z}(a))$, the existence of an inverse of the Jacobian $\mathbf{g_z}$ that is bounded along the solution guarantees (by an implicit function theorem) that $\mathbf{z}$ can be written as a function of $\mathbf{y}$, namely $\mathbf{z}(x) = \mathbf{R}(\mathbf{y}(x))$, and then $\mathbf{y}' = \mathbf{f}(\mathbf{y}, \mathbf{R}(\mathbf{y}))$, $\mathbf{y}(a)$ given, is an initial value problem for $\mathbf{y}(x)$. Thus, one way to solve index one problems is to apply your favorite integrator to $\mathbf{y}' = \mathbf{f}(\mathbf{y}, \mathbf{z})$, $\mathbf{y}(a)$ given, and every time that it needs to evaluate $\mathbf{f}(\mathbf{y}, \mathbf{z})$ for a specific $\mathbf{y}$, solve $\mathbf{0} = \mathbf{g}(\mathbf{y}, \mathbf{z})$ for the corresponding $\mathbf{z}$ and then substitute $\mathbf{y}$ and $\mathbf{z}$ in $\mathbf{f}$. This is easy enough in principle, but obviously such problems can lead to initial value problems for which the evaluation of $\mathbf{f}$ is rather expensive. DAEs that are not semi-explicit or not of index one cannot be solved so simply and "the facts of life" are quite different. A study of numerical solution of ODEs does, however, furnish a starting point for the investigation of DAEs. The monographs of Brenan, Campbell, and Petzold [1989] and Hairer, Lubich, and Roche [1989] can be consulted for more information about DAEs.

A variational description of the motion of a pendulum with a point mass of m attached to a rigid, weightless rod of length L leads to equations

$$\begin{aligned} m\ddot{x} &= -\lambda x, \\ m\ddot{y} &= -\lambda y - g, \\ 0 &= x^2 + y^2 - L^2. \end{aligned}$$

Here (x, y) are the coordinates of the mass, g is the acceleration due to gravity, and λ is the tension in the rod. This problem is not in the desired form, but differentiating twice the algebraic constraint that the rod has length L and using the differential equations leads to

$$0 = m(\dot{x}^2 + \dot{y}^2) - gy - \lambda L^2.$$

Show this. If we take as variables $y_1 = x$, $y_2 = \dot{x}$, $y_3 = y$, $y_4 = \dot{y}$, $z = L$, the problem is

$$\frac{d\mathbf{y}}{dx} = \frac{d}{dx}\begin{pmatrix} y_1 \\ y_2 \\ y_3 \\ y_4 \end{pmatrix} = \begin{pmatrix} y_2 \\ -\dfrac{z}{m} y_1 \\ y_3 \\ -\dfrac{z}{m} y_3 - \dfrac{g}{m} \end{pmatrix} = \mathbf{f}(\mathbf{y}, z),$$

$$0 = m(y_2^2 + y_4^2) - gy_3 - \lambda z^2 = g(\mathbf{y}, z).$$

In this particular instance the algebraic equation for z is easy to solve in terms of $\mathbf{y}$ and the problem can be integrated with standard codes. To make matters simple, take $m = g = L = 1$ and solve the problem for $x(0) = \dot{x}(0) = 0$, $y(0) = L$, $\dot{y}(0) = 0$. Does your numerical solution satisfy the constraint that $x^2(t) + y^2(t) = L^2$?

PROBLEM 6. Floquet theory is concerned with periodic solutions of linear equations with periodic coefficients,

$$\mathbf{y}' = P(x)\mathbf{y},$$

where the matrix P is continuous and periodic with minimal period T, i.e., $P(x + T) = P(x)$ for all x, and T is the smallest positive number for which this is true. The text of Jordan and Smith [1987] develops this theory in an elementary way in §9.2–9.4. They prove that there are non-trivial solutions $\mathbf{y}(x)$ such that for a constant μ, $\mathbf{y}(x + T) = \mu\mathbf{y}(x)$ for all x. Let

$\Phi(x)$ be a fundamental matrix for this system of equations and let E be a matrix such that $\Phi(x + T) = \Phi(x)E$. The constants μ are eigenvalues of the matrix E. If the magnitude of μ is greater than 1, the non-trivial solution grows and the differential equation is unstable. If $\mu = 1$, the solution is periodic. If the magnitude of μ is less than 1, the solution decays.

To investigate the stability of solutions of such equations, you can choose $\Phi(0) = I$ and integrate $\Phi(x)$ to $x = T$ to get $E = \Phi(T)$. The eigenvalues of E are then computed and if any is greater in magnitude than 1, there is a solution that grows without bound. This is an application of ODE solvers for which it is difficult to see how much accuracy is needed because the results of the integrations at T are used in a complicated way to get the quantity that is of real interest.

An equation that has been important in applications is Mathieu's equation,

$$y'' + (\alpha + \beta \cos x)y = 0,$$

where α and β are parameters. When written as a system

$$\frac{d}{dx}\begin{pmatrix} y \\ y' \end{pmatrix} = \begin{pmatrix} 0 & 1 \\ -\alpha - \beta \cos x & 0 \end{pmatrix}\begin{pmatrix} y \\ y' \end{pmatrix},$$

it is seen that $P(x)$ is periodic of minimal period 2π. This equation arises in the study of stability of physical systems. For example, Jordan and Smith cite A. B. Tayler's treatment of the motion of the surface of an inviscid liquid in a cylindrical tank undergoing vertical oscillations. The mathematical question for Mathieu's equation is, for which values of the parameters are solutions unstable, and for which are there periodic solutions? In the case of the example treated by Tayler, this corresponds to instability of the surface of the liquid, and he suggests that a half empty glass of beer is a suitable apparatus for experimental investigation of the phenomenon! The stability of solutions of Mathieu's equation can be analyzed by perturbation methods and explored more comprehensively by numerical means. It is known, for example, that for α near 1 and "small" β, there is a solution that is periodic of period 2π when

$$\alpha \approx 1 + \frac{1}{12}\beta^2 \quad \text{and when} \quad \alpha \approx 1 - \frac{5}{12}\beta^2.$$

For a given small β, there is at least one growing solution (the equation is unstable) if α is between these values yielding a periodic solution, i.e., α is in $(1 - 5/12\ \beta^2, 1 + 1/12\ \beta^2)$. (All solutions are bounded for α just outside this interval.) Choose a small value of β and investigate the situation numerically for an α that the perturbation analysis says will provide a stable solution and for an α that will provide an unstable solution by computing the matrix E and calculating its eigenvalues. Note that in the case of Mathieu's equation, E is only 2×2 so that the characteristic equation is a quadratic polynomial and it is easy to compute the eigenvalues. In one of your computations you should find that both eigenvalues have magnitude no greater than 1 and that one of the eigenvalues is approximately equal to 1. In the other computation you should find that one eigenvalue has magnitude greater than 1.

PROBLEM 7. Some nonlinear *Volterra integral equations* are equivalent to an initial value problem for a system of ODEs. Because effective ODE codes are readily available, some authors have sought to exploit the connection for the numerical solution of the integral equations. The survey by C.T.H. Baker [1981] describes a number of possibilities. This problem involves an equivalence vigorously pursued by Bownds in J.M. Bownds and L. Applebaum [1985] and the references cited therein. The equivalence is: If

$$u(x) = f(x) + \int_\alpha^x \sum_{i=1}^N a_i(x) g_i(t, u(t))\, dt, \qquad \alpha \le x \le \beta,$$

then with suitable smoothness assumptions about $f(x)$, the $a_i(x)$, and the $g_i(t, u)$, the solution

$$u(x) = f(x) + \sum_{i=1}^N a_i(x) y_i(x),$$

where for $i = 1, \ldots, N$,

$$\frac{d}{dx} y_i(x) = g_i\left(x, f(x) + \sum_{j=1}^N a_j(x) y_j(x)\right), \qquad y_i(\alpha) = 0.$$

Prove that the solution to the system of ODEs does provide a solution of the nonlinear Volterra integral equation in the form stated. Bownds and others solve more general integral equations by approximating their kernels by ones of this special form. This can lead to systems of ODEs with rather large N.

Solve the integral equation

$$u(x) = 1 - \int_0^x 10 e^{-(x-t)} u(t)\, dt, \qquad 0 \le x \le 10.$$

Verify the analytical solution

$$u(x) = \frac{1}{11} + \frac{10}{11} e^{-11x},$$

and use it to check your numerical results.

Initial value problems arising in this way can be stiff. An example is

$$u(x) = x^2 + \frac{x^9}{7} - \int_0^x x^2 u^3(t)\, dt, \qquad 0 \le x \le 10.$$

Verify the analytical solution $u(x) = x^2$ and use it to check your numerical solution of this equation. If you try to solve the initial value problem with a code not intended for stiff problems, you will find the solution to be expensive. To proceed further with understanding stiffness in this case, you will need to have read Chapter 8.

Introduce the column vectors of N components $\mathbf{a} = (a_1(x), \ldots, a_N(x))^T$, $\mathbf{g}(x, u) = (g_1(x, u), \ldots, g_N(x, u))^T$, and $\mathbf{g}_u = (\partial g_1/\partial u, \ldots, \partial g_N/\partial u)^T$. Show that the Jacobian of the differential equations is

$$J = \left(\frac{\partial g_i}{\partial u} a_j \right) = \mathbf{g}_u \mathbf{a}^T.$$

It is shown in Shampine [1988] that the Jacobian has at most one non-zero eigenvalue μ. If there is a non-zero eigenvalue at all, then necessarily $\mu = \mathbf{a}^T \mathbf{g}_u$, and $\mathbf{g}_u$ is the corresponding eigenvector. Prove that if $\mathbf{v}$ is any vector, then $J\,\mathbf{v} = (\mathbf{a}^T \mathbf{v})\mathbf{g}_u$. Use this to prove that if $\mathbf{a}^T \mathbf{g}_u \neq 0$, then $\mathbf{g}_u$ is an eigenvector with eigenvalue $\mu = \mathbf{a}^T \mathbf{g}_u$. For the equation that was described as stiff, show that $\mu = -3x^6$ and argue that the equation becomes increasingly stiff as the integration progresses.

A key computation when solving stiff problems is to solve linear systems of the form

$$(I - h\gamma\, J)\mathbf{z} = \mathbf{b}.$$

It is pointed out in Chapter 8 that taking account of the structure of the approximate Jacobian can greatly reduce the cost when the number of equations N is large. There is a very cheap way to solve these systems in the present circumstances. The algorithm starts by forming $\mathbf{a}^T \mathbf{g}_u$ and then verifying that $h\gamma\, \mathbf{a}^T \mathbf{g}_u \neq 1$. (This amounts to verifying that the linear system is not singular.) The step size h is to be reduced if this condition does not hold. For each vector $\mathbf{b}$, we can obtain $\mathbf{z}$ by forming $a^T \mathbf{b}$ and then

$$\mathbf{z} = \mathbf{b} + h\gamma\tau\, \mathbf{g}_u,$$

where

$$\tau = \frac{\mathbf{a}^T \mathbf{b}}{1 - h\gamma\, \mathbf{a}^T \mathbf{g}_u}.$$

Prove this last statement. Not only is this a very cheap "factorization" of the iteration matrix and solution of a linear system, but the storage required is a small multiple of N, which is just as good as in codes intended for non-stiff problems—there is no need ever to form J as a matrix. Chapter 8 discusses the formation of approximate Jacobians by differences. Here the structure is such that this can be done very cheaply; explain why.

For a single integration of a system of moderate size, it is not worth the trouble to take advantage of the remarkable structure of the Jacobian. On the other hand, if one wanted to write a quality subroutine of some generality, it would be essential to recognize that these are very special initial value problems.

PROBLEM 8. There is a class of problems called *delay-differential equations* that are closely related to the kinds of problems studied in this book. They arise when the effect of an action on the evolution of a system is not felt immediately, rather after a delay. A simple example often used to illustrate the distinctions is

$$y'(t) = y(t - 1), \qquad t \geq 0,$$
$$y(t) = 1, \qquad t \leq 0.$$

Here there is a differential equation that tells us how y changes for $t \geq 0$, but it depends on how the solution behaves one time unit in the past. In particular, it is necessary to specify not only a starting value $y(0)$, but a starting history. In this instance, specifying $y(t)$ for $-1 \leq t \leq 0$ would suffice. This is an example of a delay-differential equation with a constant delay of 1. Useful theoretical results can be obtained by applying the theory of initial value problems. In some cases the codes that are readily available for the solution of initial value problems can be used to solve delay-differential equations. The basic idea of the *method of steps* is exemplified here by solving the equation successively on the intervals [0, 1], [1, 2], [2, 3], ... If $y(0)$ is defined by continuity to be equal to 1, then the given history for $t \leq 0$ tells us that for $0 \leq t \leq 1$, $y(t)$ is the solution of $y'(t) = 1$, $y(0) = 1$, namely $y(t) = t + 1$. A glimmer of the effects of a delay is seen when we look at the first derivative of y at $t = 0$. Obviously $y'(t) \equiv 0$ for $t \leq 0$, but $y'(0+)$ is 1. Although $y(t)$ is continuous at $t = 0$, its first derivative has a jump there. Using this expression for $y(t)$ on [0, 1], we now see that on [1, 2], $y(t)$ is the solution of the initial value problem $y(1+) = y(1-) = 2$, $y'(t) = t + 1$. You should solve this problem and investigate how smoothly the solution on [0, 1] connects with the solution on [1, 2]. Continuing with this line of reasoning, the existence of a solution of the delay-differential equation can be demonstrated for all t. It is typical of delay-differential equations that discontinuities in the solution are propagated by the delay, though for many equations interesting in practice, the solution becomes smoother as t increases. Discontinuities in the history function or in the function f can also induce discontinuities in y that subsequently propagate. This has unpleasant implications we take up in a moment.

If we try to implement the method of steps numerically with our favorite code for the initial problem, we immediately encounter a number of serious difficulties. To be specific, suppose we are interested in solving a problem with a single, constant delay $\alpha > 0$:

$$\begin{aligned} y'(x) &= f(x, y(x), y(x - \alpha)), \qquad && a \le x \le b, \\ y(a) &= y_a, \qquad y(x) = \phi(x), && a - \alpha \le x < a. \end{aligned}$$

The problem is solved by solving a sequence of initial value problems, each posed on an interval of length α. Although the initial value problems have a similar form, they are different and the code must be restarted for each new subinterval. This is an example of a situation when the last step size in the integration over one subinterval should be usable as an initial step size for the next interval. All these restarts are expensive, so advantage should be taken of a guess for the initial step size that is likely to be on scale. If α is small compared to $b - a$, methods with memory are unattractive because of the expense of starting them. If α is very small, the whole approach may be impractical. When integrating over one interval, values of the solution from the previous interval are required. Where do they come from? This is a situation calling for an interpolation capability. By saving the necessary information about the solution formed when integrating over one subinterval, the solution can be evaluated anywhere in the interval by interpolation when needed as history for the next subinterval. However, if the code requires a great many steps to make it through a subinterval, a great deal of information must be retained and the whole approach may be impractical because of the storage demands. Oberle and Pesch [1981] and the references therein provide more details about the theory and practice.

Bellman [1961] suggests a way to deal with the issue of the memory that is sometimes practical. As described in the paper, he wishes to solve

$$\begin{aligned} \mathbf{x}'(t) &= \mathbf{g}(\mathbf{x}(t), \mathbf{x}(t - 1)), \qquad && t \ge 1, \\ \mathbf{x}(t) &= \mathbf{x}_0(t), && 0 \le t \le 1. \end{aligned} \tag{P.1}$$

A sequence of vector functions

$$\mathbf{x}_n(t) = \mathbf{x}(t + n), \qquad 0 \le t \le 1,$$

are introduced for $n = 1, 2, \ldots$. Then (P.1) can be written as a system of equations

$$\mathbf{x}_n'(t) = \mathbf{g}(\mathbf{x}_n(t), \mathbf{x}_{n-1}(\mathbf{t})), \qquad 0 \le t \le 1. \tag{P.2}$$

The initial condition for $\mathbf{x}_n$ is taken from the value of $\mathbf{x}_{n-1}$ at $t = 1$: $\mathbf{x}_n(0) = \mathbf{x}_{n-1}(1)$. The idea is to start with $\mathbf{x}_1(0) = \mathbf{x}_0(1)$ and integrate (P.2) for $n = 1$ from $t = 0$ to $t = 1$. The value $\mathbf{x}_1(1)$ furnishes the initial value $\mathbf{x}_2(0)$. Now equations (P.2) for *both* $\mathbf{x}_1(t)$ and $\mathbf{x}_2(t)$ are integrated from $t = 0$ to $t = 1$. This is repeated, with larger and larger systems being integrated. Obviously there is no difficulty in obtaining values from the history of $\mathbf{x}(t)$ because the $\mathbf{x}_n(t)$ are being computed simultaneously. This is an interesting example of a sequence of problems of increasing size. Obviously the approach is impractical if the delay is small compared to the length of the interval on which a solution $\mathbf{x}(t)$ is desired.

In Oberle and Pesch [1981] a model for the spread of infection due to Hoppensteadt and Waltman is treated as an example. In somewhat abbreviated form, the task is to compute $S(t)$ satisfying

$$\frac{d}{dt} S(t) = -r\, S(t) I_0(t), \qquad 0 \le t \le t_0, \tag{P.3}$$

$$\frac{d}{dt} S(t) = -r\, S(t)[I_0(t) + S_0 - e^{\mu} S(t)], \qquad t_0 \le t \le t_0 + \sigma, \tag{P.4}$$

$$\frac{d}{dt} S(t) = -r\, S(t) e^{\mu}[S(t - \sigma) - S(t)], \qquad t_0 + \sigma \le t. \tag{P.5}$$

Numerical results are provided for the data $S(0) = S_0 = 10$, $\sigma = 1$, $r = 0.5$, $\mu = 0.1\, r$, $t_0 = 1 - \sqrt{2}/2$ (≈ 0.29), and

$$\begin{aligned} I_0(t) &= 0.4(1 - t), & 0 \le t \le 1, \\ &= 0, & 1 < t. \end{aligned}$$

They find that $S(10) = 0.63020\ 89869 \times 10^{-1}$. Let us now discuss how we might solve this problem with Bellman's method. Equation (P.3) along with the initial value $S(0) = 10$ is a standard initial value problem that we can readily integrate to find $S(t_0)$. For $n = 0, 1, \ldots$, let us define the functions $S_n(t) = S(t + t_0 + n\sigma)$. (Here $\sigma = 1$, but it will be left general in the description except for the requirement that $\sigma > t_0$.) Verify that equations (P.4) and (P.5) can be restated in terms of these new functions as

$$\frac{d}{dt} S_0(t) = -r\, S_0(t)[I_0(t + t_0) + S_0 - e^{\mu} S_0(t)], \qquad 0 \le t \le \sigma,$$

$$\frac{d}{dt} S_n(t) = -r\, S_n(t) e^{\mu}[S_{n-1}(t) - S_n(t)], \qquad 0 \le t \le \sigma.$$

From the integration of (P.3) we have $S(t)$ for the interval $[0, t_0]$ and in particular, we have $S(t_0)$. With the initial value $S_0(0) = S(t_0)$, we can integrate the equation for $S_0(t)$ from $t = 0$ to $t = \sigma$ to obtain $S(t)$ for the interval $[t_0, t_0 + \sigma]$. The value $S_0(\sigma)$ furnishes the initial value $S_1(0)$. Using it, we can integrate simultaneously for $S_0(t)$ and $S_1(t)$ to extend the solution of $S(t)$ another distance of σ. The value $S_1(\sigma)$ furnishes the initial value $S_2(0)$. Using it, we can integrate simultaneously for $S_0(t)$, $S_1(t)$, and $S_2(t)$. This process is repeated to advance the solution of $S(t)$. For the case of $\sigma = 1$, if we want an answer $S(10)$, we repeat until we have available $S_0(0)$, $S_1(0), \ldots, S_9(0)$. We then integrate the 10 equations to obtain $S_9(1 - t_0) = S(1 - t_0 + t_0 + 9) = S(10)$. Carry out the computation of $S(t)$ for the data given to at least $t = 2$. It is not much more work to compute $S(10)$ for which you have the value of Oberle and Pesch as a check.

PROBLEM 9. Quite general nonlinear first-order partial differential equations can be solved using the method of characteristics. A nice development of the theory and its applications are found in Chapters 2 and 3 of Whitham [1974]. Following his treatment, we study the solution of an equation in n independent variables $\mathbf{x} = (x_1, \ldots, x_n)^T$. We seek a function $\phi(\mathbf{x})$ that satisfies the partial differential equation $H(\phi, \mathbf{p}, \mathbf{x}) = 0$. Here H is a given function and the vector $\mathbf{p}$ has components

$$p_i = \frac{\partial \phi}{\partial x_i}, \qquad i = 1, \ldots, n.$$

The idea is to find a characteristic curve $\mathscr{C}$ in the parametric form $\mathbf{x}(\lambda)$ along which ϕ can be determined by solving ordinary differential equations. Along $\mathscr{C}$,

$$\frac{d\phi}{d\lambda} = \sum_{j=1}^{n} \frac{\partial \phi}{\partial x_j}\frac{dx_j}{d\lambda} = \sum_{j=1}^{n} p_j \frac{dx_j}{d\lambda}.$$

Similarly,

$$\frac{dp_i}{d\lambda} = \frac{d}{d\lambda}\left(\frac{\partial \phi}{\partial x_i}\right) = \sum_{j=1}^{n} \frac{\partial^2 \phi}{\partial x_i \partial x_j}\frac{dx_j}{d\lambda}.$$

Differentiating the partial differential equation leads to

$$0 = \frac{\partial H}{\partial \phi}\frac{\partial \phi}{\partial x_i} + \sum_{j=1}^{n} \frac{\partial H}{\partial p_j}\frac{\partial p_j}{\partial x_i} + \frac{\partial H}{\partial x_i},$$

$$= p_i \frac{\partial H}{\partial \phi} + \sum_{j=1}^{n} \frac{\partial H}{\partial p_j}\frac{\partial^2 \phi}{\partial x_i \partial x_j} + \frac{\partial H}{\partial x_i}.$$

Comparing all these equations, it is seen that when $\mathscr{C}$ is defined by

$$\frac{dx_i}{d\lambda} = \frac{\partial H}{\partial p_i}, \qquad i = 1, \ldots, n,$$

the function ϕ and its derivatives p_i can be computed from

$$\frac{dp_i}{d\lambda} = -\, p_i \frac{\partial H}{\partial \phi} - \frac{\partial H}{\partial x_i}, \qquad i = 1, \ldots, n,$$

$$\frac{d\phi}{d\lambda} = \sum_{j=1}^{n} \mathrm{p}_j \frac{\partial H}{\partial p_j}.$$

This is a system of $2n + 1$ ordinary differential equations for ϕ and its partial derivatives along the characteristic $\mathscr{C}$. In principle the behavior of ϕ can be determined in this way from given values of ϕ and the p_i on a set of $\mathbf{x}$ that serve as initial values for the integration. In the special case of a quasilinear equation,

$$\sum_{i=1}^{n} c_i(\phi, \mathbf{x}) \frac{\partial \phi}{\partial x_i} = b(\phi, \mathbf{x}),$$

it is not necessary to introduce equations for the partial derivatives. Show that in this case the system of ODEs reduces to

$$\frac{dx_i}{d\lambda} = c_i(\phi, \mathbf{x}), \qquad i = 1, \ldots, n,$$

$$\frac{d\phi}{d\lambda} = b(\phi, \mathbf{x}).$$

The author first became acquainted with the numerical solution of ODEs by studying the text of Collatz [1960] and in honor of this seminal work, an example is taken from the chapter devoted to the solution of partial differential equations. In §4.2 of Chapter 4 an equation arising in the theory of glacier motion is solved. Conditions in the tongue of a glacier are described by an equation of the form

$$[(n + 1)\kappa u^n - a]\frac{\partial u}{\partial x} + \frac{\partial u}{\partial \mathrm{t}} = -\mathrm{a}.$$

Here $u(x, t)$ is the vertical depth of a central longitudinal section of the glacier at time t and a distance x along a slightly inclined, straight bed. For the particular example considered of a receding glacier on a very flat bed, $n = 1/3$, $\kappa = 0.075$, and $a = 1/2$. At time $t = 0$, the height in dimensionless variables is

$$u(x, 0) = 2\left(\frac{4 - x}{5 - x}\right), \qquad 0 \leq x \leq 4.$$

(This initial data suffices for this particular example because of the behavior of the physical problem, or more formally, because of the behavior of the characteristics.) What are the ODEs that you must integrate to solve this problem by the method of characteristics? Show that you can take t as the parameter λ for this problem.

You would like to know the profile $u(x, t)$ as a function of t. The initial value $u(3, 0) = 1$. If you follow the characteristic through $x = 3$, $t = 0$, you should find that at $t = 2$, the depth u is 0 and $x = 2.15$. For a set of x in the interval $[0, 4]$ compute the depths along the characteristics at times $t = 0.5, 1.0, 1.5, 2, 2.5$. If you have some graphics capability, plot the depth profile at these times.

You may find that the depth decreases to 0 at some time before $t = 2.5$. Negative values of u have no meaning, so you will want to terminate the integration then. It has been pointed out a number of times in this text that situations of this kind will present difficulties when it is not possible to continue the integration to negative values. In this instance there is no difficulty of this kind. A code with the capability of integrating to a specific time or until a solution component vanishes would be convenient. It would be almost as easy to use an interval-oriented code with intermediate output mode because one would just need to check the computed solution u at each step and if it is negative, terminate the integration.

It would be terribly misleading not to make some further comments about these partial differential equations. The function ϕ is determined along a characteristic by its initial value and the characteristic curve itself. If two characteristics were to intersect, it is clear in principle that the two values of ϕ so determined might be different. This does happen. To obtain a single valued function ϕ, the concept of solution of the partial differential equation is extended to permit solutions with discontinuities that are called "shocks" in this context. For problems of this kind, it is of great importance to locate where shocks must be introduced and how they propagate. Whitham has quite a nice discussion of the possibility and importance of shocks. The matter takes us away from illustrating the use of ODE methods so we pursue it no further. It might be remarked, however, that it is useful to compute simultaneously characteristics issuing from a set of initial **x** so that an intersection of a pair of characteristics might be recognized and special action taken. This requires monitoring the integration of a (perhaps large) set of ODEs so as to recognize if, or when, two characteristics cross in the course of a step.

PROBLEM 10. As formulated in Chapter 1, the two-body problem has one body fixed and the other moving about it. If the other body starts from rest, it moves in a straight line towards the fixed body. The equations simplify in this situation to

$$\ddot{r} = -1/r^2, \qquad r(0) = r_0, \qquad \dot{r}(0) = 0,$$

where $r(t)$ is the distance between the bodies. Daniel and Moore [1970, §7.7] discuss this problem and show that the bodies collide after a finite time t_{col}. Further, the body "passes through," so that the solution extends to times greater than t_{col}.

Use your favorite integrator in a straightforward way to integrate this problem when $r_0 = 1$. The code will "crash" somehow at the time that the bodies collide, just how depending on the integrator. Despite this, you will be able to locate t_{col} quite accurately: the body is accelerating as it approaches the origin and this means that $r(t) \to 0$ very quickly as $t \to t_{col}$. Daniel and Moore derive an analytical expression for t_{col}. There is a slip in the analytical evaluation of an integral that when corrected with Derive [1989] provides the expression

$$t_{col} = \frac{\sqrt{2}\pi r_0^{3/2}}{4}.$$

You should find that your computed value is in close agreement with this analytical result.

Approximate $r(t)$ for $t \approx t_{col}$ by an expression of the form $\alpha(t - t_{col})^{\mu}$. By substituting this expression into the differential equation, you will be able to deduce α and μ. With your numerical value for t_{col}, this expression not only allows you to approximate $r(t)$ near t_{col}, but also to obtain numerical values for starting an integration to follow the motion for times after collision. Argue that if you record values τ, $r(\tau)$, and $\dot{r}(\tau)$ for some τ "close" to t_{col}, you can start the integration from $t_{col} + (t_{col} - \tau)$ with the values $r(\tau)$ and $-\dot{r}(\tau)$. Continue the integration of the problem when $r_0 = 1$ up to $t = 3$. You will notice some interesting behavior of the distance function $r(t)$; what does this mean physically?

PROBLEM 11. A ball is thrown into the air with an initial speed v. Let the height of the ball at time t be $y(t)$. Measuring height from the ground and time from the throw, we have $y(0) = 0$ and $\dot{y}(0) = v > 0$. When the ball hits the ground at time t_1, i.e., $y(t_1) = 0$, it bounces up with speed $\dot{y}(t_1+) = -\alpha\dot{y}(t_1-)$ where $0 < \alpha < 1$ is a coefficient of restitution. With a convenient choice of units the acceleration of the ball due to gravity is $\ddot{y} = -2$. Figure P.2 shows the height of the ball when $v = 1$ and $\alpha = 0.9$. Although this seems to be a very simple problem, it is not easy for the codes because it is necessary to determine when the ball hits the ground and restart the integration at such times.

Let us denote the times the ball is at ground level as t_m with $0 = t_0 < t_1 < \ldots$. On $[t_0, t_1]$ the height $y(t)$ of the ball is the solution of the initial value problem

$$\ddot{y} = -2, \qquad t_0 \le t \le t_1,$$
$$y(t_0) = 0, \qquad \dot{y}(t_0) = v > 0.$$

For $m \ge 1$, it is the solution of the initial value problem

$$\ddot{y} = -2, \qquad t_m \le t \le t_{m+1},$$
$$y(t_m) = 0, \qquad \dot{y}(t_m+) = -\alpha\dot{y}(t_m-).$$

Verify that for $m \ge 1$,

$$t_m = \frac{1 - \alpha^m}{1 - \alpha}\, v,$$

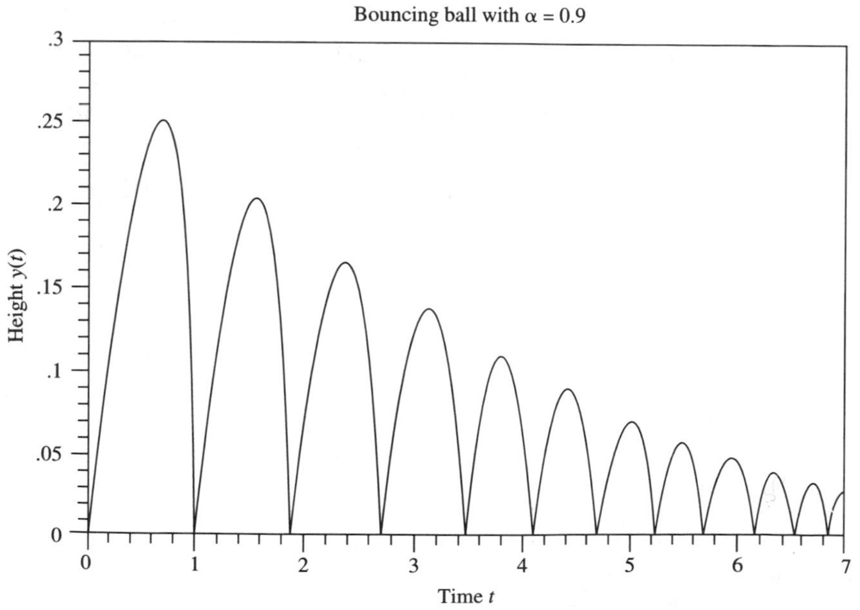

Figure P.2

and that for $m \geq 0$, the solution on $[t_m, t_{m+1}]$ is

$$y(t) = \alpha^m v(t - t_m) - (t - t_m)^2.$$

Try to reproduce the results of Figure P.2 by solving the differential equation numerically using one of the codes available to you. Obviously it would be convenient to use a code that can locate where a solution component vanishes. Discuss how the cost of starting might influence your selection of code. Some codes require a guess for an initial step size and others permit this. In the present situation you have available a reasonable guess for an initial step size at t_m for all $m \geq 1$—what is it?

PROBLEM 12. Some applications lead to a differential equation in a *complex variable z*,

$$\frac{d\mathbf{w}}{dz} = \mathbf{F}(z, \mathbf{w}).$$

The initial value problem is a little more complicated than in the real case because it is necessary to specify the path $\mathscr{C}$ that the integration is to follow from a complex initial point $\alpha + i\beta$ to a complex final point $\alpha' + i\beta'$. Software for such problems is not readily available, so it is useful to know that they can be reduced to problems involving only real variables that can be integrated with familiar codes. The reduction is particularly simple when the path of integration is parallel to the real or imaginary axis. It is accomplished by first writing all quantities in terms of their real and imaginary parts: $z = x + iy$, $\mathbf{w}(z) = \mathbf{u}(x, y) + i\mathbf{v}(x, y)$, $\mathbf{F}(z, \mathbf{w}) = \mathbf{f}(x, y, \mathbf{u}, \mathbf{v}) + i\mathbf{g}(x, y, \mathbf{u}, \mathbf{v})$. An easy calculation then shows that when integrating parallel to the real axis from $\alpha + i\beta$ to $\alpha' + i\beta$, i.e., $z = x + i\beta$ for $\alpha \leq x \leq \alpha'$, the complex valued equation is equivalent to the pair of real valued equations

$$\frac{d\mathbf{u}(x, \beta)}{dx} = \mathbf{f}(x, \beta, \mathbf{u}(x, \beta), \mathbf{v}(x, \beta)), \qquad \mathbf{u}(\alpha, \beta) = \mathrm{Re}(\mathbf{w}(\alpha, \beta)),$$

$$\frac{d\mathbf{v}(x, \beta)}{dx} = \mathbf{g}(x, \beta, \mathbf{u}(x, \beta), \mathbf{v}(x, \beta)), \qquad \mathbf{v}(\alpha, \beta) = \mathrm{Im}(\mathbf{w}(\alpha, \beta)).$$

Make this calculation. Similarly work out the equivalent real system for an integration parallel to the imaginary axis from $\alpha + i\beta$ to $\alpha + i\beta'$, i.e., $z = \alpha + iy$ for $\beta \leq y \leq \beta'$.

H.T. Davis [1962] makes an interesting use of integration in the complex plane literally to dodge a difficulty on the real axis. Chapter 9 of his book is devoted to "continuous analytic continuation," what nowadays is called Taylor series methods. Although not pointed out when we studied the methods in Chapter 4, it is clear that they can be used for integrations in the complex plane just like along the real line. Davis evaluates the first Painlevé transcendant for a number of values of the parameter by integrating the differential equation it satisfies with a fixed step size and a Taylor series method of moderate order. For one value of the parameter, the initial value problem is $y'' = 6y^2 + y, y(0) = 1, y'(0) = 0$. The difficulty is that the solution has a pole at $x \approx 1.21$. To get around the pole, Davis integrates along the real line to (0.5, 0) and then parallel to the imaginary axis to (0.5, 5.0). He continues the integration from (0.5, 5.0) to (1.5, 5.0) parallel to the real axis. Finally he integrates parallel to the imaginary axis from (1.5, 5.0) back to the real axis at (1.5, 0). He found that $y(1.5) \approx 11.8$ and $y'(1.5) \approx -79.6$. These values should be real, so the size of the imaginary parts of the computed values furnishes a measure of the accuracy of the integration. He found imaginary parts of about 1% of the size of the real parts, indicating only modest accuracy. Using your favorite code, compute more accurate values for $y(1.5)$ and $y'(1.5)$. It is not necessary to follow the path taken by Davis; you might, for example, integrate from the origin along the imaginary axis to (0, 1), then parallel to the real axis to (1.5, 1), and finally parallel to the imaginary axis back to the real axis at (1.5, 0).

Self Test

- A stable limit cycle $\mathbf{y}(t)$ of the equation $\dot{\mathbf{y}} = \mathbf{F}(\mathbf{y})$ can be computed by choosing initial values $\mathbf{s}$ that are not "too" far from $\mathbf{y}(0)$ and integrating for a "long" time. Discrete variable methods cannot be used to integrate unstable initial value problems, so how might you compute an *un*stable limit cycle?
- Memory in a formula has advantages and disadvantages—explain briefly a couple.
- The (explicit) midpoint rule

$$\mathbf{y}_{j+1} = \mathbf{y}_{j-1} + 2h\,\mathbf{F}(x_j, \mathbf{y}_j)$$

  is (weakly) stable. On the other hand, it is not absolutely stable for *any* h. (The region of absolute stability degenerates to the origin.) Obviously the two concepts of stability are different—explain briefly the difference.
- Stiffness is a complex phenomenon. Describe briefly a couple of differences between solving stiff and non-stiff problems.
- Efficiency is the usual justification offered for variation of the step size (and the estimation of local error that must accompany it), but there are other, perhaps more important, reasons. Describe briefly a couple of these reasons.
- The typical library code asks the user to specify the accuracy desired. Often users think that the code will return with a solution $\mathbf{y}_j$ that agrees

with the true solution $\mathbf{y}(x_j)$ to this accuracy. This is *not* what the codes do. Explain briefly what they do attempt.

- Explain briefly
  (a) the construction of explicit Runge–Kutta formulas by quadrature
  (b) the origin of Adams formulas
  (c) the origin of the BDFs (backward differentiation formulas)
- With a constant step size h, LMMs (linear multistep methods) have the form

$$\sum_{i=0}^{k} \alpha_i \, \mathbf{y}_{j+1-i} = h \sum_{i=0}^{k} \beta_i \, \mathbf{f}(x_{j+1-i}, \mathbf{y}_{j+1-i}).$$

  Explain the difference between an implicit LMM and a predictor–corrector pair. Include both the implementation and the properties of the formula.
- What is meant by the "(local) truncation error" of a LMM? What is the *form* of this error? State a convergence result that includes the effect of the start. Supposing that you start "accurately," how does the error behave?
- What is the "classical situation?" What is different about "stiff" problems? What factors restrict the step size when a method appropriate for non-stiff problems is applied to a stiff problem? How do we deal with these factors in codes appropriate for stiff problems?
- Consider a one-step method of order p. Define its local error. How can this error be estimated using a one-step method of order $p + 1$? Prove the validity of your estimate.
- Consider a LMM (linear multistep method) of order p. Define its local truncation error. Using a (different) explicit LMM, explain Milne's idea for estimating the local truncation error. How is it possible to estimate this error using data $\{\mathbf{y}_j\}$ that are themselves in error?
- In one of the computing exercises, an equation was integrated with AB3 and constant step size h. The rate of convergence was measured for three ways of starting the integration:

case	starting procedure	observed rate of convergence
1	forward Euler	$\mathcal{O}(h^2)$
2	second-order R–K	$\mathcal{O}(h^3)$
3	true solution	$\mathcal{O}(h^3)$

  It was also observed that although the orders in cases (2) and (3) were the same, the actual errors were different. Explain briefly why you would expect the behavior observed in this experiment.
- Suppose you wish to solve $\mathbf{y}' = \mathbf{F}(x, \mathbf{y})$, $\mathbf{y}(a) = \mathbf{A}$. What is the *form* of an explicit Runge–Kutta method? If there is a constant vector $\mathbf{v}$ such

that $\mathbf{v}^T \mathbf{F}(x, \mathbf{y}) \equiv \mathbf{0}$, then any solution $\mathbf{y}(x)$ of the initial value problem satisfies the linear conservation law $\mathbf{v}^T \mathbf{y}(x) = \mathbf{v}^T \mathbf{A}$. Recall that this was proved by observing that $\mathbf{v}^T \mathbf{y}'(x) = \mathbf{v}^T \mathbf{F}(x, \mathbf{y}(x)) = \mathbf{0}$ and integrating. Prove that any explicit Runge–Kutta method started with the correct initial value, $\mathbf{y}_0 = \mathbf{A}$, will produce $\mathbf{y}_j \approx \mathbf{y}(x_j)$ that also satisfy the conservation law, i.e., $\mathbf{v}^T \mathbf{y}_j = \mathbf{v}^T \mathbf{A}$ for all j.

- Suppose that the solution of the (scalar) problem $y' = F(x, y)$, $y(a) = A$ is a polynomial $p(x)$ of degree d. Argue that if an Adams method or a BDF (choose one possibility for your answer) of "appropriate" order is used, the local truncation error is zero. What is an appropriate order? If you were to integrate such a problem with one of the popular variable order Adams or BDF codes like STEP, LSODE, DDRIV2, . . . , you would find that the numerical solution y_j is not exactly equal to $p(x_j)$. There are two main reasons for this—explain one.
- Suppose that the equation $\dot{\mathbf{y}} = \mathbf{F}(\mathbf{y})$ has a stable limit cycle to which all nearby solution curves converge rapidly. You choose some initial values and start integrating with your favorite numerical method. After integrating over a "long" time interval, it appears that the numerical solution represents a periodic solution. The convergence theory of the numerical methods holds for fixed time intervals, so how can you expect to approximate the limit cycle?
- You need to decide which of two methods of different orders will be the more efficient in the sense that a given accuracy is obtained using fewer function evaluations. Argue that if you ask for enough accuracy, the higher order formula will always be the more efficient.
- When you specify a positive absolute error tolerance (or threshold), you are telling the code that you are not concerned about controlling the error of solution components that are smaller in magnitude than this tolerance (or threshold). Nevertheless, you are likely to find that such components are computed with some accuracy. Why is this?

References

§1 Literature

No attempt is made here to provide an exhaustive list of the literature on the numerical solution of the initial value problem for ordinary differential equations. Among the many books in this area we cite a number with particularly large bibliographies that might be consulted for this purpose. There is, of course, a considerable overlap among these bibliographies, and the newer ones will naturally contain entries not present in the older, but a number of sources are cited because some will be easier to obtain than others and none is exhaustive. In this section a chronological listing is natural.

Henrici, P., Discrete Variable Methods in Ordinary Differential Equations, Wiley, New York, 1962.

Ceschino, F. and Kuntzmann, J., Numerical Solution of Initial Value Problems, English translation by D. Boyanovitch, Prentice-Hall, Englewood Cliffs, NJ, 1966.

Willoughby, R.A., ed., Stiff Differential Systems, Plenum Press, New York, 1973.

Hall, G. and Watt, J.M., eds., Modern Numerical Methods for Ordinary Differential Equations, Clarendon Press, Oxford, 1976.

Grigorieff, R.D., Numerik gewöhnlicher Differentialgleichungen, vols. 1 and 2, B.G. Teubner, Stuttgart, 1977.

Aiken R.C., ed., Stiff Computation, Oxford University Press, New York, 1985.

Butcher, J.C., The Numerical Analysis of Ordinary Differential Equations, Wiley, New York, 1987.

§2 Cited Works

Abramowitz, M.and Stegun, I.A., eds, Handbook of Mathematical Functions, National Bureau of Standards, U.S.Government Printing Office, 1964.

Aiken, R.C., ed., Stiff Computation, Oxford University Press, New York, 1985.

Albrecht, P., A New Theoretical Approach to Runge–Kutta Methods, SIAM J.Numer. Anal., 24 (1987) pp. 391–406.

Ascher, U.M., Mattheij, R.M.M., and Russell, R.D., Numerical Solution of Boundary Value Problems for Ordinary Differential Equations, Prentice-Hall, Englewood Cliffs, NJ, 1988.

Baker, C.T.H., Volterra Integral Equations, pp. 235–246 in The Numerical Solution of Nonlinear Problems, Baker, C.T.H. and Philips, C., eds, Clarendon Press, Oxford, 1981.

Bate, R.R., Mueller, D.D., and White, J.E., Fundamentals of Astrodynamics, Dover, New York, 1971.

Bellman, R., On the Computational Solution of Differential-Difference Equations, J. Math. Anal. Applic., 3 (1961) pp. 108–110.

Bischof, C., Carle, A., Corliss, G., Griewank, A., and Hovland, P., ADIFOR—Generating Derivative Codes from FORTRAN Programs, Scientific Programming, 1 (1992) pp. 11–29.

Bogacki, P. and Shampine, L.F., A 3(2) Pair of Runge–Kutta Formulas, Appl. Math. Letters, 2 (1989a) pp. 1–9.

Bogacki, P. and Shampine, L.F., An Efficient Runge–Kutta (4, 5) Pair, Rept. 89–20, Math. Dept., Southern Methodist Univ., Dallas, TX, 1989b.

Borrelli, R.L., Coleman, C.S., and Boyce, W.E., Differential Equations Laboratory Workbook, Wiley, New York, 1992.

Bownds, J.M. and Applebaum, L., Algorithm 627: a FORTRAN Subroutine for Solving Volterra Integral Equations, ACM Trans. Math. Software, 11 (1985) pp. 58–65.

Boyce, W.E. and DiPrima, R.C., Elementary Differential Equations, 5th ed., Wiley, New York, 1992.

Brankin, R.W., Gladwell, I., and Shampine, L.F., Starting BDF and Adams Codes at Optimal Order, J. Comp. Appl. Math., 21 (1988) pp. 357–368.

Brankin, R.W., Gladwell, I., and Shampine, L.F., RKSUITE: A Suite of Runge–Kutta Codes for the Initial Value Problem for ODEs, Softreport 92-S1, Math. Dept., Southern Methodist Univ., Dallas, TX, 1992.

Brenan, K.E., Campbell, S.L., and Petzold, L.R., Numerical Solution of Initial-Value Problems in Differential-Algebraic Equations, North Holland, New York, 1989.

Brown, P.N., Byrne, G.D., and Hindmarsh, A.C., VODE: A Variable-Coefficient ODE Solver, SIAM J.Sci. Stat. Comput., 10 (1989) pp. 1038–1051.

Buchanan, J.L., MDEP Midshipman Differential Equations Program, Version 2.28, Math. Dept., U.S.Naval Academy, Annapolis, MD, 1992.

Butcher, J.C., The Numerical Analysis of Ordinary Differential Equations, Wiley, New York, 1987.

Buzbee, B.L., The SLATEC Common Mathematical Library, pp. 302–320 in Cowell, W.R., ed., Sources and Development of Mathematical Software, Prentice-Hall, Englewood Cliffs, NJ, 1984.

Byrne, G.D. and Hindmarsh, A.C., A Polyalgorithm for the Numerical Solution of Ordinary Differential Equations, ACM Trans. Math. Software, 1 (1975) pp. 71–96.

Byrne, G.D. and Hindmarsh, A.C., Stiff ODE Solvers: a Review of Current and Coming Attractions, J. Comp. Phys., 70 (1987) pp. 1–62.

Byrne, G.D., Hindmarsh, A.C., Jackson, K.R., and Brown, H.G., A Comparison of Two ODE Codes: GEAR and EPISODE, Computers and Chemical Engineering, 1 (1977) pp. 133–147.

Canuto, C., Hussaini, M.Y., Quarteroni, A., and Zang, T.A., Spectral Methods in Fluid Dynamics, Springer-Verlag, New York, 1988.

Ceschino, G. and Kuntzmann, J., Numerical Solution of Initial Value Problems, English translation by D.Boyanovitch, Prentice-Hall, Englewood Cliffs, NJ, 1966.

Christiansen, J., Numerical Solution of Ordinary Simultaneous Differential Equations of the 1st Order Using a Method for Automatic Step Change, Numer. Math., 14 (1970) pp. 317–324.

Coleman, T.F., Garbow, B.S., and Moré, J.J., Software for Estimating Sparse Jacobian Matrices, ACM Trans. Math. Software, 11 (1984a) pp. 329–345.

Coleman, T.F., Garbow, B.S., and Moré, J.J., Algorithm 618 FORTRAN Subroutines for Estimating Sparse Jacobian Matrices, ACM Trans. Math. Software, 11 (1984b) pp. 346–347.

Collatz, L., The Numerical Treatment of Differential Equations, 3rd ed., Springer, Berlin, 1960.

Cooper, G.J. and Verner, J.H., Some Explicit Runge–Kutta Methods of High Order, SIAM J.Numer. Anal., 9 (1972) pp. 389–405.

Corliss, G.F. and Chang, Y.F., Solving Ordinary Differential Equations Using Taylor Series, ACM Trans. Math. Software, 8 (1982) pp. 114–144.

Curtiss, C.F. and Hirschfelder, J.O., Integration of Stiff Equations, Proc. Natl. Acad. Sci. USA, 38 (1952) pp. 235–243.

Curtis, A.R., Solution of Large Stiff Initial Value Problems—the State of the Art, pp. 257–278 in Jacobs, D., ed., Numerical Software—Needs and Availability, Academic, London, 1978.

Curtis, A.R., Powell, M.J.D., and Reid, J.K., On the Estimation of Sparse Jacobian Matrices, J.Inst. Math. Appl., 13 (1974) pp. 117–119.

Dahlquist, G., Edsberg, L., Sköllermo, G., and Söderlind, G., Are the Numerical Methods and Software Satisfactory for Chemical Kinetics? pp. 149–164 in Hinze, J., ed., Numerical Integration of Differential Equations and Large Linear Systems, Springer, Berlin, 1982.

Daniel, J.W. and Moore, R.E., Computation and Theory in Ordinary Differential Equations, Freeman, San Francisco, 1970.

Davidenko, D.F., On a New Method for the Solution of Systems of Equations (Russian), Dokl. Akad. Nauk SSSR, 88 (1953) pp. 601–602.

Davis, H.T., Introduction to Nonlinear Differential and Integral Equations, Dover, 1962.

Derive, A Mathematical Assistant Program, 3rd. ed., Soft Warehouse, Inc., Honolulu, Hawaii, 1989.

Deuflhard, P., Recent Progress in Extrapolation Methods for Ordinary Differential Equations, Rept. 224, SFB 123, University of Heidelberg, Germany, 1983.

Deuflhard, P., Recent Progress in Extrapolation Methods for Ordinary Differential Equations, SIAM Rev., 27 (1985) pp. 505–535.

Dongarra, J.J., Bunch, J.R., Moler, C.B., and Stewart, G.W., LINPACK Users' Guide, SIAM, Philadelphia, 1979.

Dormand, J.R. and Prince, P.J., A Family of Embedded Runge–Kutta Formulae, J.Comput. Appl. Math., 6 (1980) pp. 19–26.

Dormand, J.R. and Prince, P.J., Runge–Kutta Triples, Comp. & Maths. with Applcs., 12 (1986) pp. 1007–1017.

England, R., Error Estimates for Runge–Kutta Type Solutions to Systems of Ordinary Differential Equations, Comput. J., 12 (1969) pp. 166–170.

Enright, W.H., Jackson, K.R., Nørsett, S.P., and Thomsen, P.G., Interpolants for Runge–Kutta Formulas, ACM Trans. Math. Software, 12 (1986) pp. 193–218.

Erdélyi, A., Asymptotic Expansions, Dover, New York, 1956.

Fehlberg, E., Classical Fifth-, Sixth-, Seventh-, and Eighth Order Runge–Kutta Formulas with Step Size Control, NASA Tech. Rept. 287, 1968.

Fehlberg, E., Klassiche Runge–Kutta-Formeln vierter und niedrigerer Ordnung mit Schrittweiten-Kontrolle und ihre Anwendung auf Wärmeleitungsprobleme, Computing, 6 (1970), pp. 61–71.

Forsythe, G.E., Malcolm, M.A., and Moler, C.B., Computer Methods for Mathematical Computations, Prentice-Hall, Englewood Cliffs, NJ, 1977.

Fox, P.A., Hall, A.D., and Schryer, N.L., The PORT Mathematical Subroutine Library, ACM Trans. Math. Software, 4 (1978) pp. 104–126.

Gaffney, P.W., A Performance Evaluation of Some FORTRAN Subroutines for the Solution of Stiff Oscillatory Ordinary Differential Equations, ACM Trans. Math. Software, 10 (1984) pp. 58–72.

Gear, C.W., The Automatic Integration of Ordinary Differential Equations, Comm. ACM, 14 (1971a) pp. 176–179.

Gear, C.W., Algorithm 407, DIFSUB for Solution of Ordinary Differential Equations, Comm. ACM, 14 (1971b) pp. 185–190.

Gear, C.W., Numerical Initial Value Problems in Ordinary Differential Equations, Prentice-Hall, Englewood Cliffs, NJ, 1971c.

Gladwell, I., Initial Value Routines in the NAG Library, ACM Trans. Math. Software, 5 (1979) pp. 386–400.

Gladwell, I., Shampine, L.F., and Brankin, R.W., Automatic Selection of the Initial Step Size for an ODE Solver, J.Comp. Appl. Math., 18 (1987) pp. 175–192.

Gollwitzer, H., Differential Systems User Manual, Version 3.0, Dept. Math. & Comp. Sci., Drexel Univ., Philadelphia, PA, 1991.

Gourlay, A.R., A Note on Trapezoidal Methods for the Solution of Initial Value Problems, Math. Comp., 24 (1970) pp. 629–633.

Grigorieff, R.D., Numerik gewöhnlicher Differentialgleichungen, vols. 1 and 2, B.G.Teubner, Stuttgart, 1977.

Gupta, G.K., A Note about Overhead Costs in ODE Solvers, ACM Trans. Math. Software, 6 (1980) pp. 319–326.

Gupta, G.K., Sacks-Davis, R., and Tischer, P.E., A Review of Recent Developments in Solving ODEs, ACM Computing Surveys, 17 (1985) pp. 5–47.

Haberman, R., Elementary Partial Differential Equations, 2nd ed., Prentice-Hall, Englewood Cliffs, NJ, 1987.

Hairer, E., Lubich, C., and Roche, M., The Numerical Solution of Differential-Algebraic Systems By Runge–Kutta Methods, Springer, Berlin, 1989.

Hairer, E., Nørsett, S.P., and Wanner, G., Solving Ordinary Differential Equations I Nonstiff Problems, Springer, Berlin, 1987.

Hairer, E. and Wanner, G., Solving Ordinary Differential Equations II Stiff and Differential-Algebraic Problems, Springer, Berlin, 1991.

Hall, G. and Watt, J.M., eds., Modern Numerical Methods for Ordinary Differential Equations, Clarendon Press, Oxford, 1976.

Hénon, M., On the Numerical Computation of Poincaré Maps, Physica 5D (1982) pp. 412–414.

Henrici, P., Discrete Variable Methods in Ordinary Differential Equations, Wiley, New York, 1962.

Henrici, P., Error Propagation for Difference Methods, Wiley, New York, 1963.

Hindmarsh, A.C., LSODE and LSODI, Two New Initial Value Ordinary Differential Equation Solvers, ACM SIGNUM Newsletter, 15 (1980) pp. 10–11.

Hindmarsh, A.C., ODEPACK, A Systematized Collection of ODE Solvers, pp. 55–64 in Stepleman, R.S., et al., eds., Scientific Computing, North-Holland, Amsterdam, 1983.

Hindmarsh, A.C. and Byrne, G.D., Applications of EPISODE: An Experimental Package for the Integration of Systems of Ordinary Differential Equations, pp. 147–166 in Lapidus, L. and Schiesser, W.E., eds., Numerical Methods for Differential Systems, Academic, New York, 1976.

Holodniok, M. and Kubiček, M., DERPAR: An Algorithm for the Continuation of Periodic Solutions in Ordinary Differential Equations, J.Comp. Phys., 55 (1984) pp. 254–267.

de Hoog, F. and Weiss, R., The Application of Linear Multistep Methods to Singular Initial Value Problems, Math. Comp., 31 (1977) pp. 676–690.

de Hoog, F. and Weiss, R., The Application of Runge–Kutta Schemes to Singular Initial Value Problems, Math. Comp., 44 (1985) pp. 93–103.

Horn, M.K., Fourth and Fifth-Order Scaled Runge–Kutta Algorithms for Treating Dense Output, SIAM J.Numer. Anal., 20 (1983) pp. 558–568.

van der Houwen, P.J., Explicit Runge–Kutta Formulas with Increased Stability Boundaries, Numer. Math., 20 (1972) pp. 149–164.

Hubbard, J.H. and West, B.H., Differential Equations: A Dynamical Systems Approach Part I, One Dimensional Equations, Springer, New York, 1990.

Hubbard, J.H. and West, B.H., MacMath 9.0 A Dynamical Systems Software Package for the Macintosh™, Springer, New York, 1992.

Hull, T.E. and Enright, W.H., A Structure for Programs that Solve Ordinary Differential Equations, Rept. 66, Comp. Sci. Dept., University of Toronto, Canada, 1974.

Hull, T.E., Enright, W.H., and Jackson, K.R., User's Guide for DVERK—a Subroutine for Solving Non-Stiff ODEs, Rept. 100, Comp. Sci. Dept., University of Toronto, Canada, 1975.

Hull, T.E., Enright, W.H., Fellen, B.M., and Sedgwick, A.E., Comparing Numerical Methods for Ordinary Differential Equations, SIAM J.Numer. Anal., 9 (1972) pp. 603–637.

IMSL, Math/Library User's Manual, IMSL, Houston, TX, 1989.

Isaacson, E. and Keller, H.B., Analysis of Numerical Methods, Wiley, New York, 1966.

Jackson, K.R. and Sacks-Davis, R., An Alternative Implementation of Variable Step-Size Multistep Formulas for Stiff ODEs, ACM Trans. Math. Software, 6 (1980) pp. 295–318.

Jordan, D.W. and Smith, P., Nonlinear Ordinary Differential Equations, 2nd ed., Clarendon Press, Oxford, 1987.

Juedes, D., A.Taxonomy of Automatic Differentiation Tools, pp. 315–329 in Griewank, A. and Corliss, G.F., eds., Automatic Differentiation of Algorithms: Theory, Implementation, and Application, SIAM, Philadelphia, 1991.

Kahaner, D., Moler, C., and Nash, S., Numerical Methods and Software, Prentice-Hall, Englewood Cliffs, NJ, 1989.

Keast, P. and Muir, P.H., Algorithm 688 EPDCOL A More Efficient PDECOL Code, ACM Trans. Math. Software, 17 (1991) pp. 153–166.

Klamkin, M.S., On the Transformation of a Class of Boundary Value Problems into Initial Value Problems for Ordinary Differential Equations, SIAM Rev., 4 (1962) pp. 43–47.

Klamkin, M.S., Transformation of Boundary Value into Initial Value Problems, J.Math. Anal. Appl., 32 (1970) pp. 308–330.

Koçak, H., Differential and Difference Equations through Computer Experiments, 2nd ed., Springer, New York, 1989.

Krogh, F.T., Suggestions on Conversion (with Listings) of the Variable Order Integrators VODQ, SVD1, and DVDQ, Tech. Memo. 278, Jet Propulsion Lab., California Institute of Technology, Pasadena, CA, 1971.

Kubiçek, M. and Marek, M., Computational Methods in Bifurcation Theory and Dissipative Structures, Springer, Berlin, 1983.

Lambert, J.D., Stiffness, pp. 19–46 in Gladwell, I., and Sayers, D.K., eds., Computational Techniques for Ordinary Differential Equations, Academic, London, 1980.

Lapidus, L., Aiken, R.C., and Liu, Y.A., The Occurrence and Numerical Solution of Physical and Chemical Systems Having Widely Varying Time Constants, pp. 187–200 in Willoughby, R.A., ed., Stiff Differential Systems, Plenum Press, New York, 1973.

Lastman, G.J., Wentzell, R.A., and Hindmarsh, A.C., Numerical Solution of a Bubble Cavitation Problem, J.Comp. Physics, 28 (1978) pp. 56–64.

Lighthill, J., An Informal Introduction to Theoretical Fluid Mechanics, Clarendon Press, Oxford, 1986.

Lin, C.C. and Segel, L.A., Mathematics Applied to Deterministic Problems in the Natural Sciences, SIAM, Philadelphia, 1988.

Main, I.G., Vibrations and Waves in Physics, 2nd ed., Cambridge University Press, Cambridge, 1987.

Madsen, N.K. and Sincovec, R.F., Algorithm 540 PDECOL, General Collocation Software for Partial Differential Equations, ACM Trans. Math. Software, 5 (1979) pp. 326–351.

Moler, C.B., Little, J., Bangert, S., and Kleiman, S., MATLAB User's Guide, MathWorks, Sherborn, MA, 1987.

Na, T.Y., Transforming Boundary Conditions to Initial Conditions for Ordinary Differential Equations, SIAM Rev., 9 (1967) pp. 204–210.

Na, T.Y., Computational Methods in Engineering Boundary Value Problems, Academic, New York, 1979.

NAG FORTRAN Library, NAG Ltd, Oxford.

Nayfeh, A., Perturbation Methods, Wiley, New York, 1973.

Merson, R.H., An Operational Method for the Study of Integration Processes, pp. 110-1 to 110-25 in Proc. Symp. Data Processing, Weapons Research Establishment, Salisbury, Australia, 1957.

Oberle, H.J. and Pesch, H.J., Numerical Treatment of Delay Differential Equations by Hermite Interpolation, Numer. Math., 37 (1981) pp. 235–255.

Ortega, J.M. and Rheinboldt, W.C., Iterative Solution of Nonlinear Equations in Several Variables, Academic, New York, 1970.

O'Malley, R.E., On Nonlinear Singularly Perturbed Initial Value Problems, SIAM Review, 30 (1988) pp. 193–212.

Ostrowski, A.M., Solution of Equations and Systems of Equations, 2nd ed., Academic, New York, 1966.

Prince, P.J. and Dormand, J.R., High Order Embedded Runge–Kutta Formulae, J.Comput. Appl. Math., 7 (1981) pp. 67–75.

Ralston, A., Runge–Kutta Methods with Minimum Error Bounds, Math. Comp., 16 (1962) pp. 431–437.

Ralston, A., A First Course in Numerical Analysis, McGraw-Hill, New York, 1965.

Robertson, H.H., The Solution of a Set of Reaction Rate Equations, pp. 178–182 in Walsh, J., ed., Numerical Analysis: An Introduction, Academic, London, 1967.

Rogers, C. and Ames, W.F., Nonlinear Boundary Value Problems in Science and Engineering, Academic, New York, 1989.

Rouse, H., Elementary Mechanics of Fluids, Dover, New York, 1946.

Salane, D.E., Three Adaptive Routines for Forming Jacobians Numerically, Rept. SAND86-1310, Sandia National Laboratories, Albuquerque, NM, 1986.

Salane, D.E. and Shampine, L.F., An Economical and Robust Routine for Computing Sparse Jacobians, Rept. SAND85-0977, Sandia National Laboratories, Albuquerque, NM, 1985.

Schiesser, W.E., The Numerical Method of Lines Integration of Partial Differential Equations, Academic, San Diego, CA, 1991.

Sedgwick, A.E., An Effective Variable Order Variable Step Adams Method, Rept. 53, Dept. Comp. Sci., Univ. of Toronto, 1973.

Seydel, R., From Equilibrium to Chaos Practical Bifurcation and Stability Analysis, Elsevier, Amsterdam, 1988.

Shah, M.J., Engineering Simulation Using Small Scientific Computers, Prentice-Hall, Englewood Cliffs, NJ, 1976.

Shampine, L.F., Local Extrapolation in the Solution of Ordinary Differential Equations, Math. Comp., 27 (1973) pp. 91–97.

Shampine, L.F., Limiting Precision in Differential Equation Solvers, Math. Comp., 28 (1974) pp. 141–144.

Shampine, L.F., Stiffness and Non-Stiff Differential Equations Solvers, pp. 287–301 in Collatz, L., et al., eds., Numerische Behandlung von Differentialgleichungen, ISNM 27, Birkhauser, Basel, 1975.

Shampine, L.F., Local Error Control in Codes for Ordinary Differential Equations, Appl. Math. Comp., 3 (1977) pp. 189–210.

Shampine, L.F., Limiting Precision in Differential Equation Solvers II: Sources of Trouble and Starting a Code, Math. Comp., 32 (1978) pp. 1115–1122.

Shampine, L.F., Storage Reduction for Runge–Kutta Codes, ACM Trans. Math. Software, 5 (1979a) pp. 245–250.

Shampine, L.F., Solving ODEs with Discrete Data in SPEAKEASY, pp. 177–192 in de Boor, C. and Golub, G.H., eds., Recent Advances in Numerical Analysis, Academic, New York, 1979b.

Shampine, L.F., Better Software for ODEs, Comp. & Maths. with Applcs., 5 (1979c) pp. 157–161.

Shampine, L.F., What Everyone Solving Differential Equations Numerically Should Know, pp. 1–17 in Gladwell, I. and Sayers, D.K., eds., Computational Techniques for Ordinary Differential Equations, Academic, London, 1980.

Shampine, L.F., The Step Sizes Used by One-Step Codes for ODEs, Appl. Numer. Math., 1 (1985a) pp. 95–106.

Shampine, L.F., Interpolation for Runge–Kutta Methods, SIAM J.Numer. Anal., 22 (1985b) pp. 1014–1027.

Shampine, L.F., Local Error Estimation by Doubling, Computing, 24 (1985c) pp. 179–190.

Shampine, L.F., Some Practical Runge–Kutta Formulas, Math. Comp., 46 (1986) pp. 135–150.

Shampine, L.F., Solving Volterra Integral Equations with ODE Codes, IMA J.Num. Anal., 8 (1988) pp. 37–41.

Shampine, L.F., Ill-Conditioned Matrices and the Integration of Stiff ODEs, J. Comput. Appl. Math., 48(1993) pp. 279–292.

Shampine, L.F. and Allen, R.C., Numerical Computing: an Introduction, Saunders, Philadelphia, PA, 1973.

Shampine, L.F. and Baca, L.S., A Runge–Kutta Formula with Inexpensive Error Estimate, Trans. Soc. Comp. Sim., 2 (1985) pp. 1–9.

Shampine, L.F. and Baca, L.S., Fixed vs. Variable order Runge–Kutta, ACM Trans. Math. Software, 12 (1986) pp. 1–23.

Shampine, L.F. and Gear, C.W., A User's View of Solving Stiff Ordinary Differential Equations, SIAM Rev., 21 (1979) pp. 1–17.

Shampine, L.F. and Gordon, M.K., Some Numerical Experiments with DIFSUB, SIGNUM Newsletter, 7 (1972) pp. 24–26.

Shampine, L.F. and Gordon, M.K., Computer Solution of Ordinary Differential Equations: The Initial Value Problem, Freeman, San Francisco, 1975.

Shampine, L.F. and Watts, H.A., Comparing Error Estimators for Runge–Kutta Methods, Math. Comp., 25 (1971) pp. 445–455.

Shampine, L.F. and Watts, H.A., Practical Solution of Ordinary Differential Equations by Runge–Kutta Methods, Rept. SAND 76–0585, Sandia National Laboratories, Albuquerque, NM, 1976a.

Shampine, L.F. and Watts, H.A., Global Error Estimation for Ordinary Differential Equations, ACM Trans. Math. Software, 2 (1976b) pp. 172–186.

Shampine, L.F. and Watts, H.A., Algorithm 504, GERK: Global Error Estimation for Ordinary Differential Equations, ACM Trans. Math. Software, 2 (1976c) pp. 200–203.

Shampine, L.F. and Watts, H.A., The Art of Writing a Runge–Kutta Code, Part I, pp. 257–275 in Rice, J.R., ed., Mathematical Software III, Academic, New York, 1977.

Shampine, L.F. and Watts, H.A., The Art of Writing a Runge–Kutta Code, II, Appl. Math. Comp., 5 (1979) pp. 93–121.

Shampine, L.F. and Watts, H.A., DEPAC–Design of a User Oriented Package of ODE Solvers, Sandia National Laboratories Rept. SAND79–2374, Albuquerque, NM, 1980.

Shampine, L.F. and Watts, H.A., Software for Ordinary Differential Equations, pp. 112–133 in Cowell, W.R., ed., Sources and Development of Mathematical Software, Prentice-Hall, Englewood Cliffs, NJ, 1984.

Shampine, L.F., Watts, H.A., and Davenport, S.M., Solving Nonstiff Ordinary Differential Equations—the State of the Art, SIAM Rev., 18 (1976) pp. 376–411.

Shampine, L.F. and Zhang, W., Convergence of LMM When the Solution is Not Smooth, Comp. & Maths. with Applics., 15 (1988) pp. 213–220.

Shampine, L.F. and Zhang, W., Rate of Convergence of Multistep Codes Started by Variation of Order and Stepsize, SIAM J.Numer. Anal., 27 (1990) pp. 1506–1518.

Sincovec, R.F. and Madsen, N.K., Software for Nonlinear Partial Differential Equations, ACM Trans. Math. Software, 1 (1975a) pp. 232–260.

Sincovec, R.F. and Madsen, N.K., Algorithm 494 PDEONE Solutions of Systems of Partial Differential Equations, ACM Trans. Math. Software, 1 (1975b) pp. 261–263.

Sod, G.A., Numerical Methods in Fluid Dynamics, Cambridge University Press, Cambridge, 1985.

Speckhard, F.H. and Green, W.L., A Guide to Using CSMP—the Continuous System Modelling Program, Prentice-Hall, Englewood Cliffs, NJ, 1976.

Stetter, H.J., Analysis of Discretization Methods for Ordinary Differential Equations, Springer, New York, 1973.

Strang, G., Linear Algebra and its Applications, 3rd ed., Harcourt, Brace, Jovanovich, San Diego, 1988.

Strikwerda, J.C., Finite Difference Schemes and Partial Differential Equations, Wadsworth, Pacific Grove, CA, 1989.

Watson, L.T., A Globally Convergent Algorithm for Computing Fixed Points of C2 Maps, Appl. Math. Comp., 5 (1979) pp. 297–311.

Watts, H.A., Survey of Numerical Methods for Ordinary Differential Equations, pp. 127–158 in Erisman, A.M., Neves, K.W., and Dwarakanath, M.H., eds., Electric Power Problems: the Mathematical Challenge, SIAM, Philadelphia, 1980.

Watts, H.A., Starting Step Size for an ODE Solver, J.Comp. Appl. Math., 9 (1983) pp. 177–191.

Watts, H.A. and Shampine, L.F., Smoother Interpolants for Adams Codes, SIAM J.Sci. Stat. Comput., 7 (1986) pp. 334–345.

Wheeler, D.J., Note on the Runge–Kutta Method, Computer Journal, 2 (1959) p. 23.

White, F.M., Viscous Fluid Flow, 2nd ed., McGraw-Hill, New York, 1991.

Whitham, G.B., Linear and Nonlinear Waves, Wiley, New York, 1974.

Willoughby, R.A., ed., Stiff Differential Systems, Plenum Press, New York, 1973.

Zienkiewicz, O.C. and Taylor, R.L., The Finite Element Method, Vol. I, 4th ed., McGraw-Hill, London, 1988.

Zonneveld, J.A., Automatic Numerical Integration, Mathematisch Centrum, Amsterdam, 1964.

§3 Cited Codes

Some codes are cited often enough by name in this book that it is useful to collect them here with references. See also §5 of Chapter 3 which provides some information about how to acquire software.

DDRIV2—see Kahaner, Moler and Nash [1989]

DEABM—a modernization of ODE/STEP, INTRP that appears in DEPAC

DERKF—a modernization of RKF45 that appears in DEPAC

DEPAC—see Shampine and Watts [1980]

Differential Systems—see Gollwitzer [1991]

DIFSUB—see Gear [1971a, 1971b, 1971c]

DIVPAG—one of the Gear line of codes that appears in the IMSL library

DIVPRK—a modernization of DVERK that appears in the IMSL library

DOPRI5—see Hairer, Nørsett, and Wanner [1987]

D02E??—several levels of a code that appears in the NAG library

DVERK—see Hull, Enright, and Jackson [1975]

DVDQ—see Krogh [1971]

EPISODE—see Byrne and Hindmarsh [1975]

GERK—see Shampine and Watts [1976b, 1976c]

LINPACK—see Dongarra et al. [1979]

LSODE—see Hindmarsh [1980]

LSODES—a version of LSODE for problems with sparse Jacobians. See Hindmarsh [1983]

MATLAB—see Moler et al. [1987]

METAN1—see Deuflhard [1983]

MDEP—see Buchanan [1992]

ODE/STEP, INTRP—see Shampine and Gordon [1975]

PHASER—see Koçak [1989]

RKF—see Shampine and Allen [1973]

RKF4—see Hull and Enright [1974]

RKF45—see Shampine and Watts [1976a, 1977, 1979]

SPRINT—a package written by M.Berzins and R.M.Furzeland that forms the basis for the codes in the NAG library for problems with sparse Jacobians.

VODE—see Brown, Byrne and Hindmarsh [1989]

Appendix

Some Mathematical Tools

In this appendix some of the mathematical tools that are used throughout the book are described. They are basic results in the calculus of functions of several variables, the theory of matrices, the theory of ordinary differential equations, and the theory of polynomial interpolation. Most of the results would be seen in a first course devoted to the topic, but it is useful to state them here for reference. Some of the results are deeper, but the reader need only accept them because our use of the results will be straightforward. The concept of the order of a quantity and how to work with this concept are less likely to be familiar and they are essential to the study undertaken here. The reader who is not familiar with this tool must study carefully §2 of Chapter 1. The basic results are restated here for purposes of reference.

§1 Order

Throughout this book we must talk about quantities like numerical errors tending to zero and we shall be vitally concerned with *how fast* they tend to zero. The concept of *order of convergence* or *rate of convergence* quantifies this for us.

A function $g_1(h)$ of a positive parameter h is said to be of order p, written symbolically as $\mathbb{O}(h^p)$ (and spoken as "big oh of h to the p"), if there is a constant C_1 such that for all sufficiently small h, say $0 < h \leq h_1$,

$$|g_1(h)| \leq C_1\, h^p.$$

The parameter h does not have to be positive and often it is not. The assumption then is that there is a constant h_1 such for all $0 < |h| \leq h_1$, we have $|g_1(h)| \leq C_1\, |h|^p$. In this book the order p is typically a positive integer, but this is also not necessary and there are occasions when fractional powers appear. The following properties of order follow easily from the basic definition:

If $g_1(h)$ is $\mathcal{O}(h^p)$, $g_2(h)$ is $\mathcal{O}(h^q)$, and γ is any constant, then

(1) the function $g_1(h)$ is $\mathcal{O}(h^r)$ for any $r < p$,
(2) the function $\gamma g_1(h)$ is $\mathcal{O}(h^p)$,
(3) the function $g_1(h)g_2(h)$ is $\mathcal{O}(h^{p+q})$, and
(4) the function $g_1(h) + g_2(h)$ is $\mathcal{O}(h^s)$ where $s = \min(p, q)$.

These rules say that a quantity of order p is also of order $r < p$, multiplying a quantity of order p by a constant does not change its order, multiplying quantities of orders p and q results in a quantity of order $p + q$, and adding quantities of orders p and q results in a quantity of order $s = \min(p, q)$.

Another important rule is that for a composite expression: if $g(x)$ is $\mathcal{O}(x^p)$ and x is $\mathcal{O}(\delta^q)$, then $g(\delta)$ is $\mathcal{O}(\delta^{pq})$.

We shall need some approximations obtained from series. The basic result is that if $y(x)$ is any function that is continuous along with its first p derivatives in an interval I containing x_0, then Taylor's theorem with remainder implies that for $\delta = x - x_0$,

$$y(x) = y(x_0) + y'(x_0)\,\delta + \ldots + \frac{1}{(p-1)!}\,y^{(p-1)}(x_0)\delta^{p-1} + \mathcal{O}(\delta^p).$$

Examples used throughout the book are

$$e^x = 1 + x + \mathcal{O}(x^2) = 1 + x + \frac{1}{2}x^2 + \mathcal{O}(x^3) = \ldots.$$

We say that $1 + x$ approximates $\exp(x)$ to order 2, $1, + x + ½\, x^2$ approximates $\exp(x)$ to order 3, and so forth. Another important expansion is the binomial series

$$(1 + x)^\gamma = 1 + \gamma x + \frac{\gamma(\gamma-1)}{2!}x^2 + \ldots + \frac{\gamma(\gamma-1)(\gamma-2)\cdots(\gamma-k+1)}{k!}x^k + \ldots,$$

which converges for $x^2 < 1$. We use it, for example, to approximate $\sqrt{1 + x}$ to order 2 by $1 + ½x$.

§2 Vectors, Matrices, and Norms

The set of all real numbers is denoted by R. Vectors of n real components, $\mathbf{v} \in R^n$, are indicated by printing them in bold face. They are always column vectors, but it is often convenient to write them in the form $\mathbf{v} = (v_1, v_2, \ldots, v_n)^T$ using the transpose operator T.

In order to talk about how well one vector approximates another, we must have a measure of the "size" of a vector $\mathbf{v}$. This is provided by its norm, $\|\mathbf{v}\|$. Norms are commonly taken up in introductions to linear algebra, but a person who is not very practiced in their use will have no difficulty if it is kept in mind that a norm generalizes to a vector the idea of absolute value for a scalar. A *(vector norm)* is a scalar function with one vector as its argument such that for all vectors $\mathbf{v}, \mathbf{u} \in R^n$ and all scalars $\alpha \in R$,

$$\|\mathbf{v}\| \geq 0 \quad \text{and} \quad \|\mathbf{v}\| = 0 \quad \text{if, and only if,} \quad \mathbf{v} = \mathbf{0}$$

$$\|\alpha\ \mathbf{v}\| = |\alpha|\ \|\mathbf{v}\|$$

$$\|\mathbf{v} + \mathbf{u}\| \leq \|\mathbf{v}\| + \|\mathbf{u}\|$$

Only two norms are common in the numerical solution of the initial value problem for a system of ordinary differential equations. The more familiar is the usual Euclidean length of the vector $\mathbf{v} = (v_1, v_2, \ldots, v_n)^T$:

$$\|\mathbf{v}\|_2 = \left(\sum_{i=1}^{n} |v_i|^2\right)^{1/2}.$$

The subscript 2 is used to indicate this particular norm, so this *Euclidean norm* is often referred to as the *two norm*. The other common norm is the *infinity norm*, or *max norm*,

$$\|\mathbf{v}\|_\infty = \max_{1 \leq i \leq n} |v_i|,$$

In the codes weighted norms are used to require that some component be calculated more accurately than another. These norms arise by specifying positive weights w_i, $i = 1, 2, \ldots, n$, defining the diagonal matrix $D = \mathrm{diag}\{w_1, w_2, \ldots, w_n\}$, and then defining $\|\mathbf{u}\|_w = \|\mathrm{D}\ \mathbf{u}\|$. Because these weighted norms are not qualitatively different from the unweighted norms, we need say no more about them. A common variation on the Euclidean norm is a scaled version called the *RMS norm (root-mean-square norm)*,

$$\|\mathbf{v}\| = \left(\frac{1}{n}\sum_{i=1}^{n} |v_i|^2\right)^{1/2},$$

but it is also not qualitatively different. A norm that is occasionally used is the *one norm*,

$$\|\mathbf{v}\|_1 = \sum_{i=1}^{n} |v_i|.$$

The concept of *inner product* will sometimes be convenient. An inner product on R^n is a scalar function with two vectors as its arguments such that for all vectors $\mathbf{v}$, $\mathbf{u}$, and $\mathbf{w} \in R^n$ and all scalars α and $\beta \in R$,

$$\langle \mathbf{v}, \mathbf{v} \rangle \geq 0 \quad \text{and} \quad \langle \mathbf{v}, \mathbf{v} \rangle = 0 \quad \text{if, and only if,} \quad \mathbf{v} = \mathbf{0}$$

$$\langle \mathbf{v}, \mathbf{u} \rangle = \langle \mathbf{u}, \mathbf{v} \rangle$$

$$\langle \alpha \, \mathbf{v} + \beta \, \mathbf{u}, \mathbf{w} \rangle = \alpha \langle \mathbf{v}, \mathbf{w} \rangle + \beta \langle \mathbf{u}, \mathbf{w} \rangle$$

The scalar or dot product in R^n provides the Euclidean inner product: $\mathbf{v}^T \mathbf{v} = \mathbf{v} \cdot \mathbf{v} = \langle \mathbf{v}, \mathbf{v} \rangle$. It is readily verified that if $\langle \cdot, \cdot \rangle$ is any inner product, then a norm is induced by the definition

$$\|\mathbf{v}\| = \sqrt{\langle \mathbf{v}, \mathbf{v} \rangle}.$$

In the case of the Euclidean inner product, this is the usual Euclidean norm. An important result connecting the inner product and the norm is the *Cauchy–Schwarz inequality*:

$$|\langle \mathbf{u}, \mathbf{v} \rangle| \leq \|\mathbf{u}\| \, \|\mathbf{v}\|$$

Not all norms can be derived from inner products; the max norm is sometimes inconvenient for just this reason.

A *matrix norm* is a scalar function such that for all $n \times n$ matrices A and B and all scalars $\alpha \in R$,

$$\|A\| \geq 0 \quad \text{and} \quad \|A\| = 0 \quad \text{if, and only if, } A = 0$$

$$\|\alpha A\| = |\alpha| \, \|A\|$$

$$\|A + B\| \leq \|A\| + \|B\|$$

Unlike vectors, we can multiply matrices as well as add them. In this book we work only with matrix norms that have the additional property

$$\|AB\| \leq \|A\| \, \|B\|.$$

Often the manipulations in this book involve both vectors and matrices, so it is convenient to require that the two norms be *compatible norms* in the sense that

$$\|M \, \mathbf{v}\| \leq \|M\| \, \|\mathbf{v}\|.$$

For any given vector norm, a compatible matrix norm can be defined by

$$\|M\| = \max_{\mathbf{x} \neq \mathbf{0}} \frac{\|M \, \mathbf{x}\|}{\|\mathbf{x}\|}.$$

The matrix norm defined in this way is called the *subordinate matrix norm*. In the case of the max vector norm, the subordinate matrix norm is found to be

$$\|M\|_\infty = \max_{1 \le i \le n} \sum_{j=1}^{n} |M_{ij}|,$$

and in the case of the one vector norm, the subordinate matrix norm is found to be

$$\|M\|_1 = \max_{1 \le j \le n} \sum_{i=1}^{n} |M_{ij}|.$$

The matrix norm subordinate to the Euclidean vector norm is not readily computed. A compatible norm that is easily computed is the *Frobenius norm*,

$$\|M\|_F = \left(\sum_{i=1}^{n} \sum_{j=1}^{n} |M_{ij}|^2 \right)^{1/2}.$$

A sequence of vectors $\{\mathbf{v}_m\}$ is said to converge to a limit $\mathbf{v}$ if $\|\mathbf{v}_m - \mathbf{v}\| \to 0$ as $m \to \infty$. This definition raises the question, Might a sequence converge in one norm and not in another? In R^n the answer is no, because all norms are equivalent in a certain sense. We say that two norms $\|\cdot\|$ and $\|\cdot\|_*$ are *equivalent norms* if there are constants α and β such that for all vectors $\mathbf{v} \neq \mathbf{0}$,

$$0 < \alpha\|\mathbf{v}\| \le \|\mathbf{v}\|_* \le \beta\|\mathbf{v}\|.$$

(It takes only a moment to see that this implies that

$$0 < \alpha'\|\mathbf{v}\|_* \le \|\mathbf{v}\| \le \beta'\|\mathbf{v}\|_*$$

where $\alpha' = 1/\beta$ and $\beta' = 1/\alpha$.) For example, it is easy to see that

$$0 < 1\|\mathbf{v}\|_\infty \le \|\mathbf{v}\|_2 \le \sqrt{n}\|\mathbf{v}\|_\infty.$$

Convergence in one norm implies convergence in any equivalent norm: if $\|\mathbf{v}_m - \mathbf{v}\|$ tends to 0 as $m \to \infty$, then so does $\|\mathbf{v}_m - \mathbf{v}\|_*$ because $\beta\|\mathbf{v}_m - \mathbf{v}\| \ge \|\mathbf{v}_m - \mathbf{v}\|_*$. More generally, the rules of order tell us that the rate of convergence does not depend on the norm. There is an important theorem that all vector norms on R^n are equivalent, and there is a similar result for matrix norms. One of the things this tells us is that we can prove convergence in whichever norm happens to be convenient for our manipulations and then rely upon this general result to deduce convergence in the norm that interests us. In particular, the use of weights in measuring the size of vectors does not affect their convergence. The equivalence of the Euclidean and max norms involves the dimension n of the space R^n.

This suggests that the "facts of life" might be different in an infinite dimensional space and indeed they are.

§3 Multivariate Calculus

With norms to measure the size of vectors, the usual definitions of limit, continuity, and derivative extend in a straightforward way to vector-valued functions. We shall need to work with functions $\mathbf{F}(x, \mathbf{u})$ from $R \times R^n$ to R^n. In the context of ordinary differential equations the concept of a *Lipschitz condition* is of great importance. The function $\mathbf{F}$ is said to satisfy a Lipschitz condition with constant L in a region if

$$\|\mathbf{F}(x, \mathbf{u}) - \mathbf{F}(x, \mathbf{v})\| \leq L\|\mathbf{u} - \mathbf{v}\|$$

for all $(x, \mathbf{u})$ and $(x, \mathbf{v})$ in the region. Of course the constant L is not uniquely defined, but it is usual to have in mind the smallest value possible for the region of interest. It is important to have a sense of Lipschitz conditions. First note that it implies continuity in the dependent variable $\mathbf{u}$ because if $\mathbf{v}$ is "close" to $\mathbf{u}$, then $\mathbf{F}(x, \mathbf{v})$ is "close" to $\mathbf{F}(x, \mathbf{u})$. We can get some insight by considering $\mathbf{F}$ that are linear: $\mathbf{F}(x, \mathbf{u}) = J(x)\mathbf{u} + \mathbf{g}(x)$. For such functions,

$$\|\mathbf{F}(x, \mathbf{u}) - \mathbf{F}(x, \mathbf{v})\| = \|J(x)(\mathbf{u} - \mathbf{v})\| \leq \|J(x)\|\,\|\mathbf{u} - \mathbf{v}\|$$

Such a function satisfies a Lipschitz condition, is *Lipschitzian*, if, and only if, the norm of the matrix $J(x)$ is bounded for all x in the region of interest. It is useful to observe that the norm of $J(x)$ is bounded if, and only if, the magnitudes of all its entries are bounded. The situation for more general $\mathbf{F}$ can be understood by using a mean value theorem that we take up in a moment, but first let us introduce some notation for derivatives.

In addition to the usual notation for derivative like $\partial F/\partial y$, it is often convenient to use the common subscript notation F_y. The derivative of a function $\mathbf{F} \in R^n$ with respect to a scalar y is a (column) vector with components that are the derivatives of the corresponding components in $\mathbf{F}$:

$$\frac{\partial \mathbf{F}}{\partial y} = \mathbf{F}_y = \left(\frac{\partial F_1}{\partial y}, \frac{\partial F_2}{\partial y}, \ldots, \frac{\partial F_n}{\partial y}\right)^T$$

A less common notation that will prove useful is the derivative of a function $F \in R$ with respect to a vector $\mathbf{y} \in R^m$. By this is meant the *row* vector of m components that are the derivatives of $\mathbf{F}$ with respect to each of the

m components of $\mathbf{y}$:

$$\frac{\partial F}{\partial \mathbf{y}} = F_{\mathbf{y}} = \left(\frac{\partial F}{\partial y_1}, \frac{\partial F}{\partial y_2}, \ldots, \frac{\partial F}{\partial y_m}\right)$$

The reason for introducing a row vector here rather than the more familiar notation for the gradient of F (∇F or grad(F)) is that this derivative and notation is a special case of the derivative of a function $\mathbf{F} \in R^n$ with respect to a vector $\mathbf{y} \in R^m$. This general derivative is an $n \times m$ matrix which has as its (i, j) entry the partial derivative of component i of $\mathbf{F}$ with respect to component j of $\mathbf{y}$:

$$\frac{\partial \mathbf{F}}{\partial \mathbf{y}} = \mathbf{F}_{\mathbf{y}} = \begin{matrix} \frac{\partial F_1}{\partial y_1} & \frac{\partial F_1}{\partial y_2} & \cdots & \frac{\partial F_1}{\partial y_m} \\ \frac{\partial F_2}{\partial y_1} & \frac{\partial F_2}{\partial y_2} & \cdots & \frac{\partial F_2}{\partial y_m} \\ \vdots & \vdots & & \vdots \\ \frac{\partial F_n}{\partial y_1} & \frac{\partial F_n}{\partial y_2} & \cdots & \frac{\partial F_n}{\partial y_m} \end{matrix}$$

This matrix is called the *Jacobian matrix*, or simply *Jacobian*, of $\mathbf{F}$. It will be very important in our study of the numerical solution of ordinary differential equations. Indeed, we shall use it right away to formulate mean value theorems for vector-valued functions.

The usual mean value theorem for a scalar function $F(x, u)$ states that if F has a continuous first partial derivative with respect to u in a region of interest, then

$$F(x, u) - F(x, v) = \mathcal{J}(u - v) = \frac{\partial F}{\partial y}(x, z)(u - v),$$

where $\mathcal{J} = \partial F/\partial y$ is evaluated at some point z between u and v. There is a similar result for scalar functions of vector arguments, $F(x, \mathbf{u})$. This result can be applied to the components of a function $\mathbf{F}$ to obtain a mean value theorem for vector-valued functions. For each component i of $\mathbf{F}$, there is a point $\mathbf{z^i}$ on the line between $\mathbf{u}$ and $\mathbf{v}$ such that

$$F_i(x, \mathbf{u}) - F_i(x, \mathbf{v}) = \frac{\partial F}{\partial \mathbf{y}}(x, \mathbf{z}^i)(\mathbf{u} - \mathbf{v}).$$

Assembling all the components into a vector, we obtain a *mean value theorem* for vector-valued functions of vector arguments:

$$\mathbf{F}(x, \mathbf{u}) - \mathbf{F}(x, \mathbf{v}) = \mathcal{J}(\mathbf{u} - \mathbf{v}) = \left(\frac{\partial F_i}{\partial y_j}(x, \mathbf{z}^i)\right)(\mathbf{u} - \mathbf{v}).$$

In the linear case the matrix $\mathcal{J}$ here is the Jacobian of **F**, but it is not in the nonlinear case because the partial derivatives are evaluated at different values of the dependent variables on the line between **u** and **v**. In the scalar case, expansion of the partial derivative about v leads to

$$F(x,u) - F(x,v) = \frac{\partial F}{\partial y}(x,v)(u-v) + \mathbb{O}(|u-v|^2) \approx \frac{\partial F}{\partial y}(x,v)(u-v).$$

A similar expansion in the vector case leads to

$$\begin{aligned}\mathbf{F}(x, \mathbf{u}) - \mathbf{F}(x, \mathbf{v}) &= \frac{\partial \mathbf{F}}{\partial \mathbf{y}}(x, \mathbf{v})(\mathbf{u} - \mathbf{v}) + \mathbb{O}(\|\mathbf{u} - \mathbf{v}\|^2) \\ &\approx \frac{\partial \mathbf{F}}{\partial \mathbf{y}}(x, \mathbf{v})(\mathbf{u} - \mathbf{v}).\end{aligned}$$

This form involving the Jacobian of **F** is often the more convenient.

With a mean value theorem, the meaning of a Lipschitz condition can be discussed for non-linear **F** in a way almost identical to that used for linear **F**:

$$\|\mathbf{F}(x, \mathbf{u}) - \mathbf{F}(x, \mathbf{v})\| = \|\mathcal{J}(\mathbf{u} - \mathbf{v})\| \leq \|\mathcal{J}\| \, \|\mathbf{u} - \mathbf{v}\|$$

A Lipschitz constant for F is provided by a bound on an appropriate norm of the matrix of partial derivatives that is valid for all $(x, \mathbf{u})$ in the region of interest. From the definitions of the matrix norms stated earlier, this is equivalent to a bound on the magnitudes of all the partial derivatives $\partial F_i/\partial y_j$.

§4 Initial Value Problems for Ordinary Differential Equations

A standard *existence* theorem is that if $\mathbf{F}(x, \mathbf{y})$ is continuous on an open region D of $R \times R^n$ containing the initial point $(a, \mathbf{A})$, there is *at least* one solution $\mathbf{y}(x)$ of

$$\mathbf{y}' = \mathbf{F}(x, \mathbf{y}), \qquad \mathbf{y}(a) = \mathbf{A}$$

in the region. (A region D is said to be *open* if for any point $(x, \mathbf{y})$ in D, there is an $\varepsilon > 0$ such that any point $(t, \mathbf{z})$ that is within ε of $(x, \mathbf{y})$ also lies in D.) The hypothesis that the region is open means that the initial point $(a, \mathbf{A})$ cannot be on the boundary of the region where $\mathbf{F}(x, \mathbf{y})$ is defined. Nearly all the problems seen in practice will satisfy these hypotheses, at least in subregions. The theorem goes on to say that each solution exists on a maximal interval $\alpha < x < \beta$ containing a, and as $x \to \alpha$ or $x \to \beta$, the solution tends to the boundary of the region. As simple examples taken up in Chapter 1 show, it is not necessarily the case that there is only one solution to the initial value problem.

A fundamental issue in the numerical solution of ordinary differential equations is the effect of small changes to the problem. One way to investigate this issue is by means of Gronwall's inequality. It is presented in most intermediate level books on the theory of ordinary differential equations, but because of its importance in the present context and because the proof is short, it is given here.

Gronwall's Lemma Let $g(x)$ be *continuous and non-negative for* $a \leq x \leq b$, *and let* $K \geq 0$, $L \geq 0$ *be constants. If*

$$g(x) \leq K + L \int_a^x g(t)\, dt \quad \textit{for} \quad a \leq x \leq b,$$

then

$$g(x) \leq K\, e^{L(x-a)} \quad \textit{for} \quad a \leq x \leq b.$$

Proof. Let

$$w(x) = K + L \int_a^x g(t)\, dt.$$

Then $g(x) \leq w(x)$ *and* $w'(x) = L\, g(x)$. *This implies that*

$$\begin{aligned} \frac{d}{dx}\left[w(x)e^{-L(x-a)}\right] &= w'(x)e^{-L(x-a)} - L\, w(x)e^{-L(x-a)} \\ &= e^{-L(x-a)}[L\, g(x) - L\, w(x)] \leq 0. \end{aligned}$$

This last inequality says that $w(x)\exp(-L(x-a))$ *is non-increasing for* $a \leq x \leq b$, *hence*

$$w(x)e^{-L(x-a)} \leq w(a) = K.$$

Finally,

$$g(x) \leq w(x) \leq K\, e^{L(x-a)}.$$

First let us apply the lemma to bounding the effect of a change of the initial values, or equivalently, to bounding how fast two solutions can spread apart. Suppose that in addition to the conditions of the existence theorem, $\mathbf{F}$ satisfies a Lipschitz condition with constant L. Let $\mathbf{u}(x)$ and $\mathbf{v}(x)$ be any two solutions of the differential equation:

$$\mathbf{u}'(x) = \mathbf{F}(x, \mathbf{u}(x)) \quad \text{and} \quad \mathbf{v}'(x) = \mathbf{F}(x, \mathbf{v}(x)) \quad \text{for} \quad a \leq x \leq b$$

Integrating the equations leads to

$$\mathbf{u}(x) = \mathbf{u}(a) + \int_a^x \mathbf{F}(t, \mathbf{u}(t))\, dt$$

and similarly for $\mathbf{v}(x)$. Subtraction yields

$$\mathbf{u}(x) - \mathbf{v}(x) = \mathbf{u}(a) - \mathbf{v}(a) + \int_a^x \mathbf{F}(t, \mathbf{u}(t)) - \mathbf{F}(t, \mathbf{v}(t))\, dt.$$

Taking norms and using the Lipschitz condition,

$$\|\mathbf{u}(x) - \mathbf{v}(x)\| \leq \|\mathbf{u}(a) - \mathbf{v}(a)\| + L \int_a^x \|\mathbf{u}(t) - \mathbf{v}(t)\|\, dt.$$

In Gronwall's inequality take $K = \|\mathbf{u}(a) - \mathbf{v}(a)\|$ and $g(x) = \|\mathbf{u}(x) - \mathbf{v}(x)\|$ to conclude that

$$\|\mathbf{u}(x) - \mathbf{v}(x)\| \leq \|\mathbf{u}(a) - \mathbf{v}(a)\| e^{L(x-a)} \quad \text{for} \quad a \leq x \leq b.$$

This important inequality bounds how fast two solutions of the differential equation can spread apart. In particular, it implies that an initial value problem with Lipschitzian $\mathbf{F}$ can have *at most* one solution because if two solutions agree at one point they cannot separate thereafter. This is a standard *uniqueness* theorem.

Gronwall's inequality can be used to show that if $\mathbf{F}$ is continuous on $[a, b] \times R^n$ and satisfies a Lipschitz condition there, the initial value problem

$$\mathbf{y}' = \mathbf{F}(x, \mathbf{y}), \qquad a \leq x \leq b, \qquad \mathbf{y}(a) = \mathbf{A}$$

has a unique solution that extends all the way from a to b. This differs in two ways from the general theorems just stated: The integration is started on a boundary $x = a$ of the region where $\mathbf{F}$ is defined, and it is asserted that the solution extends all the way to the boundary $x = b$ (rather than becoming unbounded at some $x < b$).

We also need to bound the effect of changing $\mathbf{F}$ on the solution of an initial value problem. We prove shortly a bound that can be obtained easily when the norm comes from an inner product as, e.g., the Euclidean norm does. Specifically, if

$$\mathbf{u}'(x) = \mathbf{F}(x, \mathbf{u}(x)) \quad \text{and} \quad \mathbf{v}'(x) = \mathbf{F}(x, \mathbf{v}(x)) + \mathbf{G}(x) \quad \text{for} \quad a \leq x \leq b,$$

then

$$\|\mathbf{u}(x) - \mathbf{v}(x)\| \leq \|\mathbf{u}(a) - \mathbf{v}(a)\| e^{L(x-a)} + \int_a^x \|\mathbf{G}(t)\| e^{L(x-t)}\, dt.$$

This bound says that in the *classical situation* that $L(b - a)$ is "not large," small changes to the initial values $\mathbf{u}(a)$ and to the equation $\mathbf{F}$ lead to small changes in the solution $\mathbf{u}(x)$.

For the discussion of stiff initial value problems it is valuable to refine the concept of a Lipschitz condition using an inner product rather than a norm. First using the Cauchy–Schwarz inequality and then a Lipschitz condition on $\mathbf{F}$, we find that

$$\begin{aligned} |\langle \mathbf{F}(x, \mathbf{u}) - \mathbf{F}(x, \mathbf{v}), \mathbf{u} - \mathbf{v}\rangle| &\leq \|\mathbf{F}(x, \mathbf{u}) - \mathbf{F}(x, \mathbf{v})\| \, \|\mathbf{u} - \mathbf{v}\| \\ &\leq L\|\mathbf{u} - \mathbf{v}\|^2 \end{aligned}$$

This can be rewritten as

$$-L\|\mathbf{u} - \mathbf{v}\|^2 \leq \langle \mathbf{F}(x, \mathbf{u}) - \mathbf{F}(x, \mathbf{v}), \mathbf{u} - \mathbf{v}\rangle \leq L\|\mathbf{u} - \mathbf{v}\|^2.$$

The pair of inequalities is a consequence of the Lipschitz condition (and the fact that the norm is induced by an inner product). Because one of the inequalities is much more important for our purposes than the other, we study $\mathbf{F}$ that satisfy

$$\langle \mathbf{F}(x, \mathbf{u}) - \mathbf{F}(x, \mathbf{v}), \mathbf{u} - \mathbf{v}\rangle \leq \mathcal{L}\|\mathbf{u} - \mathbf{v}\|^2$$

for a constant $\mathcal{L}$. We continue to assume that $\mathbf{F}$ satisfies a Lipschitz condition, so there is no point to this *one-sided Lipschitz condition* unless the constant $\mathcal{L}$ is smaller than L. By definition L is non-negative, but it is quite possible that $\mathcal{L}$ be negative. Differential equations with $\mathbf{F}$ that satisfy a one-sided Lipschitz condition with constant $\mathcal{L} < 0$ are called *dissipative*.

This refinement of the concept of a Lipschitz constant allows the bounds on the effects of changes to initial values and to the equation to be refined. Because it is not difficult to prove a result that is illuminating and the matter is not ordinarily taken up in courses on differential equations, we do so. Let us consider how much the solution of

$$\mathbf{y}' = \mathbf{F}(x, \mathbf{y}), \qquad \mathbf{y}(a) \text{ given},$$

can differ from the solution of

$$\mathbf{v}' = \mathbf{F}(x, \mathbf{v}) + \mathbf{G}(x) \qquad \mathbf{v}(a) \text{ given},$$

when $\mathbf{F}$ satisfies a one-sided Lipschitz condition with constant $\mathcal{L}$. To this end, let

$$D(x) = e^{-2\mathcal{L}x}\|\mathbf{v}(x) - \mathbf{y}(x)\|^2 = e^{-2\mathcal{L}x}\langle \mathbf{v}(x) - \mathbf{y}(x), \mathbf{v}(x) - \mathbf{y}(x)\rangle.$$

In the case of the Euclidean inner product it is easy to see that the derivative of $D(x)$ is

$$D'(x) = -2\mathcal{L}e^{-2\mathcal{L}x}\|\mathbf{v}(x) - \mathbf{y}(x)\|^2 + 2e^{-2\mathcal{L}x}\langle \mathbf{v}'(x) - \mathbf{y}'(\mathbf{x}), \mathbf{v}(x) - \mathbf{y}(x)\rangle,$$

and it is not difficult to argue that this result is true of inner products in general. Now

$$\mathbf{v}'(x) - \mathbf{y}'(x) = \mathbf{F}(x, \mathbf{v}(x)) + \mathbf{G}(x) - \mathbf{F}(x, \mathbf{y}(x)),$$

so

$$\begin{aligned}\langle \mathbf{v}'(x) - \mathbf{y}'(x), \mathbf{v}(x) - \mathbf{y}(x)\rangle &= \langle \mathbf{F}(x, \mathbf{v}(x)) - \mathbf{F}(x, \mathbf{y}(x)), \mathbf{v}(x) - \mathbf{y}(x)\rangle \\ &\quad + \langle \mathbf{G}(x), \mathbf{v}(x) - \mathbf{y}(x)\rangle \\ &\le \mathcal{L}\|\mathbf{v}(x) - \mathbf{y}(x)\|^2 \\ &\quad + \|\mathbf{G}(x)\|\,\|\mathbf{v}(x) - \mathbf{y}(x)\|.\end{aligned}$$

Then

$$\begin{aligned}D'(x) &\le 2e^{-2\mathcal{L}x}\|\mathbf{G}(x)\|\,\|\mathbf{v}(x) - \mathbf{y}(x)\| = 2\|\mathbf{G}(x)\|e^{-\mathcal{L}x}\sqrt{D(x)} \\ &\le 2\|\mathbf{G}(x)\|e^{-\mathcal{L}x}\sqrt{D(x) + \tau}\end{aligned}$$

for any $\tau > 0$. This implies that

$$\frac{d}{dx}\sqrt{(D(x) + \tau)} \le \|\mathbf{G}(x)\|e^{-\mathcal{L}x})$$

and integration leads to

$$\sqrt{D(x) + \tau} - \sqrt{D(a) + \tau} \le \int_a^x \|\mathbf{G}(t)\|\mathrm{e}^{-\mathcal{L}t}\, dt.$$

Let $\tau \to 0$ in this inequality to get

$$e^{-\mathcal{L}x}\mathbf{v}(x) - \mathbf{y}(x)\| - e^{-\mathcal{L}a}\|\mathbf{v}(a) - \mathbf{y}(a)\| \le \int_a^x \|\mathbf{G}(t)\|\, e^{-\mathcal{L}t}\, dt.$$

A little manipulation then gives the desired bound

$$\|\mathbf{v}(x) - \mathbf{y}(x)\| \le \|\mathbf{v}(a) - \mathbf{y}(a)\|e^{\mathcal{L}(x-a)} + \int_a^x \|\mathbf{G}(t)\|e^{\mathcal{L}(x-t)}\, dt.$$

This bound is especially nice for dissipative problems ($\mathcal{L} < 0$) because the exponentials are then uniformly bounded by 1. It is always possible to take $\mathcal{L} = L$ here to get a result that takes no special advantage of the extra information of a one-sided Lipschitz condition—the result stated earlier in the context of Lipschitz conditions.

In addition to bounds on the effects of changes in initial conditions and equation, we shall need to approximate the actual changes induced in the solution. Let us suppose $\mathbf{F}(x, \mathbf{y})$ satisfies the conditions given above that guarantee that the problem

$$\mathbf{y}' = \mathbf{F}(x, \mathbf{y}) \quad \text{for} \quad a \le x \le b, \quad \mathbf{y}(a) = \mathbf{s},$$

has a unique solution $\mathbf{y}(x)$. Focussing our attention on component i of the vector of initial values, let us denote this solution as $\mathbf{y}(x; s_i)$. When $\mathbf{e}_i$ is column i of the identity matrix I, the vector $\mathbf{s} + \delta\mathbf{e}_i$ is the result of adding δ to component i of $\mathbf{s}$, so the solution of the problem

$$\mathbf{y}' = \mathbf{F}(x, \mathbf{y}) \quad \text{for} \quad a \le x \le b, \quad \mathbf{y}(a) = \mathbf{s} + \delta\mathbf{e}_i$$

is $\mathbf{y}(x; s_i + \delta)$. We can then ask whether $(\mathbf{y}(x; s_i + \delta) - \mathbf{y}(x; s_i))/\delta$ has a limit as $\delta \to 0$, i.e., whether the partial derivative $\partial\mathbf{y}/\partial s_i(x; s_i)$ exists. When it does, we can make the usual approximation in applied mathematics,

$$\mathbf{y}(x; s_i + \delta) \approx \mathbf{y}(x; s_i) + \frac{\partial \mathbf{y}}{\partial s_i}(x; s_i)\delta,$$

because the difference of the two sides tends to zero as δ tends to zero. If we were to proceed formally by taking the partial derivative of the initial value problem, we would obtain

$$\frac{\partial}{\partial s_i}\left(\frac{d\mathbf{y}}{dx}\right) = \frac{d}{dx}\left(\frac{\partial \mathbf{y}}{\partial s_i}\right) = \frac{\partial \mathbf{F}}{\partial \mathbf{y}}(x, \mathbf{y}(x))\frac{\partial \mathbf{y}}{\partial s_i} \quad \text{for} \quad a \le x \le b, \quad \frac{\partial \mathbf{y}}{\partial s_i}(a) = \mathbf{e}_i.$$

Formally the partial derivative $\mathbf{v}(x) = (\partial\mathbf{y}/\partial s_i)(x)$ is the solution, along with $\mathbf{y}(x)$, of the problem

$$\begin{aligned} \mathbf{y}' &= \mathbf{F}(x, \mathbf{y}) & \mathbf{y}(a) &= \mathbf{s}, \\ \mathbf{v}' &= \frac{\partial \mathbf{F}}{\partial \mathbf{y}}(x, \mathbf{y}(x))\mathbf{v} & \mathbf{v}(a) &= \mathbf{e}_i. \end{aligned}$$

Fortunately, these formal manipulations can be justified when $\mathbf{F}$ is smooth enough. A careful study of the formal approximation provides

$$\mathbf{y}(x; s_i + \delta) = \mathbf{y}(x; s_i) + \frac{\partial \mathbf{y}}{\partial s_i}(x; s_i)\delta + \mathbb{O}(\delta^2).$$

It is convenient to assemble all the partial derivatives with respect to the components of the initial vector $\mathbf{s}$ into a matrix function $V(x) = ((\partial\mathbf{y}/\partial s_1)(x), (\partial\mathbf{y}/\partial s_2)(x) \ldots, (\partial\mathbf{y}/\partial s_n)(x))$ and then write the equations in the compact form

$$\begin{aligned} \mathbf{y}' &= \mathbf{F}(x, \mathbf{y}) & \mathbf{y}(a) &= \mathbf{s}, \\ V' &= \frac{\partial \mathbf{F}}{\partial \mathbf{y}}(x, \mathbf{y})V & V(a) &= I, \end{aligned}$$

in terms of the Jacobian of $\mathbf{F}$ and the identity matrix I. The equation for the partial derivatives is called the *equation of first variation* or the *variational equation*. Notice that it is linear. Using the partial derivatives with respect to all components of $\mathbf{s}$, we find that the effect of changing the initial vector $\mathbf{s}$ to $\mathbf{s} + \boldsymbol{\delta}$ is to change the solution from $\mathbf{y}(x; \mathbf{s})$ to $\mathbf{y}(x; \mathbf{s} + \boldsymbol{\delta})$ where

$$\mathbf{y}(x; \mathbf{s} + \boldsymbol{\delta}) = \mathbf{y}(x; \mathbf{s}) + V(x)\boldsymbol{\delta} + \mathbb{O}(\|\boldsymbol{\delta}\|^2) \approx \mathbf{y}(x; \mathbf{s}) + V(x)\boldsymbol{\delta}.$$

In addition to the effects of changing the initial conditions, we shall need to approximate the effects of changing $\mathbf{F}$. Let us suppose then that $\mathbf{F}$ depends on a number of parameters that are written conveniently as a vector $\mathbf{p}$ of m components. Again, formal differentiation of the problem leads to a variational equation for the partial derivatives of the solution with respect to the parameters. Now the solution $\mathbf{y}(x; \mathbf{p})$ and the matrix of partial derivatives $V(x) = ((\partial \mathbf{y}/\partial p_1)(x), (\partial \mathbf{y}/\partial p_2)(x), \ldots, (\partial \mathbf{y}/\partial p_m)(x))$ satisfy the initial value problems

$$\mathbf{y}' = \mathbf{F}(x, \mathbf{y}, \mathbf{p}) \qquad \mathbf{y}(a; \mathbf{p}) = \mathbf{s},$$

$$V' = \frac{\partial \mathbf{F}}{\partial \mathbf{y}}(x, \mathbf{y}, \mathbf{p})V + \frac{\partial \mathbf{F}}{\partial \mathbf{p}}(x, \mathbf{y}, \mathbf{p}) \qquad V(a) = 0.$$

The effect of changing the parameters $\mathbf{p}$ to $\mathbf{p} + \boldsymbol{\delta}$ is to change the solution from $\mathbf{y}(x; \mathbf{p})$ to $\mathbf{y}(x; \mathbf{p} + \boldsymbol{\delta})$ where

$$\mathbf{y}(x; \mathbf{p} + \boldsymbol{\delta}) = \mathbf{y}(x; \mathbf{p}) + V(x)\boldsymbol{\delta} + \mathbb{O}(\|\boldsymbol{\delta}\|^2) \approx \mathbf{y}(x; \mathbf{p}) + V(x)\boldsymbol{\delta}.$$

Note the forms of the two variational equations. Both are linear equations for the partial derivatives. One is a homogeneous equation with non-zero initial values and the other is inhomogeneous with zero initial values. In the linear approximation, the effect of changing both the initial vector and a vector of parameters is to add the effects of the separate changes.

These results about how changes to the initial conditions and to the equation affect the solution underlie (regular) *perturbation theory*. The viewpoint of perturbation theory is often useful in the present context, so a result of this kind will be stated: If $\mathbf{y}(x)$ is the solution of the initial value problem

$$\mathbf{y}' = \mathbf{F}(x, \mathbf{y}), \qquad \mathbf{y}(a) = \mathbf{A},$$

and $\mathbf{u}(x, \varepsilon)$ the solution of

$$\mathbf{u}' = F(x, \mathbf{u}) + \varepsilon\, \mathbf{g}(x), \qquad \mathbf{u}(a) = \mathbf{A} + \varepsilon\, \boldsymbol{\delta}(a),$$

then

$$\mathbf{u}(x, \varepsilon) = \mathbf{y}(x) + \varepsilon\, \boldsymbol{\delta}(x) + \mathbb{O}(\varepsilon^2)$$

where

$$\boldsymbol{\delta}' = \frac{\partial \mathbf{F}}{\partial \mathbf{y}}(x, \mathbf{y}(x))\boldsymbol{\delta} + \mathbf{g}(x), \qquad \boldsymbol{\delta}(a) \text{ given.}$$

§5 Polynomial Interpolation

The study of the numerical solution of the initial value problem for systems of ordinary differential equations builds upon the theory of polynomial interpolation as taken up in most introductions to numerical analysis. Some basic results are stated here for purposes of reference.

Let I be the finite interval $[a, b]$ and suppose that the function $y(x)$ is continuous along with its first m derivatives in this interval. If x_0 is any point in (a, b), then Taylor's theorem with remainder states that for all $x \in I$,

$$y(x) = y(x_0) + y'(x_0)(x - x_0) + \cdots + \frac{1}{(m-1)!} y^{(m-1)}(x_0)(x - x_0)^{m-1} + R(x).$$

here

$$R(x) = \frac{1}{m!} y^{(m)}(\xi_x)(x - x_0)^m \quad \text{and} \quad a < \xi_x < b.$$

The polynomial

$$T(x) = y(x_0) + y'(x_0)(x - x_0) + \cdots + \frac{1}{(m-1)!} y^{(m-1)}(x_0)(x - x_0)^{m-1}$$

is the *Taylor polynomial.* It is the unique polynomial of degree at most $m - 1$ that *interpolates* $y(x)$ and its first $m - 1$ derivatives at a, that is, $T^{(j)}(a) = y^{(j)}(a)$ for $j = 0, 1, \ldots, m - 1$. For an interval of length $h = b - a$, continuity of the derivative of order m and the form of the remainder term imply that $T(x)$ approximates $y(x)$ to $\mathcal{O}(h^m)$.

Because derivatives may not be readily available, the classic task of polynomial interpolation is to interpolate $y(x)$ at m distinct points $\{x_j\}$ in the interval I and the classic result is that there is a unique polynomial $P_m(x)$ of degree at most $m - 1$ that does this, i.e., $P_m(x_j) = y(x_j)$ for $j = 1, 2, \ldots, m$. The classic error expression for the error of approximation for all $x \in I$ is

$$y(x) - P_m(x) = \frac{1}{m!} y^{(m)}(\xi_x) \prod_{j=1}^{m} (x - x_j) \quad \text{and} \quad a < \xi_x < b.$$

It is also the case that derivatives of the interpolant approximate corresponding derivatives of $y(x)$. For example, when the nodes $\{x_j\}$ are ordered so that $a \leq x_1 < x_2 < \ldots < x_m \leq b$, it is proven in Isaacson and Keller [1966] that for any $r < m$

$$y^{(r)}(x) - P_m^{(r)}(x) = \frac{1}{(m - r)!} y^{(m)}(\xi_x) \prod_{j=1}^{m-r} (x - \xi_j) \quad \text{and}$$

$$x_j < \xi_j < x_{j-r} \quad \text{for} \quad 1 \leq j \leq m - r.$$

For an interval of length $h = b - a$, continuity of the derivative of order m and the form of the remainder term implies that $P_m^{(r)}(x)$ approximates $y^{(r)}(x)$ to $\mathbb{O}(h^{m-r})$.

Although the interpolating polynomial is unique, there are different ways of writing it that are useful in different contexts. The *Lagrangian form* is

$$P_m(x) = \sum_{k=1}^{m} y(x_k) L_k(x),$$

where the *fundamental Lagrangian interpolating polynomials* are

$$L_k(x) = \prod_{\substack{j=1 \\ j \neq k}}^{m} \left(\frac{x - x_j}{x_k - x_j} \right), \qquad k = 1, 2, \ldots, m.$$

This form is convenient because the data interpolated, the $y(x_k)$, are exposed. For some purposes it is more convenient to expose the nodes. The *Newtonian form* is

$$P_m(x) = y[x_1] + y[x_1, x_2](x - x_1) + y[x_1, x_2, x_3](x - x_1)(x - x_2)$$
$$+ \ldots + y[x_1, x_2, \ldots x_m](x - x_1)(x - x_2) \ldots (x - x_{m-1}).$$

Here the quantities $y[x_1], y[x_1, x_2], \ldots$ are the *divided differences* of order $0, 1, \ldots$. They can be expressed in terms of the data and the nodes as

$$y[x_1, x_2, \ldots, x_k] = \sum_{i=1}^{k} y(x_i) \prod_{\substack{j=1 \\ j \neq i}}^{k} \frac{1}{(x_i - x_j)}.$$

This expression is not an efficient way to compute this divided difference of order $k - 1$, but it shows the important fact that the difference depends only on the data corresponding to the first k nodes. It also shows the important fact that the difference does not depend on the order in which the k nodes are taken. The Newtonian form of the interpolating polynomial is analogous to the Taylor polynomial. The closeness of the analogy is made clearer by the result that

$$y[x_1, x_2, \ldots, x_k] = \frac{1}{(k-1)!} y^{(k-1)}(\xi)$$

for some ξ in the smallest interval containing the nodes $x_1, x_2, \ldots, x_k$. The Taylor polynomial may be thought of as a limit of the interpolating polynomial as the nodes coalesce at a point x_0. The Newtonian form shares an important qualitative property with the Taylor form. To interpolate at an additional node x_{k+1}, all that is necessary is to add a term. Specifically,

$$P_{m+1}(x) = P_m(x) + y[x_1, x_2, \ldots, x_m, x_{m+1}] \prod_{j=1}^{m} (x - x_j)$$

This fact is exploited when trying to decide how many nodes to use.

In this book we do not require any details about the computation of divided differences and the Newtonian form of the interpolating polynomial. Some details are taken up in the simpler special case of nodes that are equally spaced, namely $x_{n-j} = x_n - jh$ for $j = 0, 1, \ldots$. In this context the backward differences that are scaled versions of divided differences prove to be convenient. A *backward difference* with step size h of a vector-valued function $\mathbf{f}(x)$ is defined by

$$\nabla \mathbf{f}(x) = \mathbf{f}(x) - \mathbf{f}(x-h).$$

Higher order differences are defined recursively by

$$\nabla^q \mathbf{f}(x) = \nabla(\nabla^{q-1} \mathbf{f}(x)).$$

For example,

$$\begin{aligned}\nabla^2 \mathbf{f}(x) = \nabla(\nabla \mathbf{f}(x)) &= (\mathbf{f}(x) - \mathbf{f}(x-h)) - (\mathbf{f}(x-h) - \mathbf{f}(x-2h)) \\ &= \mathbf{f}(x) - 2\mathbf{f}(x-h)) + \mathbf{f}(x-2h)).\end{aligned}$$

In general $\nabla^q \mathbf{f}(x)$ is a linear combination of $\mathbf{f}(x)$, $\mathbf{f}(x-h), \ldots, \mathbf{f}(x-qh)$. A mean value theorem says that for each component f_i of the vector function $\mathbf{f}$, there is a ξ_i between x and $x - h$ such that $\nabla f(x) = h f'(\xi_j)$. Assembling the components as a vector, this becomes

$$\nabla \mathbf{f}(x) = h\, \mathbf{f}'(*),$$

where the asterisk is to indicate that the components are evaluated at different points between x and $x - h$. Similarly,

$$\nabla^q \mathbf{f}(x) = h^q\, \mathbf{f}^{(q)}(*),$$

where now the components are evaluated at different points between x and $x - qh$.

When the nodes are equally spaced, $x_{m-j} = x_m - jh$, $j = 0, 1, \ldots, m - 1$, the interpolating polynomial can be written in terms of backward differences as

$$P_m(x) = y(x_m) + \frac{\nabla y(x_m)}{h}(x - x_m) + \frac{\nabla^2 y(x_m)}{2!h^2}(x - x_m)(x - x_{m-1}) + \ldots + \frac{\nabla^{m-1}y(x_m)}{(m - 1)!h^{m-1}}(x - x_m)(x - x_{m-1}) \ldots (x - x_2).$$

Index